Principles of Forest Management

Principles of Forest Management

Edited by Lester Jacob

SYRAWOOD
PUBLISHING HOUSE
New York

Published by Syrawood Publishing House,
750 Third Avenue, 9th Floor,
New York, NY 10017, USA
www.syrawoodpublishinghouse.com

Principles of Forest Management
Edited by Lester Jacob

International Standard Book Number: 978-1-68286-540-8 (Hardback)

Cataloging-in-Publication Data

Principles of forest management / edited by Lester Jacob.
p. cm.
Includes bibliographical references and index.
ISBN 978-1-68286-540-8
1. Forest management. 2. Forests and forestry.
3. Ecosystem management. I. Jacob, Lester.
SD431 .P75 2018
634.92--dc23

Printed in China.

TABLE OF CONTENTS

PREFACE

Every book is a source of knowledge and this one is no exception. The idea that led to the conceptualization of this book was the fact that the world is advancing rapidly; which makes it crucial to document the progress in every field. I am aware that a lot of data is already available, yet, there is a lot more to learn. Hence, I accepted the responsibility of editing this book and contributing my knowledge to the community.

Forest management studies the social and technical aspects of forest administration. It is mainly based on conservation and the management of the resources found in forests. It is very important to understand the significance of conserving forests and managing them in a balanced manner to sustain an ecological balance. Some of the popular forest management techniques are managing the aesthetics, fire preservation, plantation, timber extraction, etc. The various studies that are constantly contributing towards advancing technologies and evolution of this field are examined in detail. Scientists and students actively engaged in this field will find this book full of crucial and unexplored concepts.

While editing this book, I had multiple visions for it. Then I finally narrowed down to make every chapter a sole standing text explaining a particular topic, so that they can be used independently. However, the umbrella subject sinews them into a common theme. This makes the book a unique platform of knowledge.

I would like to give the major credit of this book to the experts from every corner of the world, who took the time to share their expertise with us. Also, I owe the completion of this book to the never-ending support of my family, who supported me throughout the project.

Editor

1

Synergies for Improving Oil Palm Production and Forest Conservation in Floodplain Landscapes

Nicola K. Abram[1,2,3,4*], Panteleimon Xofis[5], Joseph Tzanopoulos[1], Douglas C. MacMillan[1], Marc Ancrenaz[2,4,6], Robin Chung[7], Lucy Peter[8], Robert Ong[9], Isabelle Lackman[2], Benoit Goossens[8,10,11], Laurentius Ambu[11], Andrew T. Knight[12,13]

1 Durrell Institute for Conservation and Ecology, School of Anthropology and Conservation, University of Kent, Canterbury, Kent, United Kingdom, **2** HUTAN/Kinabatangan Orang-utan Conservation Programme, Kota Kinabalu, Sabah, Malaysia, **3** Living Landscape Alliance, Reading, Berkshire, United Kingdom, **4** Borneo Futures initiative, People and Nature Consulting International, Jakarta, Indonesia, **5** Technological Education Institute of Kavala, Department of Forestry and Management of Natural Environment, Drama, Greece, **6** North England Zoological Society, Chester, Cheshire, United Kingdom, **7** C H Williams Talhar and Wong (Sabah) Sdn Bhd, Sandakan, Sabah, Malaysia, **8** Danau Girang Field Centre, c/o Sabah Wildlife Department, Kota Kinabalu, Sabah, Malaysia, **9** Forest Research Centre, Sabah Forestry Department, Sandakan, Sabah, Malaysia, **10** Organisms and Environment Division, School of Biosciences, Cardiff University, Cardiff, Wales, United Kingdom, **11** Sabah Wildlife Department, Kota Kinabalu, Sabah, Malaysia, **12** Department of Life Sciences, Imperial College London, Ascot, Berkshire, United Kingdom, **13** Department of Botany, Nelson Mandela Metrropolitan University, Port Elizabeth, Eastern Cape, South Africa

Abstract

Lowland tropical forests are increasingly threatened with conversion to oil palm as global demand and high profit drives crop expansion throughout the world's tropical regions. Yet, landscapes are not homogeneous and regional constraints dictate land suitability for this crop. We conducted a regional study to investigate spatial and economic components of forest conversion to oil palm within a tropical floodplain in the Lower Kinabatangan, Sabah, Malaysian Borneo. The Kinabatangan ecosystem harbours significant biodiversity with globally threatened species but has suffered forest loss and fragmentation. We mapped the oil palm and forested landscapes (using object-based-image analysis, classification and regression tree analysis and on-screen digitising of high-resolution imagery) and undertook economic modelling. Within the study region (520,269 ha), 250,617 ha is cultivated with oil palm with 77% having high Net-Present-Value (NPV) estimates ($\$413/ha^{-yr}$–$\$637/ha^{-yr}$); but 20.5% is under-producing. In fact 6.3% (15,810 ha) of oil palm is commercially redundant (with negative NPV of $\$-299/ha^{-yr}$-$\$-65/ha^{-yr}$) due to palm mortality from flood inundation. These areas would have been important riparian or flooded forest types. Moreover, 30,173 ha of unprotected forest remain and despite its value for connectivity and biodiversity 64% is allocated for future oil palm. However, we estimate that at minimum 54% of these forests are unsuitable for this crop due to inundation events. If conversion to oil palm occurs, we predict a further 16,207 ha will become commercially redundant. This means that over 32,000 ha of forest within the floodplain would have been converted for little or no financial gain yet with significant cost to the ecosystem. Our findings have globally relevant implications for similar floodplain landscapes undergoing forest transformation to agriculture such as oil palm. Understanding landscape level constraints to this crop, and transferring these into policy and practice, may provide conservation and economic opportunities within these seemingly high opportunity cost landscapes.

Editor: Francisco Moreira, Institute of Agronomy, University of Lisbon, Portugal

Funding: NKA thanks RICS Research Trust, U.S. Fish and Wildlife Service's Great Ape Fund and the Arcus Foundation for financial support; Planet Action and Trimble for providing satellite imagery and the Definiens eCognition License. ATK acknowledges support from the Department of Life Sciences at Imperial College London, UK, and ARC Centre of Excellence in Environmental Decisions, University of Queensland, Australia. The funders had no role in study design, data collection and analysis, decision to publish, or preparation of the manuscript.

Competing Interests: The authors have declared that no competing interests exist.

* E-mail: nicola_abram@hotmail.com

Introduction

Lowland tropical forest ecosystems contain some of the highest levels of species endemism and biological diversity globally [1,2]. However, tropical lowland forests are under increasing risk of conversion to oil palm (*Elaeis guineensis*) which drives biodiversity loss [2] and compromises forest functioning and maintenance of vital ecosystem services [3,4,5]. In 2012 over 17.1 million hectares of permanent cultivated cropland worldwide was comprised of oil palm agriculture compared with 9.97 million hectares in the year 2000 [6]. Oil palm establishment continues to increase within lowland areas of Malaysia and Indonesia, the two dominant producing countries, but also elsewhere in tropical Asia, Africa and central and South America [6,7]. Brazil alone has 2 million hectares of lowland forest identified as suitable for oil palm, and similarly the Democratic Republic of Congo, Indonesia, Peru and Columbia have an estimated 0.78 million, 0.61 million, 0.46 million and 0.42 million hectares respectively, of forest deemed as suitable for oil palm [8]. Expansion of this crop is largely due to its high opportunity costs driven by accelerating global demand for this versatile oil coupled with high palm production (i.e. yield) capacity [9].

Oil palm plantations are limited to low elevation areas and are in direct conflict with tropical lowland forests, including those found within riverine floodplains. Riverine floodplains are characterised by high levels of biodiversity and productivity [10]. These regions are low lying and subject to periodic inundation by associated rivers or streams [11]. Consequently, these landscapes present heterogenous environmental characteristics that dictate variable land suitability for oil palm cultivation, with unsuitable areas principally linked to seasonal and/or tidal inundation events. The palm *Elaeis guineensis* is flood intolerant with mortality occurring from root rot within two weeks of ground saturation and/or low oxygen levels or from saline water [12]. Palm mortalities reduce plantation yields and in-turn financial returns [13]. Nevertheless, simplistic biophysical criteria are often used by governments and agencies for agricultural zoning for oil palm that includes slope (<25°), elevation (<500 m) and soil types within suitable climatic zones [14]. These criteria may fail to capture regionalised constraints for this crop.

We investigated forest and oil palm dynamics within a floodplain system in the Malaysian context. In Malaysia it is estimated that between 1990–2005, 55–59% of Malaysia's oil palm extent replaced old-growth and secondary forests [15]. In Malaysia, Sabah (Borneo) is the largest oil palm planted state contributing 28.6% of the country's total oil palm extent [16]. In 2011, 1.43 million hectares (19.3%) of Sabah's terrestrial extent was under oil palm which could increase up to 2.1 million hectares by 2025, depending on land suitability [16,17]. It is likely that oil palm expansion will continue to target the eastern State floodplains areas that have very high yield potential. We employed the Lower Kinabatangan landscape, the largest floodplain in Sabah, as a 'model system' to understand spatial and economic components of forest conversion to oil palm in a riverine floodplain. We highlight the need for improving biophysical criteria used by governments in agricultural zoning for oil palm and emphasise the need for regional land-use planning of floodplain landscapes to promote synergies between agricultural development and biodiversity and ecosystem conservation goals. Understanding landscape level constraints to this crop, and transferring these into policy and practise, may provide conservation opportunities within these seemingly high opportunity cost landscapes.

Materials and Methods

Study Area

The study area comprised 520,269 ha of the Lower Kinabatangan floodplain region, in eastern Sabah (Figure 1(A)). The region experiences mean monthly temperatures of 21–34°C and average annual rainfall of 3,000 mm [18]. Forest types encompass those associated with mangrove, flooded forest and dry (humid) forest systems, which are threatened forest types [19]. Remaining forests have been heavily impacted from past commercial timber exploitation and extensive forest conversion resulting in significant forest loss, severe fragmentation and degradation within the forest ecosystem [20,21]. Yet, these forests remain important habitat for biodiversity harbouring 129 species of mammal, 314 species of birds, 101 species of reptiles, and 33 species of amphibians [22]. Many of these species are threatened IUCN (International Union for Conservation of Nature) Red List species such as the Bornean elephant [23], Bornean orangutan [24] and proboscis monkey [25]. These forests provide species habitat, are instrumental in connecting (at least in part) the fragmented protected areas network; as well as crucial for facilitating multiple ecological processes needed for the functioning of this ecosystem. Attempts to safeguard remaining forests has led to the gazetting of the Lower Kinabatangan Wildlife Sanctuary (LKWS, at 27,960 ha) in 2005. Other protected areas include protection forest reserves and virgin jungle forest reserves. Use-forests include the mangrove forest reserves and commercial (i.e. production) forest reserves; all herein referred to as protected areas (PA's). However, significant areas of unprotected forest remain outside the protected areas or production forests and on State land or land 'alienated' land for agriculture. Alienated land is leased (long-term) State land, under title and granted for a specific purpose such as oil palm [26]. Despite over a decade of conservation initiatives to secure these unprotected forests for conservation purposes they remain threatened largely with conversion to oil palm.

Forest Extent and Mapping Forest Types

Forest vegetation types were identified using SPOT5 10 m satellite images captured on 25/11/2007 and two on 19/06/2009; as well as two Landsat TM5 30 m images captured on 27/07/2006 and 11/08/2009 (downloaded from www.usgs.gov). Images were orthorectified and registered and indices calculated for: Normalized Difference Vegetation Index (NDVI); Normalized Difference Moisture Index (NDMI); and Soil Adjusted Vegetation Index (SAVI). Spectral transformations were performed including reflectance for SPOT5 and Tasseled Cap transformation on the Landsat TM5 Images [27]. We used a step-wise Object-Based Image Analysis (OBIA in eCognition Developer 8.7) for the classification that creates meaningful 'objects' by segmenting images into groups of spectrally similar pixels and spatial characteristics [28]. The eastern and western Kinagantagan areas were classified separately because different image capture dates prohibited the development of a common classification rule set. For training the classifier and testing the result, 1,938 ground validation points were collected in-field using a stratified random sample. The stratification was conducted on dominant regional forest types idenftified by local experts (for details see Table 1). Within a 20 m buffer around each validation point, data were recorded on: habitat type; degradation level (low, medium, high); percentage canopy cover; and dominant tree species with diameter at breast height (DBH) class. The ground truthed dataset was halved, one portion for training the classifier and the other for testing the result.

For image segmentation and object generation within eCognition, SPOT5 10 m resolution data were employed for delimitation of 'objects' within the images. Following trialing of numerous scale parameters, a scale of 15 was adopted. Colour/Shape and smoothness/compactness were set (0.9/0.1 and 0.5/0.5 respectively) [29]. All spatial data were used for ruleset development and the identification of features with the highest discrimination ability. The Support Vector Machine algorithm (SVM) within eCognition was employed to run the classification because it performed better in reproducing the training set than other trialled algorithms (i.e. Byes or classification and regression tree) [30,31]. We used a generic forest extent auxiliary layer, digitised from 2.5 m SPOT5 satellite images captured 12/12/2010 and 01/06/2011, to restrict the classification to areas of known forest (these images were not used in the classification due to inappropriate bands). Despite the high accuracy of the SVM classifications, classification refinement was undertaken on selected known misclassified objects using ground knowledge or neighborhood based features.

The eastern and western classifications were exported as shapefiles and 'merged' to form one continuous layer. An accuracy assessment was performed using an error confusion matrix method using the second portion of the ground truth dataset, and overall classification accuracy and kappa statistics were calculated for both portions [29].

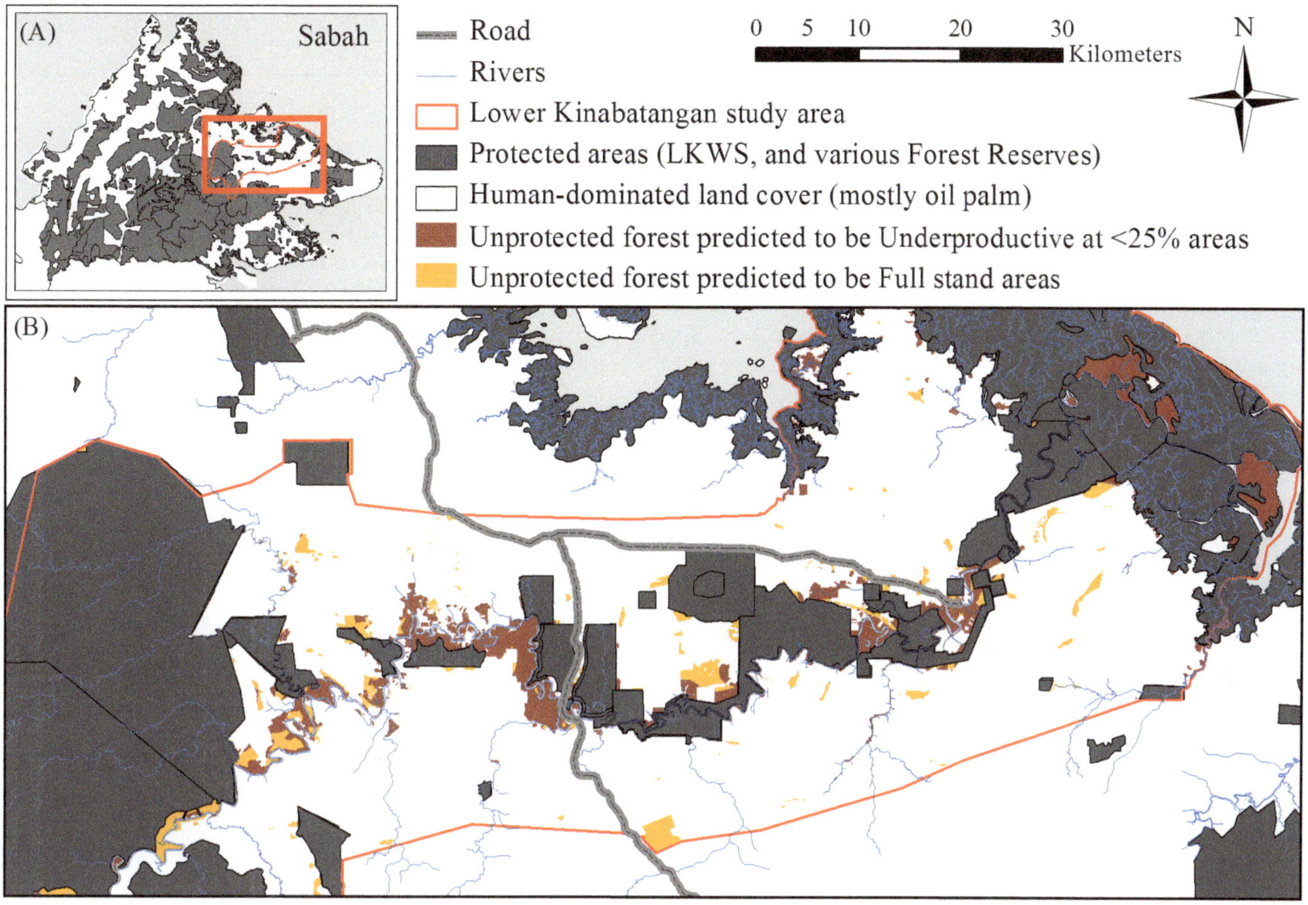

Figure 1. Study region and result from the CART analysis. (A) Map of the study region, the Lower Kinabatangan (red outline) within the context of Sabah, Malaysian Borneo. Protected areas (including all 'use' forests) are in grey. (B) Map of the Lower Kinabatangan study region with the spatial output from the Classification and Regression Tree analysis (CART) overlaid with the unprotected forest areas. Dark red demonstrates areas predicted as unsuitable for oil palm (i.e. areas that would be <25% palm capacity and commercially redundant); orange areas are predicted for full stands of oil palm (i.e. suitable areas).

Mapping Existing Oil Palm

Oil palm age and productivity classes were digitised using the 2.5 m SPOT5 images. Age classes were defined by characteristics viewable in the images that were associated with planting preparation and/or age associated features (see Table 2 for descriptions and Figure 2 for examples). To inform the visual characteristics, we used two geo-referenced maps (in ARcGIS 10) from large oil palm estates within the region (Sime Darby and Hilco Estates) that contained planting years for each oil palm 'block'. For productivity classes we based these on palm capacity. Optimal planting density for oil palm typically ranges from 128 to 148 palms per hectare depending on factors such as genotypes, environmental conditions, soil types and management practises [32]. In our study region 136 stands (i.e. palms) per hectare (SPH) is used, as recommended by the Malaysian Palm Oil Board (MPOB) and we used 136 SPH as our upper limit value for palm capacity (i.e., 136 SPH = 100% palm capacity). We classified the underproductive classes by defining the range of palm capacity within each class. Descriptions of each class can be seen in Table 2 and examples in Figure 2. In brief however, our oil palm categories were: (1) Cleared areas; (2) Planted out (0–2 years old); (3) Young mature (3–6 years, and ranging from 76–100% palm capacity); and (4) Prime mature at 'Full stand' (7–24 years, and ranging from 76–100% palm capacity); (5) Underproductive at 75% (areas that ranged from 51%–75% SPH capacity); (6) Underproductive at 50% (areas that had approximately 26–50% SPH capacity); and (7) Underproductive at ≤25% (areas that had approximately 25% SPH capacity or less).

Classification accuracy was determined through remotely obtained information as no ground-data was available. We used all recent (16/09/2009 and 24/08/2011) high-resolution tiles from Google Earth images (see http://earth.google.com) in our region to validate our oil palm age and productivity map. Although our satellite imagery was 2.5 m, which allowed individual palms to be seen, the higher resolution tiles in Google Earth enabled better interpretation ability and therefore permitted the validation of classes. We defined these tiles within a GIS (ArcGIS 10), generated 120 random points (see Figure S1 in File S1) with assigned unique identity numbers. We imported the validation points into Google Earth, assessed and allocated an oil palm class for each point. An error confusion matrix method was used to assess map accuracy and kappa statistic was calculated [29].

Economic Models for Oil Palm

We developed economic models to calculate the net present value (NPV) of oil palm for four oil palm class scenarios (see Tables S1 (a+b) to S4 (a+b) in File S2). These classes included; (1) 'Full

Table 1. Forest system and forest type classes found in the Kinabatangan with flooding periods and extents in hectares (ha) as calculated from the Object-Based Image Analysis classification.

Forest systems and forest type class	Annual flood-ing period	Total forest	Protected forest	Unprot-ected forest
		ha (%)	ha (%)	ha (%)
Mangrove forest:				
Beach forest: Occurs on sandy substrate along coastal areas. Dominant species includes *Casuarina equisetifolia.*	Tidal	5,327 (2)	4,672 (2)	655 (2)
Mangrove forest: Found in saline coastal sediments. Dominant species include *Rhizophora apiculata.*	Tidal	12,863 (5)	12,357 (6)	506 (2)
Nipah palm forest: Native type of palm (*Nypa fruticans*) found within the mangrove system either in mono-stands or mixed with *Rhizophora apiculata*	Tidal	26,618 (11)	25,399 (11)	1,219 (4)
Transitional forest: Occurs between mangrove and freshwater swamp forest. Brackish water. Dominant species of *Heritiera littoralis, Ilex cymosa, Excoecaria agallocha.*	Semi-tidal	13,849 (6)	10,567 (5)	3,282 (11)
Seasonally flooded forest:				
Freshwater swamp forest: Formed in backswamps and largely on poorly drained soil. Common species include *Dillenia excelsa, Croton oblongus, Mallotus muticus.*	>6 mths	22,284 (9)	16,721 (8)	5,563 (18)
Seasonal freshwater swamp forest: Heavy degradation thought to have occurred with many pioneer species. Common species include *Macaranga gigantea, Pterospermum elongatum, Cananga odarata.*	3–6 mths	12,501 (5)	8,253 (4)	4,248 (14)
Peat swamp forest: Oligotropic peat substrate, poorly drained forests exposed to flooding. Common treespecies include *Lophopetalum multinerviu, Baccaurea, Campnosperma coriaceum, Syzygium* and *Anisoptera costata.*	>6 mths	2,132 (1)	2,102 (1)	30 (<0)
Swamp: Open reed, swamp vegetation. Dominant tree species *Excoecaria indica*	>9 mths	2,750 (1)	2,048 (1)	702 (2)
Lowland dry forest:				
Lowland dry forest: Previous dipterocarp forest, secondary forest with species including, *Nauclea subdita, Neolamarckia cadamba, Glochidion rubrum*	<3 mths	39,008 (16)	30,693 (14)	8,315 (28)
Lowland dry dipterocarp forest: Logged lowland mixed dipterocarp forest, dominated with *Dipterocarp sp.*	Never/Rarely	101,878 (41)	100,866 (46)	1,012 (3)
Limestone forest: Gomantong substrate association of hill and ridge escarpments. Low human disturbance. Dominant species include *Dryobalanops lanceolata, Shorea pauciflora, Parashorea malaanonan* and *Dipterocarpus caudiferus.*	Never/Rarely	1,679 (1)	1,392 (1)	287 (1)
Mixed vegetation types:				
Severely degraded: areas of severe degradation with unknown previous forest types dominated by shrub/low lying vegetation.	Varied	10,511 (4)	6,159 (3)	4,352 (14)
Sum of hectares		251,400	436,295	30,173

stand' areas which incorporated Young mature and Prime mature oil palm classes that ranged from 76% (i.e., 103 SPH) to 100% (based on a maximum of 136 SPH) palm capacity (and combined enumerated 96.3% of this class extent). We also included Cleared areas and Planted out classes into 'full stand' under the assumption they will fall into this palm capacity range in the future or once planted (collectively these classes made up only 3.7% of the 'full stand' class extent); (2) 'Underproductive at 75%' which assumed 51% (69 SPH) to 75% (102 SPH) palm capacity; (3) 'Under-productive at 50%' at 26% (35 SPH) to 50% (68 SPH) palm capacity; and (4) 'Underproductive at ≤25%' which ranged from 0% (0 SPH) to 25% (34 SPH) palm capacity. We modelled the NPV of the lowest and highest palm capacity percentage range for the four classes.

Our economic models assumed a typical 25 year crop life as per industry standard [16]. For each model a yield curve was used

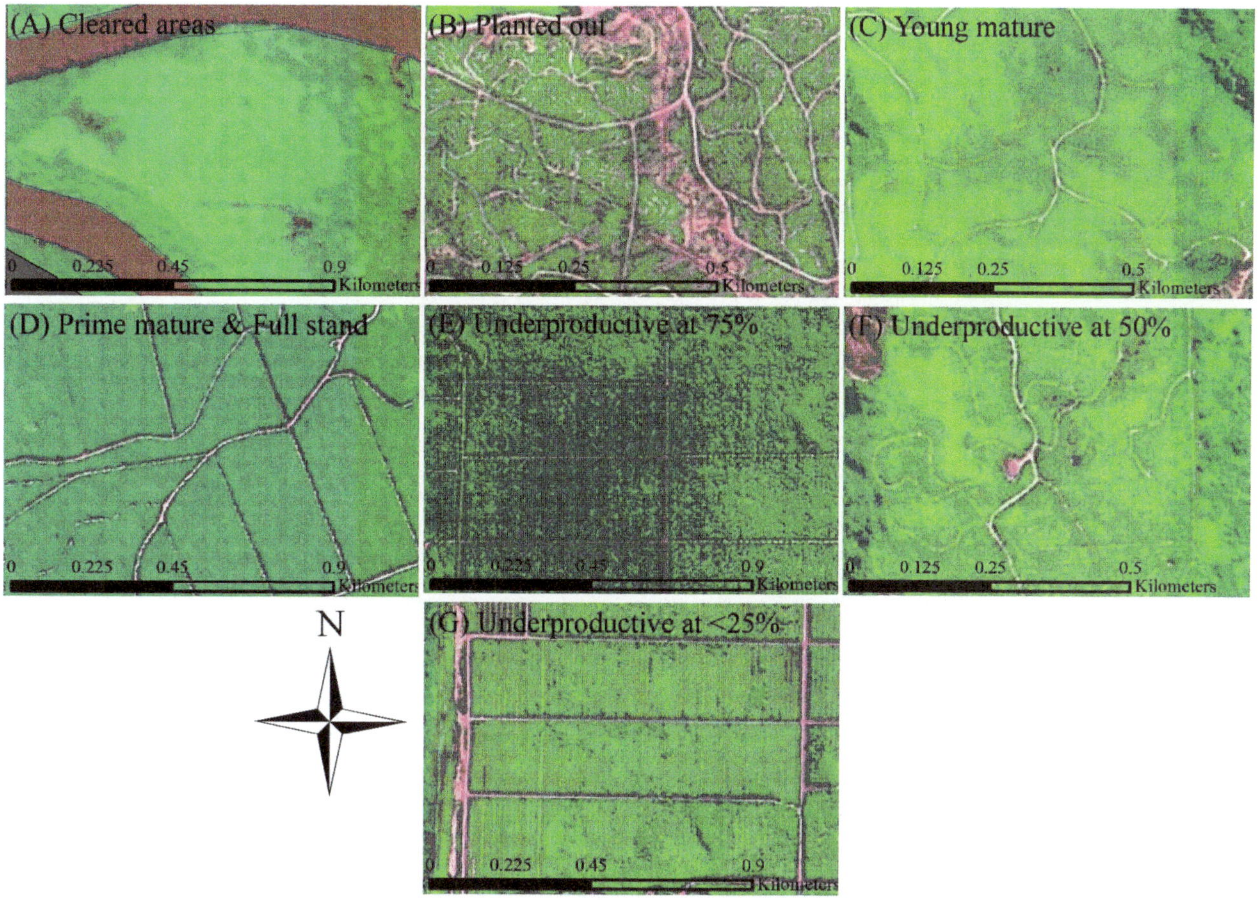

Figure 2. Examples of oil palm age and productivity classes. Images (A) to (D) show examples of the four age classes used for the oil palm mapping; and, images (E) to (G) show examples of underproductive classes, using the 2.5 m SPOT satellite images.

spanning the 25 year time frame (see oil palm yields in Figure S2 in File S2). Yields vary considerably over the lifespan of the crop with declines occurring after peak years due to factors such as palm mortality. To generate our yield estimates we used data based on actual yields from the region for full stand areas of 136 SPH (obtained from C H Williams). For the 'Full stand' at 136 SPH (100% palm capacity) NPV model, we used a maximum yield of 30 metric tons (t) of fresh fruit bunches (FFB) per hectare (ha) in peak years (years 8–11) declining to 17 t/FFB/ha in year 25, averaging at 21.92 t/FFB^{-25}. For all other models, we vary annual

Table 2. Oil palm age and productivity classes and descriptions, mapped using SPOT5 2.5 m satellite imagery.

Oil palm classes	Description
Cleared areas	Areas cleared of forest cover with bare earth or grass like vegetation cover. No roads laid nor ground prepared for oil palm planting (Figure 2A).
Planted out	Areas had roads cut and laid, ground prepared for planting, and in some areas palms planted out (palms would be approximately 2 years or younger) (Figure 2B).
Young mature	Young mature palms were visible but palm fronds did not overlap. Low-lying (leguminous) vegetation was visible. Palms in this category range from 3–6 years (Figure 2C).
Prime mature and Full stand	Palms in prime yield (7–24 years) and 'Full stand' (i.e., 76–100% palm capacity). Class had closed canopies i.e., overlapping fronds of neighbouring plants. Canopy closes at 7 years in areas with 136 palms per ha. Homogeneous texture (Figure 2D).
Underproductive at 75%	Class ranged from 51%–75% palm capacity. Class included: older palm (where mortality would naturally start to occur) or areas prone to water logging (dimpling effect can be seen in blocks) (Figure 2E).
Underproductive at 50%	Class ranged from 26–50% palms per ha capacity and were sometimes on slightly undulating areas or areas with water logging/ flooding issues (Figure 2F).
Underproductive at 25%	Class had ≤25% palms per ha and were largely associated with areas that experience annual flooding and/or daily inundation from tides (Figure 2G).

yields through the 25 year model timeframe, by calculating the proportion of yield (in t/FFB/ha) needed against our maximum 'Full stand' (100% at 136 SPH) yield values. In our models we assume a direct relationship between the number of SPH (or percentage thereof) and yield. We discuss the pitfalls and limitations of these assumptions in the discussion.

Within our models we retained FFB as our revenue unit rather than converting to crude palm oil (CPO) as for smallholders and commercial estates with no processing mills, revenue is derived from the sale of FFB. We based our model largely around 2011 data. As a result we used a constant FFB price of US$178 t/FFB, based on average 2011 values from the east coast of Sabah (calculated by C H Williams Talhar & Wong). We used a discount rate of 11% per annum in our main models, used by industry in 2011 in Sabah. However, to assess model robustness we undertook sensitivity analyses using discount rates of 5%, 8% and 14% per annum. Cost data were calculated from estates past actual costs, their budgeted future costs and typical industry average costs for 2011 (obtained from C H Talhar & Williams). These costs were cross referenced and supplemented with state wide data on costs from 2008 [33]. Costs were summarised in three categories: (1) General charges (or Joint Estate costs); (2) Field upkeep (weeding; manuring; pruning; pests and disease treatment; supplying; infrastructure etc); and (3) Harvesting and transport. Costs were constant in the four models except for 'supplying' which is the replanting of palms if palms die. We used a commercial estates approach (rather than smallholder approach) due to data availability and use in translating industry standards to build model assumptions. In regards to 'supplying' in commercial estates this is undertaken within the first two years of production to ensure homogeneous age blocks [33]. Values for supplying were adjusted for the three underproductive classes to reflect the proportion of palms needed for replanting within the initial two year period (these costs were relatively small, see Tables S1 (a+b) to S4 (a+b) in File S2). Costs excluded those associated with non-recurrent costs for establishing 'New Plantings' and those associated with costs of palm oil processing mills.

Modeling Oil Palm Suitability for Unprotected Forested Areas

To predict suitable and unsuitable areas for oil palm outside the protected areas, we used a Classification and Regression Tree (CART) analysis. The CART framework was developed by using data from existing oil palm blocks integrated with biophysical spatial information developed at 1 ha resolution. Predictor variables included: (1) Euclidian distance from main river (and main tributaries); (2) elevation; (3) aspect; (4) slope; and (5) a soil agriculture suitability layer using 4 suitability classes (unsuitable = 1; marginal = 2; suitable = 3; and very suitable = 4) [34]. Point training data were extracted randomly from known 'Full stand' and 'Underproductive at ≤25%' areas. 'Full stand' class comprised of 263 data points and 'Underproductive at ≤25%' had 211. We extracted raster values from the spatial predictors for each point using ArcGIS 10. Waikato Environment for Knowledge Analysis (WEKA) software was used to execute the CART analysis and a 10-fold cross-validation was used for tree pruning and the accuracy assessment. The software ArcPath was used to integrate the CART's classification decision tree with variables to produce a categorical map of oil palm suitability. This categorical output was imported into ArcGIS 10 and extracted for the unprotected forest extent.

Mapping Land Titles

Publicly available cadastral maps ($n = 14$) used by commercial land valuation agencies and industry were used to forecast future loss of unprotected forest by quantifying the extent of alienated land. Maps were undated but known to have been drafted in the 1990's by the Land and Survey Department. Although, these maps have been updated since the 1990's the extent of alienated land is likely underestimated. Maps were scanned at high resolution and geo-referenced to 2.5 m SPOT5 images. Land parcels were digitised, with title types and identity numbers recorded in a vector file.

Land title types included: (1) Native Title (NT), i.e., smallholdings alienated for agriculture (oil palm) for perpetuity and restricted to <40 ha in size [26]; (2) Country Land title (CL), i.e., 'alienated' State Land for commercial agriculture (oil palm) under a 99 year lease [26]; (3) Demarcated State land, with boundaries but no identity code (assumed to be under application but not alienated): and (iv) Undemarcated State land, assumed to have no title applications and not alienated (see Appendix S1 in File S1 for more details on land ordinance policies).

Results

Forest Extent and Forest Types

Overall classification accuracy was 72.2%, with a Kappa statistic of 0.65 using an error matrix method and 969 ground truth points (see Appendix S2 and Table S6 in File S1, for details), deemed as satisfactory e.g. kappa falls into 'fair' and 'good' according to differing scales outlined by Monserud [35] and Landis and Koch [36], respectively. Of the 520,269 ha study region, 48% (251,400 ha) is forested (inclusive of severely degraded forests) and comprised of 371 independent forest fragments (*mean* size = 679 ha; *sd* (standard deviation) = 5,231 ha). Of the forested extent, 12% (30,173 ha) occurs outside the protected areas (i.e. unprotected forest).

Within the unprotected forest, the Lowland dry forest system covered 32% (Table 1, Figure 3). The Seasonally flooded forest system was the largest (35%), harbouring: Freshwater swamp forest (18%); Seasonal freshwater swamp forest (14%); Peat swamp forest (<1%); and Swamp (2%) (Table 1, Figure 3). The Mangrove system comprised 19% of unprotected forest and contained: Beach forest (2%); Mangrove forest (2%); Nipah palm forest (4%); and Transitional forest (11%) (Table 1, Figure 3). These two forest systems enumerated 54% of the unprotected forest, are prone to inundation events and are likely unsuitable for oil palm. The severely degraded mixed vegetation type class totalled 14% and we predict that much of this area is prone to flooding and could also be unsuitable for oil palm.

Future Forest Loss

Of the unprotected forest, 64% had been alienated with 49% under Native or smallholder titles (i.e., 9,497 ha, either fully or partially under 1,175 titles); and 51% under Country Land or commercial titles (i.e., 9,732 ha spanning 163 titles); the remaining extent was State land (36%) (see Table 3). Of the forest systems, 75% of the Seasonally flooded forest system had been alienated for agriculture (of this NT = 46%: CL = 54%); 21% of the Mangrove forest system (of this NT = 87%: CL = 13%); 71% for the Lowland dry forest system (of this NT = 43%: CL = 57%); and 77% for the Mixed vegetation types (of this NT = 57%: CL = 43%) (see Table 4 for more details).

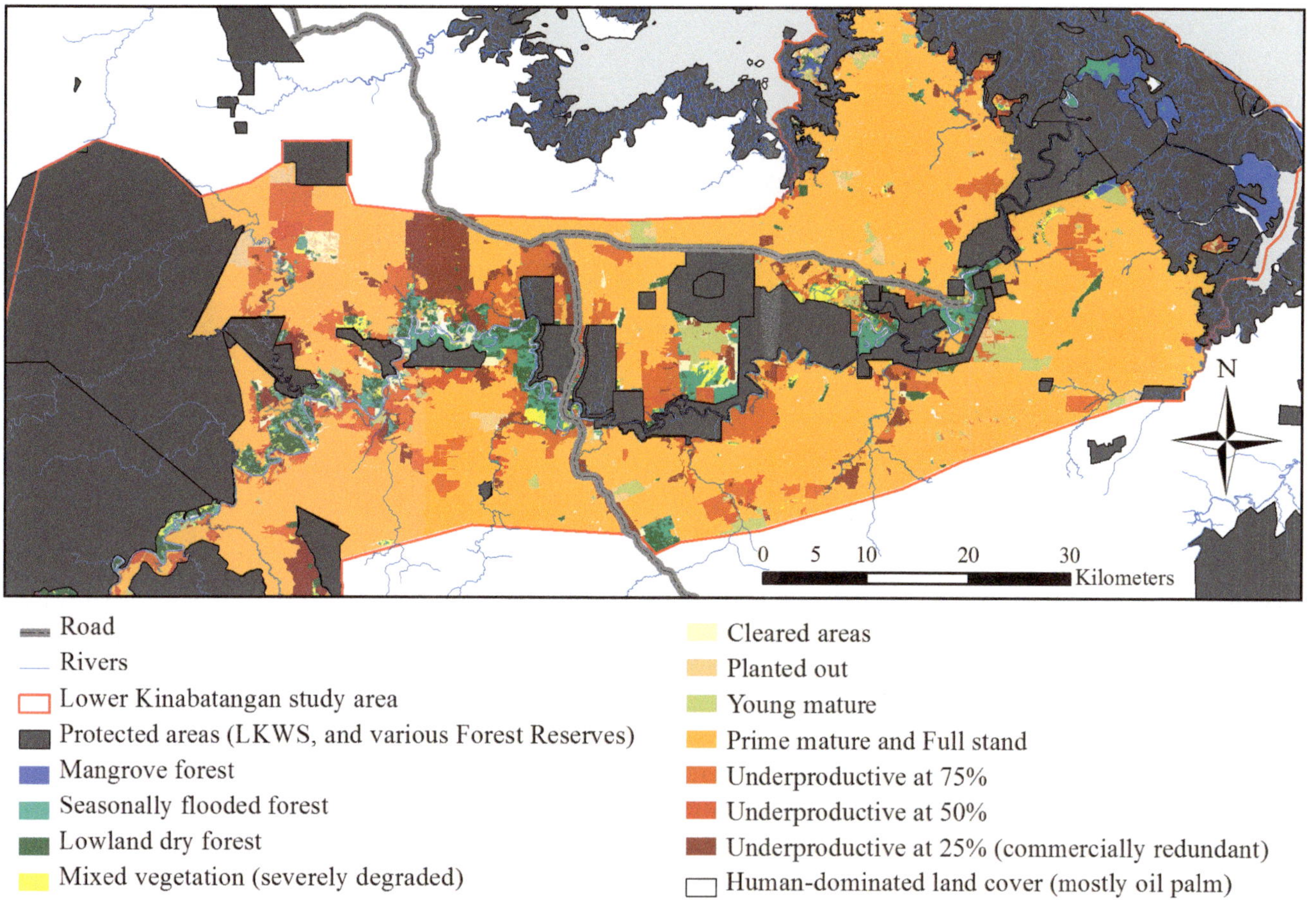

Figure 3. Map of forest systems and oil palm age and productivity classes. Map of the Lower Kinabatangan study area (red outline) showing the extent of the protected area network (grey) and the forest systems identified in the unprotected forest (as of 2010/2011 and generated through the Object-Based Image Analysis). Forest systems include: Mangrove forest (blue), Seasonal flooded forest (turquoise), Lowland dry forest (dark green) and Mixed vegetation that is severely degraded (yellow). Map also shows the oil palm age and productivity classes including: Cleared areas (cream), Planted out (beige), Young mature (olive), Prime mature and Full stand (orange), Underproductive at 75% (orange-red), Underproductive at 50% (red), and Underproductive at ≤25% (dark red), that are largely associated with areas proximal to the major river and its principal tributaries.

Oil Palm Landscape

Overall classification accuracy was 84% with a Kappa statistic of 0.816 (deemed 'very good' [35] to 'excellent' [36]) using an error matrix method with 116 test data, as four fell on cloud (see Table S7 in File S1, for details). Oil palm covered 250,617 ha (48%) of the study region (see Table 5 for ha level information and Figure 3). Seventy percent of oil palm is on commercial land titles (174,698 ha). Cleared areas covered 1.3% of the oil palm landscape; Planted out covered 1.6%; and Young mature and Prime mature and full stand collectively covered 77% of the oil palm extent. For the underproductive classes, Underproductive at 75% made up 11% and of this 39% were on NT and 59% on CL. This class largely included areas of old palm (>25 years) and/or areas of water logging both of which can increase palm mortality.

Table 3. Land title types, size in hectares (with percentages) and number of unique demarcated titles within the Kinabatangan, classified from cadastral maps for the unprotected forests.

	Unprotected forest	
Title type	**Total no. ha (%)**	**No. of unique titles**
Native Title	9,497 (31)	1,175
Country Land Title	9,732 (32)	163
State Land Demarcated	3,937 (13)	104
State Land Un-demarcated	7,009 (23)	-
TOTAL	30,173	1,338

Table 4. Land title type and extent (in ha and percentages) for: Forest systems; and, Modelled oil palm suitability (CART analysis) for suitable ('Full stand') and unsuitable ('Underproductive at ≤25%') areas.

	Land Title Type			
	Native title in ha (%)	**Country Land title in ha (%)**	**State land (demarcated) in ha (%)**	**State land (un-demarcated) in ha (%)**
Forest systems:				
Mangrove forest	1,016 (11)	147 (2)	321 (8)	4,179 (60)
Seasonally flooded forest	3,635 (38)	4,235 (44)	1,110 (28)	1,563 (22)
Lowland dry forest	2,934 (31)	3,917 (40)	1,747 (44)	1,017 (15)
Mixed vegetation types	1,912 (20)	1,433 (15)	758 (19)	250 (4)
Total number of ha	9,497	9,732	3,936	7,009
Oil palm suitability modelling:				
Full stand	3,665 (39)	3,952 (41)	1,723 (44)	1,003 (14)
Underproductive at ≤25%	5,832 (61)	5,780 (59)	2,213 (56)	6,006 (86)
Total number of ha	9,497	9,732	3,936	7,009

Underproductive at 50% totalled 3%, and spanned both NT and CL titles almost equally and were typical on slightly undulating areas or flood prone areas. Underproductive at ≤25% covered 6%, of which 28% was on NT and 66% on CL. These areas vary in age but knowledge of some areas e.g. Pontian Alljoy and Pendiros Pontian Estates, suggest conversion dates back 20 years and are largely associated with inundation.

Oil Palm Economic Models

The net present value of oil palm for each class ranged from: $413/ha^{-yr} to $637/ha^{-yr} for the 'Full stand' class; $179/ha^{-yr} to $403/ha^{-yr} for the Underproductive at 75% class; $-55/ha^{-yr} to $169/ha^{-yr} for the Underproductive at 50% class; and for those areas that were grossly underproductive at ≤25% had negative NPV of between $-299/ha^{-yr} to –$65/ha^{-yr} (see Table 6). We performed sensitivity analyses for varying discount rates for all NPV models (see Table 6; and Table S5 in File S2) and compared them to our 11% discounted base models. For Full stand areas, a 5% discount rate would increase NPV by approximately 100%, an 8% discount rate would seen a 40% approx increase in NPV; and around a 26% decrease in NPV under the 14% discount rate (Table 6). Similar percentage values were found for the Underproductive at 75% models (Table 6). For Underproductive at 50% areas the lower range models did not widely differ, but for the upper range models could see a 121% increase in NPV with a 5% discount rate, 48% NPV increase under an 8% discount rate; and decrease of about 33% for a discount rate of 14%. Models for the upper range of Underproductive at ≤25% were insensitive to changes in discount rates all having negative NPV (Table 6).

Oil Palm Suitability Modelling (CART)

The CART analysis identified distance to river, elevation and soil as strong predictors for suitable and unsuitable areas for oil palm. Aspect and slope variables were not significant. Overall classification accuracy was 75% and a Kappa statistic of 0.5, deemed as fair [35] to good agreement [36] according to differing scales. The CART decision tree identified three classification rules that dictate land feature thresholds for unsuitable areas for oil palm (Figure S3 in File S1): (1) areas ≤1,504 m from river, with elevation of ≤22.4 m above sea level (ASL); and (2) if over 1504 m from a river, ≤16 m ASL and marginal soil type; and (3) if over 1,947 m from a river, on very suitable soil and between 9.9 m–14 m ASL, with the latter two likely due to areas within lower lying depressions prone to periodic flooding in some years and/or water logged areas. Of the 30,173 ha of unprotected forest, the

Table 5. Land title type and extent (in ha and percentages) for existing oil palm age and productivity classes.

	Land Title Type			
Palm oil classes	**Total in ha (%)**	**Native title in ha (%)**	**Country land title in ha (%)**	**Un-demarcated or State land in ha (%)**
Cleared areas	3,290 (1)	1,729 (2)	1,218 (1)	343 (7)
Planted out	4,030 (2)	960 (1)	2,520 (1)	550 (11)
Young & Prime mature and full stand	191,832 (77)	49,328 (70)	140,157 (80)	2,347 (47)
Underproductive at 75%	28,081 (11)	10,959 (15)	16,563 (9)	560 (11)
Underproductive at 50%	7,575 (3)	3,489 (5)	3,754 (2)	332 (7)
Underproductive at ≤25%	15,810 (6)	4,481 (6)	10,486 (6)	843 (17)
Total area	250,617	70,945	174,698	4,974

Table 6. Range values for the main oil palm Net Present Value (NPV) (US $/ha/25 years) model discounted at 11% in the four suitability classes, as well as the outputs of the sensitivity analyses with variable discount rates of 5%, 8% and 14% (showing range NPV and percentage (%) difference from the main model at 11%).

	11% discount rate	5% discount rate	8% discount rate	14% discount rate
	lowest & highest NPV	lowest & highest NPV	lowest & highest NPV	lowest & highest NPV
Full stand	$413–$637	$813–$1,252 (101%–97%)	$579–$881 (40%–38%)	$300–$471 (−27%– −26%)
Underproductive at 75%	$179–$403	$392–$812 (119%–101%)	$263–$565 (47%–40%)	$122–$293 (−32%– −27%)
Underproductive at 50%	$-55–$169	$-48–$374 (−13%–121%)	$-52–$250 (−5%–48%)	$-57–$144 (4%– −33%)
Underproductive at ≤25%	$-299–$-65	$-505–$-66 (69%-2%)	$-380 $-65 (27%–0%)	$-242–$-64 (−19%– −2%)

CART model predicted 66% is unsuitable for oil palm (Figure 1(B)). Of these unsuitable lands, 59% has already been alienated for oil palm and 11% were probably under land applications, though may now be alienated (Table 4).

Discussion

It is likely that oil palm cultivations will continue to expand at large scales throughout the world [37], as it is estimated that global demand for palm oil will double by 2020 [38]. With such expansion careful land-use planning is needed so that multiple-benefits within landscapes can be attained and issues pertaining to biodiversity and ecosystem service loss can be mitigated [39]. We employed the Kinabatangan as a 'model system' to investigate spatial and economic patterns of land cover in a tropical forest system.

Heterogeneity of Oil Palm Landscapes

We considered the oil palm landscape to understand fine-scale variability of production within the wider floodplain extent to begin formulating ideas around conservation opportunity [40,41]. Oil palm production in many fertile floodplains may deliver high profits under optimal planting conditions [17]. Our economic models for full stand areas within the region (see Figure 3) estimated NPV from $\$413/\text{ha}^{-\text{yr}}$ to $\$637/\text{ha}^{-\text{yr}}$ using an 11% discount rate (potentially increasing by 100% with an 8% discount rate, see Table 6). However, floodplains display variable topography resulting in heterogeneous suitability for oil palm cultivation (Figure 3). Our study region is characterised by areas of inundation, particularly during the monsoon season (October to March). Our results indicated that 20.5% (51,466 ha) of the total 250,617 ha of oil palm plantations are under producing. At minimum 6.3% (15,810 ha) is likely to be commercially redundant with negative NPV estimates ranging from $\$\text{-}299/\text{ha}^{-\text{yr}}$ to $\$\text{-}65/\text{ha}^{-\text{yr}}$. These areas are largely associated with flooding in low-lying areas proximal to rivers, as predicted by our CART model (Figure 1). The economic models for the lower SPH range within the Underproductive at 50% class also demonstrated negative NPV suggesting that some of these areas could also be commercially redundant (Table 6).

Although flood mitigation measures can be implemented in flood prone areas, losses from inundation within our study region have proven to be largely ineffective and very costly [42] thereby posing a financial risk to industry. For example, in the year 2000 one company experienced palm mortalities in 5,000 hectares of immature palm, with estimated financial losses of US $3 million (equivalent to US $600/ha) due to high flood water (14 m ASL) [43]. Impacts of flood related financial losses is particularly pertinent for small scale farmers who often establish plantations using formal credit, borrow money through informal arrangements, or by investing a large proportion of their savings. Failed oil palm ventures therefore represent poor return on investment for small-scale producers [44]. Larger companies with processing mills are likely to have less associated financial risk in converting flood prone land as larger plantations may have a mosaic of land suitability thereby offsetting financial impacts. Moreover, major financial risks for some companies are generally associated with fixed costs of processing mills [45]. Nevertheless, large estates have social and environmental corporate responsibilities and conserving these forests may help companies achieve these requirements.

There were a number of assumptions in our economic models for oil palm. We assumed that the variability in yield is driven by the number of SPH. However, yield may vary due to a range of factors such as soil composition and other environmental factors [12], palm strains and management practises [32,46]; as well as pests and disease such as ganoderma [47]. In addition to factors that may affect actual yield, a number of other limitations are at play that may affect the harvesting of the fresh fruit bunches. Sabah is currently experiencing a foreign labour shortage notably of FFB harvesters, collectors and loose fruit collectors within the estates, resulting in substantial areas of oil palm remaining un-harvested [48]. Areas may also be un-harvestable if there is poor accessibility or potential health hazards such as areas with flooding. Such factors are not incorporated into our models due to a lack of data availability, as well as difficulties in up-scaling these factors to a landscape level. Nevertheless, our findings are based on sound, regional data using justifiable assumptions and provision needed values for understanding landscape economics for the region to start to formulating understanding of potential trade-offs.

Understanding Forest System Suitability for Oil Palm

Mapping natural forest systems at the agricultural frontier is fundamental to understanding opportunities for achieving biodiversity conservation goals [49]. This is particularly pertinent in forested areas zoned for oil palm that have low expected suitability. We estimated 54% (i.e., 16,207 ha of mangrove forest and flooded forests systems combined) to 68% (i.e., 20,555 ha with the inclusion of severely degraded forest) is unsuitable for oil palm production (Table 1, Figure 3), supported by our CART results. These areas experience significant inundation either daily (Mangrove system) or annually from Monsoon floods (Seasonally flooded forest system). If these forests are converted to oil palm, it is highly likely they will mirror the commercially redundant areas. Using our most conservative unsuitable forest extent estimate (16,209 ha) we estimate at minimum 56% was already alienated under smallholder (4,651 ha) and commercial (4,382 ha) titles. In parallel to existing commercially redundant areas, future investment of smallholders or companies may see no financial gain in these areas. If these areas are converted there could be over 32,000 ha of commercially redundant areas in the Kinabatangan region.

Further loss from the unprotected forests will be a significant blow for the Lower Kinabatangan ecosystem including for lowland forest types that are becoming increasingly threatened [50,51]. These forests provide a wide range of ecosystem function [52] and services such as above- and below- ground- carbon storage, essential in mitigating climate change [53], erosion control thereby mitigating severe runoff into the river system and flood attenuation, wildlife carrying capacity and species dispersal essential for the viability of globally threatened and endemic species found in the region [23,24,54,55]. Moreover the persistence of these forests and the populations of large mega-fauna species that reside in them are crucial to the flourishing international tourism industry in the region, essential to local livelihoods. Despite the value of these forests, significant forest extent is alienated and by law should be cultivated within several years of title acquisition [26]. The large extent of alienated forest suggests many landholders have not complied with these regulations. This could be due to a number of reasons including their understanding of these forests lack of suitability for oil palm. A window of opportunity for intervention schemes results from these unconverted forests. We discuss a number of plausible intervention schemes in the context of the State of Sabah.

State Policy/Management Interventions

For unprotected forests, the Sabah Government has commited to establishing corridors within the Kinabatangan to help promote the viability of State priority species e.g. the Bornean orangutan and Bornean elephant [56,57], two endangered species in the Kinabatangan. As an initial step, the Sabah Government should seek to excise State land to become incorporated into the Lower Kinabatangan Wildlife Sanctuary or as Class I Forest Reserves (i.e., protected forest status). For alienated forested land conversion may be imminent. A moratorium on conversion for these areas, regardless of forest type or level of degradation, would allow time for negotiations to be undertaken between government, the oil palm industry, and local people so that opportunities to secure these forests for conservation purposes can be sought. Solutions will need to consider land holders rights (local people and companies). Lands unsuitable for cultivation are likely to incur costs that potentially outweigh revenues and could be purchased for conservation by Government or by existing land purchase schemes by NGO's. Additionally, mechanisms such as reducing emissions from deforestation and forest degradation (REDD+) [58] may prove useful if State policies can be aligned with such mechanisms, i.e., to permit alienated land to retain standing forest (which is not permitted under current Land Ordinance Policy) [26].

For existing commercially redundant oil palm these areas could be excised to the protected area network and prioritised for restoration by either existing initiatives (e.g. HUTAN, Nestlé Project RiLeaf and MESCOT initiatives) or new reforestation projects (potentially under a carbon credit fund [59]). Companies could benefit through promoting corporate responsibility goals and securing benefits when applying for sustainable certification under the Roundtable on Sustainable Palm Oil (RSPO) [60]. Additionally, re-planting of old palm and improving yield of existing palms could help maximise returns from existing oil palm areas, offsetting need for further oil palm expansion.

Policy reform or creation of new policies could pave a more sustainable multi-use floodplain landscape. Firstly, biophysical criteria used by the State Government to demarcate areas for oil palm could be refined to account for more regional environmental constraints such as frequency and intensity of inundation. The Sabah Water Resources Enactment (1998) requires 'river reserves' to be retained in cultivated areas (up to 20 m from watercourse). This is rarely implemented by companies or enforced by authorities. River reserve widths are based on minimum needs to mitigate soil and bank erosion [61]. Revision of this policy to incorporate wildlife corridor needs or development of new 'wildlife corridor' policies should be considered. Although we strongly recommend fine-scale planning to be undertaken to promote best use of landscapes, we do suggest minimal corridor widths of 1 km for major floodplain rivers (such as the Kinabatangan which is 1.5 km wide at river mouth and 100 m wide at the most interior point of study region). This is because 'corridors' should be (multi) functional landscape features [62], resistant to issues such as edge effects that can compromise their long-term persistence [63]. Additionally, our findings suggests commercially redundant areas of oil palm within floodplains are largely within 1.5 km from (major) river banks, meaning that such corridors may not significantly impact industry profits.

In this study we highlighted prevalent issues pertaining to forest conversion to oil palm in a floodplain system. Our study provides an exemplary case study of how poor planning can result in unfavourable land cover whilst little serving either profitability for landholders or conservation agendas. Regional studies such as this one should be conducted for other floodplains undergoing land cover change. Greater understanding of opportunities and constraints in these landscapes is needed to promote informed trade-off decision making at multi-scale levels. As global palm oil demand increases, ensuring responsible utility of tropical landscapes is vital in synergising a balancing between agricultural and development needs with long-term biodiversity conservation and ecosystem functioning. This study has global significance beyond our study region as we hypothesis that biophysical criteria used by many governments and international agencies for targeting areas for oil palm cultivation are likely similar to those used in Sabah [64]. Revising biophysical criteria for agricultural zoning may better promote land use. Finally the long-term viability of the Kinabatangan lie in the hands of political will and landholders decisions. We hope that this study will provision information for better decision making within the context of the Kinabatangan landscape to ensure wise steps are made to secure this ecosystem long-term.

Supporting Information

File S1 Oil palm age and productivity map validation areas (Figure S1); Land ordinance policy under land titles (Appendix S1); Operational habitat map confusion matrix (Appendix S2); Confusion matrix table of the OBIA (Table S6); Confusion matrix table of the oil palm age and productivity (Table S7); Decision Tree from the CART analysis for oil palm suitability (Figure S3).

File S2 Excel economic models for four oil palm classes including: Full stand 100% palm capacity (136 SPH) (Table S1(a)); Full stand at 76% palm capacity (103 SPH) (Table S1(b)); Underproductive at 75% at 75% capacity (102 SPH) (Table S2(a)); Underproductive at 75% at 51% capacity (69 SPH) (Table S2(b)); Underproductive at 50% at 50% capacity (68 SPH) (Table S3(a)); Underproductive at 50% at 26% capacity (35 SPH) (Table S3(b)); Underproductive at 25% at 25% capacity (34 SPH) (Table S4(a));

Underproductive at 25% at 0% capacity (0 SPH) (Table S4(b)); Discounted summaries (Table S5); Yield curve (Figure S2). The authors are solely responsible for the content and functionality of these materials. Queries (other than absence of the material) should be directed to the corresponding author.

Acknowledgments

Thank you to D.Webber, RSPO Secretary General for comments and the staff of HUTAN/KOCP, especially A.R.Bin Saharon for his expertise, W. Bin Maharonl and H.Kler. The authors thank the two anonymous peer reviewers for their insightful comments which helped strengthen the manuscript.

Author Contributions

Conceived and designed the experiments: NKA PX JT DM MA RC IL BG LA ATK. Performed the experiments: NKA PX RC LP. Analyzed the data: NKA PX JT RC LP. Contributed reagents/materials/analysis tools: NKA PX JT MA RC LP RO IL BG. Wrote the paper: NKA PX JT DM MA RC RO IL BG ATK.

References

1. Sodhi NS, Koh LP, Brook BW, Ng PKL (2004) Southeast Asian biodiversity: an impending disaster. Trends in Ecology & Evolution 19: 654–660.
2. Fitzherbert EB, Struebig MJ, Morel A, Danielsen F, Brühl CA, et al. (2008) How will oil palm expansion affect biodiversity? Trends in Ecology & Evolution 23: 538–545.
3. Brown KA, Flynn DFB, Abram NK, Ingram JC, Johnson SE, et al. (2011) Assessing Natural Resource Use by Forest-Reliant Communities in Madagascar Using Functional Diversity and Functional Redundance Metrics. PLoS ONE 6: e24107.
4. Abram NK, Meijaard E, Ancrenaz M, Runting RK, Wells JA, et al. (2013) Spatially explicit perceptions of ecosystem services and land cover change in forested regions of Borneo. Ecosystem Services. Available: http://dx.doi.org/10.1016/j.ecoser.2013.11.004.
5. van der Werf GR, Morton DC, DeFries RS, Olivier JGJ, Kasibhatla PS, et al. (2009) CO_2 emmissions from forest loss. Nature Geoscience 2: 737–738.
6. FAO (2012) FAOSTAT online statistical service. Food and Agriculture Organization of the United Nations (FAO) Rome, Italy. Available: http://faostat.fao.org. Accessed 08-02-2014.
7. Gilbert N (2012) Palm-oil boom raises conservation concerns. Industry urged towards sustainable farming practices as rising demand drives deforestation. Nature News 12 July 2012.
8. Butler RA, Laurance WF (2009) Is oil palm the next emerging threat to the Amazon? Tropical Conservation Science 2: 1–10.
9. Feintrenie L, Chong WK, Levang P (2010) Why do Farmers Prefer Oil Palm? Lessons Learnt from Bungo District, Indonesia. Small-scale Forestry 9: 379–396.
10. Tockner K, Stanford JA (2002) Riverine flood plains: present state and future trends. Environmental Conservation 29: 308–330.
11. Junk WJ, Welcomme RL (1990) Floodplains. In: Wetlands and Shallow Continental Water Bodies, ed. B.C. Patten et al., 491–524. The Hague, The Netherlands: SPB Academic Publishers.
12. Mantel S, Wosten H, Verhagen J (2007) Biophysical Land Suitability for Oil Palm in Kalimantan, Indonesia. Wageningen, UR. Wageningen: ISRIC-World Soil Information, Alterra, Plant Research International.
13. Teoh CH (2000) Land Use and the Oil Palm Industry in Malaysia: Report of WWF Malaysia.
14. Carlson KM, Curran LM, Ratnasari D, Pittman AM, Soares-Filho BS, et al. (2012) Committed carbon emissions, deforestation, and community land conversion from oil palm plantation expansion in West Kalimantan, Indonesia. Proceedings of the National Academy of Sciences 109: 7559–7564.
15. Koh LP, Wilcove DS (2009) Oil palm: disinformation enables deforestation. Trends in Ecology & Evolution 24: 67–68.
16. MPOB (2012) Malaysian Oil Palm Statistics 2011, 31st Edition. Malaysian Palm Oil Board, Ministry of Plantation Industries and Commodities, Malaysia.
17. IDS (2007) Sabah Development Corridor, Socio-Economic Blueprint 2008–2025, Harnessing Unity in Diversity for Wealth Creation and Social Wellbeing. Institute for Development Studies, Sabah Kota Kinabalu, Sabah.
18. Ancrenaz M, Goossens B, Gimenez O, Sawang A, Lackman-Ancrenaz I (2004) Determination of ape distribution and population size using ground and aerial surveys: a case study with orang-utans in lower Kinabatangan, Sabah, Malaysia. Animal Conservation 7: 375–385.
19. Abram NK, Kandasamy Y, Sheppard S (2013) Spatial Planning for Conservation and Sustainable Development 'SPaCe': Report for Sbah Structure Plan 2033. Kota Kinabalu: Living Landscape Alliance, HUTAN, WWF-Malaysia.
20. McMorrow J, Talip MA (2001) Decline of forest area in Sabah, Malaysia: Relationship to state policies, land code and land capability. Global Environmental Change 11: 217–230.
21. Sabah Forestry Department (2007) Vegetation Assessment and Classification for Lower Kinabatangan. Kota Kinabalu: Sabah Forestry Department & WWF.
22. Lackman-Ancrenaz I, Manokaran K (2008) Kinabatangan Biosphere Reserve Nomination Dossier. HUTAN.
23. Estes J, Othman NG, Ismail S, Ancrenaz M, Goossens B, et al. (2012) Quantity and Configuration of Available Elephant Habitat and Related Conservation Concerns in the Lower Kinabatangan Floodplain of Sabah, Malaysia. PLoS ONE 7: e44601.
24. Goossens B, Chikhi L, Jalil MF, Ancrenaz M, Lackman-Ancrenaz I, et al. (2005) Patterns of genetic diversity and migration in increasingly fragmented and declining orang-utan (*Pongo pygmaeus*) populations from Sabah, Malaysia. Molecular Ecology 14: 441–456.
25. Sha JCM, Henry B, Nathan S (2008) Status and Conservation of Proboscis Monkeys (*Nasalis larvatus*) in Sabah, East Malaysia. Primate Conservation 23: 107–120.
26. Sabah Land Ordinance (ver2010) State of Sabah Land Ordinance (Sabah Cap.68). Land Ordinance.
27. Crist EP, Cicone RC (1984) Application of the Tasseled Cap Concept to simulated thematic mapper data. Photogrammetric Engineering and Remote Sensing 50: 343–352.
28. Benz UC, Hofmann P, Willhauck G, Lingenfelder I, Heynen M (2004) Multi-resolution, object-oriented fuzzy analysis of remote sensing data for GIS-ready information. ISPRS Journal of Photogrammetry and Remote Sensing 58: 239–258.
29. Heumann BW (2011) An Object-Based Classification of Mangroves Using a Hybrid Decision Tree-Support Vector Machine Approach. Remote Sensing 3: 2440–2460.
30. Foody GM, Mathur A (2006) The use of small training sets containing mixed pixels for accurate hard image classification: training on mixed spectral responses for classification by a SVM. Remote Sensing of Environment 103: 179–189.
31. Li HT, Gu HY, Han YS, Yang JH (2010) Object-orientated classification of high-resolution remote sensing imagery based on an improved colour structure and a support vector machine. International Journal of Remote Sensing 31: 1453–1470.
32. Mohamed Nazeeb AT, Tang MK, Loong SG, Syed Shahar SAB (2008) Variable density plantings for oil palms (*Elaeis guineensis*) in Peninsular Malaysia. Journal of Oil Palm Research: 61–90.
33. MPOB (2010) Palm Oil Cost of Production Malaysia 2008: A Report of the MPOB Palm Oil Cost of Production Survey 2009: MPOB.
34. Town and Regional Planning Department (unknown) Soil Suitability Map for Agriculture. Town and Regional Planning Department, Sabah.
35. Monserud RA (1990) Method for comparing global vegetation map. ASA, Laxenbury. 90–40.
36. Landis JR, Koch GC (1987) The measurement of observer agreement for catagorical data. Biometric 33: 159–179.
37. Wilcove DS, Koh LP (2010) Addressing the threats to biodiversity from oil-palm agriculture. Biodiversity Conservation 19: 999–1007.
38. UNEP (2011) Oil palm plantations: threats and oppertunities for tropical ecosystems. UNEP Global Environmental Alert Service (GEAS): Taking the pulse of the planet: connecting science with policy December 2011. www.unep.org/geas.
39. Garcia-Ulloa J, Sloan S, Pacheco P, Ghazoul J, Koh LP (2012) Lowering environmental costs of oil-palm expansion in Colombia. Conservation Letters 5: 366–375.
40. Knight AT, Cowling RM (2007) Embracing Opportunism in the Selection of Priority Conservation Areas. Conservation Biology 21: 1124–1126.
41. Knight AT, Cowling RM, Difford M, Campbell BM (2010) Mapping human and social dimensions of conservation opportunity for the scheduling of conservation action on private land. Conservation Biology 24: 1348–1358.
42. Hoh C, Ishak-Amin J (2001) Oil Palm Cultivation in the Lower Kinabatangan Floodplain: A Case Study. Forum on Making Land Use Sustainable in the Lower Kinabatangan Floodplain. Kota Kinabalu.
43. Hai TC, Ng A, Preduente C, Pang C, Yee JTC (2001) Balancing the Need for Sustainable Oil Palm Development and Conservation: The Lower ISP National Seminal 2001: Strategic Directions for the Sustainability of the Oil Palm Industry. Kota Kinabalu, Sabah, Malaysia.
44. Vermeulen S, Goad N (2006) Towards better practice in smallholder palm oil production. International Institute for Environment and Development. London, UK.
45. Levin J, Ng G, Fortes D, Garcia S, Lacey S, et al. (2012) Profitability and Sustainability in Palm Oil Production. Analysis of Incremental Financial Costs and Benefits of RSPO Compliance. WWF-US, FMO, CDC.
46. Ng PHC, Goh KJ, Mahamooth TN, Petronella GAT (2013) Challenges in fertiliser and cover crop management. Planter 89: 409–431.

47. Paterson RRM, Moen S, Lima N (2009) The Feasibility of Producing Oil Palm with Altered Lignin Content to Control Ganoderma Disease. Journal of Phytopathology 157: 649–656.
48. MPOB (2010) Oil Palm Industry Labour: Issues, Performance & Sustainability (PLIPS).
49. Cowling RM, Pressey RL (2003) Introduction to systematic conservation planning in the Cape Floristic Region. Biological Conservation 112: 1–13.
50. Butchart SHM, Scharlemann JPW, Evans MI, Quader S, Aricò S, et al. (2012) Protecting Important Sites for Biodiversity Contributes to Meeting Global Conservation Targets. PLoS ONE 7: e32529.
51. Lindermayer DB, Fischer J (2006) Habitat fragmentation and landscape change: an ecological and conservation synthesis. USA: Island Press.
52. Zavaleta ES, Pasari JR, Hulvey KB, Tilman GD (2010) Sustaining multiple ecosystem functions in grassland communities requires higher biodiversity. Proceedings of the National Academy of Sciences 107: 1443–1446.
53. Gillies CS, St Clair CC (2008) Riparian corridors enhance movement of a forest specialist bird in fragmented tropical forest. Proceedings of the National Academy of Sciences 105: 19774–19779.
54. Stark DJ (2009) Proboscis monkey (*Nasalis larvatus*) Population Viability Analysis: Reassessment and management for wild populations in Borneo.: Oxford Brookes University.
55. Ancrenaz M, Meijaard E, Ambu LN, Lackman I, Ahmad E, et al. (in press) Of pongo, palms, and perceptions – A multidisciplinary assessment of Bornean orangutans in an oil palm context. Oryx In press.
56. Sabah Wildlife Department (2012) Orangutan Action Plan 2012–2016. In: Sabah Wildlife Department MoT, Culture and Environment, editor. Kota Kinabalu.
57. Sabah Wildlife Department (2012) Elephant Action Plan 2012–2016. In: Sabah Wildlife Department MoT, Culture and Environment, editor. Kota Kinabalu.
58. den Besten JW, Arts B, Verkooijen P (2013) The evolution of REDD+: An analysis of discursive-institutional dynamics. Environmental Science & Policy.
59. Alexander S, Nelson CR, Aronson J, Lamb D, Cliquet A, et al. (2011) Opportunities and Challenges for Ecological Restoration within REDD+. Restoration Ecology 19: 683–689.
60. RSPO (2013) Roundtable For Sustainable Palm Oil: An International Multi Stakeholder Initiative Transforming Markets to Make Sustainable Palm Oil The Norm.
61. Sabah State Government (1998) Sabah Water Resources Enactment. Kota Kinabalu.
62. Roever CL, van Aarde RJ, Leggett K (2013) Functional connectivity within conservation networks: Delineating corridors for African elephants. Biological Conservation 157: 128–135.
63. Broadbent EN, Asner GP, Keller M, Knapp DE, Oliveira PJC, et al. (2008) Forest fragmentation and edge effects from deforestation and selective logging in the Brazilian Amazon. Biological Conservation 141: 1745–1757.
64. FAO (2009) FAOSTAT online statistical service. Food and Agriculture Organization of the United Nations (FAO) Rome, Italy.

2

Large-Scale Wind Disturbances Promote Tree Diversity in a Central Amazon Forest

Daniel Magnabosco Marra[1,2,3]*, Jeffrey Q. Chambers[3,4,5,6], Niro Higuchi[3], Susan E. Trumbore[2,3], Gabriel H. P. M. Ribeiro[3], Joaquim dos Santos[3], Robinson I. Negrón-Juárez[5,6], Björn Reu[1], Christian Wirth[1,2,7]

1 Spezielle Botanik und Funktionelle Biodiversität, Universität Leipzig, Leipzig, Deutschland, **2** BGC Prozesse, Max-Planck-Institut für Biogeochemie, Jena, Deutschland, **3** Laboratório de Manejo Florestal, Instituto Nacional de Pesquisas da Amazônia, Manaus, Brazil, **4** Geography Department, University of California, Berkeley, California, United States of America, **5** Climate Sciences Department, Lawrence Berkeley National Laboratory, Berkeley, California, United States of America, **6** Ecology and Evolutionary Biology Department, Tulane University, New Orleans, Louisiana, United States of America, **7** German Centre for Integrative Biodiversity Research (iDiv) Halle-Jena-Leipzig, Leipzig, Deutschland

Abstract

Canopy gaps created by wind-throw events, or blowdowns, create a complex mosaic of forest patches varying in disturbance intensity and recovery in the Central Amazon. Using field and remote sensing data, we investigated the short-term (four-year) effects of large (>2000 m^2) blowdown gaps created during a single storm event in January 2005 near Manaus, Brazil, to study (i) how forest structure and composition vary with disturbance gradients and (ii) whether tree diversity is promoted by niche differentiation related to wind-throw events at the landscape scale. In the forest area affected by the blowdown, tree mortality ranged from 0 to 70%, and was highest on plateaus and slopes. Less impacted areas in the region affected by the blowdown had overlapping characteristics with a nearby unaffected forest in tree density (583±46 trees ha^{-1}) (mean±99% Confidence Interval) and basal area (26.7±2.4 m^2 ha^{-1}). Highly impacted areas had tree density and basal area as low as 120 trees ha^{-1} and 14.9 m^2 ha^{-1}, respectively. In general, these structural measures correlated negatively with an index of tree mortality intensity derived from satellite imagery. Four years after the blowdown event, differences in size-distribution, fraction of resprouters, floristic composition and species diversity still correlated with disturbance measures such as tree mortality and gap size. Our results suggest that the gradients of wind disturbance intensity encompassed in large blowdown gaps (>2000 m^2) promote tree diversity. Specialists for particular disturbance intensities existed along the entire gradient. The existence of species or genera taking an intermediate position between undisturbed and gap specialists led to a peak of rarefied richness and diversity at intermediate disturbance levels. A diverse set of species differing widely in requirements and recruitment strategies forms the initial post-disturbance cohort, thus lending a high resilience towards wind disturbances at the community level.

Editor: Han Y.H. Chen, Lakehead University, Canada

Funding: This study was financed by the Brazilian Council for Scientific and Technological Development (CNPq) [Projeto SAWI (Chamada Universal MCTI/N° 14/2012, Proc. 473357/2012-7) and INCT - Madeiras da Amazônia], by the National Aeronautics and Space Administration's Biodiversity Program (Project 08-BIODIV-10), and by the Max-Planck-Institute for Biogeochemistry (BGC-Processes, Jena). The presented results are part of the first author's PhD thesis, which incorporate data from his Master thesis, undertaken at the INPA/CFT, with fellowship funded by the Coordination for the Improvement of Higher Level Personal (CAPES). The funders had no role in the study design, data collection and analysis, decision to publish, or preparation of the manuscript.

Competing Interests: The authors have declared that no competing interests exist.

* Email: daniel.marra@uni-leipzig.de

Introduction

Natural disturbances varying in size are suggested as major driver of tree species substitution across space and time, thus influencing vegetation structure, composition and diversity in forest ecosystems [1–4]. Disturbance dynamics in neotropical forests are considered to be dominated by small canopy gaps (<2000 m^2) from small treefall events resulting from small-scale abiotic disturbances [5–7] and biotic interactions [8]. In contrast, widespread tree mortality caused by large-scale exogenous disturbances such as forest blowdowns are often regarded as rare events.

However, remote sensing studies confirm a broad range in the size and intensity of wind disturbances representing a continuum without simple distinctions between frequent and episodic events [9–11]. In addition, widespread tree mortality associated with blowdowns is more prevalent in the Central and Western Amazon than the Eastern Amazon [12–14]. This has revived a classical question [1,5,15,16] on the ecological importance of large canopy gaps (2000 m^2) and how these influence forest dynamics and tree diversity patterns at the landscape scale [11,14,17–20].

In Central Amazon forests, there is in fact a large gradient of gap sizes created by blowdowns, which leads to a spatially complex mosaic of successional pathways [10,20,21] and gap size distribution varies locally, for example with topography. Studies outside the Amazon report that gap size distributions resulting from wind disturbance vary predictably with elevation [22–24], with a greater fraction of large contiguous blowdowns on exposed ridges and plateaus. Blowdowns can raze thousands of trees locally, and with this create a few large gaps (2000 m^2) as well as many smaller tree-fall gaps (<2000 m^2) [9,10,25].

Large blowdowns and small tree fall gaps may differ substantially in environmental conditions and thus may represent very different starting conditions for tree regeneration. In single-treefall gaps, soil and organic layer disturbance due to uprooting and snapping trees, as well as changes in nutrients and water availability can influence both mortality and recruitment rates as well as other factors vital for the maintenance of tree populations and communities [6,15,26–28]. Large blowdowns can include a wide range of variability in disturbance severity [22,29], which may affect the structure and composition in the recovering forest [30–33].

As a first order effect, gap formation by adult tree mortality is considered a key process as it increases the local availability of resources, most notably of light [5,34]. With the resulting increase in high-light microsites [26] forest patches gain the potential to promote light-demanding (pioneers) tree species specialized in colonizing and occupying gaps by efficient dispersal and rapid growth [2,34–36]. Nonetheless, the same traits that convey high growth potential for such light-demanding or pioneer species (e.g. high enzyme activity, low self-shading, and low wood density and construction costs) weaken their competitive strength under undisturbed conditions. Such traits may prevent them from enduring low light conditions, casting shade and outcompeting shade-tolerant and slower-growing (late-successional) trees in the longer term, resulting in relatively short life spans [37–41]. Light-demanding species thus depend on frequent gap formation to escape local extinction [2,5,6,34]. However, this binary mosaic view – pioneer versus late-successional species in gaps versus old growth forest – may be too simplistic for understanding the influence of a complex gradient ranging from single tree fall gaps to large blowdowns on landscape-level dynamics of highly-diverse tropical rain forests. Given the complexity of the disturbance mosaic, tropical tree species may not fall into two distinct successional groups but rather form a continuum from pioneer to late-successional strategies [37,42–44]. For example, small gaps (<2000 m^2) opened up by mortality events of one or a few trees maybe too dark for light-demanding species and colonization may depend on alternative mechanisms, such as growth-release of the understory sapling bank (advanced regeneration), resprouting of damaged trees, and lateral expansion of surviving trees. All three mechanisms may also be relevant in larger gaps (>2000 m^2) especially given distance-related limitation of seed dispersion. [31,45–48], but here high-light conditions also strongly promote the emergence of pioneer species [33,49]. However, the threshold disturbance intensity below which colonization by light-demanding species fails is not known. Moreover, it is not known whether there are specialists for intermediate levels of wind disturbance.

By combining a landscape-level approach with a detailed plot-scale analysis of high taxonomic resolution, we analyzed recruitment patterns along a large gradient of disturbance intensity and its potential relevance for maintaining tree species diversity in tropical forests. We hypothesize that the vast species pool of Amazonian forests contains species with strategies intermediate between the extremes of light-demanding and shade-tolerant species, thus enabling the forest vegetation to fully exploit the entire gradient of gap sizes and resource availability. We further hypothesize that a co-occurrence of light-demanding, shade-tolerant and intermediate species leads to a peak in diversity at intermediate disturbance intensities and gap sizes in early regeneration. To address these hypotheses we answered the following questions: *1. Are there predictable patterns of disturbance intensity distributions for large blowdowns, and do these patterns differ between topography classes (valleys, slopes, and plateaus)? 2. Do severe blowdowns exert/promote selective mortality effects at individual and species dimensions? 3. How do pre-blowdown conditions, early vegetation responses and the relative importance of the regeneration mode (establishment from seed, resprouting and growth of survivors) interact with disturbance intensity? 4. What are the implications of these effects on community composition and species coexistence at the landscape scale?*

Materials and Methods

Study sites

We conducted the research at the Estação Experimental de Silvicultura Tropical (EEST) (2°61′S, 60°20′W) of the Instituto Nacional de Pesquisas da Amazônia (INPA) and adjacent-contiguous area (ZF2), which is administered by the Superintendência da Zona Franca de Manaus (SUFRAMA) (2°56′S, 60°26′W), Amazonas, Brazil (Figure 1a). In January of 2005 storms propagating from southwest to northeast of Brazil caused large forest blowdowns across the central Amazon, including ~2500 ha of forest in the Manaus region [9]. At the ZF2, forest not known to have been previously disturbed near the Rio Cuieiras (a tributary to the Rio Negro) was heavily disturbed, as identified by field surveys and Landsat images (Figure 1).

The entire area is covered by *terra firme* forest. Prior to the 2005 disturbance there was no evidence of human intervention for at least the previous 40 years. In our Landsat chronosequence there was no signal of large natural disturbances affecting the area since 1979. Mean monthly temperature in this region is 26°C with little seasonal variation [50,51] and annual rainfall is about 2600 mm, with a distinct dry season between July and September [52,53]. The local topography is undulating with a maximum altitudinal difference of about 140 m (40–180 m a.s.l). Upland plateaus with high clay content (Oxisols) are cut by slopes and valleys dominated by soils with high sand content (Spodosols) and subject to seasonal flooding. The drainage network flows to tributaries of the Rio Negro or directly to the Rio Cuieiras. The *terra-firme* forest is characterized by a closed canopy with high tree species diversity [54–57] and a dense understory with abundant acaulescent palm species in plateaus and canopy palm species in valleys [57,58]. In this region, forest compositional and structural variations are correlated with water stress, soil and topography [54,59–61].

Forest inventory

We sampled three different patches directly affected by the 2005 blowdown (hereafter referred to as disturbed forest). To assess the entire disturbance gradient we installed three pairs of transects, measuring 200, 600 and 1000 m length by 10 m width respectively, for a total of six transects (Figure 1b – blue lines). We established 144 sub-plots of 10×25 m (total of 3.6 ha) within the 6 transects. At each sub-plot, we determined several measures of forest structure for trees with diameter at breast height (DBH) ≥10 cm. Diameters of live uprooted and damaged-snapped trees were usually measured above the DBH (1.3 m height) to avoid lesions and mechanical obstructions, such as trunks, branches, soil, etc. For species with buttressed and aerial roots, we measure the diameter just above these obstructions.

Considering previous studies [26,46,62] and field observations, we identified resprouting stimulated by mechanical injuries in individuals following uprooting and snapping (partial or total rupture of the crown). For botanical identification we collected samples from at least one individual of each species and for subsequent re-measurement we tagged all trees. We carried out

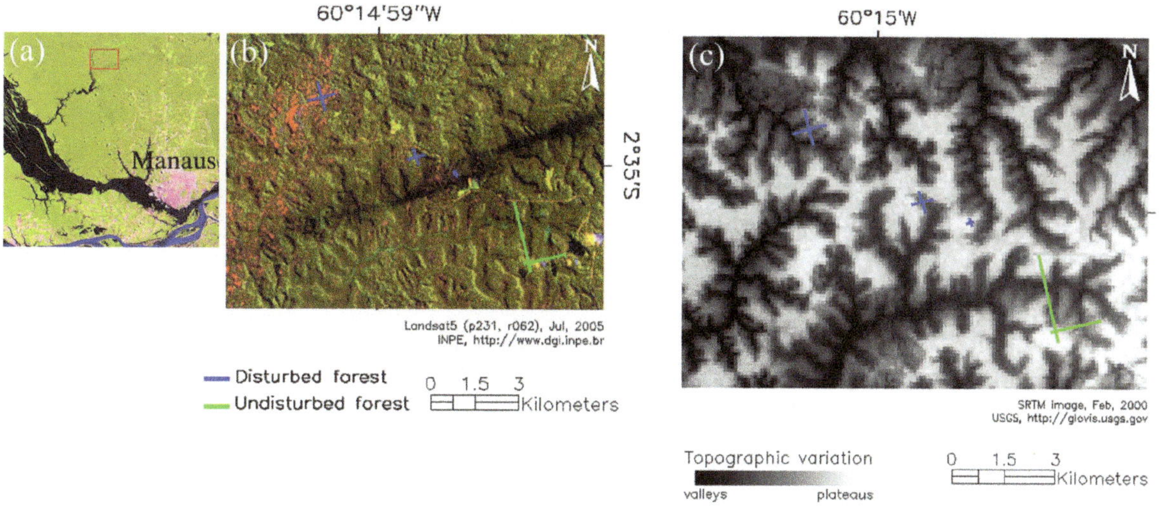

Figure 1. Study areas at the confluence of the Rio Solimões and Rio Negro, Amazonas, Brazil. Legend: (a) Landsat RGB composition of the studied areas (red inset); (b) sampled areas [short-wave infrared reflectance (red channel) indicate the 2005 blowdown tree mortality, measured by quantifying the differences on the no-photosynthetic vegetation (ΔNPV) fraction - SMA analysis]; (c) SRTM elevation model and topographic variation.

identification to the species level [63–65] when possible, by comparing the collected material with specimens at the INPA and IFAM (Instituto Federal de Educação, Ciência e Tecnologia do Amazonas) herbariums, and also with the EEST botanical collection. Reproductive material (flowers and/or fruits) was collected when available, and added to the IFAM herbarium and to the EEST botanical collection. Our research did not involve endangered or protected species and no special permits were required to access the ZF2. We carried the forest inventory between June and September of 2009. Botanical samples were collected in two campaigns, in 2009 and 2011.

We used measurements from an undisturbed contiguous area (hereafter referred to as undisturbed forest) to compare with our data for the disturbed forest (Figure 1b – green lines). In this area, two transect plots measuring 20×2.500 m (10 ha) were installed in 1996 [66] and forest structure and dynamics are monitored in 250 sub-plots (20×20 m) using measures of growth, recruitment and mortality made every two years since 2000 [52,67]. We used data from 196 (total 7.8 ha) sub-plots in an area not affected by clouds in both 2004 and 2005 Landsat images, and estimated structural parameters at these sites in 2008. We used census data from these sites for the years 2004–2006 to obtain background forest mortality rates for an area not directly affected by the 2005 blowdown (as confirmed using our Landsat ΔNPV maps) [10]. Although much of the Amazon experienced drought in 2005 with potential effects on mortality [68,69], the region we studied was not unusually dry during this period [9,53].

The transects established in disturbed and undisturbed forests were perpendicularly oriented in north-south and east-west directions (Figure 1b). These transects cross several toposequences in order to represent the floristic and structural variation between plateaus, slopes and valleys (Figure 1c). For classifying sub-plots into topographic classes we used field observations, apparent soil texture, slope and altitude [56,57,66]. Slope was measured with a clinometer and sub-plot altitude data was assessed by using a Global Positioning System (GPS) receiver (*Trimble Nomad 900*) and a digital elevation model (Shuttle Radar Topographic Mission -SRTM) (http://glovis.usgs.gov). Higher altitude areas (>80 m a.s.l.) with lower slope (~0–10°) and higher clay content soils were classified as plateaus. Areas with slope >10°, mixed sand/clay texture soils, and located in interfluve areas were classified as slopes. Areas with low elevation and relief with sandy soil and direct contact with perennial or intermittent water streams were classified as valleys (Figure 1). In the Manaus region, valleys can be also characterized by the high abundance of some palm species (e.g. *Mauritiella aculeata* (Kunth) Burret, *Oenocarpus bataua* Mart. and *Manicaria saccifera* Gaertner) and for being partially inundated during the rainy season. In the disturbed forest we classified 44 (1.1 ha), 73 (1.8 ha) and 27 (0.7 ha) sub-plots in plateaus, slopes and valleys, while in the undisturbed forest plateaus, slopes and valleys represented 49 (2.0 ha), 96 (3.8 ha) and 51 (2.0 ha) sub-plots, respectively.

In both disturbed and undisturbed forests we counted dead trees and measured their DBH. In the disturbed forest, we classified mode of tree death related to the blowdown as snapped, uprooted or standing dead according to previous studies [9,25,62]. For both disturbed and undisturbed forests, we estimated tree density (trees ha^{-1}) and basal area (m^2 ha^{-1}). For the disturbed forest, we additionally estimated mean DBH (cm) and wood density (g cm^3) for each sub-plot. We compiled species wood density from available literature [70–73] to check for possible correlations between disturbance intensity measures and wood density variation. For species for which data were not available or the identification was only possible at the genus level, we used genus-level obtained by calculating the mean wood density of local congeneric species. For six trees (snapped or without leaves), we used family-level obtained by calculating the mean wood density of the recorded species from the respective family.

Satellite data and disturbance parameters

We used Landsat-5 thematic mapper images (Path 231, Row 062), obtained from Brazil's National Institute for Space Research (INPE, http://www.dgi.inpe.br/CDSR/) and the United States Geological Survey (USGS, http://glovis.usgs.gov) to assess disturbance intensity over the disturbed area, directly affected by the 2005 blowdowns. We first georeferenced (400 control points per image) the images using the NASA Geocover data set as a basemap (https:/zulu.ssc.nasa.gov/mrsid/) and several reference points across the EEST route were collected using a GPS (*Trimble Nomad 900*); we applied a mask to remove clouds, cloud shadows, land use, and water bodies; and removed smoke and haze contamination with the Carlotto technique [74]. Further, we carried out a spectral mixture analysis (SMA) [75] to determine the fractions of green vegetation (GV), nonphotosynthetic vegetation (NPV) and shade. We normalized the pixels without shade as GV/(GV + NPV) and NPV/(GV + NPV), resulting in shade-normalized GV and NPV fractions [76]. The ΔNPV reflects changes in nonphotosynthetic vegetation (NPV, wood, dead vegetation and surface litter) before and after the storm and was calculated as NPV2005-NPV2004. We carried out SMA analysis with the Environment for Visualizing Images software (ENVI, ITT Industries, Inc, Boulder CO, USA).

We assessed disturbance intensity in the disturbed forest by estimating three disturbance parameters on a per-pixel basis (30×30 m): sub-plot mortality, mortality in the area surrounding the sub-plot (hereafter referred to as neighboring mortality) and gap size; these three disturbance parameters derived from tree mortality measures (in percentage) obtained from a mortality estimation model which has the ΔNPV as predictor [$Y = 103.22*\Delta NPV - 3.32$] ($r^2 = 0.8$ and $P<0.001$) [9]. We used the four corner coordinates (northeast, northwest, southeast and southwest) to rasterize polygons representing our sub-plots on the ΔNPV image. The sub-plot's ΔNPV was obtained by calculating the area weighted mean ΔNPV value of those pixels included in its respective rasterized polygon. To increase the accuracy of this estimation, we resampled the ΔNPV image from its native resolution of 30×30 m to a 3×3 m grid size, which allowed improved estimation of the relative area within each pixel contributing to the total sub-plot area.

We applied the sub-plot's weighted mean ΔNPV in our mortality estimation model [9] to assess tree mortality at sub-plot level in the two forests. For the disturbed forest we calculated the neighboring mortality with the same equation by applying the mean ΔNPV value of the eight pixels directly adjacent to that pixel which included the geographic coordinates of the sub-plot's central point. Additionally, we estimated gap size as the area of virtual polygons formed by groups of continguous pixels with $\Delta NPV\geq 0.16$. In these forests a $\Delta NPV\geq 0.16$ corresponds to a mortality rate of ~13% (~6 trees/pixel), which is conservatively above the minimum rate observed from smaller treefall gaps in local old growth forests (~2%) [67,77], and above the single-pixel mortality gaps obtained from field-measured mortality (~5%) [10,25]. In our mortality model [$Y = 103.22*\Delta NPV - 3.32$], $\Delta NPV<0.032$ produced negative mortality estimates, which occurred exclusively in the undisturbed forest and valleys of the disturbed forest. Because the model was adjusted only for disturbed areas, negative ΔNPV values cannot be interpreted as real gain in PV (or forest regrowth). In order to avoid negative mortality estimates, we applied a threshold ($\Delta NPV<0.032$) below which we assumed mortality was zero for sub-plots. Neighboring mortality and gap size were only estimated for the disturbed forest. We computed sub-plot and neighboring mortality by using tools from the *raster* [78] and *maptools* [79] packages implemented in the R (*version 3.0.1*) software platform [80]. We computed gap size with the SAGA software (*version 2.0.8* http://www.saga-gis.org).

Statistical analysis

Disturbance patterns. To check for recognizable relationships between disturbance patterns and topography, we compared structural (tree density and basal area) and disturbance intensity (sub-plot mortality) of the different forests. For the disturbed forest, we tested for the effects of disturbance intensity measures (sub-plot mortality, neighboring mortality and gap size) on subsequent structural and floristic variation in the different topographic classes of the disturbed forest.

To assess structural differences among topographic classes within the two forests, we classified sub-plots from the disturbed forest in two levels of disturbance. Non- and low-disturbance sub-plots ($\Delta NPV<0.16$; hereafter referred as low disturbance) [10] were compared with sub-plots that experienced disturbance above this threshold ($\Delta NPV>0.16$; hereafter referred to as high disturbance). We compared ΔNPV and structural measures of the two forests with factorial ANOVA. We used a Post-hoc Tukey test to evaluate significant differences among topographic classes. In disturbed forest, we related mode of death (snapped, uprooted and standing dead) among topographic classes with linear regressions and assessed whether mode of death is still recognizable and related to sub-plot mortality. With one-way ANOVA, we compared the mean DBH of dead trees among modes of death. Additionally, we tested for the effect of the disturbance intensity on subsequent structural variation by relating tree density and basal area to sub-plot mortality with linear regressions. The same analysis was applied to test for possible correlations between mean DBH of recorded dead trees and sub-plot mortality.

Selective mortality patterns. To address our questions related to selective mortality effects at individual and species level, we compared the size distribution of trees from the disturbed with the undisturbed forest. For the disturbed forest, we described the variation of genera's importance at low and high disturbance levels, and related species diversity measures to sub-plot mortality.

Due to the lower disturbance intensity that has been observed in valleys, and also to pre-blowdown distinct structural and floristic characteristics of the vegetation in the valleys, we only considered sub-plots established on plateaus and slopes for these analyses. To assess the influence of disturbance intensity on size distribution of trees and on the importance of genera, we also considered the classification described in section 2.4.1. We assessed the disturbance intensity effects on the size distribution of trees using Chi-squared tests comparing the diameter distribution of live and dead trees within the disturbed and undisturbed forests, and also within the levels of disturbance described above. For the disturbed forest, we additionally related the mean DBH of live trees to sub-plot mortality. We described community demographic patterns at the genus level for low and high disturbance sub-plots by estimating the Importance Value Index (IVI), calculated as the sum of relative density, frequency and dominance (basal area) of congeneric species [81]. To assess the effects of disturbance intensity on diversity patterns, we related with linear and polynomial regressions, species richness, Shannon diversity and species rarefaction to sub-plot mortality. These three measures were calculated for groups of five sub-plots, ranked and grouped by their sub-plot mortality value (23 in total). With this we avoided the tree density effect observed in those sub-plots with high mortality values, and thus few trees (e.g. with three trees); and were able to look for patterns at the community level.

Disturbance gradient effects and vegetation responses. To address our last two questions, on vegetation responses along the disturbance gradient and changes on community composition, we assessed the importance of resprouters and fast growing or pioneer species, and checked for variations in species composition with disturbance intensity measures.

With linear regressions we assessed the effect of sub-plot mortality, neighboring mortality and gap size on the fraction of resprouters; and effect of sub-plot mortality on mean wood density. We tested with one-way ANOVA for wood density and size differences between resprouters and non-resprouters. Further, we related the fraction of pioneer species basal area to sub-plot mortality and mean wood density; and related the fraction of pioneer species with neighboring mortality and gap size. We consider as pioneers, light-demanding/fast-growing species typical for the region from the genera *Cecropia*, *Conceveiba*, *Inga*, *Laetia*, *Miconia*, *Pourouma*, *Tachigali* and *Vismia*. To reduce the influence of possible pioneer trees that were established before the blowdown, we only accounted for trees in these genera with DBH≤25 cm.

To test for the effect of disturbance intensity on floristic composition, we used nonmetric multidimensional scaling (NMDS) to reduce the dimensionality of the community species composition [82,83]. Ordinations were computed with two axes from one dissimilarity matrix generated by the abundance of all recorded species. We used Mantel-tests ($P<0.001$) to test for the significance of the NMDS analysis. Finally, sub-plot mortality, neighboring mortality and gap size were used as predictors in linear regressions to assess the effects of disturbance intensity measures on the floristic composition variation detected by the first NMDS axis. Finally, to look in detail at possible variations on genera composition driven by the disturbance intensity, we fit a cubic smoothing spline function (degrees of freedom = 3) relating the abundance of the 25 most important genera (based on IVI values) to sub-plot mortality. As in our previous analyses (see section 2.4.2) only sub-plots established on plateaus and slopes were used for these last two analyses. All statistical analyses were performed in R (*version 3.0.1*) software platform [80]. We used the vegan package [84] for estimating diversity indices, species curves and to process the NMDS analysis. All other tests were custom written.

Results

Disturbance patterns

In the disturbed forest we sampled 1944 live and 363 dead trees. Our structural measures confirm a correlation between ΔNPV and tree mortality, and indicate that vegetation damage was partially controlled by topography.

Tree density and basal area in the sub-plots (250 m^2) of the disturbed forest ranged from 3 to 33 trees and from 0.08 to 2.23 m^2, respectively. Mean tree density and basal area in the undisturbed forest was higher than in the low and high disturbance sub-plots (Table 1). Mean tree density and basal area in slopes from the high disturbance was lower than in slopes from the low disturbance and undisturbed forest (Table 1).

In the disturbed forest, ΔNPV ranged from −0.213 to 0.696 with an overall mean value of 0.176±0.037 (mean±99% CI), higher than from the undisturbed forest (0.038±0.016; ranging from −0.168 to 0.202). Estimated sub-plot mortality in the disturbed forest ranged from 0 to 69.9% with an overall mean value of 16.2±3.4%, higher than from the undisturbed forest (3.7±0.8%; ranging from 0 to 30.5). In the undisturbed forest, mortality rate computed from field data (2004–2006) was 1.6±0.7% $year^{-1}$ and correlated positively but weakly with the estimated sub-plot mortality ($r^2 = 0.04$; $P = 0.004$; Pearson's $r = 0.20$). Plateaus, slopes and valleys from high disturbance sub-plots had higher mean ΔNPV and sub-plot mortality in comparison to low disturbance and undisturbed forest (Table 1). Plateaus and slopes from the high disturbance also had higher neighboring mortality and gap size than plateaus and slopes from the low disturbance. The disturbed sub-plots were set in gaps that ranged in size from 0–22.6 ha (6.5±2.3), which correlated positively with mean sub-plot mortality ($r^2 = 0.37$; $P<0.001$; Pearson's $r = 0.60$). During the analyzed period we did not find gaps larger than three pixels (2700 m^2) in the undisturbed forest.

In the disturbed forest the density of dead trees ranged from 0 to 360 trees ha^{-1} and the mean (101±22 trees ha^{-1}) varied among topographic classes ($F = 8.74$; $P<0.001$). The mean density of dead trees in plateaus (104±43 trees ha^{-1}) was similar to that for slopes (120±31 trees ha^{-1}) ($P = 0.536$), but exceeded the mean density observed in valleys (44±31 trees ha^{-1}) ($P<0.01$, both). The total number of dead trees, both snapped and uprooted, correlated positively with sub-plot mortality in slopes in plateaus. In the disturbed forest plateaus and slopes, the density of live trees and basal area were negatively correlated with sub-plot mortality, while valleys were not affected (Table 2).

Selective mortality patterns

Our data show that the blowdowns promoted subsequent changes in forest structure, genera substitution and species demography. Four years after the blowdown event, differences in size-distribution, genera importance and diversity measures still correlated with levels of disturbance.

In the disturbed forest, the DBH of recorded dead trees ranged from 10 to 120 cm, with mean value (27.5±2.3 cm) higher than from that from live trees of both disturbed (20.6±2.3 cm) and undisturbed forest (21±0.7 cm). DBH of dead trees did not vary among modes of death (ANOVA $F = 2.37$; $P = 0.125$), but for snapped and uprooted trees had a low correlation with sub-plot mortality ($r^2 = 0.02$; $P = 0.029$; Pearson's $r = 0.13$), with a non-variable mean value among modes of death (ANOVA $F = 2.37$; $P = 0.125$).

Even four years after the blowdown event, the DBH distribution comparisons indicated size distribution differences between trees of the disturbed and the undisturbed forests. The DBH distribution of dead trees in both low- and high-disturbance sub-plots (Figure 2a) did not follow that of the undisturbed forest before (2004) and after (2006) the blowdown event (Chi-squared test, $P<0.001$). While the undisturbed forest had a higher abundance of dead trees within the smaller diameter classes, low and high disturbance forests had higher abundance of large trees exhibiting a log-normal distribution. The DBH distribution under low and high disturbance was also different (Chi-squared test, $P = 0.017$). Within live trees, both disturbed and undisturbed forests had higher abundance of small trees and exhibit the typical negative exponential DBH distribution of tropical forests (Figure 2b). The DBH distribution of live trees under low disturbance was similar to that of the undisturbed forest ($P = 0.490$), while under high disturbance forest it diverged from that observed within trees from the undisturbed and low disturbance forest ($P< 0.001$). The overall mean DBH of live trees was negatively related to sub-plot mortality in plateaus and slopes of the disturbed forest (Figure 2c).

In the disturbed forest, recorded trees were distributed in 51 families, 158 genera and 324 species or morphotypes. A total of 54 species (~16%) just occurred on plateaus, while 55 (~17%) and 20 (~6%) were restricted to slopes and valleys, respectively. Fifty eight species (~18%) occurred in all topographic classes. Fabaceae (66 species), Sapotaceae (26) and Lecythidaceae (23) were the richer families in number of species and together summed ~40% of all recorded trees and ~36% of the total richness. Eighteen families were represented by just one species. 129 species were represented by only a single tree. Heavily damaged sub-plots, especially those in gaps with large amounts of necromass, often had high density of lianas. Field observations showed that gaps in which wood decay was more advanced generally had a dense understory with seedlings dominated by the genera *Casearia* and

Table 1. Summary statistics (mean±99% CI) of structure and disturbance intensity measures for different levels of wind-disturbance in a forest disturbed by a large blowdown, and an undisturbed forest, in Amazonas, Brazil.

	TC	NS	TD	BA	ΔNPV	SP	NG	GS
UN	All	196	593±28*	27.7±2.1*	0.038±0.016*	3.7±0.8		
	Plateaus	49	632±46	29.1±4.4	0.048±0.036	4.6±1.9		
	Slopes	96	590±34	26.8±2.9	0.047±0.022	4.2±1.3		
	Valleys	51	560±44	28.2±3.7	0.011±0.030	2.0±0.9		
LD	All	75	583±46*	26.7±2.4*	0.050±0.016*	4.4±0.9	5.4±1.4*	1.1±0.9*
	Plateaus	12	643±5	28.0±0.2	0.048±0.002	5.2±0.1	8.4±0.2	0.4±0.04
	Slopes	39	615±99	27.1±5.1	0.054±0.033	4.8±1.9	5.9±2.8	0.6±1.5
	Valleys	24	500±78	25.3±5.3	0.046±0.029	3.2±2.0	3.2±2.5	1.9±3.4
HD	All	69	462±62*	19.9±4.9*	0.312±0.040*	28.9±4.2*	25.0±4.2*	12.6±2.7*
	Plateaus	32	554±98	25.5±8.9	0.326±0.064¥	30.4±6.7¥	26.0±5.3¥	9.2±3.3¥
	Slopes	34	426±76¥	14.5±4.2¥	0.307±0.054¥	28.5±5.6¥	24.7±6.8¥	16.3±3.9¥
	Valleys	3	600±158	21.5±7.0	0.208±0.040¥	18.2±4.1	16.8±7.8	6.6±13.3

Legend: UN- undisturbed forest; LD- low disturbance (up to 13% of mortality); HD- high/severe disturbance (greater than 13% of mortality); TC- topographic classes; NS- number of sampled sub-plots; TD- tree density (trees ha^{-1}); BA- basal area (m^2 ha^{-1}); SP- sub-plot mortality (%); NG- neighboring mortality (%); GS- gap size (ha).

*Significant differences (ANOVA) among levels of disturbance ($P<0.001$).

¥Significant differences (Tukey HSD) between same topographic classes ($P<0.001$).

Table 2. Linear regressions and Pearson's correlation relating structural variables to mortality intensity in a forest disturbed by a large blowdown, in Amazonas, Brazil.

	Structure variables	Topographic classes[¥]	N. of trees	Regression parameters*					
				a	b	F	r^2	*P*	Pearson's *r*
Dead trees	N. of dead trees	Plateau	114	1.5191	0.0456	6.00	0.125	0.018	0.354
		Slope	219	2.1073	0.0564	13.5	0.159	<0.001	0.4
		Valley	30	0.8224	0.059	2.43	0.052	0.131	0.297
	N. of snapped trees	Plateau	51	0.2713	0.0378	9.17	0.18	0.004	0.423
		Slope	92	1.055	0.0129	2.08	0.028	0.153	0.168
		Valley	19	0.5055	0.0405	2.38	0.086	0.135	0.294
	N. of uprooted trees	Plateau	44	0.8259	0.0074	0.34	0.008	0.559	0.09
		Slope	105	0.5709	0.0548	16.45	0.188	<0.001	0.434
		Valley	7	0.138	0.0248	0.86	0.033	0.362	0.182
	N. of standing dead trees	Plateau	19	0.4218	0.0004	<0.01	−0.023	0.954	0.008
		Slope	22	0.4813	−0.0113	5.48	0.071	0.022	−0.267
		Valley	4						
Live trees	Tree density (trees ha^{-1})	Plateau	637	648.69	−3.000	2.46	0.055	0.123	−0.235
		Slope	962	658.28	−8.290	29.36	0.293	<0.001	−0.54
		Valley	345	494.97	3.3030	0.47	0.018	0.5	0.135
	Basal area (m^2 ha^{-1})	Plateau		31.57	−0.2286	2.23	0.051	0.143	−0.224
		Slope		29.14	−0.4966	36.2	0.337	<0.001	−0.582
		Valley		25.25	−0.0742	0.05	0.002	0.819	−0.046

*Model: y = a+bx+Ɛi.
[¥]Degrees of freedom for the topographic classes: plateaus (44); slopes (73); valleys (27).

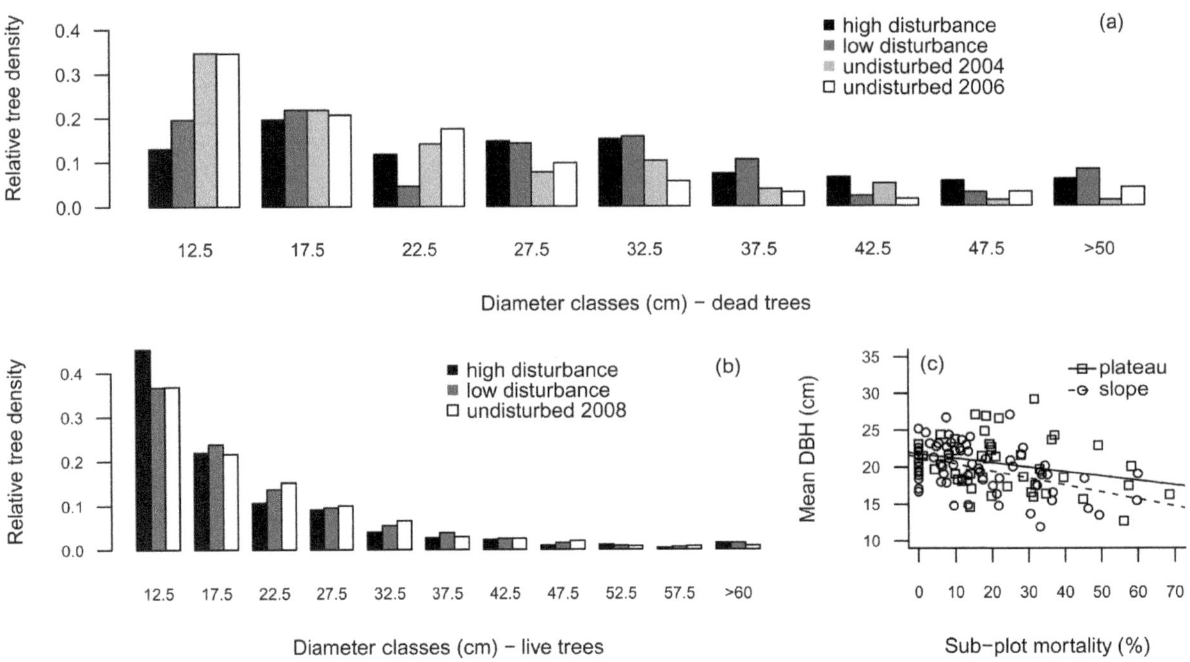

Figure 2. Tree size distribution (DBH≥10 cm) in plateaus and slopes of a forest disturbed by a large blowdown and an undisturbed forest, in Amazonas, Brazil. Legend: (a) diameter distribution of dead trees; (b) diameter distribution of live trees; (c) mean DBH of live trees in disturbed plateaus ($r^2 = 0.05$; $P = 0.06$; Pearson's $r = -0.277$) and slopes ($r^2 = 0.19$; $P < 0.001$; Pearson's $r = -0.451$) related to sub-plot mortality.

Laetia (Salicaceae), *Cecropia* and *Pourouma* (Urticaceae), *Conceveiba* and *Croton* (Euphorbiaceae), *Inga* and *Tachigali* (Fabaceae), *Miconia* (Melastomataceae), *Tapirira* (Anacardiaceae) and *Vismia* (Hypericaceae). Genera demographic analyses revealed differences between low (total of 258 species) and high disturbance (215). Ten genera among the 20 most important genera from the high and the low disturbance levels were different (Figure 3). The high disturbance level had both typical light-demanding/fast-growing (e.g. *Cecropia*, *Inga* and *Pourouma*) and shade-tolerant/slower-growing (e.g. *Brosimum*, *Pouteria*, *Sloanea* and *Vantanea*) exclusive genera within the 20 most important genera. *Eschweilera*, *Licania*, *Pouteria* and *Protium* had the highest IVI in both disturbance levels, but the contributions of the IVI estimators of these and other common genera varied between disturbance levels.

Species richness per sub-sub-plot varied from 3 to 25 species in the disturbed forest. For groups of plots, species richness was negatively related to sub-plot mortality ($r^2 = 0.39$; $P < 0.001$; Pearson's $r = -0.64$) (Figure 4a). Shannon diversity also correlated negatively with sub-plot mortality ($r^2 = 0.39$; $P = 0.002$; Pearson's $r = -0.61$). Nonetheless, at intermediate disturbance (20–50% of tree mortality), Shannon diversity was slightly lower or even similar to that from the less- and non-disturbed sub-plots (Figure 4b). Species rarefaction curve along the mortality gradient indicated that intermediate-disturbance areas can be more diverse than both heavily and undisturbed areas (Figure 4c).

Disturbance gradient effects and vegetation responses

Of the live trees recorded in the disturbed forest, 206 (~11% of total sampled) had crown or trunk injuries and 191 (~10%) had one of the mechanisms indicative of resprouting. 151 trees (~7%) belonged to light-demanding/fast-growing pioneer genera, and the mortality gradient appeared to amplify forest niches that lead to changes in species composition.

Resprouting trees were recorded in all the topographic classes and within different species. In slopes, the fraction of resprouters related positively to sub-plot mortality as well as to other disturbance intensity measures (Figure 5). In plateaus and slopes, the mean wood density correlated negatively with sub-plot mortality ($r^2 = 0.03$; $P < 0.035$; Pearson's $r = -0.19$), but resprouters and non-resprouters had no significant differences in wood density ($F = 0.262$, $P = 0.609$), but did have in DBH ($F = 4.87$, $P = 0.027$).

In general, light-demanding/fast-growing genera (e.g. *Cecropia*, *Conceveiba*, *Inga*, *Casearia*, *Miconia*, *Pourouma*, *Tachigali*, *Tapirira* and *Vismia*) contributed less to total basal area in less disturbed sub-plots, while their basal area was positively related to sub-plot mortality in plateaus ($r^2 = 0.39$; $P < 0.001$; Pearson's $r = 0.63$) and slopes ($r^2 = 0.27$; $P < 0.001$; Pearson's $r = 0.52$) (Figure 6a). In contrast, the fraction of basal area contributed by these fast-growing species was negatively related to sub-plot mean wood density, also exclusively in plateaus ($r^2 = 0.30$; $P < 0.001$; Pearson's $r = -0.56$) and slopes ($r^2 = 0.31$; $P < 0.001$; Pearson's $r = -0.57$) (Figure 6b). The fraction of these typical pioneer species was positively related to neighboring mortality in plateaus ($r^2 = 0.32$; $P < 0.001$; Pearson's $r = 0.58$) and slopes ($r^2 = 0.46$; $P < 0.001$; Pearson's $r = 0.67$) (Figure 6c). Gap size was also positively related to these group of species in plateaus ($r^2 = 0.14$; $P = 0.013$; Pearson's $r = 0.34$) and slopes ($r^2 = 0.21$; $P < 0.001$; Pearson's $r = 0.44$) (Figure 6d).

The NMDS stress value was 0.286, and the first two axes captured 34% of the floristic variation (Figure 7a and Table 3). Floristic similarities within heavily damaged sub-plots increased due to the higher abundance of light-demanding and fast-growing species. Sub-plot mortality related positively to the variations in species composition captured by the first NMDS axis (Figure 7b). The same pattern was observed for neighboring mortality and gap size, which correlated more strongly (Table 3).

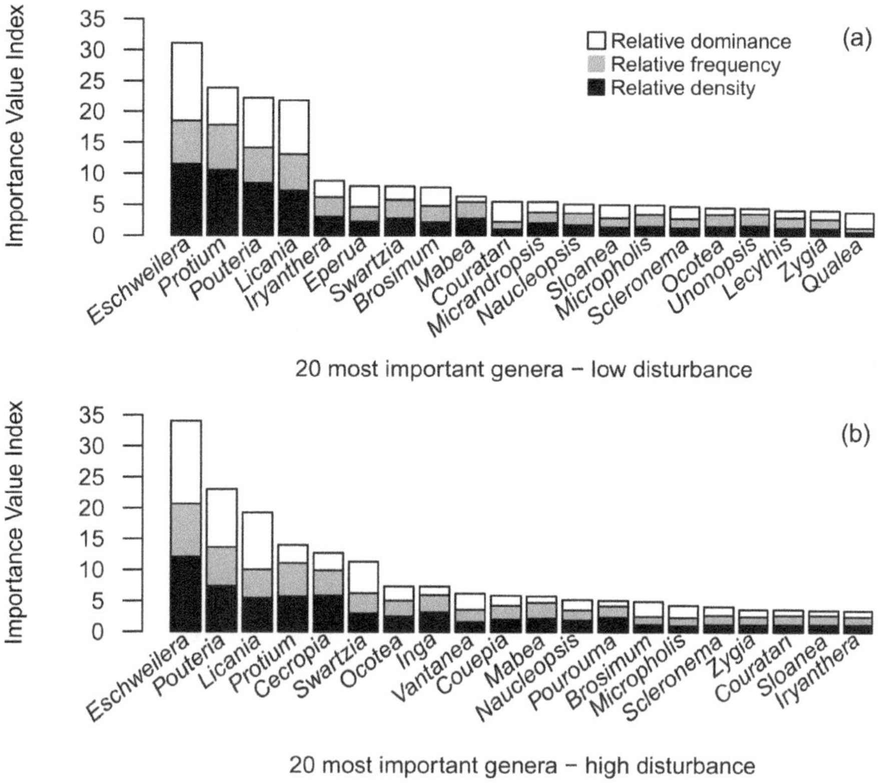

Figure 3. Genera importance ranking over plateaus and slopes of a forest disturbed by a large blowdown, in Amazonas, Brazil. Legend: (a) Importance Value Index (IVI) of the 20 most important genera recorded under low disturbance (up to 13% of tree mortality); (b) IVI of the 20 most important genera recorded under high disturbance (up to 70% of tree mortality).

The demographic patterns of the 25 most important genera in the community (identified in Figure 3) and the variations in species composition (Figure 7) corroborate our results related to genera abundance variation in respect to disturbance intensity (Figure 8). The genera abundance curves show that these 25 genera are not equally distributed along the landscape and that part of this variation was regulated by variations in sub-plot mortality. Interestingly, the abundance of these genera varied along the mortality gradient revealing five distinct groups (identified using colors in Figure 8) which had their optimum under specific disturbance intensities. This pattern indicates that there exist specialist guilds for each different level of disturbance/resource. Moreover, areas where the tree mortality driven by the blowdown ranged between 20–50% are those which provide gap space and light conditions, thus favoring a wider range of species with different requirements.

Discussion

Disturbance patterns

Mortality rate and structural measurements of the undisturbed forest were similar to those reported in other undisturbed forests nearby [52,67,77,85]. Tree mortality levels observed in the disturbed forest were in accordance with patterns described in earlier studies [9,10,25] also in Amazon and even higher than in hurricane-damaged forests [30,45,47,86].

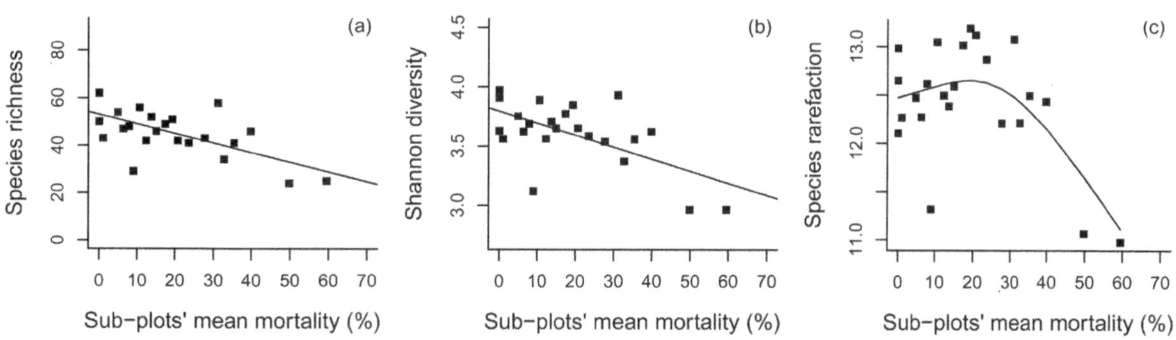

Figure 4. Species richness and diversity measures related to a mortality gradient in plateaus and slopes of a forest disturbed by a large blowdown, in Amazonas, Brazil. Legend: (a) species richness; (b) Shannon diversity; (c) species rarefaction.

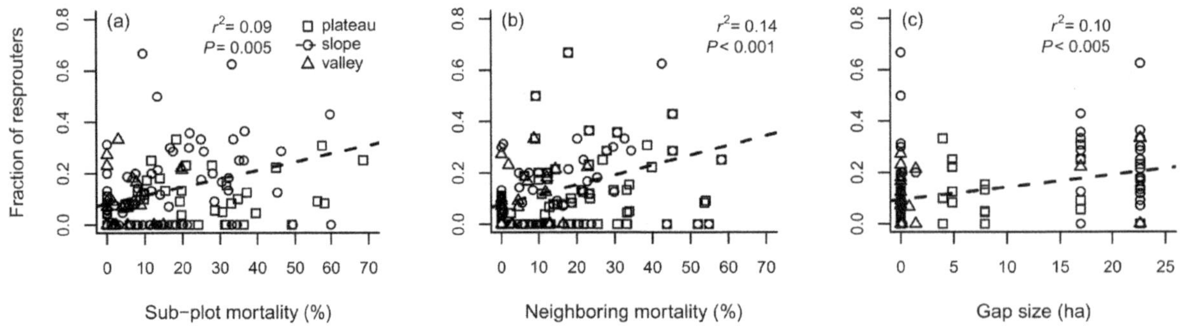

Figure 5. Fraction of resprouting tree species related to disturbance intensity measures from a forest disturbed by a large blowdown, in Amazonas, Brazil. Legend: (a) fraction of resprouters related to sub-plot mortality (slopes Pearson's $r = 0.33$); (b) fraction of resprouters related to neighboring mortality (slopes Pearson's $r = 0.38$); (c) fraction of resprouters related to gap size (slopes Pearson's $r = 0.33$).

Wind-related mortality caused by snapping and uprooting was greatest in more exposed areas, including plateaus and the top of slopes (Table 2). As a consequence of lower disturbance intensity, structural patterns were not altered by blowdowns in the valleys (Table 1). This may indicate that strong winds did not reach the valley floors, or that the vegetation in valleys is less affected by wind damage, as has been observed in other tropical and sub-tropical forests affected by cyclones [23,24,29,48,87]. We also did not observe uprooted canopy palm trees in valleys and just 14 snapped and/or standing dead ones. As there is no evidence that caulescent palm tree species that occur in these valleys are less susceptible to wind damage (see species in item 2.2), we assume that valley floors are indeed better protected from strong winds. Although in these forests vegetation damage was partially controlled by abiotic aspects such as wind characteristics and topography [22,29,88], resolution of this question requires more information on individual-species traits, such as anatomic-morphological variations among individuals and species, size distribution and populations range, and individual pre-disturbance conditions (pest-attack, biological interactions, age, etc.).

In the disturbed forest, measures of forest structure, including tree density and basal area, were lower four years following the blowdown event (Table 1). The topographic variation of measures of structure was amplified by the landscape driven differences in mortality. Immediate tree mortality effects on tree density and basal area reduction are directly dependent on the disturbance

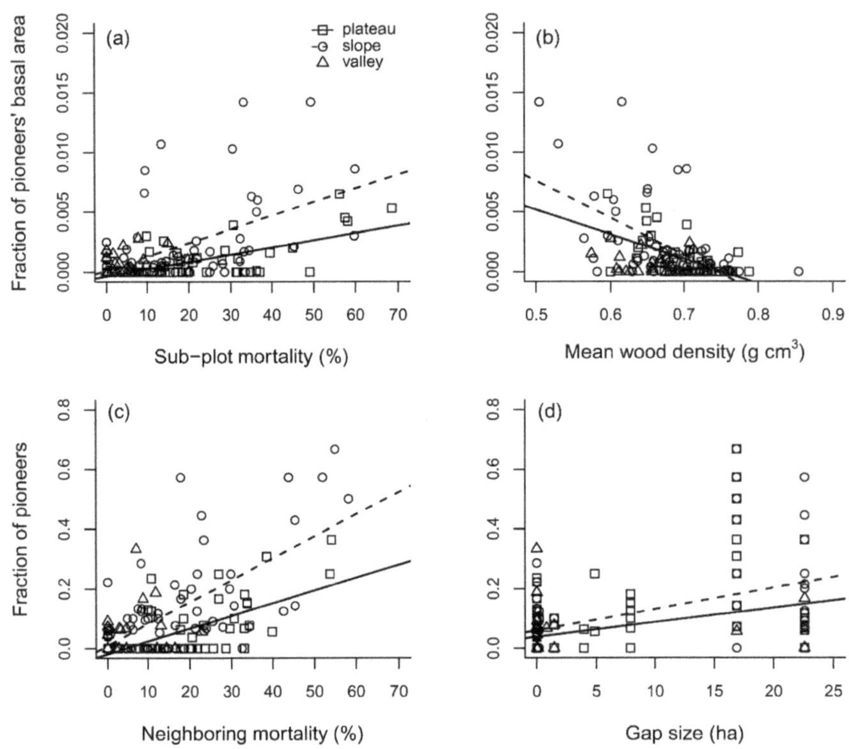

Figure 6. Pioneer (light-demanding and fast-growing) tree species importance related to disturbance intensity measures and wood density in a forest disturbed by a large blowdown, in Amazonas, Brazil. Legend: (a) fraction of pioneers' basal area related to sub-plot mortality; (b) fraction of pioneers' basal area related to sub-plot mean wood density; (c) fraction of pioneers related to neighboring mortality; (d) fraction of pioneers related to gap size.

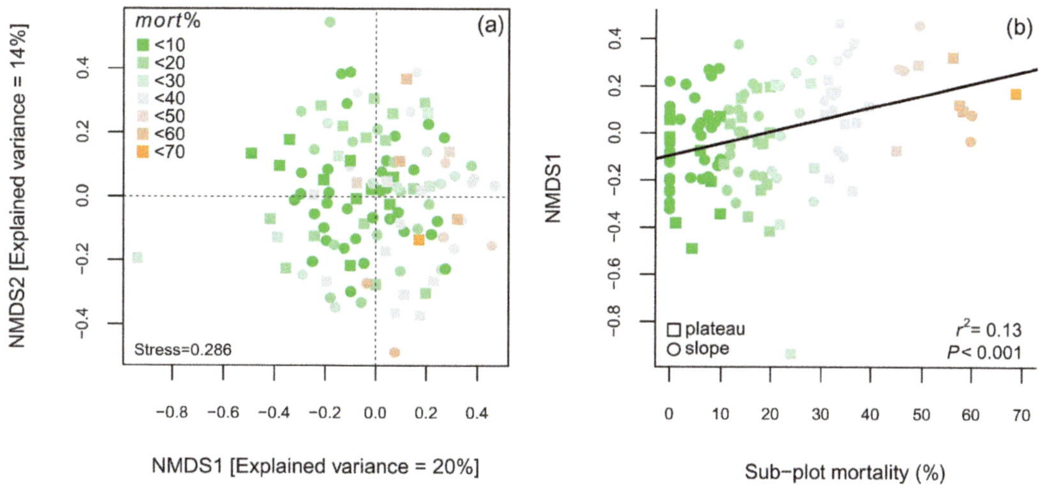

Figure 7. NMDS ordination diagram of 117 sub-plots sampled in plateaus and slopes of a forest disturbed by a large blowdown, in Amazonas, Brazil. Sub-plot mortality caused by the blowdown is highlighted by the color-scale. Legend: (a) NMDS scores computed from a dissimilarity matrix with the abundance of all recorded species; (b) first NMDS axis related to sub-plot mortality.

intensity – i.e. the fraction of trees killed by wind (Table 2). In heavily damaged areas with the highest estimated mortality rates, structural measures were lower than the early-successional stages found in other nearby and smaller disturbed forest patches [89] and anthropogenic disturbances in the same forest type [41,90]. At these sites, forest recovery will depend on species responses to the light gradient in gaps, which was supported for our results related to the third and fourth questions. Within the disturbed areas, the correlation of percent mortality with structural characteristics also showed that areas less affected by wind damage were similar to the undisturbed forest sub-plots.

The observed higher mortality within plateaus and slopes of the undisturbed forest was also related in nearby unaffected *terra-firme* forest, where under regular disturbance regimes, standing death can be expected to be more frequent in plateaus, and uprooting and snapping more frequent in slopes and in valleys [91]. Thus, competition can be more important in plateaus and exogenous disturbances dominant in slopes and valleys, which will cause differences in the size-density distribution of trees related to topographic position [92]. Despite such findings, our study indicates that for the Central Amazon forests, most notably in plateaus and slopes, tree mortality is also regulated by wind disturbance regimes such as blowdowns of varying size.

Although in the undisturbed forest we only observed small gaps, in the disturbed forest large gaps (2000 m^2) were common and changed forest structure, particularly in plateaus and slopes. Such results suggest that severe disturbances are required to form large gaps. The higher values of tree density and basal area observed in plateaus may indicate a variation in forest vulnerability due differences in species composition, or possibly higher resilience of plateaus.

Selective mortality patterns

We expected to see large emergent to be more prone to wind damage than smaller ones. While we did observe differences in DBH distribution of dead trees between undisturbed and disturbed forests that suggests lower proportional mortality for smaller trees (despite potential biases introduced because smaller dead trees decompose faster), we also observed no correlation between the DBH of dead trees and sub-plot mortality, indicating that size selective mortality did not depend on disturbance intensity.

Observed reductions in mean DBH and wood density in the live trees in highly damaged areas of the disturbed forest (Figure 2c) likely reflect fast recruitment of light-demanding and fast-growing species with shorter life span and generally lower wood density values [15,34,71]. Chambers *et al.* (2009b) compared non-disturbed and smaller blowdown gaps with ~10 years-succession and reported similar patterns, except for tree density, which was lower in our four-years disturbed plots. The substantial increase in the fraction of small trees with increasing sub-plot mortality also evidentiate the recruitment of shade-tolerant or understory species that usually do not reach the canopy, and thus demonstrates a selective mortality feedback on floristic composition.

Tree diversity in Central Amazon forests is high, and our data agree with those of other studies conducted in the same [52,67,93] and in adjacent areas [55,57,94,95]. We observed clear effects of the blowdown mortality on community demographic patterns from low to high disturbance (Figure 3). Part of the observed changes can be attributed to the fast recruitment of species with different syndromes and growth (lateral expansion) in the available gap space, such as *Cecropia sciadophylla* Mart., *Pourouma tomentosa* Mart. ex Miq., *Inga* cf. *paraensis*, *Iryanthera juruensis* Warb., *Protium hebetatum* D. C. Daly, *Mabea speciosa* Müll.Arg. and *Zygia racemosa* (Ducke) Barneby & J. W. Grimes. These species improved their importance within the high disturbance sub-plots by increasing their relative density and frequency. Such patterns may indicate success of establishment and expansion of these populations [81], which in the longer-term may increase species richness and forest diversity [89]. Common tree canopy genera in adjacent areas (e.g. *Couratari*, *Eschweilera*, *Pouteria*, *Licania*, *Micropholis*, *Ocotea*, *Sloanea*, *Swartzia* and *Vantanea*) [67,90,93], also presented high importance within the sub-plots with high disturbance (i.e. high mortality). Survivorship of shade-tolerant and slower-growing species from these genera may be attributable to the relatively high wood density (~0.79 g/cm^3) [71], small crown footprint, and presence of buttress and supporting roots [54,56]. Blowdowns are a major disturbance in these forests, thus the greater likelihood of survival and resprouting ability may be reasons for these genera to be of such high importance in the *terra firme* forests of this region [54–56,96].

Table 3. NMDS and linear regression relating measures of disturbance intensity to the floristic composition of a forest disturbed by a large blowdown, in Amazonas, Brazil.

NMDS	Mantel-test*		NMDS [Axis1] related to disturbance measures¥				
Stress	Axis 1	Axis 2	Predictors	F	r^2	P	Pearson's r
0.286	0.442	0.367	SP	18.15	0.129	<0.001	0.370
			NG	21.81	0.159	<0.001	0.400
			GS	25.56	0.175	<0.001	0.427

Legend: SP-sub-plot mortality; NG-neighboring mortality; GS-gap size.
*Related to species distance matrix and NMDS axis scores.
¥Linear regressions with NMDS (Axis1) as dependent variable and disturbance intensity measures as predictors.

Blowdown mortality, which directly affected species richness by reducing tree density at a local scale, resulted in the negative correlations between species richness and Shannon diversity with sub-plot mortality (Figure 4a and b). Nevertheless, the rarefaction curve indicate that a slightly higher number of species would be observed for any smaller subsample of individuals taken from low to intermediate disturbance areas, under the assumption of random mixing of individuals (Figure 4c). Interestingly, despite differences in species richness driven by disturbance intensity, the higher number of rare species at low and intermediate disturbance levels indicates that a short-term feedback of blowdown mortality is to promote species richness at the landscape scale, with apparent lower impacts on valleys.

Disturbance gradient effects and vegetation responses

So far we have focused on the importance of wind disturbance for providing niche space for light-demanding species requiring large canopy gaps for regeneration as a mechanism for promoting species coexistence and thus diversity at the landscape scale. Concomitantly, our data and tests related to the third and fourth questions support that the studied gaps also provide adequate niches for species with different requirements.

Disturbance and selective mortality patterns promoted immediate effects on forest structure and species composition, as evidenced by changes in tree size distribution (Figure 2), species importance between levels of disturbance (Figure 3) and species diversity along the mortality gradient (Figure 4). Nonetheless, blowdown gaps seem to produce long-term effects at the community level. The observed landscape mortality gradient also changed diversity patterns by promoting compositional changes with respect to the fraction of resprouters (Figure 5) and with light-demanding/fast-growing species (Figure 6).

Thus, a second potentially important niche axis is the ability to survive wind disturbance, which may provide a selective advantage to species that are less susceptible to wind damage either because they possess specific morphological and anatomical adaptations lending stability or because they are of small stature and are typically not exposed to wind. Both characteristics may represent a competitive disadvantage in the absence of wind disturbances. In the first case, stability conferred by e.g. high wood density is associated with high construction costs and reduced growth rates – a typical growth-defense tradeoff reported in the literature [8,71,97]. In the second case, staying short or retaining a low height/diameter ratio is certainly a disadvantage in the race for light [62].

The positive correlations between disturbance intensity measures and density of resprouters in the disturbed forest (Figure 5) confirms the importance of this regeneration mechanism in gaps created by wind disturbances [31,47,48,86,98,99]. The similarities in mean wood density between resprouters and non-resprouters (usually undamaged trees) indicate that resprouting is a regeneration pathway adopted by species with different traits and light requirements, predominantly by smaller trees (DBH≤20 cm). A similar pattern was found in a Caribbean hurricane-damaged forest [45] and indicates that, although canopy trees from higher wood density species may have higher wind resistance, resprouting is not an exclusive regeneration pathway for shade-tolerant, slower-growing or climax species. As for sub-tropical hurricane-damaged forests [46], we hypothesize that in *terra firme* forests of Central Amazon, both early secondary and understory species may benefit from wind disturbances. Finally, the decrease in the mean aggregated wood density in disturbed sub-plots is a direct consequence of the higher fraction of pioneer species which are characterized by low wood densities (Figure 6). Wood density

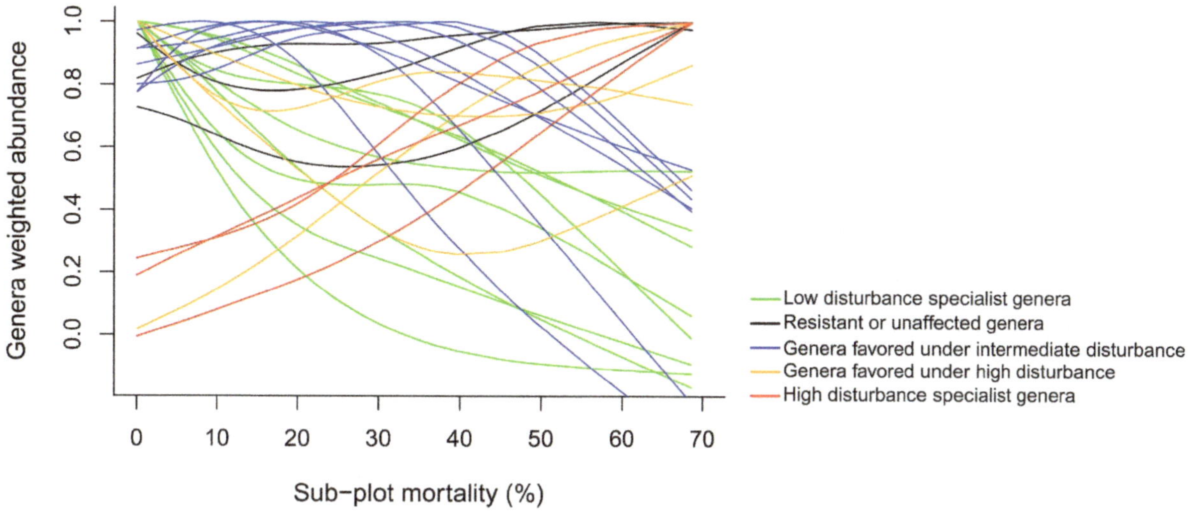

Figure 8. Genera abundance related to a mortality gradient in plateaus and slopes of a forest disturbed by a large blowdown, in Amazonas, Brazil. Legend: low disturbance specialist genera- *Brosimum, Eperua, Iryanthera, Micrandropsis, Micropholis, Pouteria, Protium* and *Unonopsis*; resistant or unaffected genera- *Couratari, Ocotea* and *Sloanea*; genera favored under intermediate disturbance- *Eschweilera, Lecythis, Licania, Naucleopsis, Swartzia, Qualea* and *Zygia*; genera favored under high disturbance- *Couepia, Mabea, Scleronema* and *Vantanea*; high disturbance specialist genera- *Cecropia, Inga* and *Pourouma.*

variation among sub-plots was partially explained by sub-plot mortality, which is significant when low wood density pioneers are observed (Figure 6a and b). This greater importance of pioneers in heavily damaged areas corroborates demographic patterns found by assessing species IVI (Figure 3b) and patterns in species composition, which was indicated by the NMDS-test and regressions (Figure 7).

Our data show that species turnover in the 20 most important genera (as quantified by IVI in Figure 3) is due to the increasing importance of both non-pioneer and pioneer species. Indirect evidence for this mechanism is provided by fact that the floristic dissimilarity between undisturbed and disturbed sub-plots seems to be driven by the admixture of light-demanding species (Figure 7a and Figure 8). There was a significant correlation between sub-plot mortality and the position of sub-plots along the first NMDS axis, the latter explaining significant part of the total variation in species composition (Figure 7b). This relationship was higher when considering neighboring-mortality and gap size values (Table 3). However, even in the plots with highest mortality rates (40–70%) there was no complete takeover by classical pioneer local genera (e.g. *Cecropia*, *Conceveiba*, *Croton*, *Goupia*, *Inga*, *Laetia*, *Miconia*, *Pourouma*, *Tachigali* and *Vismia*) or shade-tolerant common genera in undisturbed forests (e.g. *Couratari*, *Eschweilera*, *Licania*, *Pouteria*, *Scleronema*, *Sloanea* and *Swartzia*). These last, maintained up a large fraction of the IVI as indicated by their high proportion in the largest diameter class. Some of the fast growth species we have recorded, especially from the genus *Cecropia*, *Inga*, *Pourouma* and *Tachigali* were also reported as important species during early succession stages of secondary *terra-firme* forests around Manaus [40,90,100] and in the upper Rio Negro [41]. In anthropic secondary forests, under non-intensive use and without fire regimes, areas dominated by species from the genus *Cecropia* and *Pourouma* present relatively higher species richness, which hence more rapid plant succession [40,100].

The variations of genera importance in respect to mortality intensity support classical studies [2,5,6,34] that attempt to classify species into a simplified conception of pioneer species in gaps versus intermediate or late-succession species under undisturbed forest patches. These classical opposing strategies were observed within the 25 most important genera of the disturbed forest (Figure 8), as characterized as high (light-demanding and faster-growing - red lines) and low disturbance specialists (shade-tolerant and slower-growing - green lines). Additionally, our data highlighted alternative successional trajectories related to survival (resistant or unaffected - black lines), resprouting and fast-recruitment (favored under intermediate and high disturbance - blue and orange, respectively).

These alternative successional trajectories have not been observed as important regeneration mechanisms in smaller treefall events. In larger gaps (>2000 m^2), the canopy emergence of specialized guilds with an optimum at specific levels of disturbance, indicate a smooth occupation of the entire gap space and light conditions. The similar pattern within genera belonging to the same guilds may indicate similar regeneration strategies and historical life. A possible explanation for the success of species from these different genera may be the higher plasticity of traits and alternative resilience mechanisms (such as survival, resprouting and recruitment). As blowdowns are common in these forests [10,12,14], these results show that a simplified classification system is not enough to describe successional trajectories of large gaps.

Surprisingly, all the 25 most important genera in the disturbed forest (Figure 3) also figured within the 121 most abundant genera of the main portion of the Amazon [96]. *Protium*, *Pouteria* and *Eperua* characterized the low disturbance specialist guild in the disturbed forest (Figure 8) and took the second, third and seventh position in the Amazonian genera abundance ranking, respectively [96]. *Ocotea*, *Sloanea* and *Couratari* characterized the resistant or unaffected guild in the disturbed forest and took the 14°, 22° and 106° position in the Amazonian ranking, respectively. *Eschweilera*, *Licania*, *Lecythis* and *Swartzia* characterized the intermediate disturbance intensity at the disturbed forest and took the first, fourth, 10° and 17° position in the Amazonian ranking, respectively. *Inga*, *Pourouma* and *Cecropia* characterized the high disturbance specialist guild in the disturbed forest and figured at the sixth, 18° and 30° position in the Amazonian ranking.

The co-occurrence of species with a broad range of life history strategies at intermediate disturbance levels (evidentiated as higher confluence of lines - Figure 8) shows that this blowdown promoted species richness probably by amplifying niches and/or resources. We assume that large gaps (>2000 m^2) produced by blowdowns can be partly associated to increased diversity and the contrasting dominance of some species in these forests [54,55,57,67,94,96]. Variations in genera importance and species composition along the disturbance gradient indicate that gap recovery in large gaps is influenced by mortality patters, most pronounced in plateaus and slopes. Additionally, selective mortality and fast-recruitment highlight genera specialization along the mortality gradient, and indicate that large gaps contain environmental variability that together with species responses and resistance, allow high tree species diversity. In this study, intermediate-disturbance levels had higher species richness suggesting that there was a selective mortality and resulting species turnover in response to the disturbance gradient. This pattern provides evidence that blowdowns allow species richness and diversity through an interaction of wind-damage with species-resistance and resilience. Although classical successional guilds may dominate small treefall gaps, the observed pattern indicate that in large gaps (>2000 m^2) there are specialists for a wider range of disturbance and light. Thus, we assume that a diverse set of species differing widely in light requirements and recruitment strategies forms the pioneer cohort, thus lending a high resilience towards wind disturbances at the community level.

Our results add new and complementary information about succession and turnover of Western and Central Amazon forests and reinforce that recovery processes for large gaps differ from the trajectory observed in gaps formed by smaller treefall events [2,6,15,31,86] and secondary forests from anthropic activities, where fire and logging may increase biomass losses [41,90] and limit species regeneration [40,100]. Moreover, we revealed a gradient in demographic responses along the entire gradient of disturbance levels. Mortality intensity and gap size seem to influence community composition by filtering shade-tolerant survivors with resprouting ability and favoring more light-demanding and fast-growing species. As already observed in tropical and subtropical regions [26,29,32,47,98,101] wind damage depends on species composition and successional stage, which suggest that secondary forests in the Amazon might be more vulnerable and less resilient to windstorms than forests in more advanced successional stage. Thus, for better understanding the blowdown effects on species distribution, forest vulnerability, ecosystem functioning, and the ecological importance of large gaps on species maintenance it is necessary that future studies include dynamic and long-term succession data. In addition, the proliferation of lianas and pioneer tree species in severely disturbed areas may represent an important impediment to seedling establishment [102]. Previous studies in this region showed that it takes an average of ~18 years for dead trees (>10 cm DBB) to completely decay, although some can take considerably longer [103], so that coarse woody debris and greater surface litter accumulation can have a persistent effect on forest regeneration by covering the soil and acting as a physical barrier, which can favor or exclude species [104,105]. The persistence of decomposing debris may also alter nutrient and even perhaps water availability to seedlings and regenerating species, all of which will serve as filters that help determine which species recolonize large gaps.

Conclusion

Blowdown gaps larger than 2000 m^2 initiate secondary succession providing niches to establish a canopy position for species with a broad range of life history strategies and requirements. Smooth gradients in demographic responses and the existence of specialists for different levels of disturbance can perhaps help explain carbon cycle, maintenance of biodiversity and the recently reported hyperdominance of some tree and palm species in these forests [96]. If forest structure and species composition depend on the intensity and frequency of large disturbance regimes, the intensification of more extreme climate events such as convective storms, may alter forest vulnerability and resilience depending on the successional trajectories. Considering size, vegetation heterogeneity and local logistical limitations, permanent forest monitoring in Amazon must combine remote sensing methods allowing the inclusion of large natural disturbances.

Acknowledgments

We gratefully thank workers and parabotanists from the EEST for the field support, and Dr. Valdely F. Kinupp for the support at the IFAM herbarium. We also thank anonymous reviewers for their valuable contribution, Dr. Shaun Levick for helping with GIS analysis, and SUFRAMA for allowing us to research at the ZF2 area.

Author Contributions

Conceived and designed the experiments: DMM JQC NH SET JS GHPMR RIN. Performed the experiments: DMM GHPMR. Analyzed the data: DMM CW BR GHPMR RIN JQC. Contributed reagents/materials/analysis tools: SET NH JS JQC CW DMM. Wrote the paper: DMM CW SET GHPMR RIN BR JQC NH.

References

1. Clements F (1936) Nature and structure of the climax. J Ecol 24: 252–284.
2. Denslow J (1980) Patterns of plant species diversity during succession under different disturbance regimes. Oecologia 46: 18–21.
3. Sousa W (1984) The Role of Disturbance in Natural Communities. Annu Rev Ecol Syst 15: 353–391.
4. Turner M, Baker W, Peterson C, Peet R (1998) Factors Influencing Succession: Lessons from Large, Infrequent Natural Disturbances. Ecosystems 1: 511–523.
5. Grubb P (1977) The maintenance of species-richness in plant communities: the importance of the regeneration niche. Biol Rev 52: 107–145.
6. Brokaw N (1985) Gap-phase regeneration in a tropical forest. Ecology 66: 682–687.
7. Uhl C, Clark K, Dezzeo N, Maquirino P (1988) Vegetation dynamics in Amazonian treefall gaps. Ecology1 69: 751–763.
8. Coley P, Kursar T (2014) On tropical forests and their pests. Science 343: 35–36.
9. Negrón-Juárez R, Chambers J, Guimarães G, Zeng H, Raupp C, et al. (2010) Widespread Amazon forest tree mortality from a single cross-basin squall line event. Geophys Res Lett 37: 1–5.
10. Chambers JQ, Negron-Juarez RI, Marra DM, Di Vittorio A, Tews J, et al. (2013) The steady-state mosaic of disturbance and succession across an old-growth Central Amazon forest landscape. Proc Natl Acad Sci U S A 110: 3949–3954.
11. Chambers J, Negrón-Juárez R, Hurtt G, Marra D, Higuchi N (2009) Lack of intermediate-scale disturbance data prevents robust extrapolation of plot-level tree mortality rates for old-growth tropical forests. Ecol Lett 12: E22–E25.
12. Nelson B, Kapos V, Adams J, Oliveira W (1994) Forest disturbance by large blowdowns in the Brazilian Amazon. Ecology 75: 853–858.
13. Espírito-Santo F, Keller M, Braswell B, Nelson B, Frolking S, et al. (2010) Storm intensity and old-growth forest disturbances in the Amazon region. Geophys Res Lett 37: 1–6.
14. Espírito-Santo F, Gloor M, Keller M, Malhi Y, Saatchi S, et al. (2014) Size and frequency of natural forest disturbances and the Amazon forest carbon balance. Nat Commun 5: 1–6.
15. Denslow J (1987) Tropical rainforest gaps and tree species diversity. Annu Rev Ecol Syst 18: 431–451.

16. Hubbell S, Foster R, O'Brien S, Harms K, Condit R, et al. (1999) Light-Gap Disturbances, Recruitment Limitation, and Tree Diversity in a Neotropical Forest. Science 283: 554–557.
17. Gloor M, Phillips OL, Lloyd JJ, Lewis SL, Malhi Y, et al. (2009) Does the disturbance hypothesis explain the biomass increase in basin-wide Amazon forest plot data? Glob Chang Biol 15: 2418–2430.
18. Chave J, Condit R, Muller-Landau HC, Thomas SC, Ashton PS, et al. (2008) Assessing evidence for a pervasive alteration in tropical tree communities. PLoS Biol 6: e45.
19. Kellner JR, Asner GP (2009) Convergent structural responses of tropical forests to diverse disturbance regimes. Ecol Lett 12: 887–897.
20. Di Vittorio A, Negrón-Juárez R, Higuchi N, Chambers J (2014) Tropical forest carbon balance: effects of field-and satellite-based mortality regimes on the dynamics and the spatial structure of Central Amazon forest biomass. Environ Res Lett 9: 1–10.
21. Fisher J, Hurtt G, Thomas R, Chambers J (2008) Clustered disturbances lead to bias in large-scale estimates based on forest sample plots. Ecol Lett 11: 554–563.
22. Hjelmfelt M (2007) Microbursts and macrobursts: windstorms and blowdowns. In: Johnson E, Miyanishi K, editors. Plant Disturbance Ecology: The Process and the Response. Burlington, MA: Elsevier, Academic Press. 59–95.
23. Brokaw N, Grear J (1991) Forest structure before and after the hurricane Hugo at three elevations in the Luquillo Mountains, Porto Rico. Biotropica 23: 386–392.
24. Boose E, Serrano M, Foster D (2004) Landscape and regional impacts of hurricanes in Puerto Rico. Ecol Monogr 74: 335–352.
25. Negrón-Juárez R, Chambers J, Marra D, Ribeiro G, Rifai S, et al. (2011) Detection of subpixel treefall gaps with Landsat imagery in Central Amazon forests. Remote Sens Environ 115: 3322–3328.
26. Putz F (1983) Treefall pits and mounds, buried seeds, and the importance of soil disturbance to pioneer trees on Barro Colorado Island, Panama. Ecology 64: 1069–1074.
27. Denslow J, Ellison A, Sanford R (1998) Treefall gap size effects on above- and below-ground processes in a tropical wet forest. J Ecol 86: 597–609.
28. Vitousek P, Denslow J (1986) Nitrogen and phosphorus availability in treefall gaps of a lowland tropical rainforest. J Ecol 74: 1167–1178.
29. Negrón-Juárez R, Baker D, Chambers J, Hurtt G, Goosem S (2014) Multi-scale sensitivity of Landsat and MODIS to forest disturbance associated with tropical cyclones. Remote Sens Environ 140: 679–689.
30. Bellingham P, Tanner E, Healey J (1995) Damage and responsiveness of Jamaican montane tree species after disturbance by a hurricane. Ecology 76: 2562–2580.
31. Vandermeer J, la Cerda I de, Boucher D, Perfecto I, Ruiz J (2000) Hurricane Disturbance and Tropical Tree Species Diversity. Science 290: 788–791.
32. Canham C, Thompson J, Zimmerman J, Uriarte M (2010) Variation in susceptibility to hurricane damage as a function of storm intensity in Puerto Rican tree species. Biotropica 42: 87–94.
33. Uriarte M, Canham C, Thompson J, Zimmerman J, Brokaw N (2005) Seedling recruitment in a hurricane-driven tropical forest: light limitation, density-dependence and the spatial distribution of parent trees. J Ecol 93: 291–304.
34. Swaine M, Whitmore T (1988) On the definition of ecological species groups in tropical rain forests. Vegetatio 75: 81–86.
35. Dalling JW, Winter K, Hubbell SP (2004) Variation in growth responses of neotropical pioneers to simulated forest gaps. Funct Ecol 18: 725–736.
36. Clark D, Clark D (2001) Getting to the canopy: tree height growth in a neotropical rain forest. Ecology 82: 1460–1472.
37. Wright J, Muller-Landau H, Condit R, Hubbell S (2003) Gap-dependent recruitment, realized vital rates, and size distributions of tropical trees. Ecology 84: 3174–3185.
38. Schupp E, Howe H, Augspurger C, Levey D (1989) Arrival and survival in tropical treefall gaps. Ecology 70: 562–564.
39. Poorter L, Bongers F (2006) Leaf traits are good predictors of plant performance across 53 rain forest species. Ecology 87: 1733–1743.
40. Mesquita R, Ickes K, Ganade G, Williamson G (2001) Alternative successional pathways in the Amazon Basin. J Ecol 89: 528–537.
41. Saldarriaga J, West D, Tharp M, Uhl C (1998) Long-Term Chronossequence of Forest Succession in the Upper Rio Negro of Colombia and Venezuela. J Ecol 76: 938–958.
42. Rüger N, Berger U, Hubbell SP, Vieilledent G, Condit R (2011) Growth strategies of tropical tree species: disentangling light and size effects. PLoS One 6: e25330.
43. Gravel D, Canham C, Beaudet M, Messier C (2006) Reconciling niche and neutrality: the continuum hypothesis. Ecol Lett 9: 399–409.
44. Violle C, Jiang L (2009) Towards a trait-based quantification of species niche. J Plant Ecol 2: 87–93.
45. Vandecar K, Lawrence D, Richards D, Schneider L, Rogan J, et al. (2011) High mortality for rare species following hurricane disturbance in the Southern Yucatan. Biotropica 43: 676–684.
46. Putz F, Brokaw N (1989) Sprouting of broken trees on Barro Colorado Island, Panama. Ecology 70: 508–512.
47. Zimmerman J, III Everham E, Waide R, Lodge D, Taylor C, et al. (1994) Responses of tree species to hurricane winds in subtropical wet forest in Puerto Rico: implications for tropical tree life histories. J Ecol 82: 911–922.
48. Lugo A (2008) Visible and invisible effects of hurricanes on forest ecosystems: an international review. Austral Ecol 33: 368–398.
49. Foster D, Knight D, Franklin J (1998) Landscape Patterns and Legacies Resulting from Large, Infrequent Forest Disturbances. Ecosystems 1: 497–510.
50. Ranzani G (1980) Identificação e caracterização de alguns solos da Estação Experimental de Silvicultura Tropical. Acta Amaz 10: 7–41.
51. Sombroek W (2001) Spatial and temporal patterns of Amazon rainfall: consequences for the planning of agricultural occupation and the protection of primary forests. Ambio 30: 388–396.
52. Silva R da, Santos J dos, Tribuzy E, Chambers J, Nakamura S, et al. (2002) Diameter increment and growth patterns for individual tree growing in Central Amazon, Brazil. For Ecol Manage 166: 295–301.
53. Higuchi N, Santos J dos, Lima A, Higuchi F, Chambers J (2011) A floresta amazônica e a água da chuva. FLORESTA 41: 427–434.
54. Ribeiro J da S, Hopkins M, Vicentini A, Sothers C, Costa M da S, et al. (1999) Flora da Reserva Ducke: Guia de Identificação das Plantas Vasculares de uma Floresta de Terra-firme na Amazônia Central. Manaus: INPA.
55. Oliveira A de, Mori S (1999) A central Amazonian terra firme forest. I. High tree species richness on poor soils. Biodivers Conserv 8: 1219–1244.
56. Braga P (1979) Subdivisão fitogeográfica, tipos de vegetação, conservação e inventário florístico da floresta amazônica. Acta Amaz 9: 53–80.
57. Guillaumet J (1987) Some structural and floristic aspects of the forest. Experientia 43: 241–251.
58. Kahn F (1986) Life forms of Amazonian palms in relation to forest structure and dynamics. Biotropica 18: 214–218.
59. Chauvel A, Lucas Y, Boulet R (1987) On the genesis of the soil mantle of the region of Manaus, Central Amazonia, Brazil. Experientia 43: 234–241.
60. Castilho C, Magnusson W, Araújo R, Luizão R, Luizão F, et al. (2006) Variation in aboveground tree live biomass in a central Amazonian Forest: Effects of soil and topography. For Ecol Manage 234: 85–96.
61. Costa F, Guillaumet J, Lima A, Pereira O (2009) Gradients within gradients: the mesoscale distribution patterns of palms in a central amazonian forest. J Veg Sci 20: 69–78.
62. Ribeiro G, Suwa R, Marra D, Kajimoto T, Ishizuka M, et al. (2014) Allometry for juvenile trees in an Amazonian forest after wind disturbance. JARQ 48: 213–219.
63. APGIII (2009) An update of the Angiosperm Phylogeny Group classification for the orders and families of flowering plants: APG III. Bot J Linn Soc 161: 105–121.
64. APG II TAPG (2003) An update of the Angiosperm Phylogeny Group classification for the orders and families of flowering plants: APG II. Bot J Linn Soc London 141: 399–436.
65. Stevens P (2012) Angiosperm Phylogeny Website: version 12. Available: http://www.mobot.org/MOBOT/research/APweb/.
66. Higuchi N, Santos J dos, Vieira G, Ribeiro R, Sakurai S, et al. (1998) Análise Estrutural da Floresta Primária da Bacia do Rio Cuieiras, ZF-2, Manaus-AM, Brasil. In: Higuchi N, Campos M, Sampaio P, Santos J dos, editors. Pesquisas Florestais para Conservação da Floresta e Reabilitaçao de Áreas Degradadas da Amazônia. Manaus, Amazonas, Brazil: INPA. 51–81.
67. Vieira S, Camargo P de, Selhorst D, Silva R da, Hutyra L, et al. (2004) Forest structure and carbon dynamics in Amazonian tropical rain forests. Oecologia 140: 468–479.
68. Phillips OL, Aragão LEOC, Lewis SL, Fisher JB, Lloyd J, et al. (2009) Drought sensitivity of the Amazon rainforest. Science 323: 1344–1347.
69. Marengo J, Nobre C, Tomasella J, Oyama M, Oliveira G de, et al. (2008) The drought of Amazonia in 2005. J Clim 21: 495–517.
70. Fearnside P (1997) Wood density for estimating forest biomass in Brazilian Amazonia. For Ecol Manage 90: 59–87.
71. Chave J, Muller-Landau H, Baker T, Easdale T, ter Steege H, et al. (2006) Regional and phylogenetic variation of wood density across 2456 neotropical tree species. Ecol Appl 16: 2356–2367.
72. Nogueira EM, Nelson BW, Fearnside PM (2005) Wood density in dense forest in central Amazonia, Brazil. For Ecol Manage 208: 261–286.
73. Nogueira EM, Fearnside PM, Nelson BW, França MB (2007) Wood density in forests of Brazil's "arc of deforestation": Implications for biomass and flux of carbon from land-use change in Amazonia. For Ecol Manage 248: 119–135.
74. Carlotto M (1999) Reducing the effects of space-varying, wavelength-dependent scattering in multispectral imagery. Int J Remote Sens 20: 3333–3344.
75. Adams J, Sabol D, Kapos V, Filho R, Roberts D, et al. (1995) Classification of multispectral images based on fractions of endmembers: Application to land-cover change in the Brazilian Amazon. Remote Sens Environ 52: 137–154.
76. Adams J, Gillespie A (2006) Remote sensing of landscapes with spectral images. A physical modeling approach. Cambridge University Press.
77. Higuchi N, Chambers J, Santos J dos, Ribeiro R, Pinto A, et al. (2004) Dinâmica e balanço do carbono da vegetaçao primária da Amazônia Central. Floresta 34: 295–304.
78. Hijmans R, van Etten J (2012) raster: Geographic data analysis and modeling. R package. Available: http://cran.r-project.org/.
79. Lewin-Koh N, Bivand R (2013) maptools: Tools for reading and handling spatial objects. R package. Available: http://cran.r-project.org/.
80. R Core Team (2013) R: A language and environment for statistical computing. Available: http://www.r-project.org.

81. Curtis J, McIntosh R (1950) The interrelations of certain analytic and synthetic phytosociological characters. Ecology 31: 434–455.
82. Anderson M, Crist T, Chase J, Vellend M, Inouye B, et al. (2011) Navigating the multiple meanings of β diversity: a roadmap for the practicing ecologist. Ecol Lett 14: 19–28.
83. Legendre P, Legendre L (2012) Numerical ecology. 3rd ed. Amsterdam: Elsevier.
84. Oksanen J, Blanchet F, Kindt R, Legendre P, Minchin P, et al. (2013) vegan: Community Ecology Package. Available: http://cran.r-project.org/.
85. Lewis S, Phillips O, Sheil D, Vinceti B, Baker T, et al. (2004) Tropical forest tree mortality, recruitment and turnover rates: calculation, interpretation and comparison when census intervals vary. J Ecol 92: 929–944.
86. Walker L (1991) Tree damage and recovery from Hurricane Hugo in Luquillo experimental forest, Puerto Rico. Biotropica 23: 379–385.
87. Xi W, Peet RK, Decoster JK, Urban DL (2008) Tree damage risk factors associated with large, infrequent wind disturbances of Carolina forests. Forestry 81: 317–334.
88. Garstang M, White S, Shugart H, Halverson J (1998) Convective cloud downdrafts as the cause of large blowdowns in the Amazon rainforest. Meteorol Atmos Phys 67: 199–212.
89. Chambers J, Robertson A, Carneiro V, Lima A, Smith M, et al. (2009) Hyperspectral remote detection of niche partitioning among canopy trees driven by blowdown gap disturbances in the Central Amazon. Oecologia 160: 107–117.
90. Lima A, Teixeira L, Carneiro V, Santos J dos, Higuchi N (2007) Análise da estrutura e do estoque de fitomassa de uma floresta secundária da região de Manaus AM, dez anos após corte raso seguido de fogo. Acta Amaz 37: 49–54.
91. Toledo J de, Magnusson W, Castilho C, Nascimento H (2012) Tree mode of death in Central Amazonia: Effects of soil and topography on tree mortality associated with storm disturbances. For Ecol Manage 263: 253–261.
92. Toledo J de, Magnusson W, Castilho C (2013) Competition, exogenous disturbances and senescence shape tree size distribution in tropical forest: evidence from tree mode of death in Central Amazonia. J Veg Sci 24: 651–663.
93. Carneiro V, Higuchi N, Santos J dos, Pinto A, Teixeira L, et al. (2005) Composição Florística e Análise Estrutural da Floresta de Terra Firme na Região de Manaus, Estado do Amazonas, Brasil. Actas das Comunicações - Inventário, Modelação e Gestão. Viso, Portugal: V Congresso Florestal Nacional A Floresta e as Gentes. Available: http://www.esac.pt/cernas/cfn5/docs/T2-40.pdf.
94. Prance G, Rodrigues W, Silva M da (1976) Inventário florestal de um hectare de mata de terra firme km 30 da Estrada Manaus-Itacoatira. Acta Amaz 6: 9–35.
95. Oliveira A, Nelson B (2001) Floristic relationships of terra firme forests in the Brazilian Amazon. For Ecol Manage 146: 169–179.
96. Ter Steege H, Pitman NC a, Sabatier D, Baraloto C, Salomão RP, et al. (2013) Hyperdominance in the Amazonian tree flora. Science 342: 1243092.
97. Muller-Landau H (2004) Interspecific and inter-site variation in wood specific gravity of tropical trees. Biotropica 36: 20–32.
98. Brokaw N, Walker L (1991) Summary of the effects of Caribbean hurricanes on vegetation. Biotropica 23: 442–447.
99. Burslem D, Whitmore T, Brown G (2000) Short-term effects of cyclone impact and long-term recovery of tropical rain forest on Kolombangara, Solomon Islands. J Ecol 88: 1063–1078.
100. Massoca P, Jakovac A, Bento T, Williamson G, Mesquita R (2012) Dynamics and trajectories of secondary succession in Central Amazonia. Bol Mus Para Emílio Goeldi 7: 235–250.
101. Thompson J, Brokaw N, Zimmerman J, Waide RB, Everham EM III, et al. (2002) Land use history, environment, and tree composition in a tropical forest. Ecol Appl 12: 1344–1363.
102. Schnitzer S, Dalling J, Carson W (2001) The impact of lianas on tree regeneration in tropical forest canopy gaps: evidence for an alternative pathway of gap-phase regeneration. J Ecol 88: 655–666.
103. Chambers J, Higuchi N, Teixeira L, Santos J dos, Laurance S, et al. (2004) Response of tree biomass and wood litter to disturbance in a Central Amazon forest. Oecologia 141: 596–611.
104. Clark D, Clark D (1991) The impact of physical damage on canopy tree regeneration in tropical rain forest. J Ecol 79: 447–457.
105. Scariot A (2006) Seedling Mortality by Litterfall in Amazonian Forest Fragments1. Biotropica 32: 662–669.

3

Merging National Forest and National Forest Health Inventories to Obtain an Integrated Forest Resource Inventory – Experiences from Bavaria, Slovenia and Sweden

Marko Kovač[1]*, Arthur Bauer[2], Göran Ståhl[3]

1 Department of Forest and Landscape Planning and Monitoring, Slovenian Forestry Institute, Večna pot 2, Ljubljana, Slovenia, **2** Department of Food, Agriculture and Forestry Cham – Division of Forestry, Bavarian State Ministry of Food, Agriculture and Forestry, Ölbergstraβe 3, Waldmünchen, Germany, **3** Department of Forest Resource Management, Swedish University of Agricultural Sciences (SLU), Skogsmarksgränd, Umeå, Sweden

Abstract

Backgrounds, Material and Methods: To meet the demands of sustainable forest management and international commitments, European nations have designed a variety of forest-monitoring systems for specific needs. While the majority of countries are committed to independent, single-purpose inventorying, a minority of countries have merged their single-purpose forest inventory systems into integrated forest resource inventories. The statistical efficiencies of the Bavarian, Slovene and Swedish integrated forest resource inventory designs are investigated with the various statistical parameters of the variables of growing stock volume, shares of damaged trees, and deadwood volume. The parameters are derived by using the estimators for the given inventory designs. The required sample sizes are derived via the general formula for non-stratified independent samples and via statistical power analyses. The cost effectiveness of the designs is compared via two simple cost effectiveness ratios.

Results: In terms of precision, the most illustrative parameters of the variables are relative standard errors; their values range between 1% and 3% if the variables' variations are low (s%<80%) and are higher in the case of higher variations. A comparison of the actual and required sample sizes shows that the actual sample sizes were deliberately set high to provide precise estimates for the majority of variables and strata. In turn, the successive inventories are statistically efficient, because they allow detecting the mean changes of variables with powers higher than 90%; the highest precision is attained for the changes of growing stock volume and the lowest for the changes of the shares of damaged trees. Two indicators of cost effectiveness also show that the time input spent for measuring one variable decreases with the complexity of inventories.

Conclusion: There is an increasing need for credible information on forest resources to be used for decision making and national and international policy making. Such information can be cost-efficiently provided through integrated forest resource inventories.

Editor: Ben Bond-Lamberty, DOE Pacific Northwest National Laboratory, United States of America

Funding: The Slovene part of research was financed by the project JGS 4 (MAE; http://www.mko.gov.si/en/areas_of_work/forestry/) and the research program PS 1001 (SRA; http://www.arrs.gov.si/en/dobrodoslica.asp). The Bavarian and Swedish parts of research were part of regular research activities. The funders had no role in study design, data collection and analysis, decision to publish, or preparation of the manuscript.

Competing Interests: The authors have declared that no competing interests exist.

* Email: marko.kovac@gozdis.si

Introduction

The principle of sustainability [1], suggesting the use of forest resources in such a way as to continuously supply industry and households with wood assortments and fuel wood, was the main driving force that compelled European countries to begin inventorying their forestlands [2], [3]. Although the principle was understood similarly by the countries, different forest management traditions, practices and information needs compelled forest sciences and practices to develop forest inventories in two different directions.

The bottom-up stand-wise inventorying concept, promoted in Central Europe [4–7] and in other continents [8], was based on the idea that the data collected in small-sized allotments (viz. forest stands, compartments and forest management units) could be compiled upwards to support the information needs at larger planning scales (viz. forest management units, forest regions, nations). However, despite this theoretical possibility, in most cases, limited human and financial resources preventing the collection of detailed data in all management units, different data-collection techniques (e.g. full enumeration, purposive and probabilistic sampling, visual estimation) and the differences between the reference data hindered the derivation of statistically credible estimates at planning scales larger than forest management units [9]. Moreover, for the same reasons, these inventories could not provide reliable estimates on trends in time and space.

In contrast, the top-down inventory concept, developed in Nordic countries [10], [11], aimed to provide the basic estimates on forest resources such as the area, growing stock volume, increment, and so forth for large forest complexes. Unlike the first group of inventory concepts, this concept fully met the requirements of probabilistic sampling and consequently provided estimates with known levels of precision [3]. However, because of low sampling intensities, the usability of these estimates remained limited to large areas. Despite this deficiency, the theory behind these inventories was essential to all modern statistically-based forest inventories, especially to national forest inventories (hereafter: NFI).

Despite the shortcomings and rather poor sets of variables that aimed to describe forest sites, stands and forest resources (e.g. growing stock volume, increment), both inventory systems managed to provide information for formulating national forest policies, for international reporting [12], [13] and for creating forest management plans [9].

An important event that had implications for the development of the presently well-known integrated national forest resource inventories (especially in terms of comparable designs and multitudes of ecological variables) was the endorsement of the Convention on Long Range Trans-boundary Air Pollution [14]. Because the care for forest health was an essential element, the signatory states were invited to participate in the International Co-operative Program - Forests (hereafter: ICP Forests) and to build a large-scale cross-boundary forest health monitoring system. The system was to be used for detecting changes in tree and forest conditions over time and space and for better understanding the cause-effect relationships between the forest conditions and the factors affecting forest stands [15]. Although ICP Forests managed to build a joint inventory network covering the forests of all signatory states and to define the common sets of target variables along with the measurement protocols [15], it has neither succeeded in establishing close collaboration with the NFIs, nor in including their data in its analyses. Furthermore, the Life+ project [16], which, after the expiration of the Forest Focus scheme [17], was a kind of substitute for forest health monitoring, and whose actions were focused on the creation of a common sample grid and to the testing of methods for adapting data from NFIs, also had only partial success. Its greatest achievement was perhaps a scientific meeting in Florence in March 2009 (http://www.aisf.it/futmon/20.htm) at which a number of scientists described the progress, obstacles and future work in the field of inventory integrations.

A second event that potentially could contribute to the unification of forest inventory systems was the COST E43 action "Harmonization of National Forest Inventories in Europe: Techniques for Common Reporting" [18], [19], which was launched in 2004. Its objectives were (i) to improve and harmonize the existing national forest resource inventories in Europe, (ii) to support national teams in designing inventories to meet national, European and global requirements in supplying up-to-date, harmonized and transparent forest resource information, and (iii) to promote the use of scientifically sound and validated methods. The action supported three modules: Definitions and Measurement Practices, Estimation Procedures for carbon pools and carbon pool changes, and Indicators and Estimation Procedures of forest biodiversity [20].

Despite the general needs for integrated multi-resource forest inventories, which would not only bring the integrated data needed for complex statistical evaluations (viz. forest growth and yield, health and vitality [15], sustainable forest management [21], the assessment of the conservation status of forest habitat types and habitats [22]), but would also bring financial savings, the processes have only occasionally dealt with the questions of inventory integrations and have not, despite the many advantages of integrated inventories, compelled European countries to abandon the still prevailing single-purpose forest inventorying. Nevertheless, in addition to Switzerland and Austria, which integrated their inventories decades ago, some countries, among them the German state of Bavaria, Slovenia and Sweden, have managed to merge their independent, single-purpose forest inventory systems into integrated forest resource inventories.

In this article, we aim to present the methodological and organizational approaches that were used in designing the integrated forest resource inventories. We further aim to present the statistical features and the cost effectiveness of the three inventories. Finally, we discuss the state-of-the-art in the field of integrated forest resource inventories in Europe and North America, from which the latest innovations come.

Materials, Methods, Integration Patterns

Ethics Statement

All the results come from public data. No ethic issues have been violated.

Material, Working Methods

The presented results come from the NFI and the national forest health inventory datasets of Bavaria, Germany, Slovenia and Sweden [23–25]. The same datasets are used as the official data-sources for national and international reporting [26] and for other purposes.

The statistical efficiencies, related to the statistical precisions and to the abilities of the three designs to detect changes, were investigated by analyzing three variables (Table 1): growing stock volume (hereafter: GSV), share of damaged trees (trees defoliated more than 25%; hereafter: ShDT), and deadwood volume (hereafter: DWV). These variables were chosen because i) they have been monitored by all the three inventory systems, ii) they have been recognized as the core indicators by several international processes [21], [15], [27], and iii) they generally have rather different variability, which definitely affects the efficiencies of integrated forest resource inventories. Apart from the share of damaged trees that has been defined by ICP Forests [28], the other two variables have remained non-harmonized.

The statistical parameters of the three variables (i.e. means, variances and confidence intervals) were derived by using the estimators for the given inventory designs [2], [10], [29], [30], [42], [66]. The required sample sizes were derived by:

i) the confidence interval, also known as the general formula for non-stratified independent samples [2], [30] $n = (t*s\%/E\%)^2$

whereby t = t statistics, s% = coefficient of variation, E% = relative standard error (also relative precision or relative allowable error), and also through

ii) statistical power analyses [31]. The latter approach was used to further explore the sample sizes of the variables at given different values of statistical power. The power analysis was employed in three ways:

a) the choice "independent sample t-test H0: M1 = M2" (whereby H0 = null hypothesis; M1 = sample mean at the first occasion; M2 = sample mean at the second occasion) was used for calculating the sample sizes of the independent variables (e.g. cut time series);

Table 1. Core variables and their definitions as used in the NFIs of Bavaria, Germany, Slovenia and Sweden.

Country	Variable	Definition	Thresholds	Method/ Reference	Inventory year/ cycle	Remarks
Bavaria/ Germany	GSV (m^3/ha)	Standing alive trees, volume over bark, stump included	DBH≥10 cm	[23]	1987	
			DBH≥7 cm	[41]	2002, 2012	
Slovenia	GSV (m^3/ha)	Standing alive trees, volume over bark, stump included	DBH≥10 cm	[42]	1985, 1991, 1995, 2000	
			DBH>0 cm		2007	
Sweden	GSV (m^3/ha)	Standing alive trees, volume over bark, stump excluded	DBH>0 cm	[10]	Continuous inventory	Estimates calculated on 5-year averages.
Bavaria/ Germany	ShDT(%)	% of damaged trees	Trees with an assessed defoliation of 25% or more	[28]		
Slovenia	ShDT(%)	% of damaged trees	Trees with an assessed defoliation of 25% or more	[28]	1985, 1991, 1995, 2000, 2007	Continuous inventory on the 16×16 km grid
Sweden	ShDT(%)	% of damaged trees	Trees with an assessed defoliation of 25% or more	[28]	Continuous inventory	Estimates calculated on 5-year averages.
Bavaria/ Germany	DWV (m^3/ha)	Dead standing and downed trees, snags, stumps, coarse woody debris	Standing deadwood: DBH≥20 cm	[41]	2002	
			Stumps: D≥50 cm or H≥50 cm			
			Snags: D≥20 cm; L without threshold			
Slovenia	DWV (m^3/ha)	Dead standing and downed trees, snags, stumps, coarse woody debris	Standing deadwood: DBH≥10 cm	[42]	2007, 2011	
			Stumps: D≥10 cm and H≥20 cm			
			Snags and coarse woody debris: D≥10 cm, L≥50 cm			
Sweden	DWV (m^3/ha)	Dead standing an downed trees, snags and coarse woody debris; stumps excluded	Standing deadwood: DBH>4 cm	[10]	Continuous inventory	Estimates calculated on 5-year averages.
			Lying dead wood: DBH>10 cm and 1.3 m long			

GSV = growing stock volume; ShDT = share of damaged trees; DWV = deadwood volume; DBH = diameter at breast height.

b) the choice "dependent sample t-test H0: M1 = M2" (whereby H0 = null hypothesis; M1 = sample mean at the first occasion; M2 = sample mean at the second occasion) was used for calculating the sample sizes of the paired variables (viz. GSV and ShDT as continuous time series); and

c) the choice "one mean t-test H0: M1 = M0" (whereby H0 = null hypothesis; M1 = sample mean; M0 = hypothesized mean) was used for calculating the sample sizes of the variables that had been measured or assessed only once (e.g. the first measurement of deadwood).

In judging the overall statistical efficiencies of the designs, a single variable approach was used [29].

The cost effectiveness of the designs was compared via simple cost effectiveness ratios, such as CER (cost effectiveness ratio) and ICER (incremental cost effectiveness ratio) [32], [33] as follows:

$$\text{CER} = (\text{costs of forest inventorying})/(\text{number of assessed variables in inventorying}), \text{ and}$$

Table 2. Historical development of large-scale forest resource inventory in Bavaria, Germany.

Year	Inventory type	Grid density (km)	Number of clusters/ plots (Type of plots)	Major variables	Major improvements/Comments
1983–2005	NFHI	4×4	Approx. 1700 clusters (angle- count plots)	Characteristics of forest stands; defoliation; damages; mortality;	6 to 8 plots within forest stands; Trees chosen by angle-count-sampling [35]. No data assessment in 1990 due to heavy damages caused by the storms Vivian and Wiebke.
	FHI_ICP	16×16	123 plots (angle-count plots)		
1986–1988	NFI I	4×4; in some areas 2.83×2.83 and 2×2	3279 clusters (cluster 150×150 m with four angle-count plots)	Growing stock, regeneration, forest openness, site characteristics (site form, site gradient, site aspect);	Introduction of NFI grid (4×4 km) and NFI clusters in parallel (200 m westward and northward) to the old NFHI grid [43].
2001–2002	NFI II	4×4; in some areas 2.83×2.83	Approx. 2700 clusters (cluster 150×150 m with four angle-count plots)	Growing stock, regeneration, site characteristics (site form, site gradient, site aspect), deadwood, naturalness;	Cluster with the side lengths of 150 m; Sample trees are chosen by angle-count sampling at each plot of the cluster.
1987	NSI I	8×8	424 clusters (ten sub-plots distributed over the forest stand)	Soil types, acidity, carbon, soil water, humus, nitrate, base saturation, heavy metal;	Sample plots are located next to the plots of ICP-plots; the center of the plots is unmarked [36].
2006–2008	NSI II	8×8	386 clusters (five sample squares/cluster)	Soil types, acidity, carbon, soil humidity, humus, nitrate, base saturation, heavy metals;	
2006	IFRI III	8×8	386 clusters (four angle-count and four 6-tree plots)	Growing stock, regeneration, site characteristics (site form, site gradient, site aspect), deadwood, naturalness; soil types, acidity, carbon, soil humidity, humus, nitrate, base saturation, heavy metals;	The ICP and the soil sample plots are installed on the southwest corner of the NFI-cluster; The ICP plots are installed according to the ICP-Forests Manual [28].

FHI_ICP = Forest health inventory for the needs of the ICP (only variables related to forest health); NFHI = National Forest Health Inventory (only variables related to forest health); NSI = National Forest Soil Inventory (soil variables); NFI = National Forest Inventory (conventional forest management variables - without forest health variables); IFRI = Integrated forest resource inventory (all variables viz. forest health, forest area, forest growth and yield, biodiversity, carbon-sequestration, forest soil, etc.); CPSP = circular permanent sample plot.

$$\text{ICER} = \frac{(\text{costs of integrated forest resource inventorying} - \text{costs of forest health inventorying})}{(\text{number of assessed variables in integrated forest resource inventorying} - \text{number of assessed variables in forest health inventorying})}$$

where by as the costs of inventorying were used the average total time inputs taken from the field-forms of arbitrary inventorying (viz. forest health, integrated forest resource inventorying) and, as the number of assessed variables, the numbers of variables assessed by each type of inventorying were used. The average total time input includes: transportation from the office to a cluster, walk from the car to the cluster and back, walk between the plots in the cluster, field measurements, transportation from one cluster to another and transportation back to the office.

The CER index was used for ranking or comparing the cost-effectiveness of mutually independent inventories, and the ICER index was used for a comparison of the cost effectiveness of mutually exclusive forest inventories. The latter approach can also be understood as the approach of marginal costs.

All computation was carried out by using the Statsoft (v.10) statistical modules [34].

The presentation of the historical developments of the inventory designs required no specific methodology. As the basic sources, we used the relevant published literature.

Historical Development of the Inventory Designs

Bavaria (Germany). Bavaria began statistically-based large-scale forest inventorying in 1983 (Table 2), when it launched (along with the neighboring state of Baden-Wuerttemberg) its forest health inventory. In terms of statistical design, it was a systematic cluster-sampling inventory, whereby the systematically arranged grid of 4×4 km defined forest stands, which were sampled with clusters of six to ten (Figure 1) systematically arranged angle-count plots [35]. In concert with the ICP Forests Manual [28], only the trees belonging to the Kraft classes 1, 2 and 3 were selected. The computations were based on the time series of the tree data, meaning that harvested and naturally died trees, as well as the trees unsuitable (e.g. broken, bent) for assessments, were continuously replaced with the trees closest to the center of a plot [35].

The same inventory design was also used at the occasion of First Bavaria's Forest Soil Inventory in 1986, whose aim was to detect the radioactive fallout caused by the Chernobyl nuclear power plant disaster as well as to determine the rates of forest sites'

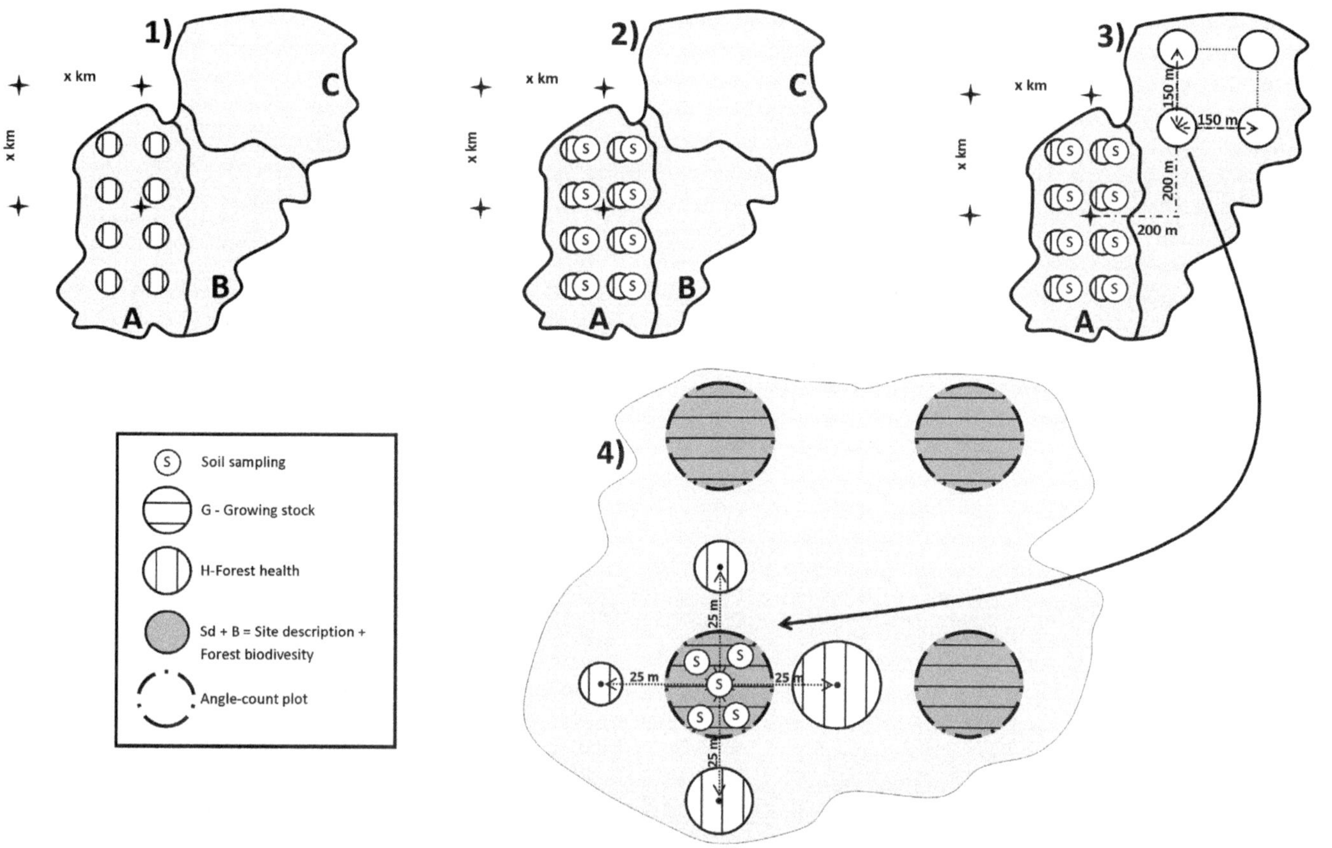

Figure 1. Historical development of Bavaria's integrated forest resource inventory. 1) 1983–1985: National Forest Health inventory. The inventorying was performed in angle-count plots, evenly distributed across forest stands. 2) 1986: National Forest Health and Forest Soil Inventory. Soil plots, used for soil sampling, were distributed in parallel to the forest health plots. 3) 1986–2005. National Forest Inventory. The introduction of the NFI clusters, consisting of four angle-count plots, distributed across forest complexes. 4) 2006–2008. Integrated Forest Resource Inventory (IFRI). Forest soil sampling and forest health assessment, performed on the subsamples of the NFI/IFRI plots (southwestern plot on the 8×8 km grid), have become part of the inventory design. The majority of variables has been collected on the angle-count plots (NFI/IFRI plots), while forest health has been assessed on the 6-tree plots.

nutrition and contamination [36]. As shown in the figure, the soil plots were located in parallel to the forest health plots.

A shift from the described development of the national forest monitoring system occurred in 1986 when the nation, including Bavaria, underwent the first statistically-based national forest inventorying [66]. Despite the existing forest health permanent sample plot (hereafter PSP) system, Bavaria did not use it, but rather installed a parallel system of clusters, each consisting of four angle-count plots. The main reason that a new system of clusters was introduced (200 m eastward and northward of the existing forest health grid, Figure 1) was the possibility of biased results due to the visibly marked sample trees used in forest health monitoring. This forest inventorying was conducted until 2005.

Another change of the inventorying system occurred in 2006 when the German states set up the Second National Forest Soil Inventory. Although the inventory should have been a repetition of the first inventory of 1986, Bavaria moved the clusters of PSP for soil sampling to the NFI grid. The main reason was the fact that the soil sampling plots were not marked at the occasion of the first inventory. Finally, the last improvement came with the latest design for forest health inventory which followed the ICP-Forests proposal [28]. The clusters were composed of four 6-tree plots [37] and were installed on the subset of the NFI grid. All the previously used plots were abandoned.

Slovenia. Until the 1980s, Slovenia was obtaining forest resources information at a national level through the bottom-up inventorying concept [9]. A statistically sound large-scale forest health inventory design (cluster sampling with clusters consisting of four 6-tree plots [37]) was first introduced in 1985 (Table 3), when the country entered the UN/ECE ICP Forests program [38]. In addition to the required tree health data [28], the surveying also provided data on forest vegetation types, site and stand conditions, tree-species composition and so forth. In 1995, this inventory design was improved with angle-count plots [37] that were superimposed over the 6-tree plots (Figure 2). Both estimates (derived by angle-count sampling and by 6-tree sampling method) enabled the derivation of an objective estimate of the growing stock volume for the first time [39].

However, because 6-tree plots had been rather inconvenient for the long-term tracking of changes (due to the variability of plot sizes, different inclusion probabilities of trees and the exchange of trees on the plots [65]) and would have shortly become unrepresentative due to visibly marked trees, the design was changed in the year 2000 [40], by adding invisibly-marked circular permanent sample plots (hereafter: CPSP) to the existing clusters. The main features of this design are: i) compliance with the principles of random sampling; ii) open clusters [30] requiring one day's work of a crew; iii) possibility of tracking the historical

Table 3. Historical development of large-scale forest resource inventory in Slovenia.

Year	Inventory type	Grid density (km)	Number of clusters/plots (Type of plots)	Major variables	Major improvements/ Comments
Since 1985	FHI_ICP	16×16	40–43 clusters (four 6-tree plots/cluster)	Forest site, tree-species, defoliation, mortality, damages;	
1985 and 1991	NFHI	4×4	700–800 clusters (four 6-tree plots/cluster)	Forest site, stand features, tree-species, defoliation, mortality, damages, DBH;	Growing stock volume and increment volume were not computed.
1995	NFI I	4×4	712 clusters (four 6-tree plots/ cluster+four angle-count plots superimposed over 6-tree plots)	Forest site, stand features, tree-species, defoliation, mortality, damages, DBH, height, lichens;	Update of the number of clusters Introduction of angle-count plots (overlaid over 6-tree plots); Calculation of growing stock and increment volume.
2000	IFRI I	4×4	712 clusters (one CPSP and two 6-tree plots/cluster)	Forest site, stand features, tree-species, defoliation, mortality, damages, DBH, height, age-estimate, lichens, deadwood (without stumps);	Update of the number of clusters; Introduction of CPSP; Change of the clusters' standpoints.
2007	IFRI II	4×4	737 clusters; (one CPSP and two 6-tree plots/cluster)	Forest site, stand features, tree-species, defoliation, mortality, damages, DBH, height, age -estimate, lichens, deadwood (with stumps), small trees;	Calculation of increment volume; Introduction of soil variables.
		Soil sampling: 8×8	Sample squares	Organic horizon, mineral part of soil, pH, C/N, bulk density;	
2012	IFRI III	4×4 Non-forestland: sample from the 1×1 km grid	a) 762 clusters+509 non-forest clusters (one CPSP/cluster)	Forest site, stand features, tree-species, mortality, DBH, height, age-estimate, deadwood (with stumps), small trees; Organic horizon, mineral part of soil, pH, C/N, bulk density; Non-forest area: Wooded area, DBH, height, organic horizon, mineral part of soil, pH, C/N, bulk density;	Updated clusters; Extension of inventorying to non-forest area (other wooded land, orchards, grasslands).

FHI_ICP = Forest health inventory for the needs of the ICP (only variables related to forest health); NFHI = National Forest Health Inventory (only variables related to forest health); NSI = National Forest Soil Inventory (soil variables); NFI = National Forest Inventory (conventional forest management variables - without forest health variables); IFRI = Integrated forest resource inventory (all variables viz. forest health, forest area, forest growth and yield, biodiversity, carbon-sequestration, forest soil, etc.); CPSP = circular permanent sample plot.

development of the CPSPs; iv) possibility of using various statistical techniques; v) ability to use the time series from the past, and; vi) openness to future modifications.

Despite improvements, the 2007 inventory has not fully covered the needs of end-users. Consequently, the surveying carried out in 2012 took account of non-forestland to provide estimates on the biomass outside forestland. Many improvements are yet to come; hence, future modifications will have to address the optimization of the sample grid to provide a variety of estimates of different variables for the strata and regions [44], as well as the temporary sample plots to control the representativeness of the installed CPSPs, indicators of ecosystem services and so forth.

Sweden. NFIs have a long tradition in Sweden (Table 4). They emerged in the 1920s as regional inventories, proceeding county by county until the entire country was covered [45]. Later, the importance of timely information about the state of and changes in forestry conditions for the entire country was recognized, and the NFIs were rearranged into a system in which a sample of the entire country was selected each year. Statistical developments of this process were documented by Matérn [46]. From the 1950s onwards, the basic set-up of Swedish NFIs has remained fairly constant (Figure 3), although slight modifications are made every five to ten years. The main features of the Swedish system are: i) compliance with the principles of random sampling; ii) employment of clusters of plots, corresponding to one day's work for a field crew; iii) adaptation to varying conditions in different parts of the country, partly through a stratified design in which details of the design are regionally optimized; iv) a combination of permanent (visited every five years) and temporary plots; v) openness to modifications of the plot-level measurement protocol.

In the 1990s, the ICP Forests Level I plots were installed as a subset of the NFI permanent plots. In this way, 770 plots were randomly distributed throughout the country; plots were selected (one plot per selected cluster) so that a fairly systematic grid of Level I plots was obtained across the country. Until 2006, the Level I plots were visited each year, and registrations of crown condition and tree damage parameters were made. Basic data about site conditions, growing stock, etc., were taken from the NFI measurements.

From 2007 onwards, the annual measurements of the Level I plots were abandoned and crown condition data were instead only taken from the ordinary NFI plots. On all the NFI plots, certain (sub-) sample trees are selected for more careful measurements. These trees are selected through probability to size (basal area) sampling, and normally about two to five trees per plot are

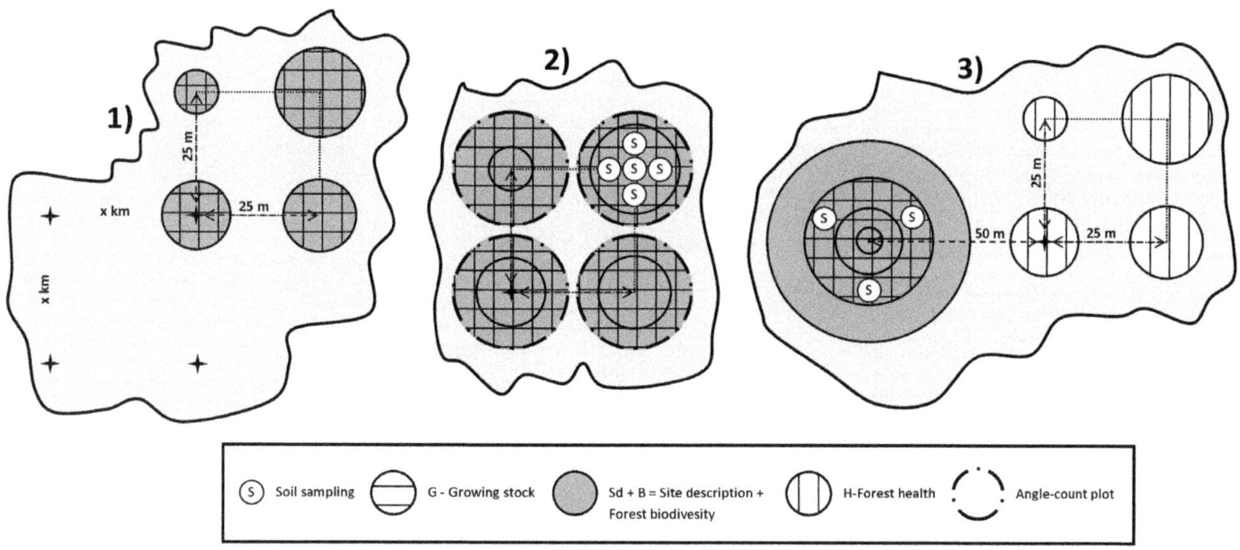

Figure 2. Historical development of Slovenia's integrated forest resource inventory. 1) 1987–1994: National Forest Health Inventory. The inventorying was performed in clusters, consisting of four 6-tree plots. 2) 1995: Integrated Forest Resource Inventory (IFRI). Forest soil sampling and the assessment of growing stock became part of inventorying. To achieve good estimates on growing stock, angle-count plots (superimposed over 6-tree plots) were introduced. 3) 2000 onwards: Integrated Forest Resource Inventory. Introduction of invisible concentric permanent sample plots with fixed areas. The plots have been used for collecting data and information on forest sites, growing stock and increment, forest health, biodiversity, ecosystem services, average tree cut (harvest). Forest health has been assessed on 6-tree plots and on concentric permanent sample plots.

Table 4. Historical development of large-scale forest resource inventory in Sweden.

Year	Inventory type	Grid density (km)	Number of clusters/plots (Type of plots)	Major variables	Major improvements/ Comments
1923–1929	NFI I	Strip inventory	Different in different regions	Growing stock, stand and site characteristics;	
1935–1952	NFI II	Mix of strips and plots	Different in different regions	Growing stock, stand and site characteristics;	Introduction of sample plots instead of sample strips.
1953–1982	NFI III-V	Unique for each of 5 different strata	Unique for each of 5 different strata	Growing stock, stand and site characteristics;	Introduction of sample plot clusters, so-called tracts.
1983–1992	NFI VI	Unique for each of 5 different strata	Unique for each of 5 different strata	Growing stock, stand and site characteristics, as well as soil classification and detailed vegetation variables;	Introduction of permanent plots and a specific soil survey.
1993–2002	NFI VII	Unique for each of 5 different strata	Unique for each of 5 different strata	Growing stock, stand and site characteristics, as well as soil classification, detailed vegetation variables, crown condition, and biodiversity indicators;	Introduction of several new variables to capture biodiversity and forest damage.
1994–2006	FHI_ICP	16×16 km	770 plots	Crown condition and forest damage; annual assessments;	ICP Forests plots selected as a subset of permanent plots of the NFI.
2003–2013	IFRI VIII	Unique for each of 5 different strata	About 10000 plots measured annually (allocation different in each of 5 different strata)	Growing stock, stand and site characteristics, as well as soil classification detailed vegetation variables, crown condition, forest damage, and biodiversity indicators;	ICP Forests Level I and NFI merged From 2007 onwards.

FHI_ICP = Forest health inventory for the needs of the ICP (only variables related to forest health); NFHI = National Forest Health Inventory (only variables related to forest health); NSI = National Forest Soil Inventory (soil variables); NFI = National Forest Inventory (conventional forest management variables - without forest health variables); IFRI = Integrated forest resource inventory (all variables viz. forest health, forest area, forest growth and yield, biodiversity, carbon-sequestration, forest soil, etc.); CPSP = circular permanent sample plot.

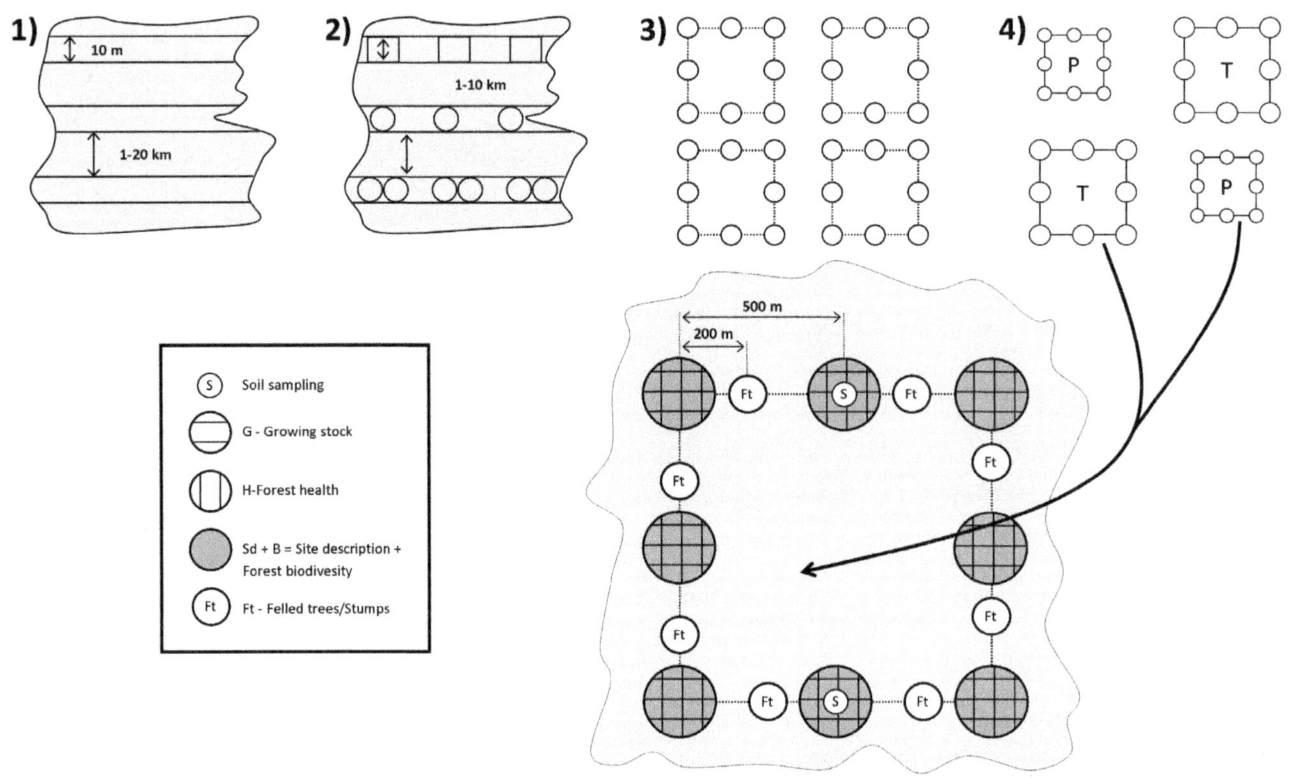

Figure 3. Historical development of Sweden's integrated forest resource inventory. 1) 1923–1929: County-by-county forest inventory. Full callipering in strips. 2) 1938–1952: County-by-county forest inventory. Introduction of squared and circular sample plots. 3) 1953–1982: NFI. Introduction of sample clusters (tracts). 4) 1983 onwards: Integrated Forest Resource Inventory (IFRI). Introduction of permanent (P) and temporary (T) clusters (tracts) of different sizes, used for inventorying forest areas, growth, biodiversity, soil condition, forest health. 5) 1983–2006. National Forest Health Inventory. Forest health assessment was performed on one NFI/IFRI plot of selected permanent tracts. 6) 2007 onwards: Forest health assessment has been performed on all NFI/IFRI plots. Trees have been selected via angle-count method.

selected. For these trees, in addition to traditional measurements, such as height, crown height, stem quality, etc., assessments of crown condition variables are made. On the basis of these measurements, statistical estimates of crown condition parameters are computed.

Although the basic cluster design remains the same throughout the country, several regional differences exist. In southern Sweden, dense grids with small clusters are used, whereas in northern Sweden sparse grids with larger (more plots) clusters are applied. The smallest clusters in southern Sweden comprise four plots, whereas the largest clusters in northern Sweden comprise sixteen plots. The clusters generally are squares with the side lengths ranging from about 200 meters to 1.600 meters. The plots are concentric, with most of the tree-related measurements conducted within a 10-meter (permanent plots) or 7-meter (temporary plots) radius. Slightly more than 10.000 plots are sampled every year; about two thirds of these are permanent.

Results

Tables 5 and 6 provide an overview of the statistical efficiencies of the three inventory designs. Among the many parameters of the variables presented in Table 5, the relative standard errors are perhaps the most illustrative; their values range between 1% and 3% if the variations of the variables are low (s%<80%) and are much higher if the variables have larger variations. Furthermore, a comparison of the actual (n_{plots}) and the required sample sizes (n_{gf}) leads to the conclusion that the actual sample sizes (especially of Bavaria and Sweden) are being set deliberately high to provide precise estimates for all variables (not only for target ones), and for diverse strata.

A more detailed overview of the statistical efficiencies is provided in Table 6, which presents the results of the power analysis. The first finding is that the successive inventories have been statistically very efficient, because the actual mean changes of the given variables have been detected with statistical powers higher than 90% (column P_c). Also informative are the simulations of mean changes of the three variables (set to 5%, 10% change etc.), which, at the set powers of 90% and 99% respectively, reveal that the existing designs aim to be multi-resourceful as they enable detecting the mean changes of the variables with a highly diverse variation; in the case of GSV, all the three designs allow the detection of 5% and larger mean changes at the power of 90% and the mean changes of 10% and larger at the power of 99%. The Swedish inventory is even more efficient, as it allows detecting a 2% mean change at the power of 90%. Less precise is the detection of mean changes of the variable ShDT; the power of 90% mostly allows the detection of mean changes to the extent of 10% and more, while the power of 99% enables detecting the mean changes to the extent of 15% and more. Similar effectiveness of the designs is also experienced at detecting changes in DWV. However, the exploration of changes of this variable suggests improvements. This is especially so for the Slovenian inventory,

Table 5. Basic statistics of core variables.

Old inventory system									Current inventory system								Remarks
C/R	**Var**	**Y**	**n**	**M**	**s**	**s%**	**E**	**E%**	**n**	**M**	**s**	**s%**	**E**	**E%**	**$E\%_{all}$**	**n_{gf}**	
BavG	GSV (m^3/ha)	1987							3279	333.00	267.00	80.20	4.66	1.40	5	988	(including bark)
BavG	GSV (m^3/ha)	2002							2622	403.00	206.00	51.10	4.02	1.00	5	401	(including bark)
SLO	GSV (m^3/ha)	1995	712	283.17	152.69	53.92	5.72	2.02									[39]
SLO	GSV (m^3/ha)	2000							712	281.39	173.48	61.65	6.50	2.31	5	584	[47]
SLO	GSV (m^3/ha)	2007							737	323.60	193.98	59.94	7.14	2.21	5	552	[47]
SWE	GSV (m^3/ha)	2000							8450	108.80	90.01	82.73	0.98	0.90	5	1052	
SWE	GSV (m^3/ha)	2007							6930	114.30	85.64	74.93	1.03	0.90	5	863	
BavG	ShDT (%)	2005	202	22.70	14.20	56.40	1.00	4.40									
BavG	ShDT (%)	2006							386	22.70	12.90	55.80	0.66	2.89	5	478	
SLO	ShDT (%)	1995	680	24.69	18.37	74.40	0.70	2.84									
SLO	ShDT(%)	2000	677	24.72	20.56	83.17	0.79	3.20	683	22.19	17.52	78.96	0.67	3.02	5	958	
SLO	ShDT(%)	2007							726	39.63	25.00	63.09	0.93	2.35	5	611	
SWE	ShDT(%)	2000	770	29.30	30.00	102.39	1.08	3.69									Spruce
SWE	ShDT(%)	2007							1386	34.30	30.00	87.46	0.79	2.30	5	1175	Spruce
SWE	ShDT(%)	2000	770	14.30	25.00	174.83	0.90	6.29									Pine
SWE	ShDT(%)	2007							1386	15.00	25.00	166.66	0.67	4.48	5	4268	Pine
BavG	DWV (m^3/ha)	2002							1280	12.90	5.60	43.40	0.16	1.24	5	289	
SLO	DWV (m^3/ha)	2007							737	18.64	36.36	195.05	1.34	7.19	10	1461	
SLO	DWV (m^3/ha)	2007													15	650	
SWE	DWV (m^3/ha	2000							8450	6.50	11.95	183.85	0.13	2.00	5	5193	
SWE	DWV (m^3/ha	2007							6930	8.30	13.82	166.51	0.17	2.00	5	4260	

C/R = country/region; BavG = Bavaria_Germany; SLO = Slovenia; SWE = Sweden; Var = Variable; Y = year; n = number of plots; M = mean value; s = sample standard deviation; s% = coefficient of variation; E = standard error; E% = relative standard error; $E\%_{all}$ = allowable relative standard error; n_{gf} = n computed by general formula for non-stratified sampling $n = (1.96 \times s\%/E\%_{all})^2$; See also abbreviations in Table 1.

Table 6. Statistical efficiencies of the designs.

InvC	Var	Y	n	M	P_c (%)	Required n at given power (P_{set}) and % of mean change							InvPair	Test type
						P_{set}	±2%	±5%	±10%	±15%	±20%	±25%		
BavG_1	GSV	1987	3279	333.00										
BavG_2a	GSV	2002	2622	403.00	100	0.99		3842	962*	428*			BavG_1-2a	2M IDP
BavG_2b	GSV	2002	2622	403.00	100	0.90		2198*	551*	246*			BavG_1-2b	2M IDP
SLO_1	GSV	2000	712	281.40										
SLO_2	GSV	2007	737	323.60	100	0.99		529*	134*	61*			SLO_1-2	2M DEP
SWE_1	GSV	2000	8450	108.80										
SWE_2a	GSV	2006	6930	114.30	100	0.99	7879	1285*	323*				SWE_1-2a	2M DEP
SWE_2b	GSV	2006	6930	114.30	100	0.90	4507*	736*	186*				SWE_1-2b	2M DEP
BavG_1a	ShDT	2006	386	22.70	93	0.99		2529	580	167*			BavG_1a	1M
BavG_1b	ShDT	2006	386	22.70	93	0.90		1447	333*	154*			BavG_1b	1M
SLO_1	ShDT	2000	683	22.20										
SLO_2a	ShDT	2007	726	39.60	100	0.99		3620	907	418*			SLO_1-2a	2M DEP
SLO_2b	ShDT	2007	726	39.60	100	0.90		2071	520*	240*			SLO_1-2b	2M DEP
SWE_1	ShDT	2000	770	29.30										
SWE_2a	ShDT	2007	1386	34.30	96	0.99		5724	1433	638*			SWE_1-2a	2M DEP
SWE_2b	ShDT	2007	1386	34.30	99	0.90		3275	820*	366*			SWE_1-2b	2M DEP
SWE_1	ShDT	2000	770	14.30										
SWE_2a	ShDT	2007	1386	15.00		0.99			5106	2271	1278*		SWE_1-2a	2M DEP
SWE_2b	ShDT	2007	1386	15.00		0.90			2921	1300*	732*		SWE_1-2b	2M DEP
BavG_1a	DWV	2002	1280	12.90	98	0.99		1603	343*	162*	153*		BavG_1a	1M
BavG_1b	DWV	2002	1280	12.90	98	0.90		918*	197*	94*	89*		BavG_1b	1M
SLO_1a	DWV	2007	737	18.60	93	0.99				3107	1781	1126	SLO_1a	1M
SLO_1b	DWV	2007	737	18.60	93	0.90			3859	1778	1017	645*	SLO_1b	1M
SWE_1	DWV	2000	8450	6.50										
SWE_2a	DWV	2007	6930	8.30	100	0.99		10062	2517*	1120*			SWE_1-2a	2M DEP
SWE_2b	DWV	2007	6930	8.30	100	0.90		5755*	1441*	642*			SWE_1-2b	2M DEP

InvC = inventory code; Var = Variable; Y = Year; n = number of plots; M = mean; P_c(%) = computed power for independent and dependent samples and for the hypothesized mean; P_{set} = set power; Required n at given power (P_{set}) and % of mean change = required sample size at 2%, 5%, 10%, 15% …change in mean; InvPair = inventory pair; 2M IDP = two mean test for independent samples; 2M DEP = two mean test for dependent samples; 1M = one mean test (of hypothesized mean);

* = actual sample size successfully detects the differences in mean change. See also abbreviations in Table 1.

Table 7. Cost-effectiveness of the inventory designs.

C/R	Inventory	n var	Total time input (mts/cluster)	CER (mts/1var)	ICER (mts/1 var)	Inventory pair
BavG	FHI_{ICP}	35	300	8.57		
BavG	NFI_{2002}	150	840	5.60	4.70	FHI_{ICP}–NFI_{2002}
SLO	FHI_{ICP}	32	344	10.75		
SLO	NFI_{1995}	65	494	7.60	4.54	FHI_{ICP}–NFI_{1995}
SLO	$IFRI_{2000}$	55	418	7.60	3.21	FHI_{ICP}–$IFRI_{2000}$
SLO	$IFRI_{2007}$	76	480	6.31	3.09	FHI_{ICP}–$IFRI_{2007}$
SWE	FHI_{ICP}	45	150	3.33		
SWE	IFRI	250	500	2.00	1.71	FHI_{ICP}–IFRI

C/R = Country/Region; Inventory = inventory name; n var = number of variables; total time input = time needed for transportation, walking and measurements in a cluster; CER = cost effectiveness ratio = average time needed for measuring one variable; ICER = incremental cost effectiveness ratio = average time needed for measuring the additional variables; See also abbreviations in Tables 2, 3 and 4.

which needs a denser grid to become more effective in change detection.

Table 7 presents two cost effectiveness indicators for two different types of inventories. The first indicator, CER, clearly shows that the time input spent for measuring one variable decreases with the complexity of inventories; the more variables to be recorded, the smaller time input per variable needed. Additionally, the second indicator, ICER, shows that by implementing integrated (merged) multi-purpose instead of single-purpose inventories (e.g. forest health versus integrated forest resources inventorying), there is an improvement in the time input efficiencies. Unlike single-purpose inventories, integrated forest resources inventories are more effective due to the less time input spent for the measurement of remaining (marginal) variables.

Discussion

National forest inventories, integrated national forest resource inventories and environmental monitoring in general are becoming increasingly important tools for decision makers in planning a sustainable future. Information is needed to assess resource availability and environmental impacts; however, the very same information also assists in forestry scenario analysis, whereby the outcome of goods and services at given different management options can be compared, and optimal management schemes can be selected [48].

Whereas conventional forest inventories traditionally have been the responsibilities of individual countries, an increasing number of international agreements highlight the need for international collaboration and for comparable information from all collaborating parties [18]. The already mentioned ICP Forests, operating within the LRTAP convention [14] and occasionally supported by EU legislation [17], was an early example of multi-lateral collaboration. Despite some methodological imperfections [18], controversies associated with air pollution and forest decline [49] and remaining challenges, such as developing indicators for assessing forest biodiversity, as well as the improvement of measurement protocols to be used in monitoring ecosystem processes, ICP Forests has proved to be successful in many areas. Its greatest achievement was perhaps the promotion and support to the participatory countries in building the geographically largest and most complete forest health database in the world [50]. Additionally, with the collected tree defoliation and other data, over the years it has managed to illuminate some of the many effects of tree die-back, some of the relationships between the crown condition and the stress factors [51], as well as the processes related to soils, local climate, local light conditions, etc. Finally, many of the CLRTAP's International co-operative programs have assisted in bringing to light the CLRTAP's amended protocols (e.g. Helsinki protocol, Gothenburg protocol [14]), aiming to improve the environment through emission reductions, control of gaseous inorganic and organic compounds, etc.

Perhaps even more collaboration is needed in the implementation of the Conventions on Climate Change [27] and Biodiversity [52]. Because assessing carbon pools and favorable conservation statuses of habitats, habitat types and species, introduced by Habitats Directive [22], requires measuring the sets of variables, much effort will be needed for their harmonization. Consequently, diverse discussions about how to harmonize forest data and how forest inventories should be organized to meet all major user needs [53], [18], [50] are underway.

An additional reason that harmonizing the data is necessary is the use of the same data at different political and spatial levels. As reported [13], a large number of countries already cover the information needs for national policy decisions with specially designed forest inventories and monitoring systems. The very same information is afterwards used by the FAO, which is responsible for preparing regular compilations of the state and development of forests (known as global forest resource assessments) at the global level [26]. At regional levels, processes such as the Montreal Process [54] and Forest Europe [21] compile regional estimates similar to that which the FAO does at the global level.

In addition to improvements due to data-harmonization, our study demonstrates that significant improvements can also be achieved through inventory program integrations. Bavaria, Slovenia and Sweden, for instance, have managed to merge independent national forest and forest health inventories to obtain integrated national forest resource inventories and to end independent, single-purpose inventorying. As described and graphically presented, these integrations were achieved in three different ways:

- by introducing a new NFI grid (in parallel to the old national forest health inventory grid) and the gradual establishment of new sets of subplots for soil-sampling and tree-defoliation assessments in Bavaria (Figure 1);
- by introducing a new set of concentric permanent sample plots (CPSPs) into the existing forest health clusters in Slovenia (Figure 2); and

- by introducing tree-defoliation assessments on the subsets of trees on the cluster's NFI plots in Sweden (Figure 3).

The results, presented for the core variables, demonstrate that the merging of the inventories has not undermined the precision of the estimates. Moreover, the results are very promising as they allow the detection of rather small differences (e.g. considering the inventory period, 1%–2% change annually) in mean changes and trends in time, which is certainly one of the most important demands imposed on modern inventorying and monitoring [55–59]. Furthermore, because all the three integrated forest resource inventories supply data for variables describing growth and yield, site conditions, biodiversity, forest soils, forest health etc., the data are also being successfully used for exploring the relationships and the causalities between the variables [40], [60] and for supporting the international processes along with the FAO reporting. Finally, as shown in Table 7, these integrated inventories also bring financial savings.

However, merging inventories does not necessarily bring only advantages. As Ferretti [50] pointed out and as this study has also demonstrated, the inventory integrations almost always result in cuts of time series. Bavaria, for instance, cut the soil and forest defoliation time series, Slovenia the NFI time series and Sweden cut the forest defoliation time series. Another disadvantage of inventory integration is that the scope of inventories may become too broad and consequently may increase the risk of measurement and assessment errors [61]. This is especially probable if the field crews are not properly trained or they are inexperienced in collecting certain variables.

In the view of the aforementioned ongoing processes, it is also worth considering at what level large-scale inventories should be (if at all) merged. Because forest policy in the EU is the responsibility of individual countries, it seems that the EU nations themselves should develop and run forest assessment systems that meet their specific needs. In concert with this policy, almost all European countries currently perform national forest inventories [13]. However, because the EU also needs forest information for general purposes and for the reporting to the international agreements to which it is a party, this type of information can presently be obtained from the individual countries, provided that the information is properly harmonized.

Naturally, there are also rather different practices around the world. For instance, after long-lasting statewide forest inventorying, the group working at the US Forest Inventory and Analysis Program has managed to merge statewide forest inventories and a nationwide forest health inventory into a common system for the entire country [62], [63]. A similar direction was followed Canada; despite the fact that forest management and policy are to a great extent regulated by the provinces and territories, it managed to create a sampling design across the whole nation [64].

We conclude that presently there is an increasing need for trustworthy information about forests to be used by decision makers within individual countries as well as for international collaboration. This type of information can be cost-efficiently provided if inventories are integrated within individual countries as demonstrated in our study and properly harmonized so that the results can be readily compared between countries and applied to obtain regional and global figures.

Acknowledgments

We direct our thanks to Jure Žlogar, who designed the uniform figures. Our thanks go also to Dr. Sören Wulff for having provided us with valuable comments to previous versions of the manuscript, and to Terry T. Jackson for having improved the manuscript's wording.

Author Contributions

Conceived and designed the experiments: MK AB GS. Performed the experiments: MK AB GS. Analyzed the data: MK AB GS. Contributed reagents/materials/analysis tools: MK AB GS. Wrote the paper: MK AB GS.

References

1. Bernasconi A (1996) Von der Nachhaltigkeit zu nachhaltigen Systemen. Forstliche Planung als Grundlage nachhaltiger Waldbewirtschaftung. Beiheft zur Schweiz Z Forstwes 76: 176 p.
2. Loetsch F, Haller KE (1964) Forest inventory. Volume I Statistics of forest inventory and information from aerial photographs. München, Basel, Wien: BLV Verlagsgesellschaft München. 436 p.
3. Schreuder HT, Gregoire TG, Wood GB (1993) Sampling Methods for Multiresource Forest Inventory. New York: John Wiley & Sons. 446 p.
4. Braun Von R (1975) Verfahrensmerkmale der Österreichischen Forstinventuren. Allg Forst Jagdztg 146: 2–6.
5. Tzschupke W (1986) Die Geschichte der deutschen Forsterhebungen. Allg Forst Zeitschrift 41: 790–791.
6. Gschwantner T, Gabler K, Schadauer K, Weiss P (2010) Austria. In: Tomppo E, Gschwantner T, McRoberts RE, editors. National Forest Inventories: Pathways for Common Reporting. Heidelberg: Springer. pp. 57–71.
7. Cerný M, Kučera M, Cienciala E, Beranova J (2010) Czech Republic. In: Tomppo E, Gschwantner T, McRoberts RE, editors. National Forest Inventories: Pathways for Common Reporting. Heidelberg: Springer. pp. 145–156.
8. Gillis MD (2001) Canada's National Forest Inventory (responding to current information needs). Environ Monit Assess 67: 121–129.
9. Kovač M, Hočevar M (2010) Short Description of Forest Inventories and Control Sampling Method Around the World and in Slovenia. In: Planinšek Š, Kovač M, Kušar G, Hočevar M, Simončič P, et al., editors. Control Sampling Method in Slovenia - History, Characteristic and Use. Ljubljana: Slovenian Forestry Institute. Studia forestalia Slovenica 136: 7–10.
10. Axelsson AL, Ståhl G, Söderberg U, Petersson H, Fridman J, et al. (2010) Sweden. In: Tomppo E, Gschwantner T, McRoberts RE, editors. National Forest Inventories: Pathways for Common Reporting. Heidelberg: Springer. pp. 541–553.
11. Tomppo E, Tuomainen T (2010) Finland. In: Tomppo E, Gschwantner T, McRoberts RE, editors. National Forest Inventories: Pathways for Common Reporting. Heidelberg: Springer. pp. 185–206.
12. SLU_NFI (2013) Swedish national forest inventory. Available: http://www.slu.se/nfi. Accessed 2013 Jul 3.
13. Tomppo E, Schadauer K, McRoberts RE, Gschwantner T, Gabler K, et al. (2010) Introduction. In: Tomppo E, Gschwantner T, McRoberts RE, editors. National Forest Inventories: Pathways for Common Reporting. Heidelberg: Springer. pp. 1–18.
14. CLRTAP (1979) The 1979 Geneva Convention on Long-range Transboundary Air Pollution. Available: http://www.unece.org/env/lrtap/lrtap_h1.html. Accessed 2013 Jul 14.
15. ICP Forests (2013) ICP Forests. Available: http://icp-forests.net/page/programme-1. Accessed 2013 Jul 14.
16. Futmon (2012) Futmon project home page. Available: http://www.futmon.org/structure-and-organisation. Accessed 2012 Dec 25.
17. Forest Focus (2003) Regulation (EC) No 2152/2003 of the European Parliament and of the Council of 17 November 2003 concerning monitoring of forests and environmental interactions in the Community (Forest Focus). Available: http://eur-lex.europa.eu/LexUriServ/LexUriServ.do?uri = CELEX:32003R2152:EN:NOT. Accessed 2012 Dec 27.
18. Tomppo E, Schadauer K (2012) Harmonization of National Forest Inventories in Europe: Advances under COST Action E43. Forest Science 58: 191–200.
19. Memorandum (2004) Memorandum of understanding for the implementation of a European Concerted Research Action designated as COST E43 "Harmonisation of National Forest Inventories in Europe: Techniques for Common Reporting". Available: http://www.metla.fi/eu/cost/e43/cost-e43-mou.pdf. Accessed 2013 Jul 14.
20. COST E43 (2007) COST E43 HOME PAGE. Harmonisation of National Forest Inventories in Europe: Techniques for Common Reporting. Available: Accessed 2013 Jul 14.
21. Forest Europe (2013) Improved Pan-European Indicators for Sustainable Forest Management as adopted by the MCPFE Expert Level Meeting 7–8 October 2002. Vienna, Austria. Available: http://www.foresteurope.org/docs/ELM/2002/Vienna_Improved_Indicators.pdf. Accessed 2013 Jul 14.
22. Habitats Directive (1992) Council Directive 92/43/EEC of 21 May 1992 on the conservation of natural habitats and of wild fauna and flora. Available: http://

eur-lex.europa.eu/LexUriServ/LexUriServ.do?uri = CELEX:31992L0043:EN:html. Accessed 2013 Jul 14.
23. BMELV (2013) Die Bundeswaldinventur. Available: http://www.bundeswaldinventur.de/enid/f046c24539097007df607c40b13ff2f5,51519f6d6f6465092d09/2.html. Accessed 2013 Jun 27.
24. FECS (2012) Forests and forest ecosystems condition survey - field data. Ljubljana: Slovenian Forestry Institute.
25. Skogsdata (2012) Forest statistics 2012. Official Statistics of Sweden. Umeå: Swedish University of Agricultural Sciences. 134 p. Available: http://pub.epsilon.slu.se/9266/1/SkogsData2012_webb.pdf. Accessed 2013 Jul 3.
26. FAO (2010) Global Forest Resources Assessment 2010. Main report. FAO Forestry Paper 163. Rome: FAO. Available: http://www.fao.org/docrep/013/i1757e/i1757e.pdf. Accessed 2013 Jan 4.
27. Penman J, Gytarsky M, Hiraishi T, Krug T, Kruger D, et al, editors (2003) Good Practice Guidance for Land Use, Land-Use Change and Forestry. Kanagawa, Japan: Institute for Global Environmental Strategies, Intergovernmental Panel on Climate Change, National Greenhouse Gas Inventories Programme. Available: http://www.ipcc-nggip.iges.or.jp/public/gpglulucf/gpglulucf.html. Accessed 2013 Jun 20.
28. Eichhorn J, Roskams P, Ferretti M, Mues V, Szepesi A, et al. (2010) Visual Assessment of Crown Condition and Damaging Agents. Manual Part IV. In: Manual on methods and criteria for harmonized sampling, assessment, monitoring and analysis of the effects of air pollution on forests. UNECE ICP Forests Programme Co-ordinating Centre Hamburg. 1–49. Available: http://www.icp-forests.org/pdf/FINAL_Crown.pdf. Accessed 2013 Jul 14.
29. Cochran WG (1977) Sampling Techniques. New York, Chichester, Brisbane, Toronto, Singapore: John Wiley & Sons. 428 p.
30. DeVries PG (1986) Sampling Theory for Forest Inventory. Berlin, Heidelberg, New York, London, Paris, Tokyo: Springer-Verlag. 399 p.
31. Bernstein S, Bernstein R (1999) Elements of Statistics II. Inferential statistics. Schaum's outline series. New York, San Francisco, Washington DC, Auckland, Bogota: McGraw-Hill. 451 p.
32. Detsky AS, Naglie IG (1990) A Clinician's Guide to Cost-Effectiveness Analysis. Annals of Internal Medicine 113: 147–154.
33. Phillips C (2009) What is cost-effectiveness? Available: www.whatisseries.co.uk. Accessed 2012 Nov 28.
34. StatSoft Inc (2011) STATISTICA (data analysis software system), version 10.
35. Kennel E (1983) Waldschadensinventur Bayern 1983: Verfahren und Ergebnisse. Forstliche Forschungsberichte München 57.
36. Gulder HJ, Kölbel M (1993) Waldbodeninventur in Bayern. Forstliche Forschungsberichte München 132: 243 p.
37. Zöhrer F (1980) Fortstinventur. Ein Leitfaden für Studium und Praxis. Hamburg, Berlin: Verlag Paul Parey. 207 p.
38. Šolar M, Jurc D (1985) Poškodbe gozdov zaradi onesnaženja zraka. (Varstvo okolja in splošno koristnih funkcij gozdov). Ljubljana: RSS 1 zvezek.
39. Hočevar M (1997) Possibilities and reliability of the growing stock and increment estimation based on the 1995 forest decline inventory. In: Robek R, editor. Proučevanje propadanja gozdov v Sloveniji v obdobju 1985–1995. Research reports Forestry and Wood Science and Technology, special issue 52: 93–118.
40. Hočevar M, Mavsar R, Kovač M (2002) Forest Condition in Slovenia in the Year 2000. Research reports Forestry and Wood Science and Technology 67: 119–157.
41. Schnell A, Bauer A (2005) Die zweite Bundeswaldinventur 2012: Ergebnisse für Bayern. Berichte der Bayerischen Landesanstalt für Wald und Forstwirtschaft. LWF Wissen 49: 102 p. Available: http://www.lwf.bayern.de/publikationen/daten/wissen/p_33202.pdf. Accessed 2013 Jul 3.
42. Kušar G, Kovač M, Simončič P (2010) Slovenia. In: Tomppo E, Gschwantner T, McRoberts RE, editors. National Forest Inventories: Pathways for Common Reporting. Heidelberg: Springer. 505–526.
43. Krüger S, Mössmer R, Bäumler A (1994) Der Wald in Bayern. Ergebnisse der Bundeswaldinventur 1986–1990. Berichte aus der Bayerischen Landesanstalt für Wald und Forstwirtschaft No 1.
44. Hladnik D, Žižek Kulovec L (2012) Forest area assessment in the Slovenian forest inventory design. Research reports Forestry and Wood Science and Technology 97: 31–42.
45. Thorell KE, Östlin EO (1931) The national forest survey of Sweden. Journal of Forestry 4: 585–591.
46. Matérn B (1960) Spatial variation. Stohastic models and their application to some problems in forest surveys and other sampling investigations. Meddelanden Från Statens Skogsforskningsinstitut 49 (5): 144 p.
47. Hočevar M (2003) Stanje in simulacija trajnostnega razvoja gozdnih fondov v Sloveniji. In: Bončina A, editor. Območni gozdnogospodarski načrti in razvojne perspektive slovenskega gozdarstva. Conference proceedings. Ljubljana: Biotechnical Faculty, Department of Forestry and Renewable Forest Resources. pp. 103–122.
48. Barth A, Lind T, Petersson H, Ståhl G (2006) A framework for evaluating data acquisition strategies for analyses of sustainable forestry at national level. Scandinavian Journal of Forest Research 21: 94–105.
49. Kandler O (1992) Historical declines and diebacks of central European forests and present conditions. Environmental Toxicology and Chemistry 11: 1077–1093.
50. Ferretti M (2010) Harmonizing forest inventories and forest condition monitoring - the rise or the fall of harmonized forest condition monitoring in Europe? iForest 3: 1–4. doi: 10.3832/ifor0518-003.
51. Müller-Edzards C, De Vries V, Erisman JW, editors (1997) Ten years of monitoring forest condition in Europe. Studies on temporal development, spatial distribution and impacts of natural and anthropogenic stress factors. Technical background report. Brussels, Geneva: EC-UN/ECE. 386 p.
52. CBD (1992) Convention on Biological Diversity. Available: http://www.cbd.int. Accessed 2012 Dec 28.
53. Chirici G, McRoberts RE, Winter S, Bertini R, Brändli U-B, et al. (2012) National Forest Inventory Contributions to Forest Biodiversity. Forest Science 58: 257–268.
54. Montreal process (2013) The Montreal Process. Available: http://www.montrealprocess.org. Accessed 2013 Jan 4.
55. Schieck J (2002) Chapter 6. Statistical Power in the Alberta Forest Biodiversity Monitoring Program. Vegreville, Alberta: Alberta Research Council Inc. 28. p.
56. Taylor LB, Gerrodette T (1993) The Uses of Statistical Power in Conservation Biology: The Vaquita and Northern Spotted Owl. Conservation Biology 7: 489–500.
57. Foster JR (2001) Statistical power in forest monitoring. Forest Ecology and Management 151: 211–222.
58. Kovač M (2010) Statistical Power of the Control Sampling Method - The Case of Forest and Forest Ecosystem Condition Survey. In: Planinšek Š, Kovač M, Kušar G, Hočevar M, Simončič P, et al., editors. Control Sampling Method in Slovenia - History, Characteristic and Use. Studia forestalia Slovenica 136: 23–29.
59. Christensen P, Hedström Ringvall A (2013) Using statistical power analysis as a tool when designing a monitoring program: experience from a large-scale Swedish landscape monitoring program. Environ Monit Assess 185: 7279–7293. doi: 10.1007/s10661-013-3100-z.
60. Kovač M, Hladnik D (2009) Monitoring poškodovanosti jelke na Slovenskem v obdobju 1987–2007. In: Diaci J, editor. Ohranitveno gospodarjenje z jelko: zbornik razširjenih povzetkov predavanj. XXVII. gozdarski študijski dnevi, Dolenjske Toplice 2. in 3. april 2009. Ljubljana: Biotehniška fakulteta, Oddelek za gozdarstvo in obnovljive gozdne vire. pp. 46–49.
61. Allegrini M-C, Canullo R, Campetella G (2009) ICP-Forests (International Co-operative Programme on Assessment and Monitoring of Air Pollution Effects on Forests): Quality Assurance procedure in plant diversity monitoring. J Environ Monit 11: 782–787.
62. Bechtold WA, Patterson PL, editors (2005) The Enhanced Forest Inventory and Analysis Program - National Sampling Design and Estimation Procedures. USDAF, Forest Service, Southern Research Station, GTR SRS-80: 85 p.
63. McRoberts RE, Hansen MH, Smith BW (2010) United States of America (USA). In: Tomppo E, Gschwantner T, McRoberts RE, editors. National Forest Inventories: Pathways for Common Reporting. Heidelberg: Springer. pp. 567–582.
64. Gillis MD, Boudewyn P, Power K, Russo G (2010) Canada. In: Tomppo E, Gschwantner T, McRoberts RE, editors. National Forest Inventories: Pathways for Common Reporting. Heidelberg: Springer. pp. 97–112.
65. Schreuder HT (2004) Sampling Using a Fixed Number of Trees Per Plot. USDAF, Forest Service, RMRS-RN-17, 4 p.
66. Polley H, Schmitz F, Hennig P, Kroiher F (2010) Germany. In: Tomppo E, Gschwantner T, McRoberts RE, editors. National Forest Inventories: Pathways for Common Reporting. Heidelberg: Springer. pp. 223–243.

4

Combining Aesthetic with Ecological Values for Landscape Sustainability

Dewei Yang[1,2], Tao Luo[1,2*], Tao Lin[1,2], Quanyi Qiu[1,2], Yunjian Luo[1,2]

1 Key Lab of Urban Environment and Health, Institute of Urban Environment, Chinese Academy of Sciences, Xiamen, China, **2** Xiamen Key Lab of Urban Metabolism, Xiamen, China

Abstract

Humans receive multiple benefits from various landscapes that foster ecological services and aesthetic attractiveness. In this study, a hybrid framework was proposed to evaluate ecological and aesthetic values of five landscape types in Houguanhu Region of central China. Data from the public aesthetic survey and professional ecological assessment were converted into a two-dimensional coordinate system and distribution maps of landscape values. Results showed that natural landscapes (i.e. water body and forest) contributed positively more to both aesthetic and ecological values than semi-natural and human-dominated landscapes (i.e. farmland and non-ecological land). The distribution maps of landscape values indicated that the aesthetic, ecological and integrated landscape values were significantly associated with landscape attributes and human activity intensity. To combine aesthetic preferences with ecological services, the methods (i.e. field survey, landscape value coefficients, normalized method, a two-dimensional coordinate system, and landscape value distribution maps) were employed in landscape assessment. Our results could facilitate to identify the underlying structure-function-value chain, and also improve the understanding of multiple functions in landscape planning. The situation context could also be emphasized to bring ecological and aesthetic goals into better alignment.

Editor: Ben Bond-Lamberty, DOE Pacific Northwest National Laboratory, United States of America

Funding: This study was supported by the National Natural Science Foundation of China (41371535, 41001098, 41371192 and 40971111) and the Fujian Provincial S&T Programme (2013Y0083 and 2011J01280). The funders had no role in study design, data collection and analysis, decision to publish, or preparation of the manuscript.

Competing Interests: The authors have declared that no competing interests exist.

* Email: tluo@iue.ac.cn

Introduction

Various landscapes continually provide human society with multiple benefits, which are derived from ecological services and aesthetic attractiveness [1–5]. However, few researches gave consideration to aesthetic preferences in ecological assessment of landscapes [6,7], and thus efforts towards landscape sustainability encounter the dilemma of making eco-aesthetically appealing landscapes [8–12].

Studies on landscape aesthetics have revealed two contrasting paradigms, which were either from an objective perspective (intrinsic attribute of the landscape) or a subjective point of view (human preferences) [1,13,14]. The employed methods varied from formalized indicators to aesthetic surveys [13,15]. In the urban and regional planning, it still remains challengeable to construct eco-aesthetically appealing landscapes that meet multiple human needs [7,10,11,16–19].

In attempt to combine aesthetic with ecological values, a number of models and/or indicators have been employed in landscape assessment [7,20,21]. Some studies have found that people appreciate aesthetically appealing landscapes that tend to be natural in ecological assessment [1–3], while some studies have indicated that ecologically sound landscapes may not be aesthetically pleasing [2,22,23]. This contradiction poses a challenge for efforts to combine aesthetic preferences and ecological service using a reliable assessment framework [7,24,25]. Moreover, the scale and spatial patterns of various landscape types need to be taken into account for both ecologically vital and aesthetically attractive.

Land use/cover change is a key socio-economic footprint of human activity and potentially affects aesthetic and ecological functions of landscapes [24,26,27]. The ecological services vary with different land-use types [28,29], while the spatial patterns of land-use types give rise to various aesthetic experiences. Thus, to what extent do landscape types meet human needs in terms of ecological services and aesthetic attractiveness [3,17,23,30,31]? This would be useful to determine what measures of both ecological and aesthetic landscapes are potentially positive [20,25].

Therefore, the aims of the present study included that: (1) a hybrid framework, which combine aesthetic and ecological landscape values, were proposed to assess integrated functional performances of different landscape types; (2) the framework was used to examine integrated functional performances and interrelationships of different landscape types in the Houguanhu Region of Wuhan City, central China; and (3) potential strategies for landscape planning were discussed. Our results could encourage an attempt of interdisciplinary research in a shared vision of landscape sustainability.

Methodology

Case study

The Houguanhu Region (113°41′–114°13′E, 30°15′–30°41′N) lies within the southwestern Wuhan city of Hubei Province in central China. Houguanhu Region covers an area of approximately 148.18 km^2, and is mainly comprised of lakes and flatlands with a height of less than 100 m. The average annual temperature is 16.5°C and the total annual precipitation is 1,100–1,450 mm. Houguanhu Region has a population of approximately 63,300 people living in 78 settlements, and 79% of residents are classified as rural population. Regional planning for Houguanhu currently classifies it as an emerging "Ecological Livable Area" and "Tourist Resort", and this region is experiencing rapid changes in landscape structures [5].

The Houguanhu Region has a variety of diverse landscape resources and a rich historical culture. Five general landscape types, consisting of forest, farmland, grassland, water body and non-ecological land (i.e., desert in [28]), are shown in Figure 1, where the classification criteria follow Costanza et al. [28]. Non-ecological land includes rural settlement, built-up areas, roads, mines, industry zones and historic areas. The water body covers 30.56% of the total area of the Houguanhu Region, where the Zhiyin Lake and the Houguan Lake are the two largest lakes in this region. Farmland and scattered non-ecological land cover 46.75% and 15.98% of total area, respectively. Forest mainly in hilly areas and scattered grasslands occupy 0.21% of the total area. Forest and grasslands are mainly distributed in the south region, while farmland and non-ecological land are scattered across the entire area. The water body is mainly located in the central region.

In recent years, diverse landscapes have attracted a large number of tourists from outside the region. At the same time, however, rapid population growth and urban expansion have exerted intensive pressures on the natural environment and landscape. In response, since 2009 the local government has developed a series of regional landscape planning for sustainable development [5,32]. Since cultural functions of landscapes usually receive less attention in landscape planning in China, this study may also benefit ongoing landscape planning in other similar regions.

Field survey

The perception-based aesthetic landscape value evaluation method [1,33] was employed in this study. A revised survey instrument of Von Haaren [33] was used to evaluate the landscape aesthetic value. The survey instrument consists of basic information, an aesthetic landscape value score and a viewshed map. Basic information (i.e. survey date, survey site coordinate, and site description) provided a landscape background for the 146 sample sites. The aesthetic landscape value score ($Vt = V1 - V2 + V3$) for each site included a positive score ($V1$), negative score ($V2$), and positive impact score of surrounding landscapes ($V3$). In each site, the biophysical landscape components (e.g., lakes, grassland, forest, hedgerows, electricity pylons, mines, and unsightly isolated buildings) were considered in the evaluation process. Landscape features (e.g., naturalness, openness, uniqueness, diversity, accessibility and visual comfort) also contributed to evaluation scores. The aesthetic landscape value was coded from 1 (lowest) to 5 (highest) on a 5-point rating scale. The viewshed map was sketched on each site for further revision using ArcGIS 9.2. Each viewshed map would be assigned an aesthetic landscape value score.

A total of 146 grid samples of 1 km^2 were established within the study area using the uniform-grid-square sampling method. Basic maps (e.g., topographical map, landscape type map, administrative map) and socio-economic data (e.g., regional planning documents, and regional eco-economic statistic materials) were collected for spatial analysis using ArcGIS 9.2 tools.

The site survey of aesthetic landscape value was conducted from June 9th to 15th, 2010. Six surveyors from Institute of Urban Environment participated in a 3-hour field training program, and then conducted the formal survey as three two-person groups. During the survey, surveyors compared each other's score and discussed until an agreement on a final score. After a 7-day survey, 141 valid viewshed maps with corresponding aesthetic scores were obtained. Blind viewshed areas without field aesthetic scores and five invalid survey samples were given aesthetic scores using spatial interpolation and joint method in ArcGIS 9.2. Consequently, a

Figure 1. Landscape type map in Houguanhu Region. This Figure and following Figure 3a are redrawn according to Luo et al. [5] and Yang et al. [32].

distribution map of aesthetic landscape value with 146 assigned samples was established. In this map, the study area was divided into graded aesthetic landscape value units based on a 5-point rating scale, with 5 being the highest value. The higher value will be more aesthetic attractive. Yang et al. [32] discussed in detail the survey instrument and data processes.

Aesthetic landscape value coefficients

The landscape type map and the aesthetic landscape value distribution map were overlaid in ArcGIS 9.2 to identify the aesthetic value of different landscape types. Then the area of each landscape type was extracted for each graded aesthetic landscape value region. The aesthetic landscape value coefficient, which refers to the aesthetic landscape value of unit area per landscape type, was calculated as:

$$Q_i = \frac{\sum P_i}{A_i} = \frac{\sum M_i \bullet N_i}{A_i} \tag{1}$$

Where Q_i is the aesthetic landscape value coefficient (point/hm^2) of the ith land type (forest, grassland, farmland, water body, and non-ecological land); P_i is the aesthetic landscape value of the ith landscape types; M_i is the aesthetic landscape value score of the ith landscape type; N_i is the area of the ith landscape type in 5-grade aesthetic landscape value units; A_i is total area of the ith landscape type.

Ecological landscape value coefficients

According to the ecological service valuation framework of Costanza et al. [28], ecological service values of Chinese terrestrial ecosystems were evaluated by export judgment, which were developed by Xie et al. [29]. Ecological landscape value coefficients were calculated for five landscape types based on the sum of the eight ecosystem service values (see Table 1). The ecological landscape value coefficients were added up, where the recreation and cultural service were excluded, because they were inappropriate to simply add and subtract when comparing with monetary value. Based on the ecological value coefficients and the areas of the five landscape types, an ecological landscape value distribution map was calculated using Equation (2):

$$H_i = U_i \cdot V_i \tag{2}$$

Where H_i is the ecological landscape value (\$/hm^2) of the ith landscape type (forest, grassland, farmland, water body and non-ecological land); U_i is the ecological landscape value of the ith landscape type on a scale of 1–5; V_i is the area of the ith landscape type.

Integrated evaluation of landscape values

The coefficients of ecological and aesthetic landscape value were normalized to evaluate integrated landscape values. The ecological and aesthetic landscape value coefficients of five landscape types were converted into values from −2 to 2 using Equation (3):

$$K = \frac{X - X_{min}}{4(X_{max} - X_{min})} - 2 \tag{3}$$

Where K is the normalized aesthetic or normalized ecological landscape value coefficient of each landscape type; X is the aesthetic or ecological landscape value coefficient of each landscape type; X_{max} and X_{min} are the maximum and minimum value of the aesthetic or ecological landscape value coefficient.

We assumed that the aesthetic and ecological landscape values contributed equally to the integrated landscape value. In order to quantify the functional roles of different landscape types in the integrated landscape values, the two normalized coefficients were visually displayed in a two-dimension coordinate system. The distribution maps of the aesthetic and ecological landscape value were overlapped into an integrated one of landscape value in ArcGIS 9.2, which were used to identify spatial characteristics of landscapes.

Results

Landscape value coefficients

Functional performances and value contribution of the five landscape types were quantified using a 5-point Likert Scale with 5 being the highest value. Aesthetic landscape value per unit area was in decreasing order: grassland, water body, forest, farmland, and non-ecological land (Table 2). The natural landscapes performed better than the artificial landscapes in terms of aesthetic functional performance.

The normalized aesthetic and ecological value coefficients were obtained using Equation (3) (Table 3). The aesthetic landscape value per unit area of water bodies was highest, followed by forest, grassland, farmland, and non-ecological land. Similarly to the aesthetic landscape values, natural landscapes had higher ecological values than cultivated or man-made landscapes.

Combination of Landscape values

We identified aesthetic, ecological and integrated functional performances of the five landscape types using a two-dimension coordinate system (Figure 2). In the two-dimension coordinate system, the five landscape types contributed unevenly, or even in opposite directions, to both landscape values. Water body was the only type where both landscape values were positive. Grassland was aesthetically attractive, but it was not ecologically valuable. The other three landscape types (i.e. forest, farmland, and non-ecological land) showed a positive relationship on aesthetic and ecological landscape value, respectively, but made a negative contribution to aesthetic and ecological landscape value.

Distribution of landscape values

We drew the aesthetic landscape value distribution map after the aesthetic landscape survey (Figure 3a), and the ecological landscape value distribution map based on Equation (2) (Figure 3b). The integrated landscape value distribution map was obtained from the overlay of the aesthetic and ecological landscape value distribution maps in ArcGIS 9.2 (Figure 4). The landscape value scores ranged from 1 (lowest) to 5 (highest).

The aesthetic, ecological and integrated landscape values varied with the human activity intensity, and gradually decreased from the external water body to the transitional farmland and central non-ecological land. Central water body and adjacent areas received the highest values, the lowest was in northeastern built-up areas, and the moderate scores were in most other areas.

The distribution of aesthetic landscape value showed a large-scale spatial continuity, with the exception of the northeastern region. The central water body with visual characteristics of broad sight, good order and naturalness obtained higher aesthetic landscape value scores, while scattered non-ecological land and farmland with disorderliness, high-intensity buildings and fragmentation received lower evaluation scores. In contrast, the distribution of ecological landscape values exhibited a mosaic distribution with an unclear spatial trend. However, water body and their adjacent areas received higher ecological value scores.

Table 1. Ecosystem service values of Chinese terrestrial ecosystems (based on Xie et al. [29] and ecological landscape value coefficients of the five landscape types.

Ecosystem services	Forest ($/hm^2)	Grassland ($/hm^2)	Farmland ($/hm^2)	Water body ($/hm^2)	Non-ecological land ($/hm^2)
Atmospheric regulation	463.62	105.97	66.23	0.00	0.00
Climate regulation	357.65	117.12	115.81	59.85	0.00
Water conservation	423.88	104.10	78.07	2651.94	3.90
Soil formation and protection	516.60	253.75	189.99	1.30	2.60
Waste treatment	173.53	170.47	213.41	2365.68	1.30
Biodiversity conservation	431.83	141.84	92.38	324.01	44.24
Food production	13.25	39.04	130.13	13.01	1.29
Raw materials	344.40	6.50	13.01	1.29	0.00
Ecological landscape value coefficients	2724.76	938.80	899.04	5417.10	53.33

Discussion

Metrics for evaluating landscape values

The functional performances and interrelationships between the five landscape types were depicted as the integrated landscape value in a two-dimension coordinate system and displayed in landscape value distribution maps. In contrast to objective aesthetic approaches [3,13,24,34], the subjective landscape aesthetics in this study reflects the intrinsic perception of the forms and spatial configuration of landscape elements. Compared with conceptual models [7,35,36], our metrics (i.e., survey, coefficients, the two-dimension coordinate system and scopes) may quantitatively capture the functional performance of different landscape types.

Landscape functions/values and landscape attributes

Our results showed that landscape values were strongly related to landscape attributes and the human activity intensity. Unlike semi-natural and human-induced landscapes, such as farmland and non-ecological land, the natural landscapes (i.e., water bodies and forest) contributed positively to both aesthetic and ecological functions. The controversial relationships (i.e., coincidence or disjuncture) between aesthetic performances and ecological quality have raised wide concerns [2,7,8,20,37]. These relationships can be quantified from the two-dimension coordinate system and distribution maps in this study, e.g., the negative or positive contribution level of landscape types in landscape values.

Human activities can result in the decrease of landscape values by changing natural landscapes into semi-natural and artificial landscapes. It was found that the landscape values increase from the human-dominated and semi-natural landscapes to more natural landscapes. The underlying structure-function-value chain can be induced from the modified landscape gradient influenced by human activities, which has been discussed in some landscape assessments [6,36,38,39].

Implications for landscape planning

The functional performances and distribution of landscape values were identified in present study, which may help optimize landscape planning in terms of developing priorities and intensity. The two-dimension coordinate system indicated that replacing one landscape type with another may result in a positive or negative change in the comprehensive landscape value. For example, more value would decrease if the water body was replaced by the non-ecological land than by the farmland. Thus, in order to bring aesthetic needs and ecological goals into better alignment, a

Table 2. Aesthetic landscape value scores (P, point$\cdot$hm^2) of five landscape types.

Aesthetic landscape value units	Forest	Grassland	Farmland	Water body	Non-ecological land
1	0.00	0.00	34.32	4.80	20.82
2	0.00	0.00	313.94	112.48	160.72
3	144.76	0.82	2239.91	670.59	1350.75
4	2672.88	47.30	16046.76	8326.75	5069.18
5	1238.10	98.48	9884.85	10807.28	2743.78
Total	4055.74	146.60	28519.77	19921.89	9345.24
Aesthetic landscape value coefficients	4.21	4.61	4.12	4.40	3.95

Table 3. Normalized aesthetic and ecological landscape value coefficients.

	Forest	Grassland	Farmland	Water body	Non-ecological land
Ecological landscape value coefficients ($/hm^2)	2724.76	938.80	899.04	5417.10	53.33
Normalized ecological value coefficients	−0.01	−1.34	−1.37	2.00	−2.00
Aesthetic landscape value coefficients (point/hm^2)	4.21	4.61	4.12	4.40	3.95
Normalized aesthetic value coefficients	−0.44	2.00	−0.97	0.73	−2.00

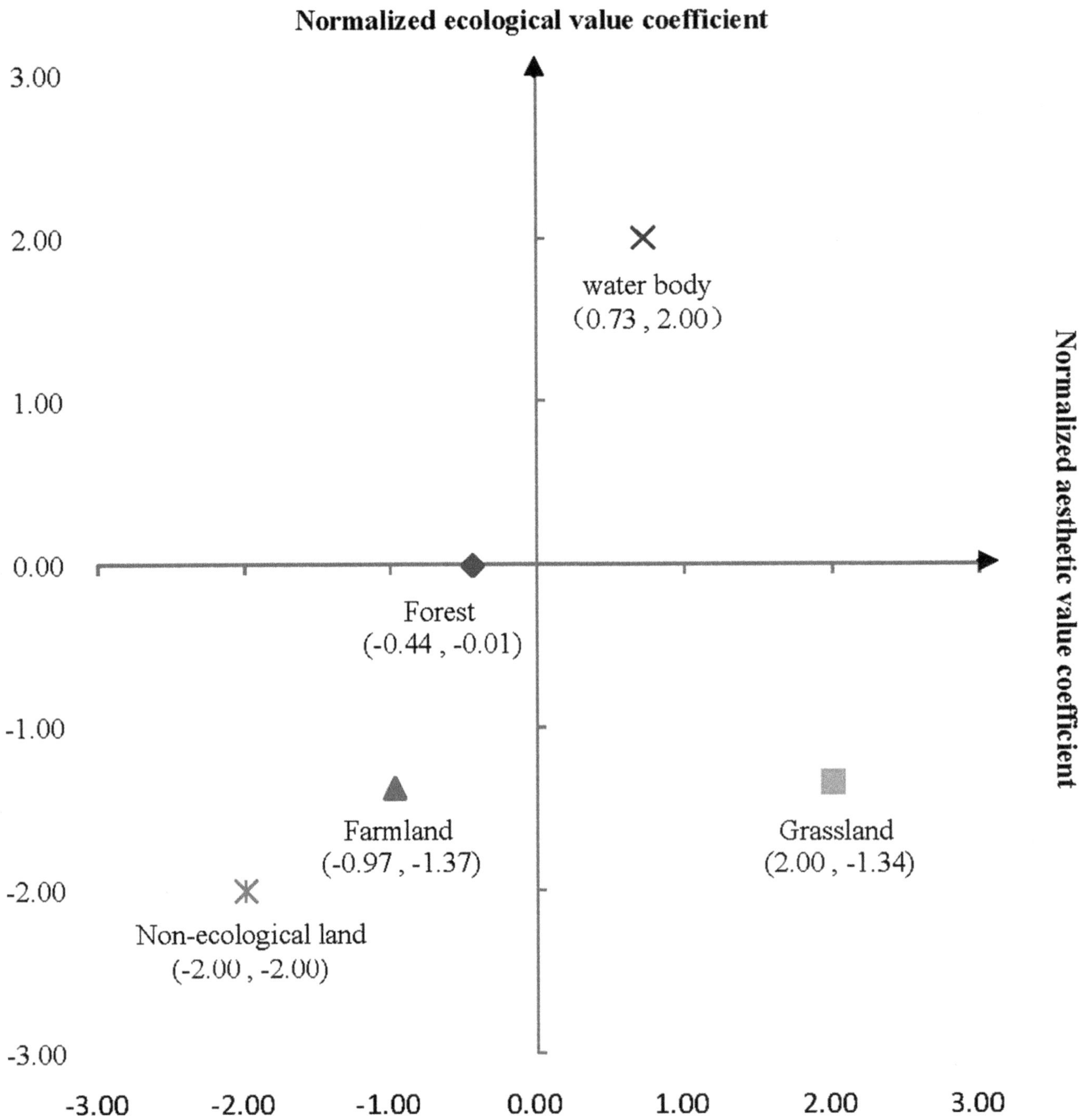

Figure 2. Two-dimensional coordinate system for landscape values.

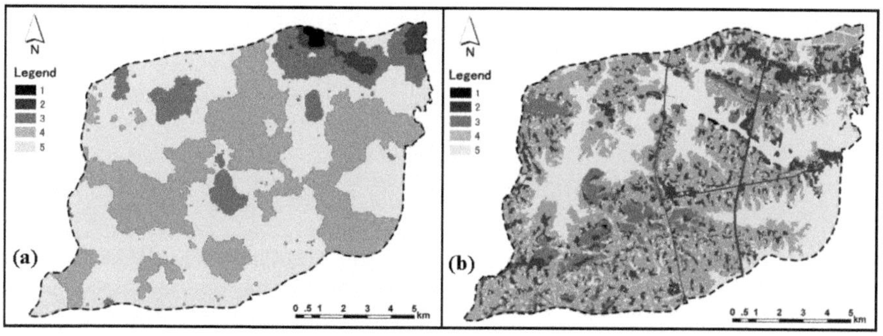

Figure 3. (a) Aesthetic landscape value distribution map and (b) ecological landscape value distribution map in Houguanhu Region.

functional compromise may encounter in landscape planning. In the landscape value distribution maps, the gradients of inherent landscape values and the concerns of underlying landscape value can be used to identify appropriate strategies, such as protection, restoration, and reconstruction. The protection and restoration measures may be taken to expand natural landscape spaces, while human-influenced landscapes should be reconstructed in an orderly way. The reconstruction measures (e.g., building hedgerow buffer zones, reducing the cutting effect of a road network on the overall landscape, or increasing accessibility between landscapes) may be involve in landscape planning.

The situational context (e.g. the social phase, indigenous culture and stakeholder demands) should also be incorporated into landscape planning in response to various concerns regarding the landscape functions [2,24,40]. Tourists are likely to care more about farmland aesthetics, while farmers put a higher value on

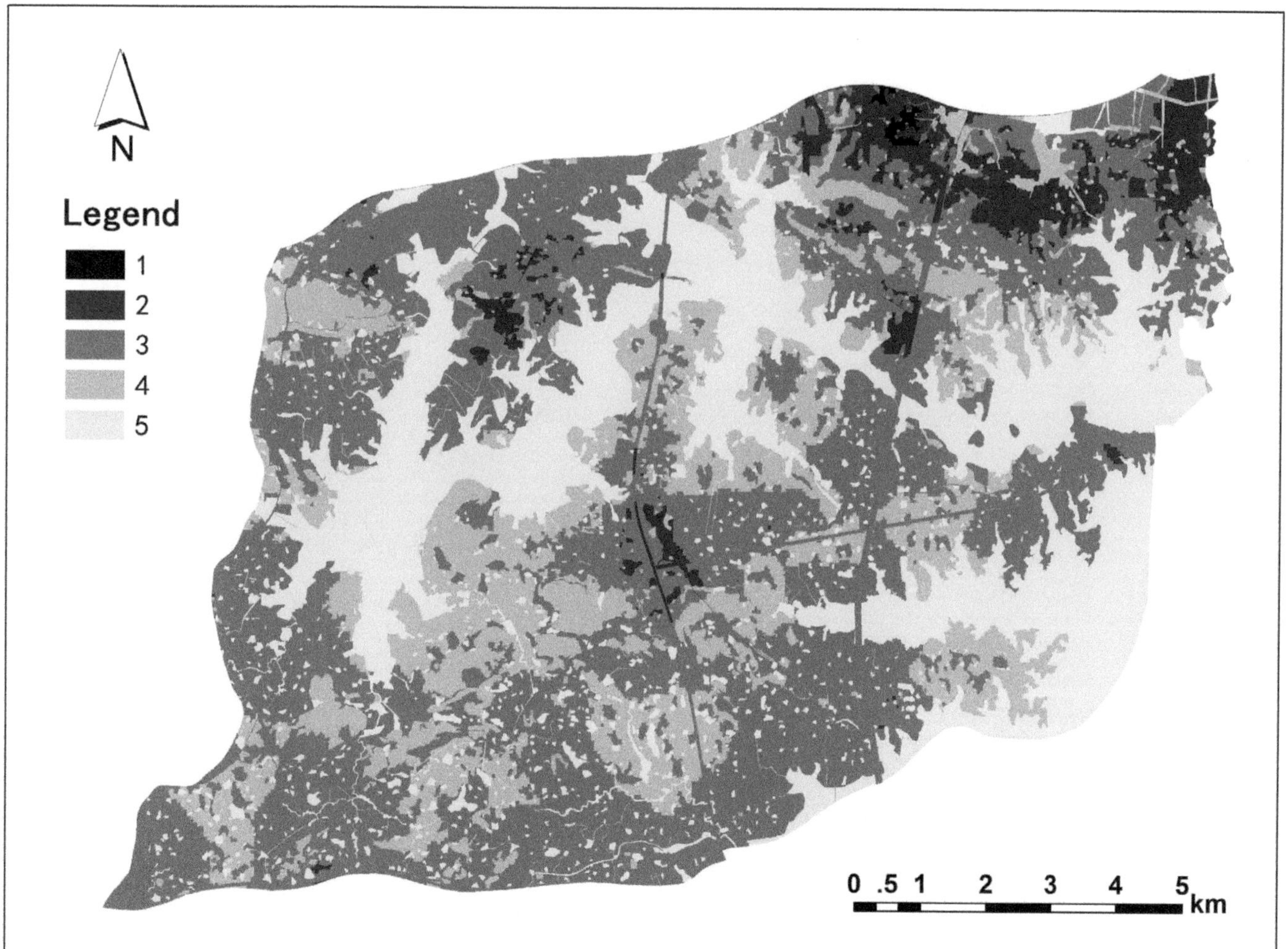

Figure 4. Integrated landscape value distribution map.

agricultural productivity. However, both farmland attractiveness enjoyed by tourists and production features valued by farmers are likely to undermine ecological functions. Therefore, it was recommended for policy makers to align aesthetic features to better support ecological health. It was also recommended that the disorderly non-scenic regions which were important to the overall landscape (e.g. rural settlements, roads, and historic sites) should be carefully enhanced using green corridors and an ordered appearance to foster a more positive aesthetic experience.

Limitations and future improvements

Some limitations and possible improvements of this study included:

(1) In order to make the survey easier, site aesthetic evaluation was kept simple, and only five major landscape types were identified. It is recommended that the subdivisions of landscapes should be adopted for future research to convey more detailed information. In addition, surveyors from a variety of cultural backgrounds could be involved to incorporate a greater diversity of perspectives into landscape planning.
(2) More comprehensive aspects of cultural landscape values (e.g. knowledge systems, social relations, and aesthetic values) could be incorporated into future studies to address the overall performance of landscape resources in the decision-makings.
(3) Subjective aesthetic landscape coefficients can only convey information related to the relative importance of landscape values of the five landscape types. This may limit the use of aesthetic landscape coefficients when comparing with ecological evaluations.

Conclusions

Due to the difficulties in integrating different data types, quantitative models and interdisciplinary studies, it was a far-reaching challenge to connect subjective (aesthetic) and objective (ecological) aspects in an integrated landscape assessment. Nevertheless, the landscape sustainability requires a careful investigation of the relationship between ecology and aesthetics. Compared to conventional conceptual frameworks, the metrics (i.e. field survey, normalized landscape value coefficients, and a two-dimensional coordinate system) were employed in this study in order to quantify the functional performances of landscape types and to present an integrated landscape value. The results bring aesthetic and ecological goals into a better alignment in landscape planning. The interdisciplinary approach may lead to a revolution in landscape designing and planning, because it balances ecological and aesthetic functions for landscape sustainability.

Acknowledgments

We would like to express our gratitude to the people and institutions in Wuhan who assisted us during the field survey. We sincerely thank the editor and anonymous reviewers for their useful comments and suggestions on our paper. We are grateful to Yan Zhu and Jonathan Vause for improving the written quality of the paper.

Author Contributions

Conceived and designed the experiments: DY T. Luo. Performed the experiments: DY T. Luo. Analyzed the data: DY. Contributed reagents/materials/analysis tools: DY T. Luo T. Lin QQ YL. Wrote the paper: DY.

References

1. Daniel TC (2001) Whither scenic beauty? Visual landscape quality assessment in the 21st century. Landscape Urban Plan 54: 267–281.
2. Gobster PH, Nassauer JI, Daniel TC, Fry G (2007) The shared landscape: What does aesthetics have to do with ecology? Landscape Ecol 22: 959–972.
3. Tyrväinen L, Silvennoinen H, Kolehmainen O (2003) Ecological and aesthetic values in urban forest management. Urban Forestry Urban Green 1: 135–149.
4. Klein LR (2013) Quantifying relationships between ecology and aesthetics in agricultural landscapes. Washington State University. 70–98.
5. Luo T, Wang H, Qiu QY, Liu J, Yang ZY (2010) Assessment of non-ecological landscape functions in environmental impact assessment for plan. Res Environ Sci 23(10): 1320–1327. (in Chinese).
6. Wu J (2010) Landscape of culture and culture of landscape: does landscape ecology need culture? Landscape Ecol 25: 1147–1150.
7. Musacchio LR (2009) The ecology and culture of landscape sustainability: emerging knowledge and innovation in landscape research and practice. Landscape Ecol 24: 989–992.
8. Parsons R (1995) Conflict between ecological sustainability and environmental aesthetics: conundrum, canärd or curiosity. Landscape Urban Plan 32: 227–244.
9. Tress B, Tress G, Décamps H, D'Hauteserre AM (2001) Bridging human and natural sciences in landscape research. Landscape Urban Plan 57: 137–141.
10. Naveh Z (2007) Landscape ecology and sustainability. Landscape Ecol 22: 1437–1440.
11. Wu J (2013) Landscape sustainability science: ecosystem services and human well-being in changing landscapes. Landscape Ecol 28: 999–1023.
12. Porto M, Correia O, Beja P (2014) Optimization of landscape services under uncoordinated management by multiple landowners. PLoS ONE 9(1): 1–16.
13. Frank S, Fürst C, Koschke L, Witt A, Makeschin F (2013) Assessment of landscape aesthetics–Validation of a landscape metrics-based assessment by visual estimation of the scenic beauty. Ecol Indic 32: 222–231.
14. Lothian A (1999) Landscape and the philosophy of aesthetics: is landscape quality inherent in the landscape or in the eye of the beholder? Landsc Urban Plan 44: 177–198.
15. Roth M, Gruehn D (2012) Visual landscape assessment for large areas–using GIS, internet surveys and statistical methodologies in participatory landscape planning for the Federal State of Mecklenburg-Western Pomerania, Germany. Latvian Acad Sci A: Humanit Soc Sci 129–142.
16. Forman RTT (2008) The urban region: natural systems in our place, our nourishment, our home range, our future. Landscape Ecol 23: 251–253.
17. Mozingo LA (1997) The aesthetics of ecological design: seeing science as culture. Landscape J 16: 46–59.
18. Nassauer JI (1997) Cultural sustainability: aligning aesthetics and ecology. In: Nassauer JI (Eds.) Placing Nature: Culture and Landscape Ecology. Washington, DC: Island Press.
19. Mikusiński G, Blicharska M, Antonson H, Henningsson M, Göransson G, et al. (2013) Integrating ecological, social and cultural dimensions in the implementation of the landscape convention. Landscape Res 38(3): 384–393.
20. Junker B, Buchecker M (2008) Aesthetic preferences versus ecological objectives in river restorations. Landscape Urban Plan 85: 141–154.
21. Norton LR, Inwood H, Crowe A, Baker A (2012) Trialing a method to quantify the 'cultural services' of the English landscape using Countryside Survey data. Land Use Policy 29: 449–455.
22. Hands DE, Brown RD (2002) Enhancing visual preference of ecological rehabilitation sites. Landscape Urban Plan 58: 57–70.
23. Williams KJH, Cary J (2002) Landscape preferences, ecological quality, and biodiversity protection. Environ Behav 34: 257–274.
24. Fry G, Tveit MS, Ode Å, Velarde MD (2009) The ecology of visual landscapes: Exploring the conceptual common ground of visual and ecological landscape indicators. Ecol Indic 9: 933–947.
25. Hale BW, Steen-Adams MM, Predick K, Fisher N (2005) Ecological conservation through aesthetic landscape planning: A case study of the Lower Wisconsin State Riverway. Environ Manage 35: 381–395.
26. Bastian O, Lütz M (2006) Landscape functions as indicators for the development of local agri-environmental measures. Ecol Indic 6: 215–227.
27. Zhao JW, Luo PJ, Wang RH, Cai YL (2013) Correlations between aesthetic preferences of river and landscape characters. J Environ Eng Landscape Manag 21(2): 123–132.
28. Costanza R, d'Arge R, Groot DR, Farber S, Monica G, et al. (1997) The values of the world's ecosystem services and natural capital. Nature 387: 253–260.
29. Xie GD, Lu CX, Leng YF, Zheng DU, Li SC (2003) Ecological assets valuation of the Tibetan Plateau. J Nat Resour 18: 189–196. (in Chinese).
30. Nassauer JI (2004) Monitoring the success of metropolitan wetland restorations: cultural sustainability and ecological function. Wetlands 24: 756–765.
31. Purcell AT, Lamb RJ (1998) Preference and naturalness: an ecological approach. Landscape Urban Plan 42: 57–66.
32. Yang ZY, Luo T, Zhu ZC, Wang H, Song Y, et al. (2012) Landscape aesthetic assessment based on experiential paradigm assessment technology. Environ Sci Technol 35(3): 187–205. (in Chinese).
33. Von Haaren C (2004) Landschaftsplanung. Stattgart: Eugen GmbH & Co. 245–272. (in German).

34. Dandy N, Van Der Wal R (2011) Shared appreciation of woodland landscapes by land management professionals and lay people: An exploration through field-based interactive photo-elicitation. Landscape Urban Plan 102: 43–53.
35. Gómez-Sal A, Belmontes JA, Nicolau JM (2003) Assessing landscape values: a proposal for a multidimensional conceptual model. Ecol Model 168: 319–341.
36. Termorshuizen JW, Opdam P (2009) Landscape services as a bridge between landscape ecology and sustainable development. Landscape Ecol 24: 1037–1052.
37. Yang B, Brown TC (1992) A cross-cultural comparison of preferences for landscape styles and landscape elements. Environ Behav 24: 471–507.
38. Yang DW, Kao WTM, Zhang GQ, Zhang NY (2014) Evaluating spatiotemporal differences and sustainability of Xiamen urban metabolism using emergy synthesis. Ecol Model, 272: 40–48.
39. Yang DW, Qiu QY, Liu WJ (2012) Complementary zoning frameworks for sustainable management of eco-economic capital: a case study of Lijiang, China. J MT Sci, 9: 760–769.
40. Aretano R, Petrosillo I, Zaccarelli N, Semeraro T, Zurlini G (2013) People perception of landscape change effects on ecosystem services in small Mediterranean islands: A combination of subjective and objective assessments. Landscape Urban Plan 112: 63–73.

5

Increased Drought Impacts on Temperate Rainforests from Southern South America: Results of a Process-Based, Dynamic Forest Model

Alvaro G. Gutiérrez[1,2*¤], Juan J. Armesto[3,4], M. Francisca Díaz[5], Andreas Huth[1]

1 Department of Ecological Modeling, Helmholtz Centre for Environmental Research (UFZ), Leipzig, Germany, **2** Forest Ecology Group, Institute of Terrestrial Ecosystems, Department of Environmental Sciences, Swiss Federal Institute of Technology (ETH Zürich), Zürich, Switzerland, **3** Instituto de Ecología y Biodiversidad (IEB), Santiago, Chile, **4** Departamento de Ecología, Facultad de Ciencias Biológicas, Universidad Catolica de Chile, Santiago, Chile, **5** Departamento de Ciencias Biológicas, Facultad de Ciencias Biológicas, Universidad Andrés Bello, Santiago, Chile

Abstract

Increased droughts due to regional shifts in temperature and rainfall regimes are likely to affect forests in temperate regions in the coming decades. To assess their consequences for forest dynamics, we need predictive tools that couple hydrologic processes, soil moisture dynamics and plant productivity. Here, we developed and tested a dynamic forest model that predicts the hydrologic balance of North Patagonian rainforests on Chiloé Island, in temperate South America (42°S). The model incorporates the dynamic linkages between changing rainfall regimes, soil moisture and individual tree growth. Declining rainfall, as predicted for the study area, should mean up to 50% less summer rain by year 2100. We analysed forest responses to increased drought using the model proposed focusing on changes in evapotranspiration, soil moisture and forest structure (above-ground biomass and basal area). We compared the responses of a young stand (YS, ca. 60 years-old) and an old-growth forest (OG, >500 years-old) in the same area. Based on detailed field measurements of water fluxes, the model provides a reliable account of the hydrologic balance of these evergreen, broad-leaved rainforests. We found higher evapotranspiration in OG than YS under current climate. Increasing drought predicted for this century can reduce evapotranspiration by 15% in the OG compared to current values. Drier climate will alter forest structure, leading to decreases in above ground biomass by 27% of the current value in OG. The model presented here can be used to assess the potential impacts of climate change on forest hydrology and other threats of global change on future forests such as fragmentation, introduction of exotic tree species, and changes in fire regimes. Our study expands the applicability of forest dynamics models in remote and hitherto overlooked regions of the world, such as southern temperate rainforests.

Editor: Bruno Hérault, Cirad, France

Funding: This study was supported by grants from Millennium Scientific Initiative (P05-002) and CONICYT PFB-23 to the Institute of Ecology and Biodiversity. AGG was funded by a DAAD doctoral fellowship, by a Marie Curie Intra European Fellowship within the 7th European Community Framework Programme (PIEF-GA-2010-274798) and CONICYT-PAI grant (82130046). Additional funding was provided by the ERC advanced grant 233066. The funders had no role in study design, data collection and analysis, decision to publish, or preparation of the manuscript.

Competing Interests: The authors have declared that no competing interests exist.

* Email: bosqueciencia@gmail.com

¤ Current address: Instituto de Conservación Biodiversidad y Territorio, Facultad de Ciencias Forestales y Recursos Naturales, Universidad Austral de Chile, Valdivia, Chile

Introduction

Climate and forests are dynamically linked through the spatial and temporal variability of soil moisture [1], with climate system effects on ecological processes which are still poorly understood. Forest dynamics models, particularly those based on interactions among individual trees (i.e. gap models [2]), provide a simple, and general framework to assess the impacts of climate on forest dynamics. These models simulate the fate of single trees on the basis of species' life-history traits and limited resource availability (e.g. soil moisture), thereby facilitating the analysis of climate-forest interactions [3].

Forest gap models use a variety of approaches to model forest hydrology. While some gap models use a simple bucket water balance model [3,4], others include physiology-based representations of plant and soil controls on water uptake and evapotranspiration [5,6]. Regardless of the level of detail used to model forest hydrology, it seems necessary that forest gap models address water availability (i.e. soil moisture) as an integrating factor, with effects on canopy transpiration [7]. Changes in rainfall regimes, summarized by changes in the duration and frequency of periods of water stress during the year, should influence soil moisture dynamics limiting plant productivity [8]. Introducing dynamic linkages of ecological processes with soil moisture variation in gap models will contribute to predict drought-induced changes in forest dynamics. Such model improvements are increasingly relevant to understanding how forests can adapt to climate change [6,9].

Forest gap models have successfully simulated the dynamics of a variety of forest types including temperate rainforests of the southern hemisphere [10,11]. In southern South America (SSA, 37–43°S), the progressive loss, fragmentation and subsequent degradation of temperate rainforests due to unsustainable logging and fire is threatening the integrity of ecosystem functions [12,13] and modifying their hydrological balance [14,15]. Annual precipitation has decreased in the same region by about 40% in the last century (time period 1901–2005, [16]) and summer rainfall is expected to decrease up to 50% by the year 2100 [17,18]. SSA forests share similar structural characteristics with temperate rainforests of the Pacific Northwest of North America, Tasmania, and New Zealand [19]. In addition, SSA forests represent the largest area of temperate forest remaining in the southern hemisphere [20]. Floristic richness is the highest among evergreen temperate rainforests worldwide and the high concentration of endemism has given this region a unique global conservation value [20,21]. The global relevance of SSA forests and climate trends predicted for the coming decades make it urgent to expand model applications into this region, as a tool to predict temperate rainforest responses to impending declines in rainfall.

This study introduces a forest gap model specifically designed for assessing the responses of temperate rainforests in southern South America to increased drought. The model provides accurate estimates of forest water fluxes and incorporates dynamical linkages among rainfall regimes, soil moisture, and individual tree growth. We assessed model performance by comparing the results with detailed field measurements of water cycling in a stand located on northern Chiloé Island, Chile (41°50′S). We also conducted a sensitivity analysis of the response of current forests to drought, i.e. when rainfall is decreased. Model predictions of forest hydrology (evapotranspiration and soil moisture) and structure (above-ground biomass and basal area) under increased drought predicted for 2100 in the study area were compared for a young-secondary (YS) and an old-growth (OG) forest stand to analyze differences in responses to expected changes in rainfall.

Materials and Methods

Study area

The study was conducted on northern Chiloé Island, Chile (41°50′ S, Fig. 1) at the private protected area *Estación Biológica Senda Darwin* (EBSD), with permission granted by the owner. Fragments of secondary and primary forests occur over rolling hills of low altitude (50–100 m) dispersed in a matrix of bogs, shrublands and grazing pastures. The present landscape has been shaped by a history of widespread use of fire to clear land for pastures since the late 1800s, followed by selective logging of remaining forest patches [22]. Soils are generally thin (<0.5 m), originated from moraine fields and outwash plains from the last glaciation, often with poor drainage [23]. Soils have high organic matter content, soil texture loam to silty loam, and a 2–4 mm thick iron silicate layer or hardpan (found at ca. 52 cm depth), where roots cannot penetrate [24]. The prevailing climate is wet-temperate with strong oceanic influence [25]. Rainfall occurs throughout the year, with an annual average of 2158 mm (25% occurring in summer). Mean annual temperature is 9.1°C. Maximum and minimum monthly temperatures are 13.9°C (January) and 4.2°C (July) [26].

Floristically, forests of the study area belong to the North Patagonian temperate rainforest type [27]. The canopy is dominated by evergreen trees, mainly *Podocarpus nubigena* (Podocarpaceae), *Drimys winteri* (Winteraceae) and *Nothofagus nitida* (Nothofagaceae), with the common presence of *Tepualia stipularis* (Myrtaceae) and several Myrtaceae tree species in the understory. Ferns (e.g. *Hymenophyllum* spp., *Hymenoglossum cruentum, Polypodium feullei)* and angiosperms (e.g. Gesneriaceae and Bromeliaceae) growing epiphytically are frequent. Detailed descriptions of structure and dynamics of this forest type have been previously published [26,28]. The study did not involve endangered or protected tree species.

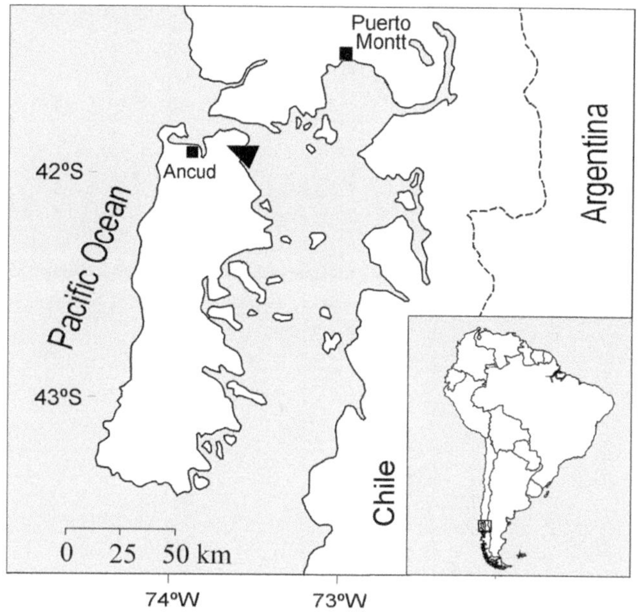

Figure 1. Location of study site (triangle) on northern Chiloé Island, Chile.

The forest model

Here, we introduce an individual-oriented dynamic forest model (FORMIND-CL v.1.0) that includes calculations of hydrologic balance. The model is based on FORMIND, a forest model comprehensively tested to simulate the dynamics of temperate rainforests in SSA [11,13]. FORMIND is a generalized forest growth model that simulates the spatial and temporal dynamics of uneven-aged, mixed species forest stands [29–31]. The model simulates forest dynamics (in annual time steps, *t*) as a mosaic of interacting forest patches of 20×20 m, which is the approximate crown size of a large mature tree in the forest. Within these patches, stand dynamics is driven by competition for light and space following the gap model approach [2]. For the explicit modeling of the competition for light, each patch is vertically divided into height layers of 0.5 m, where leaf area is summed up and the light environment under the canopy is calculated via a light extinction law. The carbon balance of each individual tree is modeled explicitly, including the main physiological processes (photosynthesis and respiration [13]). Allometric functions and geometrical relations are used to calculate above-ground biomass, tree height, crown diameter and stem volume from the stem diameter at 1.3 m height of the tree (*dbh*). Tree mortality can occur either through self-thinning in densely populated stands, tree senescence, gap formation by large falling trees, slow tree grow, or external disturbances (e.g. windthrow). Gap formation links neighboring forest areas. Tree regeneration rates are formulated as maximum rates of recruitment of small trees at *dbh* threshold of 1 cm, with seed loss through predation and seedling mortality being incorporated implicitly [13]. Maximum recruitment rates are reduced by shading. Nutrient availability is considered to be homogeneous at the stand scale. A description of the core model and its equations is given elsewhere [11,13]. We focus below on the extensions added to incorporate forest hydrology.

Table 1. Structure and composition of the young-secondary North Patagonian forest stand (YS) in Senda Darwin Biological Station, compared to values reported for other secondary forest stands, elsewhere in Chiloé Island, Chile.

	Study site		Literature*			
Species	Density (trees ha^{-1})	Basal area (m^2 ha^{-1})	Density (trees ha^{-1})		Basal area (m^2 ha^{-1})	
Drimys winteri	1550	32.3	1800	4532	24.7	61.4
Eucryphia cordifolia	175	0.9	112	480	1.4	14.7
Myrtaceae†	1014	3.5	252	827	2.3	4.4
Nothofagus nitida	713	17.7	4	260	0.1	4
Podocarpaceae¥	38	0.1	80		0.5	
Weinmannia trichosperma	88	0.3				
Others	287.5	1.9				
Total	3575	54.8	2431	7950	46	86.9

*Source: [26,77–80] only forest stands <100 years old. Only listed tree species present in YS.
†*Tepualia stipularis, Myrceugenia spp.* and *Amomyrtus spp.;*
¥*Podocarpus nubigena, Saxegothaea conspicua.*

The hydrologic submodel

Soil moisture dynamics is described at a daily timescale, treating soil as a reservoir with an effective storage capacity that is intermittently filled by rainfall events. Soil water losses occur via transpiration, interception by the forest canopy, and drainage below the root zone. We neglected lateral water flow, thus the model applies mainly to flat terrains. This is a reasonable assumption in forests of the study area because during the rainy season soils tend to be saturated and accumulated water cannot infiltrate the soil.

Soil moisture s (dimensionless, $0 \leq s \leq 1$), vertically averaged over the soil depth z (mm), was considered as central state variable [8]. Thus, the water balance equation for a given point in the forest can be expressed as [1]:

$$n \cdot z \cdot \frac{ds}{dd} = Pnet_d - Tr_d - Q(s,d) \quad (1)$$

where d is the Julian day of the year, n is the porosity (volume of voids/total volume of soil, i.e. dimensionless, vertically averaged); $Pnet_d$ is the net precipitation falling to the soil surface (mm day^{-1}); Tr_d is the transpiration rate (mm day^{-1}); and $Q(s,d)$ is the soil drainage (mm day^{-1}). Both n and z are assumed to be time-invariant parameters [1]. The volumetric water content (θ, m^3 water/m^3 soil, i.e. dimensionless) can be calculated as follows [1]:

$$\theta = s \cdot n \quad (2)$$

The normalized version of equation (1) is used through the text where all terms are divided by $n \cdot z$. Both the local vertical and horizontal variability of soil moisture are considered negligible at the daily timescale, assuming an equal propagation of the wetting front and equal soil moisture redistribution over the rooting zone [8,32].

Net precipitation. Daily net precipitation falling to the soil surface ($Pnet_d$) is described by,

$$Pnet_d = P_d - Ec_d \quad (3)$$

where, Ec_d is the canopy interception (mm day^{-1}), defined here as the total daily rainfall (P_d, mm day^{-1}) that is retained by the canopy and is evaporated so that it does not reach the ground. Following [33], we assumed that Ec_d asymptotically approaches the canopy retention capacity and can be modeled at daily intervals as:

$$Ec_d = S_t \cdot (1 - e^{-\alpha_h \cdot P_d}) \quad (4)$$

where S_t is the canopy water retention capacity of the stand at year t and α_h is a parameter describing the slope of the saturation curve. The parameter α_h represents, in a simplified terms, the complex process of water partitioning into throughfall and stem flow [34]. S_t depends on leaf area index of the forest patch at simulated year t (LAI_t) and is calculated by the expression [34,35]:

$$S_t = \frac{LAI_t}{LAI_{\max}} \cdot f_h \cdot \log(1 + LAI_t) \quad (5)$$

where, LAI_{max} is the maximum leaf area index of the forest and f_h is a shape parameter. We avoided unrealistic canopy interception values in the model by setting $Ec_d = P_d$ when $Ec_d > P_d$.

Table 2. Parameter descriptions and parameterization methods used for running simulations in FORMIND-CL v.1.0.

	Description	Value	Units	Method	Reference
Weather generator					
$1/\lambda$	Mean interval time between rainfall events[¥]	†	days	a	This study
η	Mean depth of rainfall events[¥]	†	mm day^{-1}	a	This study
T_{μ}	Mean daily temperature[¥]	†	°C	a	This study
T_{σ}	Standard deviation daily temperature[¥]	†	°C	a	This study
Rg_{μ}	Daily global radiation above canopy[¥¶]	†	µmol (photons) m^{-2} s^{-1}	a	This study
Rg_{σ}	Standard deviation of Rg_{μ} [¥¶]	†	µmol (photons) m^{-2} s^{-1}	a	This study
Hydrologic submodel					
n	Vertically averaged porosity of the soil	0.757	-	b	[53]
z	Soil depth	520	mm	b	[14]
k_{soil}	Saturated hydraulic conductivity	4	mm day^{-1}	b	[36,54]
α_h	Slope of the canopy saturation curve	0.7	-	b, c	[34]
f_h	Parameter of the relationship LAI and canopy storage capacity	3	mm day^{-1}	c	[35]
LAI_{max}	Maximum LAI of the studied forest	5.5	m^2 m^{-2}	b	[81]
WUE	Water-use efficiency	9	g CO_2 $kg^{-1}H_2O$	e	[14]
γ	Psychometer constant	65	Pa K^{-1}	b	[43,44]
L	Latent heat of vaporization of water	2.56×10^6	J kg^{-1}	e	[43,44]
Δ	Rate of change of saturated vapor pressure with temperature	c	Pa K^{-1}	d	[43,44]
θ_{wp}	Wilting point of the soil	0.125	-	b	[36]
θ_{fc}	Field capacity of the soil	0.3	-	b	[36]

Method refers to a: calculated from daily meteorological data from *Senda Darwin Biological Station*, period 1998–2009, b: from literature, c: calibrated with field data, d: calculated, e: calibrated using literature.
†values indicated in Table 3.
¥per season.
¶calculated for dry ($Pd<1$ mm) and wet days ($Pd\geq1$ mm).

Soil moisture modeling. Drainage out of the root zone ($Q(s,d)$) was modeled according to [1]. When the soil is saturated ($s=1$), soil water is permitted to percolate at a rate equivalent to the saturated hydraulic conductivity of the soil (k_{soil}, mm day^{-1}, [36]). Runoff occurs when the soil is saturated and no more water can be held in place. The excess of water is assumed to leave the system, which is reasonable to assume given the large rainfall intensity in the study area. When $s<1$, soil deep percolation rate is calculated using the empirical relationship of Neilson [37],

$$Q(s,d)=k_{soil}\cdot s^2 \tag{6}$$

Transpiration. Water-use efficiency describes the proportion of water used for the assimilation of a unit of carbon in the photosynthesis [38,39]. This concept can be used to estimate transpiration of trees (Tr, mmol H_2O m^{-2} s^{-1}) from:

$$Tr=\frac{PB}{WUE} \tag{7}$$

where, PB is the gross biomass production of the tree (µmol carbon dioxide m^{-2} s^{-1}), and WUE is a parameter denoting water-use efficiency at stand level. PB is obtained from the rate of single-leaf photosynthesis following [40], which is integrated over the total LAI of the tree to account for self-shading [13]. The resulting photosynthetic rate is then multiplied by the crown area of the tree to obtain PB (see also equations in Appendix S1 and [13]). Daily transpiration (Tr_d) of trees is obtained from equation 8 and dividing Tr by the length of the active photosynthetic period per year.

The daily potential evapotranspiration (PET_d, mm day^{-1}) describes a physical limit for the amount of water that can be held and transported away from the canopy under given climatic conditions. Evaporation is neglected in the model; therefore, it is assumed that maximum water losses by vegetation are limited by the difference between PET_d and the canopy interception of the day (Ec_d), as follows:

$$PET_d-Ec_d\geq Tr_d \tag{8}$$

PET_d is calculated using a modified Penman-Monteith expression in case of aerodynamic conductance [41,42] and determined by the variation of the daily net radiation flux (Rn_d, J m^{-2} day^{-1}):

$$PET_d=(\frac{\Delta}{\Delta+\gamma})Rn_d/L \tag{9}$$

where γ is the psychometric constant (ca. 65 Pa K^{-1}, slightly depends on temperature), L is the latent heat of vaporization of

Table 3. Parameter values used to run the weather generator under different climatic scenarios.

	Current climate			Future scenarios		
Rainfall	Seasonal sum (mm, average ± sd)	1/λ (days)	η (mm day^{-1})	Seasonal sum (mm, average)	1/λ (days)	η (mm day^{-1})
DJF	284.6±132.7	0.9	8.3	238–131	0.99–1.36	4.2–7.5
MAM	543.6±138.6	0.55	12.8	813.4	0.55	12.8
JJA	813.4±197.2	0.29	16	543.6	0.29	16
SON	424.5±144.7	0.54	9.5	382–212	0.59–0.8	4.7–8.5
Annual sum (mm, average ± sd)	2094.8±353.8					
Temperature	T_μ (°C)	T_σ (°C)			T_μ (°C)	
DJF	12.5	2.6			16.5	
MAM	10.1	3.6			13.1	
JJA	8.4	4.4			10.4	
SON	9.8	3.3			11.8	
Radiation	Rg_μ (Rg_σ)	Rg_μ (Rg_σ)				
	[P_d<1 mm]	[P_d≥1 mm]				
DJF	1413.9 (317.0)	986.9 (378.2)				
MAM	701.2 (294.1)	398.8 (261.6)				
JJA	408.9 (159.1)	229.7 (139.4)				
SON	1065.5 (348.7)	640.8 (320.1)				

Current climate indicate parameters used to run the model under current climate based on instrumental records (weather station at *Senda Darwin Biological Station*, period 1998–2009). Radiation describes parameters daily global radiation Rg_μ and Rg_σ (the latter in brackets, μmol(photons) m^{-2} s^{-1}). Temperature is mean daily air temperature. *Future scenarios* are the range of climatic parameters that were varied to run the model under increased drought (36 scenarios in total, see *Methods* for details). *DJF:* December to February (austral summer, growing season); *MAM*: March to May (austral autumn); *JJA*: June to August (austral winter), *SON*: September to November (spring, growing season). sd: standard deviation.

water (ca. 2.56×10^6 J kg^{-1} slightly depends on temperature). The rate of change of saturated vapor pressure with temperature (Δ, Pa K^{-1}) is calculated as [43,44]:

$$\Delta = \frac{2.503x10^6 e^{17.269\frac{T_d}{237.3+T_d}}}{(237.3+T_d)^2} \quad (10)$$

Rn_d was calculated from latitude, day of the year, sunshine hours and daily air temperature (T_d, °C) following [42,45].

Soil moisture impact on tree biomass production. The dependence of water uptake for tree biomass production on soil moisture is described by a function representing a reduction factor due to water scarcity ($\omega(s)$, $0\leq \omega(s) \leq 1$ [46]). This factor accounts indirectly for the impact of water demand on potential photosynthetic production (i.e. possible to achieve under competition for light). $\omega(s)$ is implemented as a daily reduction factor due to water scarcity by,

$$\omega(s)=0 \text{ if } s<\theta_{wp}$$

$$\omega(s)=\frac{s-\theta_{wp}}{\theta_{msw}-\theta_{wp}} \text{ if } \theta_{wp}<s\leq\theta_{msw}$$

$$\omega(s)=1 \text{ if } s>\theta_{msw} \quad (11)$$

where, θ_{wp} is the wilting point, and θ_{msw} represents a threshold when enough soil moisture is available for potential tree biomass production. We calculated θ_{msw} from:

$$\theta_{msw}=\theta_{wp}+\frac{1}{3}(\theta_{fc}-\theta_{wp}) \quad (12)$$

where, θ_{fc} is the soil field capacity. θ_{wp}, θ_{msw}, and θ_{msw} are expressed as normalized soil moisture. In the model, the wilting point (θ_{wp}) determines the minimum soil moisture content necessary for tree biomass production. Thus, we assumed a linear reduction of biomass production when soil water content was between θ_{msw} and θ_{wp}. The water required for biomass production of trees is completely removed from the soil compartment when soil moisture reaches θ_{msw} (i.e., $\omega(s)=1$), after the calculation of maximum possible transpiration of trees. Both biomass production and water supply are reduced until the water needed for biomass production corresponds with θ_{wp}. The calculated rate of biomass production influences tree respiration rate through maintenance and growth respiration, which are calculated subsequently in the model (see [13] for equations and Fig. S1 for a diagram). All calculations are performed for every tree and pooled together to calculate the stand-level values.

Weather generator. Rainfall time series, representing the frequency and depth of rainfall events, were constructed as series of random numbers generated by probability distributions. The interval between rainfall events, τ (day) can be expressed as an exponential distribution given by [47].

$$f_T(\tau)=\lambda\cdot e^{-\lambda\tau}, \text{ for } \tau\geq0 \quad (13)$$

Table 4. Model estimates of water balance components for a young secondary forest (YS, 60 years-old) and an old-growth North Patagonian forest (OG, >500 years-old) located in northern Chiloé Island, Chile, under current climate.

Variable	This study		Literature	
	mm year^{-1}± sd	%	Value	Source*
Young secondary stand (YS)				
Canopy Interception	381.5±30.3	20.0	20–40%	1
Deep percolation	980.7±156.8	49.9	47%	1
ET	573.6±35.1	30.3	45.2%	1
Net precipitation	1617.2±434.6	80.0	60–80%	1
Runoff	722.9±397.7	33.3		
Soil moisture	-	62.5		
Transpiration	192.1±18.8	10.3	22%	1
Old-growth stand (OG)				
Canopy Interception	378.3±33.6	19.9	17.8%	2, 3
Deep percolation	907.4±142.4	46.5	66.5%	2
ET	648.0±44.1	34.3	19.9–33.3%	2
Net precipitation	1591.8±368.6	80.1		
Runoff	665.2±321.9	31.8	30–55%	4
Soil moisture	-	55.3		
Transpiration	269.7±26.1	14.4		
Potential evapotranspiration	769±4		576–724 mm year^{-1}	5

Model results are the average of 100 simulations per stand (see *Methods* for details), with annual sum of rainfall averaging 1970–2000 mm year^{-1}. Literature refers to values reported by independent studies in Chilean temperate rainforests and comparable forests elsewhere. sd: standard deviation; %: percentage of total annual rainfall; ET: Evapotranspiration (sum of canopy interception and transpiration). sd: standard deviation.
*(1) Data for other broad-leaved evergreen forests, ca. 200 years old. Annual rainfall 2500 mm year^{-1} [82,83]. (2): Mixed deciduous-broad-leaved old-growth forest. Annual rainfall 2400 mm year^{-1}[12]. (3) Mixed broad-leaved and conifer forest, ca. 200 years old in New Zealand. Annual rainfall 3400 mm year^{-1}[84]. (4) Annual rainfall 1700–4500 mm year^{-1}, data from evergreen, broad-leaved forests with 90% cover [85]. (5) Annual rainfall of 2427–3991 mm year^{-1} [86], weather stations of Castro and Punta Corona.

where $1/\lambda$ is the mean time interval between rainfall events (days). Total daily rainfall (P_d) depends on the amount of rain of each event (h, mm day^{-1}), which is also assumed to be an independent random variable, expressed by an exponential probability density function [47]:

$$P_d(h) = \frac{1}{\eta} e^{-\frac{1}{\eta}h}, \text{ for } h \geq 0 \tag{14}$$

where η is the mean depth of rainfall events (mm day^{-1}). Both $1/\lambda$ and η parameters are calculated for each season of the year. We obtained daily global radiation (Rg_d) from EBSD instrumental records (period from May 1998 to February 2009). Rg_d varied among seasons in relation to daily rainfall (T-test, $p<0.001$). Therefore, in the model, Rg_d was distributed as a Gaussian variable whose mean and standard deviation depended on P_d (mean, Rg_μ, and standard deviation, Rg_σ). Values of Rg_μ and Rg_σ were obtained from instrumental records and varied depending on a threshold value of 1 mm of P_d and the season of the year. Daily temperature (°C, T_d) was simulated by a Gaussian random variable with parameters (mean, T_μ, and standard deviation, T_σ) that varied according to season of the year.

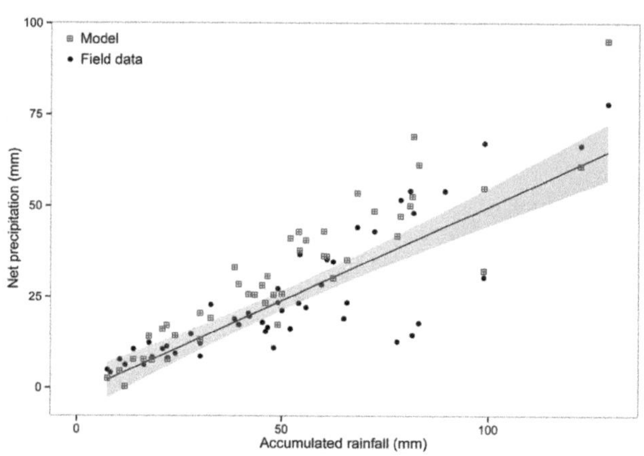

Figure 2. Comparison between measured and modeled net precipitation for 50 rain events recorded in a young secondary stand in northern Chiloé Island, Chile, for the time period 2007–2010. The line represents a linear regression between field measured net precipitation and accumulated rainfall during each event. The gray area represents 0.95 confidence intervals. Model results are for a forest patch are of 400 m^2, with a LAI = 5.

Field data

Stand structure. We characterized stand structure in terms of tree species dominance (basal area, m^2 ha^{-1}) and size (*dbh*) distribution in a young secondary and an old-growth North Patagonian forest stand (hereafter YS and OG respectively) found in a flat forested area at EBSD (Fig. 1). The YS stand was initiated from a stand-replacing fire about 60 years ago and currently presented no evidence of logging. In 2007, we set up two 20×20 m

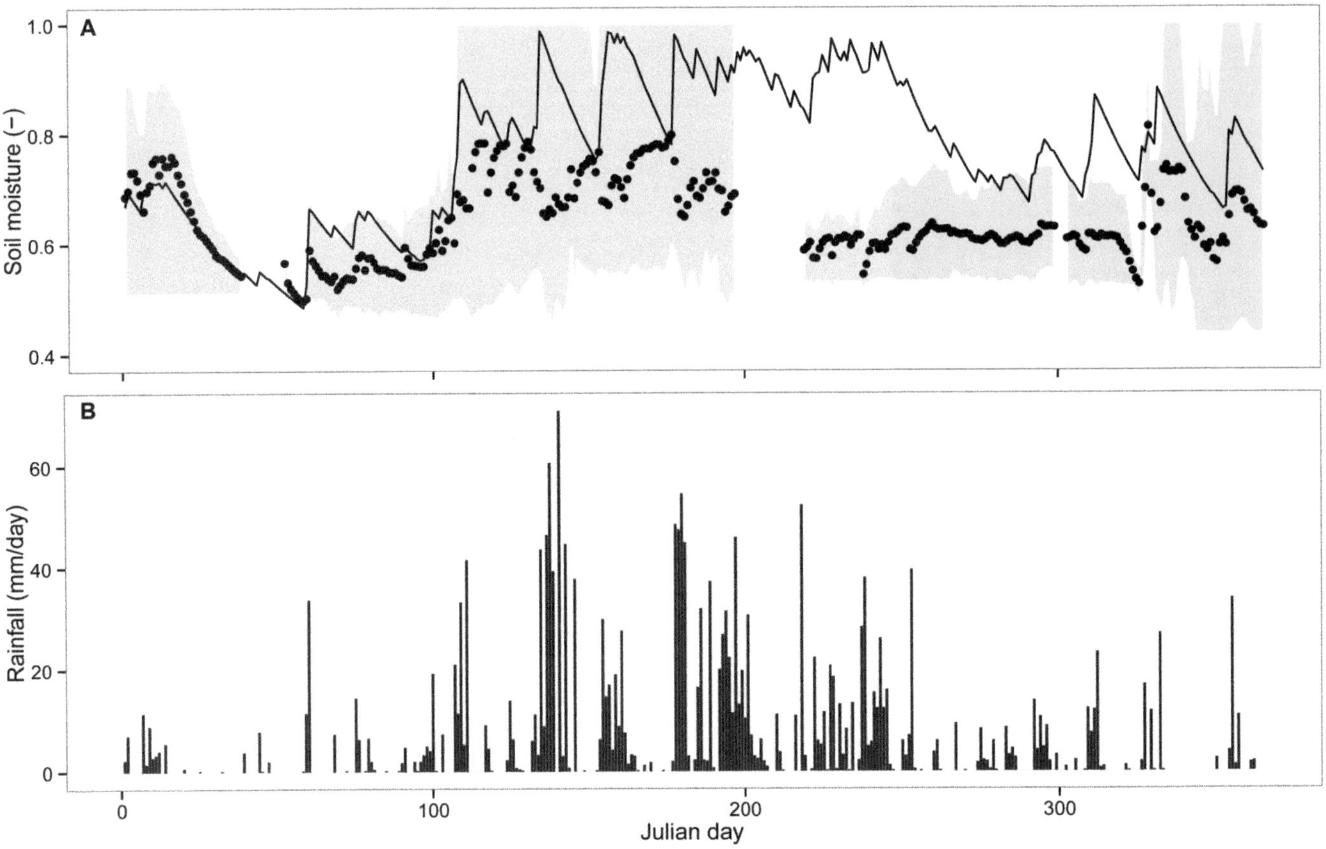

Figure 3. Variation in soil moisture during 2008 in a young secondary, North Patagonian forest stand (YS) in northern Chiloé Island, Chile. a) Comparison between observed and simulated soil moisture (normalized data, dimensionless). Observed data were obtained from five sensors randomly placed inside a 400-m^2 plot in YS (daily means denoted by filled dots, gray area showing the range of data). Simulated soil moisture (line) is for one-hectare forest with a successional age of 60 years (LAI = 4.5). Soil parameters are the same as described in Table 2 (see also *Methods*). b) Rain events recorded instrumentally during 2008.

plots to measure the hydrologic balance and stand structure. All trees rooted within each plot with stems >1.3 m height and > 5 cm *dbh* were permanently marked with numbered aluminum tags, identified to species, and their *dbh* measured to the nearest cm. Structure and composition of YS is comparable to that described for young-secondary stands elsewhere on northern Chiloé Island (Table 1). OG was an unmanaged forest stand >590 years old, without evidence of recent human disturbance, and representative of old-growth North Patagonian forest on Chiloé Island and elsewhere in the region [26]. Sampling methods, stand history, species composition and structure of OG is described by [28]. OG had total basal area of 72 m^2 ha^{-1} and density of 2610 trees ha^{-1}, with a mixed dominance of Podocarpaceae species, *D. winteri, N. nitida* and Myrtaceous tree species in the understory [28].

Hydrologic measurements. We estimated net precipitation using throughfall measurements taken in two YS forest plots (i.e. rainfall falling through canopy gaps plus canopy drip), adding stemflow (water running down the stems). We conducted these measurements of volume accumulated during rainfall events occurred between June 2007 and December 2010. During this time period, we also analyzed hourly records of rainfall from the meteorological station at EBSD to obtain daily incident rainfall above the canopy. Rainfall events considered in the analysis occurred with a separation of at least two hours without rain to allow for full drip from the forest canopy. Stemflow collectors consisted of a 2 mm thick smooth polycarbonate sheet molded around the stem to form a funnel. A hose led from the lowest point of the funnel to a 25 l polythene container, where the stemflow volume was collected after each rain event. Stemflow collectors were placed in 10 randomly selected trees of the two main canopy species, *D. winteri* and *N. nitida* (*dbh* >10 cm), in each plot. We eliminated two trees from our stemflow estimates that died during the study period. We converted the volume of collected water to millimeters of rain assuming that the surface of the collectors equals the projected tree crown area. Crown area was approximated by the area of an ellipse. Throughfall collectors, 0.12×2 m long (0.7 m^2 total area per plot) gutters were held, with a slight inclination, 0.5 m above the ground at three different locations within each plot. Collectors were connected with a funnel to a 25 l polythene container. Soil matric potential was measured every 30 minutes with four sensors per plot (*WatchDog Data Loggers* 450 and 800) placed approximately in every quarter of each plot, beneath the canopy and at ca. 15 cm soil depth. Continuous soil moisture measurements were obtained for the period January 2007 to March 2009.

Model parameterization

We used a previous model parameterization for North Patagonian forests including all main canopy tree species (11 tree species) occurring in the studied forests. The calibration, validation and robustness of this parameter set to reproduce forest stand

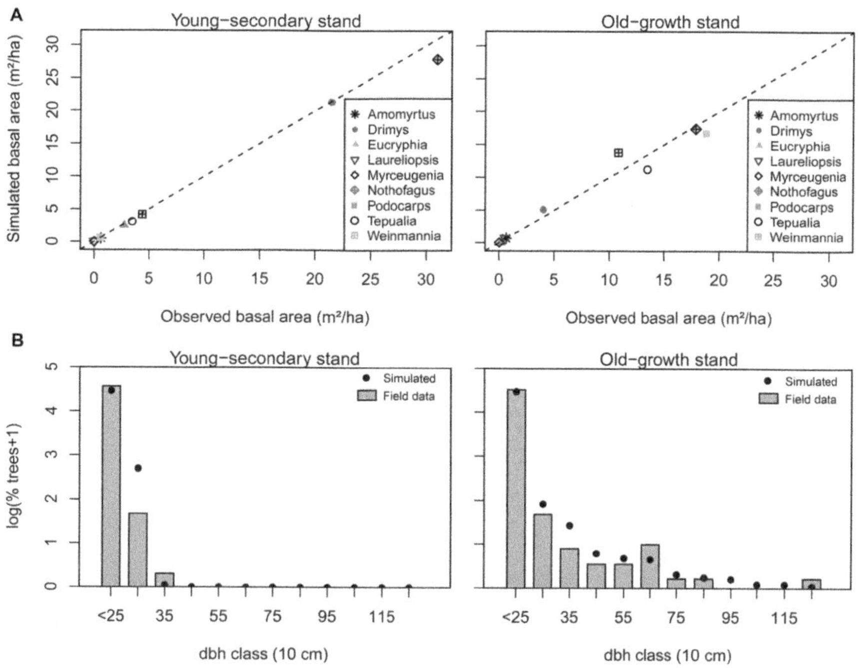

Figure 4. Comparison of forest structure between (a) observed (field data) and simulated basal area of tree species (Spearman's r^2>0.9, p<0.01 in both cases) and (b) *dbh* distributions for the young secondary (YS) and old-growth (OG) North Patagonian forest stands studied in Chiloé Island, Chile. Simulated OG structure was obtained initializing the model with inventory data.

structure is discussed in detail by [11]. Here, we describe calibration of parameters related to the inclusion of hydrologic balance into the model. New parameters needed to run `FORMIND-CL v.1.0` and their values are shown in Table 2.

The parameter f_h describing the relationship between leaf area index and canopy water storage capacity was calibrated following [34] and assuming that storage capacity reaches 4.9 mm day^{-1} at a leaf area index of 5.0 as measured by [14]. LAI_{max} was set to 5.5 following the maximum value observed in other Chilean temperate rainforests [48]. The slope of the saturation curve of the canopy rain retention capacity (α_h) was set according to common values for broad-leaved temperate trees [34]. To the best of our knowledge no estimation exists for water-use efficiency at stand scale in forests of the study area. Therefore we calibrated *WUE* using transpiration estimates of Díaz *et al.* [14] in Chiloé Island and the potential canopy photosynthetic rate estimated by the model for the study area under current climate ($Tr = 296$ mm year^{-1} and $PB = 32.9$ tC ha^{-1}. The selected *WUE* was then confirmed by comparison with reported values from other temperate rainforests [49–52]. Soil characteristics (porosity and depth) followed field descriptions from Chiloé Island [53]. We set water-retention and percolation properties of the soil (parameters θ_{wp}, θ_{fc} and k_{soil}) to average values [36,54] using texture classes (loam to silty loam) described for soils in the study area [53]. Daily records of rainfall from the EBSD weather station (60 m a.s.l, period from May 1998 to February 2009) were directly used to calculate rainfall parameters for the current climate simulations (Table 3). EBSD is the nearest and most representative weather station for the climate at the study site. We calculated the mean time interval between rainfall events from the duration (days) of rain events occurring in each season. The mean depth of rainfall events was calculated by dividing seasonal rainfall sum by the amount of wet days (P_d >0). These calculations were done only for seasons with >85 daily records. We avoided the potential overestimation of annual rainfall by normalizing predicted seasonal rainfall sum by prescribed seasonal rainfall averages (Table 3). The weather generator reproduced well the seasonal fluctuations in rainfall, temperature and radiation during the year with no significant departures from observed climatic records (Fig. S2, S3). Calibrated climatic parameters were assumed representative for growing conditions of North Patagonian forests on Chiloé Island and neighboring regions in the mainland.

Analyses

Model verification. We compared field measurements of net precipitation with model predictions at daily temporal scale. For this analysis we selected 50 rain events, representing field measurement intervals <20 days long, for which accumulated rainfall during the event (i.e. sum of daily rainfall from first to last day of the event) was correctly measured by EBSD weather station, and for which <50% of containers were filled completely during the rain event. Hydrologic parameters used for model estimation of net precipitation are indicated in Table 2. LAI measurements were unavailable for YS, thus we set LAI to comparable, averaged values reported by Diaz *et al.* [14] for the same forest type. We qualitatively compared daily variation in soil moisture produced by the model with field observations. For this analysis soil matric potential obtained for an entire year was transformed to soil moisture contents using a water retention curve [55]. We set parameters of the water retention curve following [56] for loam to silty loam texture classes and particle density of the soil type under study (ca. 2.0 g/cm^3, [53]). Model simulations were run using rainfall data for the same period of field measurements conducted over a whole year, i.e. 2008. Note that for both analyses spatial scale of model results (>1 ha) differed from the scale of field measurements (400 m^2 plots). We also compared model results with field measurements of hydrologic

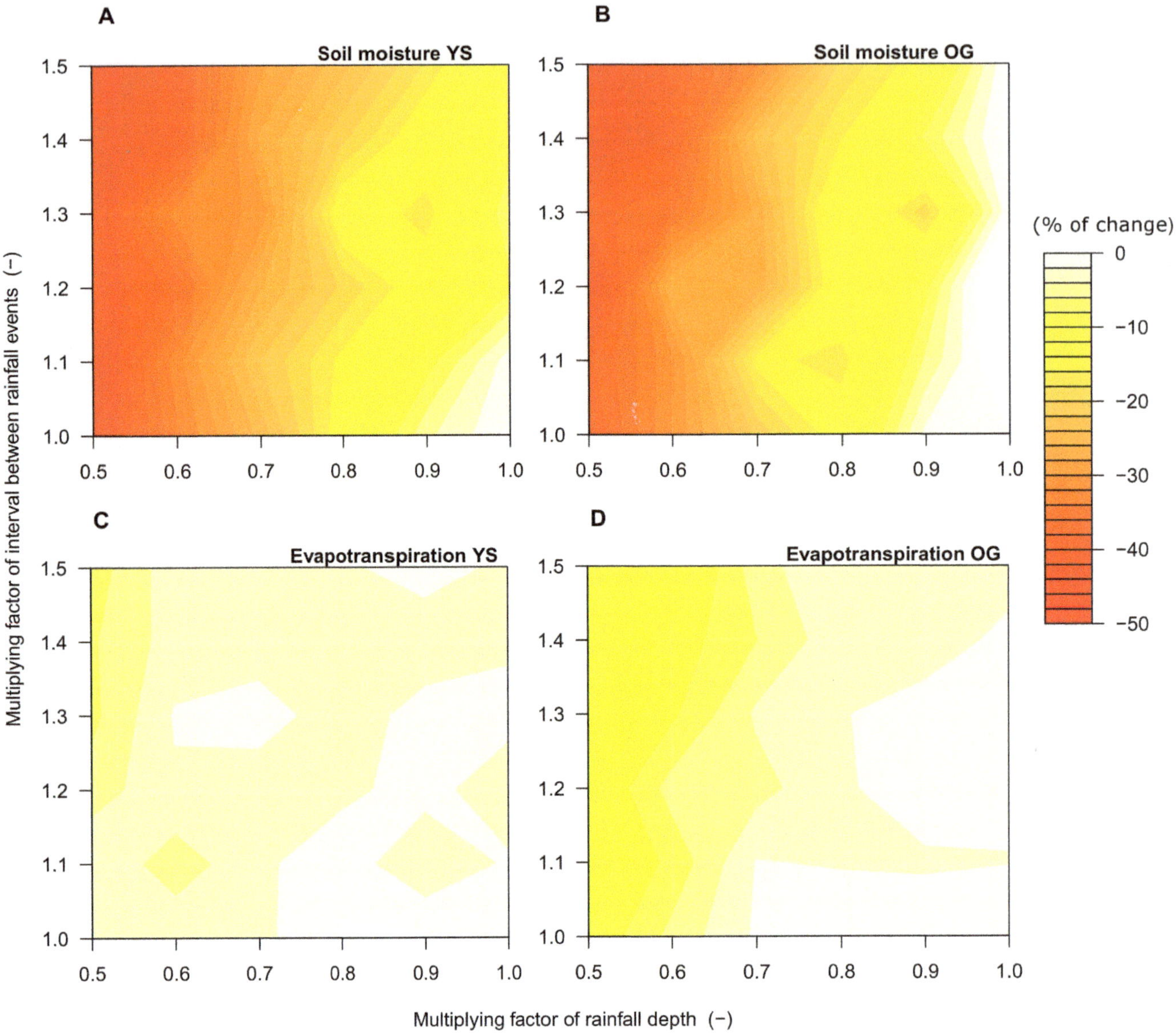

Figure 5. Sensitivity of soil moisture (a, b) and evapotranspiration (c, d) predicted by the model under increased drought in a young-secondary (YS) and an old-growth (OG) North Patagonian forest. Results represent the difference between the average under current climate (indicated in Table 4) and the average under future scenarios. A value of 0% indicates no change. The y-axis represents the variation of mean interval time between rainfall events (parameter $1/\lambda$) and the x-axis represents the variation in the mean depth of rainfall events (η) under increased drought. To represent increased drought scenarios, the parameter $1/\lambda$ was multiplied by a factor ranging from 1 (current climate) to 1.5 whereas the parameter η was multiplied by a factor ranging from 1 (current climate) to 0.5. The axes of the figures correspond to these multiplying factors of rainfall parameters. Results are the averages of 30 simulations per scenario for YS and OG.

balance at yearly temporal scales in temperate rainforests elsewhere in Chile (i.e. independent studies).

We tested model performance to reproduce forest structure under current climate. For this analysis, we compared tree basal areas and stand *dbh* distributions predicted by the model with measured structure of both YS and OG North Patagonian stands. Model comparison for the YS was performed after 60 years of succession, with succession initiated from a treeless state. To compare OG forest structure, we initiated simulations with stand inventory data and run the model for 1000 years to allow the simulated stand to reach dynamic equilibrium. We compared data at the end of the simulations with the known OG structure [28]. For each stands, we ran 100 simulations of 1 ha (i.e. 25×0.04 ha forest patches, 2500 patches in total) using current climate parameters and model parameters listed in Table 2 and 3. Demographic and species parameters were taken from [11] (site Tepual). To assess the consequences of the hydrologic submodel for the simulated forest composition, we compared species-specific basal areas reported by [11] with data from the simulations using the model version described in this study. Simulations were conducted following methods outlined by [11] and results were compared at the corresponding successional ages of YS and OG.

Simulations under increased drought. We tested the sensitivity of model predictions for forest structure and hydrological balance to changes in rainfall regimes. Climatic scenarios were selected to represent the potential range of expected climate

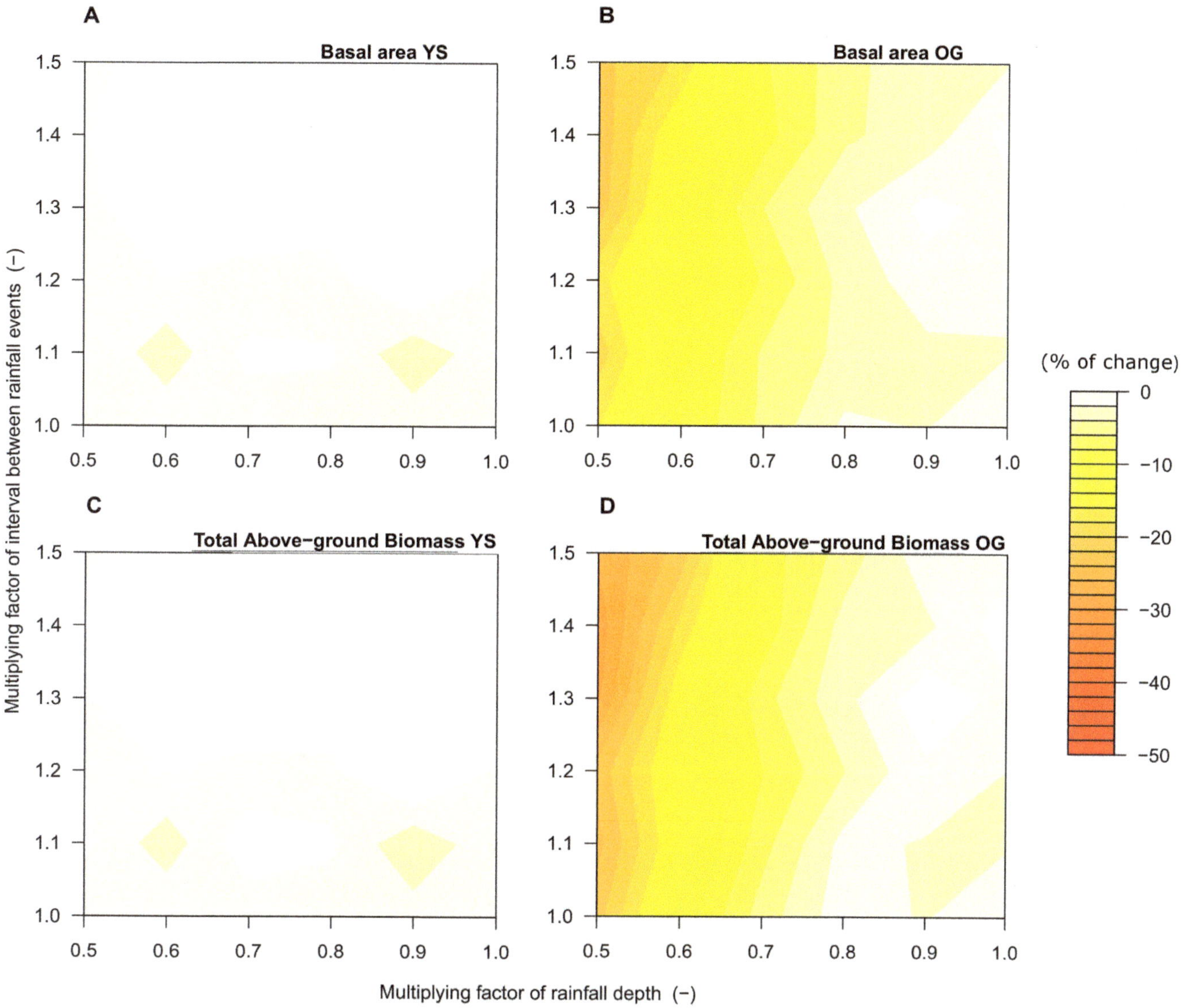

Figure 6. Sensitivity of total basal area (a, b) and above-ground biomass (c, d) predicted by the model under increased drought in a young-secondary (YS) and an old-growth (OG) North Patagonian forest. Results are the percentage of change between the average under current climate (indicated in Table 4) and the average under future scenarios. Results are the average of 30 simulations per scenario for YS and OG. The axes of the figures are as in Figure 5.

change (here increased drought) predicted for this century in the study area. To this end, we used a regional climate model downscaled for Chilean landscapes (PRECIS-DGF model, [17]). The business-as-usual scenario provided by PRECIS-DGF suggests a 50% decrease in rainfall during the growing season by year 2100. We used the daily scale output of the PRECIS-DGF model for year 2100 to calculate seasonal climatic parameters for year 2100. We linearly interpolated this seasonal rate of change between 1998–2009 and 2100. We used this scenario to set the limit of change in rainfall by year 2100 and developed climatic scenarios covering the change from current climate to the business-as-usual scenario. Climate scenarios were developed by gradually changing current climate parameters $1/\lambda$ and η until they reached the estimated value for year 2100, i.e. 50% of the current value. First, we reduced η (mean depth of rainfall events) multiplying current values by 0.9 to 0.5 in steps of 0.1. Then, we increased the parameter $1/\lambda$ (interval between events) multiplying current values by 1.1 to 1.5 in steps of 0.1. These two parameters were first varied separately (keeping the second parameter constant) and then both together. Parameter variations produced a total of 36 climate scenarios, i.e. six levels of each of the two rainfall parameters, including the current climate scenario (Table 3). In addition, we assessed the impact of warming trends on potential evapotranspiration, and its influence on other hydrologic components and forest structure. To this end, we ran an additional scenario including temperature changes expected for year 2100, with parameters $1/\lambda$ and η kept at their current values (Table 3). We considered four output variables computed at yearly temporal scale: total basal area, above-ground biomass, evapotranspiration (computed as the sum of canopy transpiration and interception) and soil moisture. The latter is dynamically linked to forest processes such as annual gross biomass production of each tree (*PB*, cf. Eq. 8, see also Fig. S1). We described changes in forest structure at different successional stages based on simulations for

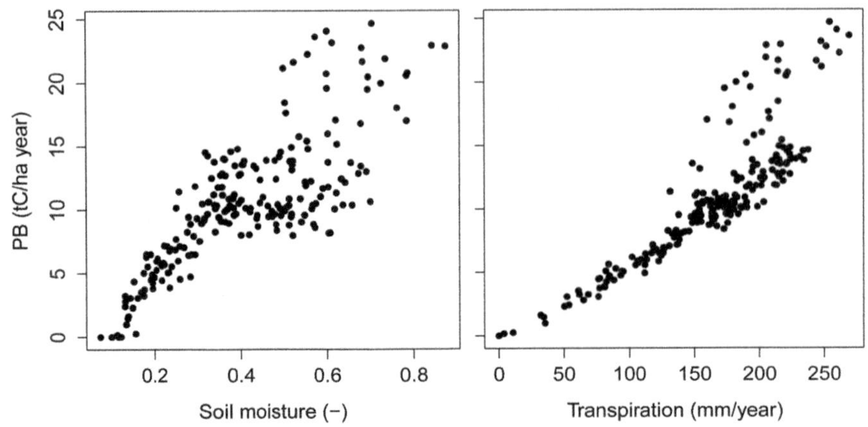

Figure 7. Shifts in annual gross biomass production (*PB*) of the YS North Patagonian forest stand in response to soil moisture and canopy transpiration changes in northern Chiloé Island, Chile. Results are simulations for one hectare of forest with a successional age of 60 years.

both YS and OG forest stands. Simulations were initialized and ran under the same conditions described in section *Model verification*. We ran 30 simulations per scenario and for each forest stand. To assess the impacts of increased drought on forests we compared the differences between means of the studied variable under current climate and means of the same variable under climate change scenarios. Finally, we checked soil moisture influence on *PB* and *Tr* by running simulations under different soil moisture conditions. We ran 216 simulations, at a scale of one ha, by monotonically varying current rainfall depth parameter ($1/\lambda$) from 0 to 100% of the current value (Table 2). All other parameters were kept constant. We assessed results at the corresponding successional ages of YS and OG, with succession initiated from a treeless state. All statistical analyses were done in R [57].

Results

Model results under current climate

Model predictions for the major components of forest hydrologic balance at a yearly temporal scale were similar to reported values for broad-leaved temperate rainforests in Chile and elsewhere (Table 4, independent studies). At a daily scale, the hydrologic model captured a large portion of net precipitation variability measured in the field ($N = 50$, $r^2 = 0.8$, $P<0.001$, Fig. 2). However, for rain events accumulating >100 mm, the model predicted higher net precipitation than recorded in the field (Fig. 2). Modeled daily variation in soil moisture resembled the daily pattern of soil matric potential measured in the field during the year (Fig. 3, $r^2 = -0.65$, $p<0.001$). Soil moisture increased during austral fall and winter (Fig. 3a, 59< julian day <242) due to a higher frequency and depth of rainfall events and reduced transpiration of trees. In contrast, during the growing season, soil moisture gradually decreased according to the model and field data (Fig. 3). Forest structure simulated by the model with the inclusion of the hydrologic submodel was qualitatively similar to field data (Fig. 4a), but with some departures from observed data for small dbh-classes in YS (<35 cm, Fig. 4b). Total basal area of YS simulated by the model was similar to basal area observed in the field but total basal area of OG was underestimated (Fig. S4). However, model predictions for forest structure including the hydrologic submodel varied in the same manner across species as in the original model version and resembled field data for both stands ($r^2 >0.97$, $P<0.001$, Fig. S4).

Net precipitation and runoff predictions under current climate did not differ between YS and OG stands ($P>0.12$, two-sample Wilcoxon test, Table 4). Water loss through deep percolation was lower in OG than in YS ($P<0.001$, two-sample Wilcoxon test, Table 4). Evapotranspiration was higher in OG than YS, mainly due to higher canopy transpiration rates in OG (Table 4, $P<0.001$, two-sample Wilcoxon test). Modeled soil moisture was lower in OG than in YS (0.55 and 0.63 respectively, $P<0.001$, two-sample Wilcoxon test, Table 4).

Simulations under increased drought

The model predicted changes in hydrological components and forest structure when simulations were run under expected trends of increasing drought (Fig. 5, 6). In both forest stands, YS and OG, we found similar responses of soil moisture to changes in rainfall parameters, with declines of soil moisture up to 50% (Fig. 5a, b). Reducing η to values <80% of the current value (Table 2, e.g. multiplying factor of 0.8) consistently reduced soil moisture by 20% (Fig. 5a, b). The influence of $1/\lambda$ (mean interval time between rainfall events) was negligible when η was kept constant at its current value (Fig. 5a, b). The model consistently predicted less evapotranspiration (hereafter ET) relative to current values (Fig. 5c, d). In YS, decreases in ET under drier climate were < 50 mm year^{-1} (<8% reduction relative to current values, Fig. 5c). In OG forest, ET was reduced up to 94.4 mm year^{-1} from the current value (15% reduction relative to current values Fig. 5d). Such decreases in ET occurred when current value of η was multiplied by 0.6 and $1/\lambda$ was multiplied by 1.2 (Fig. 5d). An increase in PET was predicted by the model when simulations included warming trends ($P<0.001$, two-sample Wilcoxon test, Fig. S5). Only transpiration and ET in YS changed in response to warming trends and increased drought ($P<0.05$, two-sample Wilcoxon test, Fig. S5). Moreover, in both forest stands, changes in PET due to warming trends did not transfer to changes in forest structure ($P>0.05$, two-sample Wilcoxon test, see also Fig. S5).

We did not find distinct differences in basal area and aboveground biomass (AGB) in YS attributable to changes in rainfall parameters (Fig. 6). In contrast, in OG forest increased drought produced decreases in basal area by 21% of the current value (Fig. 6b, current basal area of 63.8 m^2 ha^{-1}) and decreases in AGB by 27% of the current value (Fig. 6d, current AGB of

309.5 tC ha^{-1}). Main changes in basal area and AGB of OG stand were predicted when current η and $1/\lambda$ parameters were multiplied by 0.6 and 1.3, respectively (Fig. 6b, d).

Discussion

Model performance

We developed and evaluated the performance of an individual- and process-based dynamic forest model that incorporates detailed calculations of water cycling for temperate rainforests of southern South America. The model allows for the investigation of dynamic linkages between rainfall trends and forest processes at stand scale. Parameters selected for this study were taken from the literature or calibrated using our own field data (Table 2), thus they can be considered as empirically based. The model incorporates the main known hydrological controls on forest processes in temperate rainforests of southern South America (Table 2, 4).

The accuracy of model predictions regarding forest composition and structure were comparable to previous results (Fig. S4), which were performed using a previous model version without hydrologic balance calculations [11]. Model prediction of total basal area of forests decreased with the inclusion of the hydrologic module, mainly due to fewer large trees (>100 cm dbh) predicted by the model in OG stands (Fig. 4b). However, model predictions of forest structure were considered realistic at small spatial scales (< 0.2 ha) used for field sampling, compared to model results (25 ha, see also discussion in [11]). Some departures between observed and predicted daily values of net precipitation could also be attributed to canopy heterogeneity operating at different spatial scales of model results and empirical data. These discrepancies may appear when comparing forest hydrologic balance estimated for 1 ha against net precipitation measured within considerably smaller areas (<1 m^2). Further, model predictions of daily net precipitation were based on average estimated LAI values for YS because of the lack of field measurements. LAI is a relevant variable to understanding biogeochemical cycles [58], and it should therefore be incorporated in future hydrologic analyses. Additionally, sampling errors during field measurements cannot be disregarded. Despite these limitations, we considered model predictions of forest hydrology acceptable at daily temporal scale for YS (Fig. 2, 3).

At yearly temporal scale, the model reproduced the main hydrologic components reported for similar evergreen, broad-leaved temperate rainforests in southern South America and elsewhere (Table 4, independent studies). Differences in deep percolation predicted by the present model compared to values reported in Table 4 might be accounted for specific physical characteristics of glacially originated soils on northern Chiloé Island. Deep percolation in the model is mainly depending on k_{soil}, which was calibrated specifically for soils from northern Chiloé Island. Applying this model to forests developing on other soil types in SSA (e.g. volcanic originated soils) will require a site-specific calibration of soil parameters (Table 2). Moreover, we compared model results against a different forest type (Table 4), with shared dominance by evergreen and deciduous tree species, and differing in LAI and annual net precipitation. Despite these broad differences, Table 4 suggests that selected hydrologic parameters (e.g. α_{soil}, f_h, and LAI_{max}, see Table 1) yielded reasonable values for canopy water retention capacity of broad-leaved temperate rainforests in SSA.

Forest responses to increased drought

According to predictions from our model, North Patagonian forests are likely to be altered by increased drought predicted for this century by climate change models. In our modeling study, the simulation of 50% reduction in summer rainfall predicted for the study area (business-as-usual scenario, [17]) can induce changes in both hydrological balance (up to ca. 100 mm year^{-1} decrease in ET, Fig. 5) and forest structure (up to 83 tC ha^{-1} decrease in current AGB, Fig. 6), even without considering the potential ecological effects of concomitant global warming. A direct interpretation of changes in rainfall regimes is possible because the model accounts explicitly for changes in frequency and depth of rainfall events [8,32,47]. Decreasing in the depth of rainfall alone can induce some structural changes in the studied forest type (Fig. 6), but simultaneous changes in the frequency and depth of rainfall produced the strongest changes in hydrology and structure of stands (Fig. 6). These results highlight the impact on forest structure and growth of the duration and frequency of water limitation periods.

Reductions of basal area due to increased drought (Fig. 6) conform to empirical findings that drought increases the likelihood of mortality of large trees [59–61]. Our model formulation implies that trees assimilating greater biomass will have increasing demands of water for growth (Eq. 7, see also [62]). For example, annual gross biomass production increased with greater soil moisture availability during the year, a mechanism triggered by increased canopy transpiration (Fig. 7). Consequently, big trees that occur primarily in old-growth forests (Fig. 4b) experience increased stress-induced mortality due to greater hydrologic limitations during dry years. Trees that die under increased drought produce a decrease in both stand basal area and above-ground biomass (AGB, Fig. 6). Moreover, the model predicted that the OG forest has a higher PB than the young secondary stand (31.6±2.3 vs. 18.2±1.2 tC ha^{-1} year^{-1}, both obtained under current climate), and consequently a higher water demand for biomass production. Under increased drought, water demand for biomass production in OG forest is not fully covered by soil moisture supply, which causes the predicted decline of AGB and basal area (Fig. 6). We propose that the contrasting ET and structural patterns between YS and OG predicted for the coming decades under increasing drought are mainly due to significant limitation of available soil moisture for biomass production in OG, with lower impact on YS.

Research needs and model application

Here, we focused on developing an accurate model for assessing the influence of hydrologic processes involved in forest dynamics. Using the model, we quantitatively demonstrated the relative importance of soil moisture on forest structure (Fig. 6). We excluded other hydrologic processes to keep model formulation simple and results tractable with empirical information available. Processes such as water table dynamics, root dynamics, increased run off in slopes, and soil moisture dynamics across multiple soil layers can be incorporated as more empirical data becomes available for model calibration and validation. However, we strongly suggest that future model applications prioritize processes known to have an influence on the system under study [63]. In our study area, variations in the height of the water table may interfere with ecological processes such as tree establishment and mortality [14,24]. Our model provides a convenient starting point to incorporate water table dynamics into the analysis of climate change impacts, and to explore its effects on long-term forest dynamics. Our results also highlight the need of further fieldwork and experimental research on less known, mechanistic soil parameters (e.g. k_{soil}).

The rise of atmospheric carbon dioxide concentration and ensuing climate change are influencing water-use efficiency of forests [64,65]. WUE under future climate is likely to differ among forests [66], local scales [67], and species [68]. Forest water-use efficiency is a sensitive parameter in our model (e.g. Fig. S6) and illustrates the need for detailed studies of the expected variations in this parameter under drought, and their connection to gas exchange capacity of trees in different topographic settings and for a large set of species in SSA. It was beyond the scope of the present study to discuss model behavior along drought-to-moist gradients operating at regional scales because such analyses require an accurate and quantitative assessment of species-specific water-use efficiency [4]. In SSA, WUE variations have been experimentally tested in few study sites and for only three tree species included in our study [49,69]. Additional experimental research should address the complex interaction between photosynthetic carbon assimilation and water loss via transpiration with declining water supply [70]. Future research can address this question by studying changes in photosynthetic parameters along climatic gradients. Incorporating WUE as a species-specific parameter or a state dependent variable in our model is straightforward based on further empirical information.

Here, we deliberately excluded the influence of expected regional changes in temperature on forest processes (e.g. tree growth) to rather emphasize direct impacts of drought on model results. However, the model suggests that changes in PET due to expected warming trends in the study region are negligible compared to the strong impacts of increased drought (Fig. S5). Our simulated scenarios have been done using the common assumption of vegetation dynamic models that climate-forest interactions under inter-annual variation of climatic conditions can be used as proxy also for impacts of long-term climatic variations. In the later case it would be possible that some tree species would show adaptations effects. However, tree species adaptation to climate variability is still poorly understood in SSA. Moreover, warming can modify individual tree growth by affecting photosynthesis and both plant and soil respiration [38] [71], nutrient dynamics [72] and tree establishment [73]. An undergoing study analyzes the combined effect of temperature and rainfall changes on tree demography in the study area [74].

To the best of our knowledge, this is the first application of a forest gap model in temperate rainforests of SSA that integrates dynamic calculations of forest hydrology. The present work uses the best information available to ensure that climate patterns were directly comparable to hydrologic field measurements used for model calibration. However, our results on climate change impacts should be interpreted with caution because our baseline climate is constrained to a short period (1998–2009) within a long-term trend of rainfall (time period 1901–2005). Long-term monitoring of forest hydrology and dynamics can corroborate our results. To date, long-term monitoring (>10 years) of forest hydrology is lacking in forests of SSA. As more empirical data becomes available, the model can be revised and updated. The model developed here allows for the analysis of multiple environmental factors driving forest dynamics. For example, our model can help us understand the long-term responses of regional forest to drought events induced by El Niño Southern Oscillation that amplify background tree mortality rates in *Nothofagus* forest of SSA [75]. Moreover, increased regional drought is likely to interact with other drivers of global change such as changes in fire regimes, massive introduction of exotic forestry species, and forest fragmentation. To date, the implications of these drivers in the context of a changing climate remain poorly understood in SSA. These interacting growing threats demand from ecologists to understand and integrate multiple dimensions of global change on forest functioning. The model presented here is a particularly suitable tool for analyzing broad global change questions in forests of SSA because it also includes logging and fragmentation submodels [13,76]. Model-based experiments can also contribute to develop sound management strategies that anticipate forest responses to increasing drought and other drivers of climate change.

Conclusions

We developed and applied a forest dynamic model to analyze the impact of climate-driven increased drought on ecological and hydrological processes. The developed model was accurate for depicting forest hydrology at stand scales (i.e. <100 ha) and allowed the analysis of the dynamical linkages among rainfall regimes, soil moisture variation, and individual tree growth. Using the model we demonstrated that evergreen, broad-leaved temperate rainforests in southern South America are expected to be highly sensitive to future climate change, particularly increases in drought during parts of the year. Increased summer drought predicted for this century will likely impair biomass carbon accumulation, and amplify background tree mortality rates in this region. The developed model expands the range of applicability of gap models to assess climate change impacts in remote and understudied regions of the world, such as temperate forests of the southern hemisphere. It also represents an advance in the development of simple, general models to account for complex and dynamical processes operating at multiple spatial scales in forests.

Supporting Information

Figure S1 A diagram of the hydrologic submodel of FORMIND-CL v.**1.0.** Interaction between processes and variables in the hydrologic submodel, and their respective time scales of calculations. Arrows indicate whether the results of a model calculation influence the calculations of another submodel. Blue boxes represent analyzed variables of this study. All calculations are done in yearly time steps in the model, excepting the ones indicated in the dashed box. Variable notations follow the text. AGB: Above-ground biomass, LAI: leaf area index.

Figure S2 Weather generator results. Density functions of daily rainfall and daily mean temperature predicted by the weather generator compared to observed weather records from EBSD weather station.

Figure S3 Weather generator results. Comparison between simulated and observed climatic patterns during the year. Simulations were run for 100 years using parameters in Table 3. Daily data were averaged by seasons (mean daily temperature and daily radiation). Rainfall is the amount of rainfall during each season. Observed weather data are from EBSD weather station and seasons according to table 3.

Figure S4 Forest composition predictions. Model results for forest composition using different model versions compared to field data. Simulations run under the same conditions detailed in Methods section.

Figure S5 Drought induced simulations with warming included. Changes in hydrologic components and forest structure when warming and increased drought was considered. PET: Potential evapotranspiration (mm $year^{-1}$), T: transpiration (mm year^{-1}), Ec: canopy interception (mm year^{-1}), ET: evapotranspiration (mm year^{-1}), BAT: total basal area (m^2 ha^{-1}), BT: Total biomass (tC ha^{-1}). Result of a two-sample Wilcoxon test is shown on the upper right of each panel. Pink lines, drought induced simulations with warming included, blue lines drought induced simulations without warming, circles represent the values of simulation results. Note different scales for the axes.

Figure S6 Sensitivity of evaporatranspiration. Changes in evapotranspiration (ET) of the old-growth stand under current climate when using different water-use efficiency values (WUE). Simulations run under the same conditions detailed in Methods section.

Acknowledgments

We appreciate the useful comments offered by Hans Pretzsch, and anonymous reviewers in previous versions of the manuscript. This is a contribution to the Research Program of the Chilean LTSER network.

Author Contributions

Conceived and designed the experiments: AGG AH JJA MFD. Performed the experiments: AGG JJA MFD. Analyzed the data: AGG AH MFD. Contributed reagents/materials/analysis tools: AGG JJA AH. Wrote the paper: AGG.

References

1. Rodriguez-Iturbe I, Porporato A, Ridolfi L, Isham V, Cox DR (1999) Probabilistic modelling of water balance at a point: the role of climate, soil and vegetation. Proceedings of the Royal Society a-Mathematical Physical and Engineering Sciences 455: 3789–3805.
2. Botkin DB, Wallis JR, Janak JF (1972) Some Ecological Consequences of a Computer Model of Forest Growth. Journal of Ecology 60: 849–872.
3. Bugmann H (2001) A review of forest gap models. Climatic Change 51: 259–305.
4. Bugmann H, Cramer W (1998) Improving the behaviour of forest gap models along drought gradients. Forest Ecology and Management 103: 247–263.
5. Bugmann HKM, Wullschleger SD, Price DT, Ogle K, Clark DF, et al. (2001) Comparing the performance of forest gap models in North America. Climatic Change 51: 349–388.
6. Reynolds JF, Bugmann H, Pitelka LF (2001) How much physiology is needed in forest gap models for simulating long-term vegetation response to global change? Challenges, limitations, and potentials. Climatic Change 51: 541–557.
7. Asbjornsen H, Goldsmith GR, Alvarado-Barrientos MS, Rebel K, Van Osch FP, et al. (2011) Ecohydrological advances and applications in plant-water relations research: a review. Journal of Plant Ecology-Uk 4: 3–22.
8. Porporato A, Daly E, Rodriguez-Iturbe I (2004) Soil water balance and ecosystem response to climate change. American Naturalist 164: 625–632.
9. Vose JV, Sun G, Ford CR, Bredemeier M, Otsuki K, et al. (2011) Forest ecohydrological research in the 21st century: what are the critical needs? Ecohydrology 4: 146–158.
10. Hall GMJ, Hollinger DY (2000) Simulating New Zealand forest dynamics with a generalized temperate forest gap model. Ecological Applications 10: 115–130.
11. Gutierrez AG, Huth A (2012) Successional stages of primary temperate rainforests of Chiloe Island, Chile. Perspectives in Plant Ecology Evolution and Systematics 14: 243–256.
12. Echeverría C, Newton AC, Lara A, Benayas JMR, Coomes DA (2007) Impacts of forest fragmentation on species composition and forest structure in the temperate landscape of southern Chile. Global Ecology and Biogeography 16: 426–439.
13. Rüger N, Gutiérrez AG, Kissling WD, Armesto JJ, Huth A (2007) Ecological impacts of different harvesting scenarios for temperate evergreen rain forest in southern Chile - A simulation experiment. Forest Ecology and Management 252: 52–66.
14. Díaz MF, Bigelow S, Armesto JJ (2007) Alteration of the hydrologic cycle due to forest clearing and its consequences for rainforest succession. Forest Ecology and Management 244: 32–40.
15. Little C, Lara A, McPhee J, Urrutia R (2009) Revealing the impact of forest exotic plantations on water yield in large scale watersheds in South-Central Chile. Journal of Hydrology 374: 162–170.
16. Trenberth KE, Jones PD, Ambenje P, Bojariu R, Easterling D, et al. (2007) Observations: Surface and Atmospheric Climate Change. In: Solomon S, Qin D, Manning M, Chen Z, Marquis M et al., editors. Climate Change 2007: The Physical Science Basis Contribution of Working Group I to the Fourth Assessment Report of the Intergovernmental Panel on Climate Change. Cambridge, United Kingdom and New York, NY, USA.: Cambridge University Press. 336
17. DGF CONAMA (2006) Estudio de la variabilidad climática en Chile para el siglo XXI. Santiago: Departamento de Geofisica. Universidad de Chile. 63 p.
18. Christensen JH, Hewitson B, Busuioc A, Chen A, Gao X, et al. (2007) Regional Climate Projection. In: Solomon S, Qin D, Manning M, Chen Z, Marquis M et al., editors. Climate Change 2007: The Physical Science Basis Contribution of Working Group I to the Fourth Assessment Report of the Intergovernmental Panel on Climate Change. Cambridge, United Kingdom and New York, NY, USA: Cambridge University Press. 94
19. Alaback PB (1991) Comparative Ecology of Temperate Rain-Forests of the America Along Analogous Climatic Gradients. Revista Chilena de Historia Natural 64: 399–412.
20. Armesto JJ, Smith-Ramírez C, Carmona MR, Celis-Diez JL, Díaz I, et al. (2009) Old-growth temperate rain forests of South America: Conservation, plant-animal interactions, and baseline biogeochemical processes. In: Wirth C, Gleixner G, Heimann M, editors. Old-growth forests: Function, fate and value. Berlin, Heidelberg: Springer New York. 367–390
21. Armesto JJ, Rozzi R, Smith-Ramírez C, Arroyo MTK (1998) Conservation targets in South American temperate forests. Science 282: 1271–1272.
22. Willson MF, Armesto JJ (1996) The natural history of Chiloé: on Darwin's trail. Revista Chilena de Historia Natural 69: 149–161.
23. Holdgate MW (1961) Vegetation and soils in the South Chilean Islands. Journal of Ecology 49: 559–580.
24. Díaz MF, Armesto JJ (2007) Physical and biotic constraints on tree regeneration in secondary shrublands of Chiloe Island, Chile. Revista Chilena De Historia Natural 80: 13–26.
25. di Castri F, Hajek E (1976) Bioclimatología de Chile. Santiago: Universidad Católica de Chile.
26. Gutiérrez AG, Armesto JJ, Aravena JC, Carmona M, Carrasco NV, et al. (2009) Structural and environmental characterization of old-growth temperate rain-forests of northern Chiloe Island, Chile: Regional and global relevance. Forest Ecology and Management 258: 376–388.
27. Veblen TT, Schlegel FM, Oltremari JV (1983) Temperate broad-leaved evergreen forest of South America. In: Ovington JD, editor. Temperate Broad-Leaved Evergreen Forest. Amsterdam: Elsevier Science Publishers. 5–31
28. Gutiérrez AG, Armesto JJ, Aravena JC (2004) Disturbance and regeneration dynamics of an old-growth North Patagonian rain forest in Chiloe Island, Chile. Journal of Ecology 92: 598–608.
29. Köhler P (2000) Modelling anthropogenic impacts on the growth of tropical rain forests-using an individual-oriented forest growth model for the analyses of logging and fragmentation in three case studies [PhD thesis http://hdl.handle.net/10013/epic.15101].Osnabrück, Germany: Center for Environmental Systems Research and Department of Physics. University of Kassel. Der Andere Verlag.
30. Köhler P, Chave J, Riera B, Huth A (2003) Simulating the long-term response of tropical wet forests to fragmentation. Ecosystems 6: 114–128.
31. Köhler P, Huth A (1998) The effects of tree species grouping in tropical rainforest modelling: Simulations with the individual-based model FORMIND. Ecological Modelling 109: 301–321.
32. Kumagai T, Katul GG, Saitoh TM, Sato Y, Manfroi OJ, et al. (2004) Water cycling in a Bornean tropical rain forest under current and projected precipitation scenarios. Water Resources Research 40: W01104.
33. Rutter AJ, Morton AJ (1977) Predictive Model of Rainfall Interception in Forests.3. Sensitivity of Model to Stand Parameters and Meteorological Variables. Journal of Applied Ecology 14: 567–588.
34. Wattenbach M, Hattermann F, Weng R, Wechsung F, Krysanova V, et al. (2005) A simplified approach to implement forest eco-hydrological properties in regional hydrological modelling. Ecological Modelling 187: 40–59.
35. Rey JM (1999) Modelling potential evapotranspiration of potential vegetation. Ecological Modelling 123: 141–159.
36. Maidment DR (1993) Handbook of Hydrology; McGraw-Hill, editor: McGraw-Hill.
37. Neilson RP (1995) A Model for Predicting Continental-Scale Vegetation Distribution and Water-Balance. Ecological Applications 5: 362–385.
38. Lambers H, Chapin FS, Pons TL (1998) Plant physiological ecology. New York: Springer. 540 p.
39. Bazzaz FA (1979) Physiological ecology of plant succession. Annual Review of Ecology and Systematics 10: 351–371.

40. Thornley HMJ, Johnson IR (1990) Plant and Crop Modelling – A mathematical approach to plant and crop physiology. Oxford, UK.: Clarendon Press.
41. Gerten D, Schaphoff S, Haberlandt U, Lucht W, Sitch S (2004) Terrestrial vegetation and water balance - hydrological evaluation of a dynamic global vegetation model. Journal of Hydrology 286: 249–270.
42. Venevsky S, Maksyutov S (2007) SEVER: A modification of the LPJ global dynamic vegetation model for daily time step and parallel computation. Environmental Modelling & Software 22: 104–109.
43. Prentice IC, Sykes MT, Cramer W (1993) A Simulation-Model for the Transient Effects of Climate Change on Forest Landscapes. Ecological Modelling 65: 51–70.
44. Haxeltine A, Prentice IC (1996) BIOME3: An equilibrium terrestrial biosphere model based on ecophysiological constraints, resource availability, and competition among plant functional types. Global Biogeochemical Cycles 10: 693–709.
45. Prentice IC, Cramer W, Harrison SP, Leemans R, Monserud RA, et al. (1992) A Global Biome Model Based on Plant Physiology and Dominance, Soil Properties and Climate. Journal of Biogeography 19: 117–134.
46. Dingman SL (2002) Physical hydrology. New Jersey: Prentice Hall. 646 p.
47. Laio F, Porporato A, Ridolfi L, Rodriguez-Iturbe I (2001) Plants in water-controlled ecosystems: active role in hydrologic processes and response to water stress - II. Probabilistic soil moisture dynamics. Advances in Water Resources 24: 707–723.
48. Lusk CH (2001) When is a gap not a gap? Light levels and leaf area index in bamboo-filled gaps in a Chilean rain forest. Gayana Botanica 58: 25–30.
49. Piper FI, Corcuera LJ, Alberdi M, Lusk C (2007) Differential photosynthetic and survival responses to soil drought in two evergreen Nothofagus species. Annals of Forest Science 64: 447–452.
50. Zuñiga R, Alberdi M, Reyes-Diaz M, Olivares E, Hess S, et al. (2006) Seasonal changes in the photosynthetic performance of two evergreen Nothofagus species in south central Chile. Revista Chilena de Historia Natural 79: 489–504.
51. Cunningham SC (2005) Photosynthetic responses to vapour pressure deficit in temperate and tropical evergreen rainforest trees of Australia. Oecologia 142: 521–528.
52. Brodribb T, Hill RS (1998) The photosynthetic drought physiology of a diverse group of southern hemisphere conifer species is correlated with minimum seasonal rainfall. Functional Ecology 12: 465–471.
53. Janssen I, Kruemmelbein J, Horn R, Ellies AS (2004) Physical and hydraulic properties of the ñadi soils in south Chile - Comparison between untilled and tilled soil. Revista de la Ciencia del Suelo y Nutrición Vegetal 4: 14–28.
54. Sitch S, Smith B, Prentice IC, Arneth A, Bondeau A, et al. (2003) Evaluation of ecosystem dynamics, plant geography and terrestrial carbon cycling in the LPJ dynamic global vegetation model. Global Change Biology 9: 161–185.
55. van Genuchten MT (1980) A closed-form equation for predicting the hydraulic conductivity of unsaturated soils. Soil Science Society of America Journal 44: 892–898.
56. Seki K (2007) SWRC fit-a nonlinear fitting program with a water retention curve for soils having unimodal and bimodal pore structure. Hydrology and Earth System Sciences Discussions 4: 407–437.
57. R-Development-Core-Team (2012) R: A language and environment for statistical computing. Vienna, Austria.: R Foundation for Statistical Computing.
58. Breda NJJ (2003) Ground-based measurements of leaf area index: a review of methods, instruments and current controversies. Journal of Experimental Botany 54: 2403–2417.
59. Condit R, Hubbell SP, Foster RB (1995) Mortality-Rates of 205 Neotropical Tree and Shrub Species and the Impact of a Severe Drought. Ecological Monographs 65: 419–439.
60. Allen CD, Macalady AK, Chenchouni H, Bachelet D, McDowell N, et al. (2010) A global overview of drought and heat-induced tree mortality reveals emerging climate change risks for forests. Forest Ecology and Management 259: 660–684.
61. Phillips OL, van der Heijden G, Lewis SL, Lopez-Gonzalez G, Aragao LEOC, et al. (2010) Drought-mortality relationships for tropical forests. New Phytologist 187: 631–646.
62. Larcher W (2001) Ökophysiologie der Pflanzen. Stuttgart: 6th edition, Verlag Eugen Ullmer.
63. Grimm V, Railsback SF (2005) Individual-based modeling and ecology. New Jersey: Princeton University Press.
64. Penuelas J, Canadell JG, Ogaya R (2011) Increased water-use efficiency during the 20th century did not translate into enhanced tree growth. Global Ecology and Biogeography 20: 597–608.
65. Keenan TF, Hollinger DY, Bohrer G, Dragoni D, Munger JW, et al. (2013) Increase in forest water-use efficiency as atmospheric carbon dioxide concentrations rise. Nature 499: 324–+.
66. Silva LCR, Anand M (2013) Probing for the influence of atmospheric CO_2 and climate change on forest ecosystems across biomes. Global Ecology and Biogeography 22: 83–92.
67. Penuelas J, Hunt JM, Ogaya R, Jump AS (2008) Twentieth century changes of tree-ring delta C-13 at the southern range-edge of Fagus sylvatica: increasing water-use efficiency does not avoid the growth decline induced by warming at low altitudes. Global Change Biology 14: 1076–1088.
68. Levesque M, Saurer M, Siegwolf R, Eilmann B, Brang P, et al. (2013) Drought response of five conifer species under contrasting water availability suggests high vulnerability of Norway spruce and European larch. Global Change Biology 19: 3184–3199.
69. Figueroa JA, Cabrera HM, Queirolo C, Hinojosa LF (2010) Variability of water relations and photosynthesis in *Eucryphia cordifolia* Cav. (Cunoniaceae) over the range of its latitudinal and altitudinal distribution in Chile. Tree Physiology 30: 574–585.
70. Chaves MM, Pereira JS, Maroco J, Rodrigues ML, Ricardo CPP, et al. (2002) How plants cope with water stress in the field. Photosynthesis and growth. Annals of Botany 89: 907–916.
71. Ryan MG, Law BE (2005) Interpreting, measuring, and modeling soil respiration. Biogeochemistry 73: 3–27.
72. Aerts R, Chapin FS (2000) The mineral nutrition of wild plants revisited: A re-evaluation of processes and patterns. Advances in Ecological Research, Vol 30 30: 1–67.
73. Hobbie SE, Chapin FS (1998) An experimental test of limits to tree establishment in Arctic tundra. Journal of Ecology 86: 449–461.
74. Gutiérrez AG (2010) Long-term dynamics and the response of temperate rainforests of Chiloé Island (Chile) to climate change [Phd Thesis]. Freising, Germany: Technische Universität München. 170 p.
75. Suarez ML, Ghermandi L, Kitzberger T (2004) Factors predisposing episodic drought-induced tree mortality in Nothofagus-site, climatic sensitivity and growth trends. Journal of Ecology 92: 954–966.
76. Groeneveld J, Alves LF, Bernacci LC, Catharino ELM, Knogge C, et al. (2009) The impact of fragmentation and density regulation on forest succession in the Atlantic rain forest. Ecological Modelling 220: 2450–2459.
77. Donoso PJ, Soto DP, Bertin RA (2007) Size-density relationships in *Drimys winteri* secondary forests of the Chiloe Island, Chile: Effects of physiography and species composition. Forest Ecology and Management 239: 120–127.
78. Navarro C, Donoso C, Sandoval V (1999) Los renovales de Canelo. In: Donoso C, Lara A, editors. Silvicultura de los bosques nativos de Chile. Santiago, Chile: Editorial Universitaria.
79. Naulin P (2002) Estimacion de la biomasa en un renoval de canelo (Drimys winteri J.R. et Foster) en la Comuna de Ancud. Santiago: Universidad de Chile. 45 p.
80. Aravena JC, Carmona MR, Perez CA, Armesto JJ (2002) Changes in tree species richness, stand structure and soil properties in a successional chronosequence in northern Chiloe Island, Chile. Revista Chilena De Historia Natural 75: 339–360.
81. Lusk CH (2001) Leaf life spans of some conifers of the temperate forests of South America. Revista Chilena De Historia Natural 74: 711–718.
82. Huber A, Iroume A (2001) Variability of annual rainfall partitioning for different sites and forest covers in Chile. Journal of Hydrology 248: 78–92.
83. Oyarzun CE, Godoy R, Staelens J, Donoso PJ, Verhoest NEC (2011) Seasonal and annual throughfall and stemflow in Andean temperate rainforests. Hydrological Processes 25: 623–633.
84. Barbour MM, Hunt JE, Walcroft AS, Rogers GND, McSeveny TM, et al. (2005) Components of ecosystem evaporation in a temperate coniferous rainforest, with canopy transpiration scaled using sapwood density. New Phytologist 165: 549–558.
85. Lara A, Little C, Urrutia R, McPhee J, Alvarez-Garreton C, et al. (2009) Assessment of ecosystem services as an opportunity for the conservation and management of native forests in Chile. Forest Ecology and Management 258: 415–424.
86. CIREN (1994) Delimitación y descripción de microregiones para la transferencia tecnológica de INDAP. Santiago: CORFO. 99 p.

6

Effects of Litter Manipulation on Litter Decomposition in a Successional Gradients of Tropical Forests in Southern China

Hao Chen[1,5], Geshere A. Gurmesa[2,5,6], Lei Liu[1,3], Tao Zhang[1,4], Shenglei Fu[1], Zhanfeng Liu[1], Shaofeng Dong[1], Chuan Ma[1], Jiangming Mo[1]*

1 Key Laboratory of Vegetation Restoration and Management of Degraded Ecosystems, South China Botanical Garden, Chinese Academy of Sciences, Guangzhou, China, **2** Department of Geosciences and Natural Resource Management, University of Copenhagen, Copenhagen, Denmark, **3** State Key Laboratory of Urban and Regional Ecology, Research Center for Eco-Environmental Sciences, Chinese Academy of Sciences, Beijing, China, **4** Institute of Tropical Pratacultural Science, Zhanjiang Normal University, Zhanjiang, China, **5** University of Chinese Academy of Sciences, Beijing, China, **6** Sino-Danish Center for Education and Research, Aarhus, Denmark

Abstract

Global changes such as increasing CO_2, rising temperature, and land-use change are likely to drive shifts in litter inputs to forest floors, but the effects of such changes on litter decomposition remain largely unknown. We initiated a litter manipulation experiment to test the response of litter decomposition to litter removal/addition in three successional forests in southern China, namely masson pine forest (MPF), mixed coniferous and broadleaved forest (MF) and monsoon evergreen broadleaved forest (MEBF). Results showed that litter removal decreased litter decomposition rates by 27%, 10% and 8% and litter addition increased litter decomposition rates by 55%, 36% and 14% in MEBF, MF and MPF, respectively. The magnitudes of changes in litter decomposition were more significant in MEBF forest and less significant in MF, but not significant in MPF. Our results suggest that change in litter quantity can affect litter decomposition, and this impact may become stronger with forest succession in tropical forest ecosystem.

Editor: Ben Bond-Lamberty, DOE Pacific Northwest National Laboratory, United States of America

Funding: This study was founded by National Key Basic Research 973 Programme (2011CB403204) and National Natural Science Foundation of China (No: 41273143). The funders had no role in study design, data collection and analysis, decision to publish, or preparation of the manuscript.

Competing Interests: The authors have declared that no competing interests exist.

* E-mail: mojm@scib.ac.cn

Introduction

Litter decomposition is a key process that regulates nutrient recycling in ecosystems, influences net ecosystem carbon (C) storage, and is the first step in the formation of soil humus [1]. Litter decomposition can be affected by many environmental factors including the physical environment (e.g. temperature, moisture, and soil pH), nutrients availability and activities of decomposers in the soil [2]. On the other hand, litter quantity can alter microclimate, number and dynamics of decomposer organisms and nutrient availability in forest floor and mineral soil [3]. Thus, the quantity of the litter itself has an impact on litter decomposition. It has been reported that global climate changes due to rising atmospheric CO_2 concentration and temperature can increase net primary production (NPP) and consequently litter production in forest ecosystems [4–6]. In contrast, litter inputs are also likely decreased due to extensive deforestation and cultivation [7]. Therefore, evaluating the effects of these changes in litter inputs on litter decomposition is crucial for our understanding of ecosystem nutrient supplies and future global C cycle.

Despite the substantial number of studies on decomposition in a wide range of ecosystems, the influence of a sustained change in litter quantity on litter decomposition has not been well addressed [3]. Existing evidence on the effects of litter quantity on the leaf litter decomposition is indirect, and the results are inconsistent and incomparable. For example, comparison of litter decomposition before and after hurricanes showed that the large inputs of litter following hurricanes would cause accelerated or slower decomposition rate [8,9]. Similarly, results from the labile-C (glucose, cellulose, or root exudates) addition experiments have shown that labile-C addition could have both positive and negative effect on leaf litter decomposition [10–13]. Moreover, potentially decreased litter quantity after understory removal also showed both decreasing [14,15] and increasing [16] effects on litter decomposition.

Compared to the methods mentioned above, litter manipulation (removal or addition) experiment is considered to be a direct way for studying effects of litter quantity on ecosystem processes [17]. Using this method, recent studies have reported that increased litter input would accelerate soil C release and decreased soil C content in tropical forests due to the "priming effect" [18–21]. However, most of these studies assessed the effect of litter quantity on the decomposition of soil organic matter and only few studies addressed the effect on decomposition of fresh leaf litter. In addition, more studies focused on the effect of litter removal [22,23] and to our knowledge only one studied effects of litter addition on litter decomposition in forest ecosystem [3], highlighting the need for more complete comparative study on effects of both litter addition and removal on leaf litter decomposition.

While estimates of decomposition after litter manipulation are commonly reported in single forest type, forests at different successional stages are likely to present distinct responses. The Dinghushan Biosphere Reserve (DHSBR) consists of three typical forest types in southern China at different forest successional stages in terms of age and exposure to human disturbances; hence it provides an excellent opportunity to study the response of litter decomposition to litter manipulation along a forest successional gradient. The three forest types are a pioneer masson pine forest (MPF), a transitional coniferous and broad-leave mixed forest (MF) and a climax monsoon evergreen broad-leaved forest (MFBF) [24]. Previous studies showed that these forests have large variation in litter production, nutrient status in plant and soil, and some other environment factors [24–26]. This study presents the results of litter decomposition experiment where we quantitatively measured the effects of litter removal and addition on litter decomposition in three forests in DHSBR. Our objectives were to examine the effects of litter removal and addition on litter decomposition in tropical forests and to compare these effects among forests at different successional stages. We hypothesized that (1) litter removal would decrease litter decomposition but litter addition would increase litter decomposition in each forest types due to the priming effect; and (2) response of litter decomposition to litter removal/addition would vary among the three forests.

Materials and Methods

Ethics Statement

No specific permits were required for the described field studies. This research station (DHSBR) belongs to South China Botanical Garden, Chinese Academy of Sciences, which also supported the study. We confirmed that the location is not privately owned. We also confirmed that the field studies did not involve endangered or protected species. Data will be made available upon request.

Study Site

The study was conducted in Dinghushan Biosphere Reserve (DHSBR), which is in the middle part of Guangdong province, southern China (112°33' E and 23°10' N). Average annual precipitation is 1927 mm, with 75% occurring from March to August and only 6% from December to February [27]. Mean annual relative humidity is 80%. Mean annual temperature is 21.0°C with the lowest and highest average temperature being in January (12.6°C) and July (28.0°C) [27]. Soil type is lateritic red earth (Ultisols in the USDA soil taxonomy or Acrisols in the FAO soil classification) formed from sandstone [28].

There are three main forest types in the reserve; a pioneer masson pine forest (hereafter named as MPF), a coniferous and broad-leave mixed forest (hereafter named as MF) and a monsoon evergreen broad-leaved forest (hereafter named as MEBF) [24]. The three forest types form distinct successional gradients [24,29]. The MPF belongs to the first stage of the successional processes and occurs in the transition zone (periphery) of the reserve at an elevation of about 200m. *Pinus massoniana* Lamb is the dominant species in MPF (Table S1), which was planted in the 1930s. Top soil (0–30 cm) texture in MPF is medium gravel-medium loam (Ultisols in the USDA soil taxonomy), and the capacity of field moisture and wilting coefficient in the soil expressed as gravimetric water content are 26.0% and 10.9%, respectively [30]. The MF is distributed on areas next to the MPF and towards the core areas of the reserve at an elevation of about 200–300 m. It was developed by a gradual invasion of the originally planted MPF by some pioneer broadleaf species through natural succession [31]. The plant composition in MF has greatly been changed. Dominant tree species in the mixed forest are *Pinus massoniana* Lamb, *Schima superba* Chardn. & Champ., *Cryptocarya chinensis* Hance, *Craibiodendron kwangtungense* S. Y. Hu, *Lindera metcalfiana* Allen, and *Cryptocarya concinna* Hance (Table S1). The texture of top soil in MF is medium gravel-heavy loam with field capacity and wilting point at 25.3% and 8.2% gravimetric water content, respectively [30]. The MEBF is distributed in the core area of the reserve at an elevation varying from 250 to 300 m. It has been protected from direct human impact for more than 400 years [28]. Major species in MEBF are *Castanopsis chinensis* Hance, *Schima superba* Chardn. & Champ., *Cryptocarya chinensis* (Hance) Hemsl., *Machilus chinensis* (Champ. ex Benth.) Hemsl., *Syzygium rehderianum* Merr. & Perry in the canopy layer (Table S1) and *Calamus rhabdicladus* Burret, *Ardisia quinquegona* B1. and *Hemigramma decurrens* (Hook.) Copel in the understory layer. The texture of top soil in MEBF is light gravel-heavy loam. The water retention capacity of top soil is the highest with field moisture capacity and wilting coefficient of 34.6% and 11.4% gravimetric water content, respectively [30]. Litter layers are different among three forests. Litter layer is thin in the MEBF due to the faster decomposition rate, but thick in the MF and MPF. The three forests also vary in litter production, nutrient status in soil and plant, and some other environmental factors. See Table 1 for information on selected site characteristics of the three forest types.

Experimental Design

In each forest, we used a randomized complete block design with five blocks (i.e., n = 5). In each block, we set up three 1 m×1 m plots to be used as a control (CT), a litter removal (L−), and a litter addition (L+) plots making up a total of 15 plots in each forest. There was at least 3 m buffer zone between two adjacent plots to avoid overlapping effects of different treatments. The litter decomposition experiment started in February, 2007, by placing litterbags on the surface of litter layer in the plots. We removed all litter every month from the L− plots and added it to the corresponding L+ plots to cover the litterbags. Litter removal from the L− plots caused low average litter input but did not cause entire absence of litter input to the plots because the plots was receiving litter in the following days of each mouth. The controls received the normal input of aboveground litter.

Litter Collection and Initial Nutrients Analysis

Leaf litter was collected using litter traps and nylon mesh placed on the forest floor under the trees in the study sites in May and June 2006, during which litterfall is peak [32]. Leaf litter of *Schima superba* Chardn. & Champ. (SS) and *Castanopsis chinensis* Hanc (CC) was collected in MEBF and MF. Leaf litter of SS, CC and *Pinus. massoniana* Lamb (PM) was collected in MPF. These species were chosen in this study because they are the dominant tree species and contribute 40–91% the total leaf litter fall in these forests [33]. All litter was air-dried to a constant weight and six sub-samples from each kind of litter were analyzed for initial nitrogen (N) and phosphorus (P) concentrations (Table S2). N concentration was determined by the semimicro-Kjeldahl digestion method followed by the detection of ammonium with a Wescan ammonia analyzer, while total P concentration was analyzed colorimetrically after acidified ammonium persulfate digestion [34].

Litter Decomposition Experiment

Litter decomposition was determined by placing fresh litter in mesh bags in the plots. A total of 2625 litter bags were prepared from 25×25 cm polyvinyl screen with 0.5×0.5 mm mesh in the bottom and 2×2 mm in the top. Leaf litter of each species was mixed before filling in the mesh bags. Each bag was filled with

Table 1. Comparisons of litter production, leaf litter N and P concentration and selected soil properties among MEBF, MF, and MPF.

Forest types	MEBF	MF	MPF
Litter production (Mg ha^{-1} yr^{-1})‡	8.3 (0.64)^{a}	8.5 (0.62)^{a}	3.3 (0.57)^{b}
Leaf litter N (mg g^{-1})†	17.5 (0.12)^{a}	15.0 (0.10)^{b}	9.6 (0.07)^{c}
Leaf litter total P (mg g^{-1})†	0.53 (0.01)^{a}	0.32 (0.01)^{b}	0.36 (0.01)^{b}
Soil N (mg g^{-1})	1.99 (0.18)^{a}	0.93 (0.08)^{b}	1.15 (0.16)^{b}
Soil organic matter (%)	7.3 (0.8)^{a}	3.7 (0.2)^{b}	5.2 (0.2)^{b}
Soil C/N ratio	21.0 (0.6)^{a}	23.8 (2.2)^{a}	28.1 (3.9)^{b}
Soil total P (mg g^{-1})	0.49 (0.03)^{a}	0.38 (0.01)b	0.44 (0.01)^{ab}
Soil available P (mg kg^{-1})	2.2 (0.5)^{ab}	1.5 (0.5)^{a}	2.9 (0.2)^{b}
Soil moisture (%)	22.6 (1.1)^{a}	16.4 (1.9)^{b}	15.3 (1.1)^{b}
Soil temperature (°C)	21.8 (0.36)^{a}	22.6 (0.37)^{b}	23.41 (0.39)^{c}

‡from Zhou et al. (2011); †leaf litter chemical characteristics were average values of main litter species in the each forest, which were measured in 2012 in the control plots. Other values were measured in August 2007 in the control plots. Values are means, standard error in parentheses, n = 5, means not sharing the same superscript letter were statically different at *P*-value of 0.05.

10 g air-dried mass of litter. Only one litter type was put in each bag and the litter bags were evenly distributed among each plot. Litter bags were retrieved at about 3, 6, 9, 12, and 18 months (hereafter named as first, second, third, fourth and last sampling date, respectively) after the start of the study. Five litter bags of each species (a total of 525 bags on each collection) were randomly selected and collected from each plot at each sample date. The average of the 5 litterbags of each species per plot at each sampling date was used in the statistical calculations. After removing roots, soil and other extraneous materials, the leaf residue in each litter bag was oven dried at 45°C for 48 h and weighed. Litter from the last sampling date was measured for N and P concentrations using the method as described above.

Soil and Microbial Sampling

Soil sampling was conducted in August 2008 (1.5 years after litter manipulation). From each plot, 5 soil cores with 2.5 cm inner diameter were collected at random points from a 0–10cm soil depth and combined to create one composite sample per plot. Changes in soil microbial biomass was studied only in MEBF using phospholipid fatty acid (PLFA) analysis as described in Bossio and Scow [35], because samples from the other two forests were polluted during transportation of the samples from fields.

Calculations and Data Analysis

To determine litter decomposition rates we used the following decomposition model [36]: $y = e^{(-kt)}$ (exponential model), where y is the fraction of mass remaining at a specific time t (years), "e" the base of natural logarithm, k the decomposition coefficient ($year^{-1}$). To determine changes in decomposition due to treatment effect, k value changes were given by $(k_{trt}-k_{con})/k_{con}$, where k_{trt} is the decomposition coefficient in treatments (L− or L+), k_{con} the decomposition coefficient in control. Nutrient content was calculated by multiplying the nutrient concentration by the mass remaining. Nutrient content was then reported as a proportion of the initial leaf content [37].

One-way Analysis of Variance (ANOVA) was used to test the difference in decomposition rate (k) among forests or litter species in controls, and to test the difference in PLFAs among treatments in MEBF. Two-way ANOVA was used to determine the interaction effect of litter manipulation treatments and litter types on litter decomposition rate and nutrient remaining in each forest. Repeated measure ANOVA was used to determine the difference in litter mass remaining among forests, litter manipulation treatments and litter species. In all ANOVAs, the block was included as random factors. In addition, data was log transformed to fulfill the requirements of normality and homogeneity of variance. All analyses were conducted using SAS software (SAS Institute Inc., Cary NC, USA). Statistically significant differences were set with P values <0.05 unless otherwise stated. Mean values ± standard errors were reported in the text.

Results

Litter Decomposition Rate in Control Plots

The leaf litter decomposition rates in control plots varied among forest types and tree species (Fig. 1, Table 2). Leaf litter decomposition rate (i.e., k value) was significantly higher in MEBF than in MF and MPF for all possible species comparison. However, no significant difference in litter decomposition rate was found between MF and MPF for all species. Average litter decomposition rates (k) were 1.97, 0.92, and 0.73 for MEBF, MF, and MPF, respectively. When compared within forest type, litter decomposition rate was significantly lower in CC leaf litter than in SS in the MEBF and MF (Table 2, $P = 0.037$ and 0.021, respectively). However, there was no significant difference among the selected three species in the MPF (Table 2, $P = 0.261$).

Effects of Litter Manipulation on Litter Decomposition

Responses of litter decomposition rates to litter removal/ addition varied with forest types (Table 2). In MEBF, litter removal significantly decreased litter decomposition rates for both SS and CC ($P = 0.045$ and 0.030, respectively). Repeated measure ANOVA with Turkey' HSD test showed that mass remaining was significantly higher in L− than in control plots in the second, third, fourth and fifth sampling dates for SS leaf litter. For CC leaf litter, significant difference in litter decomposition between L− and control plots was observed only after the last sampling date (Fig. 1). In contrast, litter addition significantly increased litter decomposition rate for both SS and CC ($P = 0.018$ and 0.005, respectively). However, such a response of leaf litter decomposition rates to litter addition showed temporal variation between the two

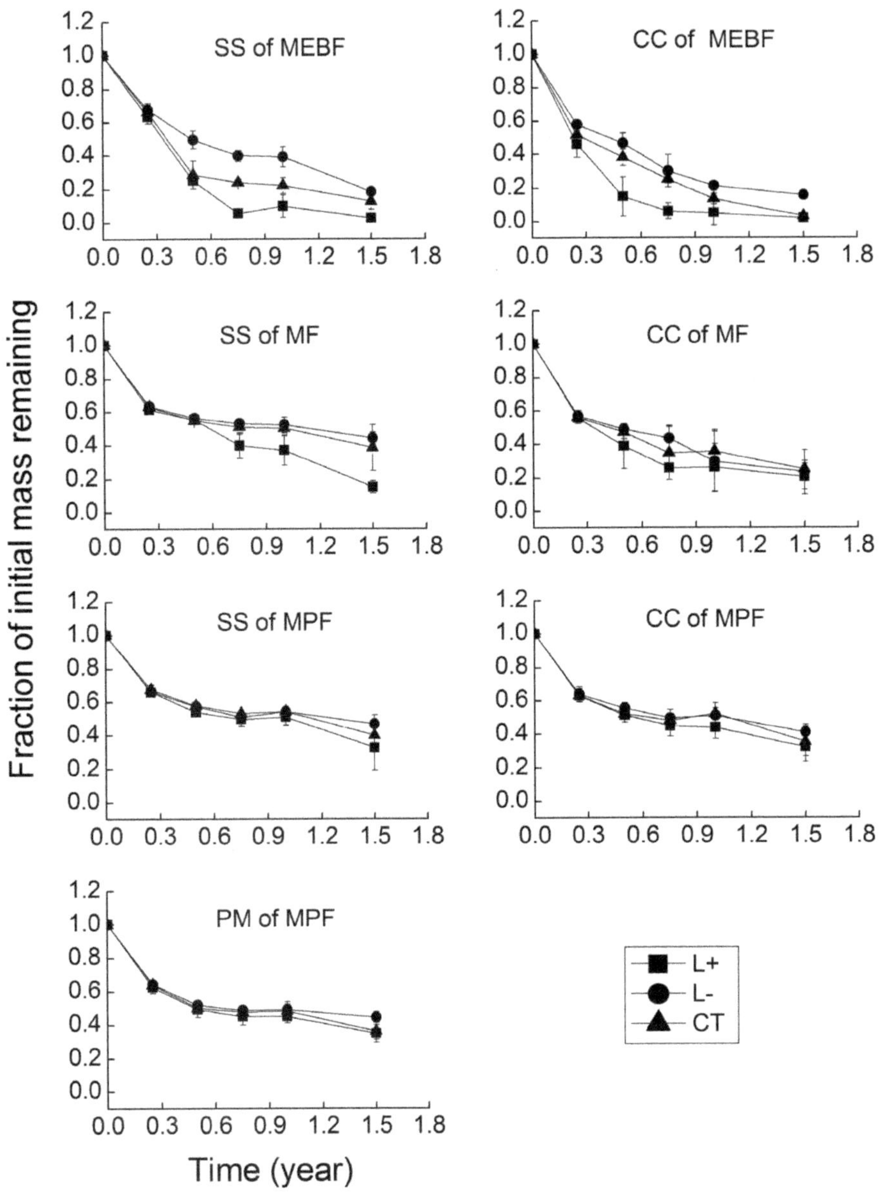

Figure 1. Mass loss of decomposing leaf litter of three dominant tree species in various litter manipulation treatments in the MEBF, MF, and MPF.

species. For SS, significant difference between control and L+ plots was found in the second, third, and fourth sampling dates, whereas the reported response for CC was observed in the third and fourth sampling dates (Fig. 1).

In MF, SS and CC leaf litter showed similar response to litter addition. Litter decomposition rate increased significantly after litter addition for both species (Table 2, $P=0.019$ and 0.012 for SS and CC, respectively). However, the effect of litter removal on litter decomposition varied between the two species (Table 2). Litter removal significantly decreased litter decomposition for SS ($P=0.033$), but not for CC ($P=0.842$). Repeated measure ANOVA with Turkey' HSD test indicated that litter addition significantly increased decomposition of SS leaf litter in the third, fourth and fifth sampling dates, and significantly increased decomposition of CC leaf litter in the second and third sampling dates (Fig. 1). Litter removal significantly decreased SS leaf litter decomposition in the last sampling date, but no significant change in leaf litter decomposition was observed for CC leaf litter decomposition at all sampling dates (Fig. 1).

In MPF, decomposition rate increased slightly with litter addition and decreased slightly with litter removal for all species (Table 2), but one-way ANOVA showed that this increase/decrease was not significant among treatments for all species. Repeated measure ANOVA also showed no significant difference in mass remaining among treatments over the entire study period for all species. There are no significant interaction effects of litter species and treatments on litter decomposition rate in all the three forests ($P>0.05$ for all).

The relative change in k values in response to the treatments decreased in the order: MEBF>MF>MPF (Fig. 2). Average decrease in litter decomposition rate as a result of litter removal was 27% in the MEBF, which was significantly higher than the

Table 2. Comparisons of decomposition rates (*k* values) between treatments and litter species in each forest.

Species	Treatments	MEBF		MF		MPF	
		K	R^2	K	R^2	k	R^2
SS	CT	1.74 (0.36)^{a}	0.84 (0.04)	0.80 (0.05)^{a}	0.60 (0.09)	0.68 (0.04)^{a}	0.71(0.06)
	L−	1.23 (0.19)^{b}	0.86 (0.04)	0.66 (0.02)^{b}	0.57 (0.07)	0.64 (0.02)^{a}	0.54 (0.09)
	L+	2.80 (0.24)^{c}	0.86 (0.06)	1.13 (0.05)^{c}	0.81(0.04)	0.84 (0.09)^{a}	0.69 (0.07)
CC	CT	2.19 (0.06)^{c}	0.75 (0.06)	1.04 (0.06)^{cd}	0.68 (0.10)	0.78 (0.05)^{a}	0.62 (0.09)
	L−	1.63 (0.29)^{a}	0.80 (0.04)	1.01 (0.05)^{cd}	0.77 (0.07)	0.72 (0.04)^{a}	0.60 (0.06)
	L+	3.30 (0.25)^{d}	0.84 (0.04)	1.35 (0.12)^{d}	0.85 (0.09)	0.89 (0.08)^{a}	0.75 (0.07)
PM	CT	–		–		0.80 (0.03)^{a}	0.64 (0.04)
	L−	–		–		0.71 (0.02)^{a}	0.51 (0.06)
	L+	–		–		0.84 (0.05)^{a}	0.64 (0.09)

Notes: *k* values and coefficients of determination (R^2) are based on a single negative exponential model. Two-way ANOVA with SNK test was used in each forest respectively to determine the effect of litter treatments and litter types on *k* values. Values are means, standard error in parentheses, n = 5. Under each forest type and for each variable column, means not sharing the same superscript letter were statically different at *P*-value of 0.05.

respective values in MF (10%) and MPF (8%) ($P<0.001$). On the other hand, litter addition increased leaf litter decomposition by 55% in the MEBF, which is significantly larger than the average percent increase in leaf litter decomposition rates in MF (36%) and MPF (14%) ($P<0.001$).

Litter Nutrient Remaining

Nutrient (N and P) remaining in all leaf litter types was measured after the last sampling date and it varied depending on forests and species (Fig. 3). In MEBF, litter removal significantly increased N and P remaining in decomposing leaf litter for both SS and CC. In contrast, litter addition decreased N and P remaining in both species leaf litter, but the decrease in N and P remaining was significant only for SS leaf litter. In MF, litter removal significantly increased N and P remaining and litter addition significantly decreased N and P remaining in SS leaf litter. However, none of the treatments caused significant changes in N and P remaining in decomposing CC leaf litter when compared to the control plots, but both litter removal and litter addition tended to decrease N and P remaining in the CC leaf litter (all $P>0.05$). In MPF, litter removal tended to increase N and P remaining and litter addition tended to decrease N and P remaining in all species, but the magnitude of these changes varied among the treatments and species. For example, litter removal significantly increased N remaining in PM ($P=0.005$), P remaining in SS and CC ($P=0.024$ and $P=0.033$, respectively).

Soil Microbial Biomass

Litter manipulation had no significant effect on soil microbial biomass in the MEBF (Fig. 4, $P=0.074$). The average total PLFAs were 55.40 (6.91), 57.94 (7.9), and 45.34 (6.1) nmol g^{-1} in control, L−, and L+ plots, respectively. Similarly, there was no significant difference among litter manipulation treatments for the Bac PLFAs, Fun PLFAs, and F:B (Fig. 4).

Discussion

In the present study, decomposition rates (*k* values) of leaf litter observed in the controls (ranging from 0.68 to 2.19) were similar to those found in subtropical forests of south China [33,38–39] and other subtropical/tropical forests [40–42], but were slightly lower than that of some other tropical forests with high rainfall [43]. However, the *k* values in our study were higher than those reported from temperate forests [44,45], indicating higher decomposition rate in these humid sub-tropical forests of southern china. In addition, decomposition rate increased with forest succession: MEBF>MF>MPF, which was in agreement with results reported by previous studies in adjacent forests [33,38].

Compared to control plots which received normal litter input, litter removal significantly decreased litter decomposition in the MEBF. Similar effect of litter removal on litter decomposition was reported by a study in an old-growth forest in Panama [3]. Based on available evidence from literature, there are several possible mechanisms that could explain our observation. Litter removal changes the microclimate (e.g. moisture and temperature) in the forest floor [46,47], causes direct nutrient losses and changes in soil physical and chemical characteristics [48,49], and decreases decomposer biomass and activity [3,50]. Even though we did not measure plot microclimate, previous studies have reported that litter removal usually affects soil microclimate, mainly decreasing moisture and greater fluctuations in temperature in forest floor [46,47]. On the other hand, a previous study observed in adjacent forests showed that litter removal had no significant effect on the soil temperature in all the three forests, and significantly decreased

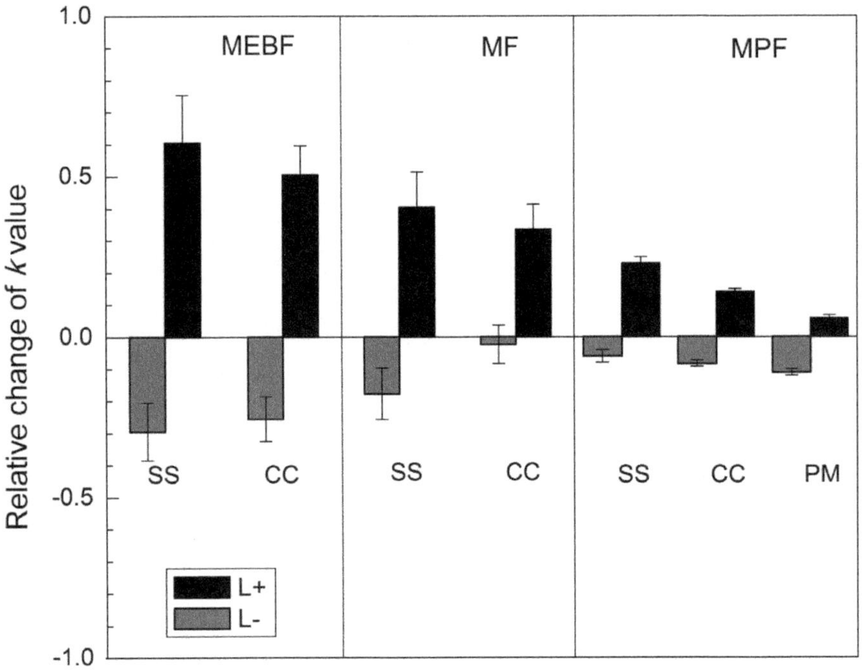

Figure 2. Comparisons of relative change of *k* value between different litter manipulation treatments, litter species and forest types.

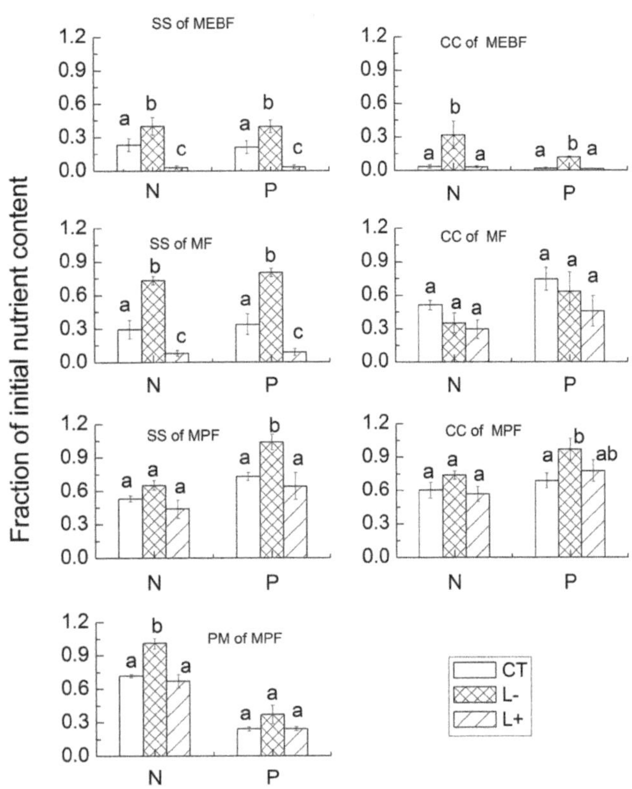

Figure 3. Nutrients remaining in the last stage of the decomposition. Values are means, standard error in parentheses, n = 5, means not sharing the same superscript letter were statically different at *P*-value of 0.05.

soil moisture only in MF [26]. However, the soil temperature and moisture were measured in mineral soil at 5 cm depth [26]. We expect more changes in forest floor microclimate following the litter manipulation, but such data are missing in our study.

Decreased nutrient supply for decomposer microbial communities after litter removal is another likely explanation for the slower decomposition in the L− plots. In this study, litter was removed once a month in L− plots, which could remove 44–73% of litter nutrient input [49]. This decreased nutrient supply might decrease microbial activities in the L− plots. We measured the litter nutrients remaining in the last sampling date and found that the content of N and P increased in L− plots compared to those in controls for all litter. This might indicate that more N and P were immobilized in L− plots, possibly because the microbes need more N and P in L− treatment due to reduced litter input of these nutrients making the microbes nutrient limited as suggested by [36].

However, we did not observe significant and clear changes in soil microbial biomass and microbial community as indicated by PLFA result (Fig. 4). Similar results were also reported from temperate forests, where change in the soil microbial biomass or activity in litter removal treatment were not significant [51]. It has been suggested that absence of expected increase in soil microbial biomass could be due to soil pools buffering the effect of litter manipulation from aboveground [51,52]. However, our data on microbial biomass included only mineral soil because clear forest floor layer in MEBF is usually absent due to fast turn-over of organic materials. Even though data on the forest floor is not available in this study, previous study reported that litter removal decreased decomposers biomass more in forest floor than in mineral soil [3]. The study also showed that litter removal reduced the abundance of meso-arthropods in forest floor because meso- and micro-arthropods, which play significant role in decomposing organic materials, mostly inhabit top forest floor layer. Though our data only for mineral soil might not be very conclusive

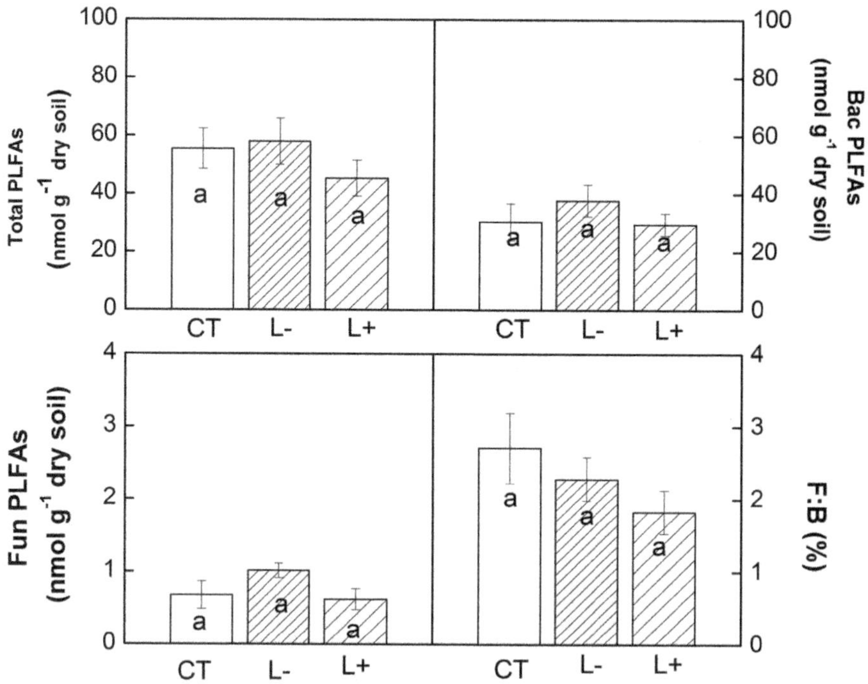

Figure 4. Comparisons of soil microbial PLFAs between treatments in the MEBF. Data from August 2008. Bac PLFAs: Bacterial PLFAs; Fun PLFAs: Fungal PLFAs; F:B (%): the percentage of fungal to bacterial PLFAs. Values are means, standard error in parentheses, n = 5, means sharing the same superscript letter were not statically different (P-value ≥0.05).

especially when data is absent for the more active and dynamic litter layer, it showed that microbial activities might not be affected by short term litter manipulation. However, more comprehensive studies including response of microbial dynamics to litter manipulation both in mineral soil and forest floor are needed for further understanding of the subject.

By contrast, litter addition significantly accelerated the decomposition of leaf litter in the MEBF. Our result is inconsistent with results reported by Sayer *et al.* [3], which showed that litter addition did not affect leaf litter decomposition but significantly increased wood litter decomposition. However, our result partially supports the "priming effect" hypothesis which suggests that the addition of fresh organic matter (leaf litter) can stimulate decomposition of the organic matter [53,54]. The reason for this response is not clear in this study. We observed decreased N and P reaming in L+ plots compared to control plots indicating that increased nutrient availability from the additional litter could stimulate the microbe to release litter nutrients, which partially contribute to increased decomposition of leaf litter. In addition, we did not find significant changes in soil microbial biomass and microbial community after the litter addition (Fig. 4), but we still cannot rule out the possibility of the change of microbial biomass and microbial communities in litter layer. For example, Sayer *et al.* [3] found increased abundance of meso-arthropode in forest floor after doubling litter input resulting in increased mass loss of leaf litter. However, we could not attribute our observation to an increase in meso-arthropodes because the size of the mesh we used (0.5mm×0.5 mm) might have reduced the contribution macro-decomposer communities despite possible increases in their abundance. Comparative studies of effects of litter manipulation on litter decomposition with different mesh sizes might provide better understanding on possible confounding effects of mesh size in litter decomposition studies.

The effect of litter manipulation experiment on leaf-litter decomposition in our study varied significantly among the three forest types and exhibited clear pattern with successional gradients. The effect was highest in the climax MEBF and lowest in the MPF whereas MF showed intermediate response (Fig. 2). The reasons for this response pattern are currently not clear. However, several explanations could be suggested. Litter production was found to differ among these three forests, amounting 83, 85 and 33 Mg ha^{-1} yr^{-1} in MEBF, MF, and MPF, respectively (Table 1). Similarly, nutrient concentration and quality of leaf (C/N ratio, lignin content etc.) differs among the three forests. Foliar N and P concentrations in MF and MPF were significantly lower than in MEBF (Table 1). This suggests that more C and other nutrients in MEBF would be removed or input after litter manipulation compared to those in MF and MPF, and thus resulting in more significant response in MEBF. Another possible explanation could be different responses in soil microclimate after litter treatments among the forests, especially for soil moisture. Soil moisture is a more important factor than soil temperature for the fastest decomposition rate in the MEBF because Table 1 showed that the MEBF had the highest moisture and the lowest temperature in soil than other two forests (Table 1). Because of this higher initial moisture, we believe that the thorough litter removal/addition may change the soil moisture of the MEBF in a larger extent compared to other forests with lower initial moisture. In addition, for the MF and MPF, the long-term history of relatively more exposure to human disturbance, lower litter production and less canopy cover also might have reduced the responses to our short-term litter manipulation experiment.

Conclusion

Our results showed that litter removal decreased litter decomposition and litter addition increased litter decomposition in three tropical forests at different successional stages. However, the effect was highest in the climax MEBF and lowest in the MPF whereas MF showed intermediate response, suggesting that the change in litter quantity can affect litter decomposition in tropical forests and this impact may become stronger with forest succession in the studied tropical forests. Our results can provide relevant information on how future climate changes modify accumulation of organic matter (C) in tropical forest and consequently affect nutrient cycle, and for any sudden events in the forest such as hurricane and tree diseases which also changes quantity of litter input to the forest floor. However, recent mechanisms are still far from clear, thus we also suggest similar studies in temperate and other tropical forest ecosystems to further strengthen the findings.

Acknowledgments

We thank Xiaoming Fang and Wei Zhang for their skilful assistance in field work, Guohua Liang for his assistance in laboratory work, and Drs. Per Gundersen and Weixing Zhu for their invaluable suggestions in experiment and earlier version of the manuscript.

Author Contributions

Conceived and designed the experiments: JMM HC SLF ZFL. Performed the experiments: CM SFD LL TZ HC. Analyzed the data: HC GAG. Contributed reagents/materials/analysis tools: SFD CM HC. Wrote the paper: HC. English editing: GAG.

References

1. Hobbie SE, Vitousek PM (2000) Nutrient limitation of decomposition in Hawaiian forests. Ecology 81: 1867–1877.
2. Chapin FS, Matson PA, Mooney HA (2002) Principles of terrestrial ecosystem ecology. New York, USA: Springer.
3. Sayer EJ, Tanner EVJ, Lacey AL (2006) Effects of litter manipulation on early-stage decomposition and meso-arthropod abundance in a tropical moist forest. Forest Ecology and Management 229: 285–293.
4. Hickler T, Smith B, Prentice IC, MjÖFors K, Miller P, et al. (2008) CO_2 fertilization in temperate FACE experiments not representative of boreal and tropical forests. Global Change Biology 14: 1531–1542.
5. Norby RJ, DeLucia EH, Gielen B, Calfapietra C, Giardina CP, et al. (2005) Forest response to elevated CO_2 is conserved across a broad range of productivity. Proceedings of the National Academy of Sciences of the United States of America 102: 18052–18056.
6. Raich JW, Russell AE, Kitayama K, Parton WJ, Vitousek PM (2006) Temperature influences carbon accumulation in moist tropical forests. Ecology 87: 76–87.
7. Holmes KW, Chadwick OA, Kyriakidis PC, de Filho EPS, Soares JV, et al. (2006) Large-area spatially explicit estimates of tropical soil carbon stocks and response to land-cover change. Global Biogeochemical Cycles 20: GB3004.
8. Ostertag R, Scatena FN, Silver WL (2003) Forest floor decomposition following hurricane litter inputs in several Puerto Rican forests. Ecosystems 6: 261–273.
9. Sullivan NH, Bowden WB, McDowell WH (1999) Short-Term Disappearance of Foliar Litter in Three Species Before and After a Hurricane. Biotropica 31: 382–393.
10. Barantal S, Schimann H, Fromin N, Hattenschwiler S (2012) Nutrient and Carbon Limitation on Decomposition in an Amazonian Moist Forest. Ecosystems 15: 1039–1052.
11. Boberg J, Finlay RD, Stenlid J, Nasholm T, Lindahl BD (2008) Glucose and ammonium additions affect needle decomposition and carbon allocation by the litter degrading fungus *Mycena epipterygia*. Soil Biology & Biochemistry 40: 995–999.
12. Chigineva NI, Aleksandrova AV, Tiunov AV (2009) The addition of labile carbon alters litter fungal communities and decreases litter decomposition rates. Applied Soil Ecology 42: 264–270.
13. Kuzyakov Y, Hill PW, Jones DL (2007) Root exudate components change litter decomposition in a simulated rhizosphere depending on temperature. Plant and Soil 290: 293–305.
14. Liu ZF, Wu JP, Zhou LX, Lin YB, Fu SL (2012) Effect of understory fern (*Dicranopteris dichotoma*) removal on substrate utilization patterns of culturable soil bacterial communities in subtropical Eucalyptus plantations. Pedobiologia 55: 7–13.
15. Wu JP, Liu ZF, Wang XL, Sun YX, Zhou LX, et al. (2011) Effects of understory removal and tree girdling on soil microbial community composition and litter decomposition in two Eucalyptus plantations in South China. Functional Ecology 25: 921–931.
16. Matsushima M, Chang SX (2007) Effects of understory removal, N fertilization, and litter layer removal on soil N cycling in a 13-year-old white spruce plantation infested with Canada bluejoint grass. Plant and Soil 292: 243–258.
17. Sayer EJ (2006) Using experimental manipulation to assess the roles of leaf litter in the functioning of forest ecosystems. Biological Reviews 81: 1–31.
18. Crow SE, Lajtha K, Bowden RD, Yano Y, Brant JB, et al. (2009) Increased coniferous needle inputs accelerate decomposition of soil carbon in an old-growth forest. Forest Ecology and Management 258: 2224–2232.
19. Fontaine S, Bardoux G, Abbadie L, Mariotti A (2004) Carbon input to soil may decrease soil carbon content. Ecology Letters 7: 314–320.
20. Leff JW, Wieder WR, Taylor PG, Townsend AR, Nemergut DR, et al. (2012) Experimental litterfall manipulation drives large and rapid changes in soil carbon cycling in a wet tropical forest. Global Change Biology 18: 2969–2979.
21. Sayer EJ, Heard MS, Grant HK, Marthews TR, Tanner EVJ (2011) Soil carbon release enhanced by increased tropical forest litterfall. Nature Climate Change 1: 304–307.
22. Kranabetter JM, Chapman BK (1999) Effects of forest soil compaction and organic matter removal on leaf litter decomposition in central British Columbia. Canadian Journal of Soil Science 79: 543–550.
23. Ruan HH, Li YQ, Zou XM (2005) Soil communities and plant litter decomposition as influenced by forest debris: Variation across tropical riparian and upland sites. Pedobiologia 49: 529–538.
24. Zhou GY, Guan LL, Wei XH, Zhang DQ, Zhang QM, et al. (2007) Litterfall production along successional and altitudinal gradients of subtropical monsoon evergreen broadleaved forests in Guangdong, China. Plant Ecology 188: 77–89.
25. Huang WJ, Zhou GY, Liu JX (2012) Nitrogen and phosphorus status and their influence on aboveground production under increasing nitrogen deposition in three successional forests. Acta Oecologica-International Journal of Ecology 44: 20–27.
26. Tang X, Liu S, Zhou G, Zhang D, Zhou C (2006) Soil-atmospheric exchange of CO_2, CH_4, and N_2O in three subtropical forest ecosystems in southern China. Global Change Biology 12: 546–560.
27. Huang ZF, Fan ZG (1982) The climate of Dinghushan. Tropical and subtropical Forest Ecosystem. Tropical and Subtropical Forest Ecosystem 1: 11–23.
28. Mo HM, Brown S, Peng SL, Kong G (2003) Nitrogen availability in disturbed, rehabilitated and mature forests of tropical China. Forest Ecology and Management 175: 573–583.
29. Peng SL, Wang BS (1993) Forest succession at Dinghushan, Guangdong, China. Botanical Journal of South China 7: 34–42.
30. Zhang BG, Zhuo MN (1985) The physical properties of soil under different forest types in Dinghushan Biosphere Reserve. Tropical and Subtropical Forest Ecosystem Research 3: 1–10 (in Chinese with English abstract).
31. Fang YT, Zhu WX, Mo JM, Zhou GY, Gundersen P (2006) Dynamics of soil inorganic nitrogen and their responses to nitrogen additions in three subtropical forests, south China. Journal of Environmental Sciences 18: 752–759
32. Zhang DQ, Ye WH, Yu QF, Kong GH, Zhang YC (2000) The litter-fall of representative forests of successional series in Dinghushan. Acta Ecologica Sinica 20: 938–944 (in Chinese with English abstract).
33. Mo JM, Brown S, Xue J, Fang Y, Li Z (2006) Response of litter decomposition to simulated N deposition in disturbed, rehabilitated and mature forests in subtropical China. Plant and Soil 282: 135–151.
34. Anderson JM, Ingram JSI (1989) Tropical Soil Biology and Fertility: A Handbook of Methods. Oxford, England: CAB International Wallingford.
35. Bossio DA, Scow KM (1998) Impacts of Carbon and Flooding on Soil Microbial Communities: Phospholipid Fatty Acid Profiles and Substrate Utilization Patterns. Microbial Ecology 35: 265–278.
36. Olson JS (1963) Energy Storage and the Balance of Producers and Decomposers in Ecological Systems. Ecology 44: 322–331.
37. McGroddy ME, Silver WL, de Oliveira RC (2004) The effect of phosphorus availability on decomposition dynamics in a seasonal lowland Amazonian forest. Ecosystems 7: 172–179.
38. Fang H, Mo JM, Peng SL, Li ZA, Wang H (2007) Cumulative effects of nitrogen additions on litter decomposition in three tropical forests in southern China. Plant and Soil 297: 233–242.
39. Chen H, Dong S, Liu L, Ma C, Zhang T, et al. (2013) Effects of experimental nitrogen and phosphorus addition on litter decomposition in an old-growth tropical forest. PLOS ONE (in press).

40. Alhamd L, Arakaki S, Hagihara A (2004) Decomposition of leaf litter of four tree species in a subtropical evergreen broad-leaved forest, Okinawa Island, Japan. Forest Ecology and Management 202: 1–11.
41. Cusack DF, Chou WW, Yang WH, Harmon ME, Silver WL, et al. (2009) Controls on long-term root and leaf litter decomposition in neotropical forests. Global Change Biology 15: 1339–1355.
42. Loranger G, Ponge JF, Imbert D, Lavelle P (2002) Leaf decomposition in two semi-evergreen tropical forests: influence of litter quality. Biology and Fertility of Soils 35: 247–252.
43. Cleveland CC, Reed SC, Townsend AR (2006) Nutrient regulation of organic matter decomposition in a tropical rain forest. Ecology 87: 492–503.
44. Gholz HL, Wedin DA, Smitherman SM, Harmon ME, Parton WJ (2000) Long-term dynamics of pine and hardwood litter in contrasting environments: toward a global model of decomposition. Global Change Biology 6: 751–765.
45. Melillo JM, Aber JD, Muratore JF (1982) Nitrogen and Lignin Control of Hardwood Leaf Litter Decomposition Dynamics. Ecology 63: 621–626.
46. Ponge JF, Arpin P, Vannier G (1993) Collembolan response to experimental perturbations of litter supply in a temperate forest ecosystem. European Journal of Soil Science 29: 141–153.
47. Villalobos-Vega R, Goldstein G, Haridasan M, Franco A, Miralles-Wilhelm F, et al. (2011) Leaf litter manipulations alter soil physicochemical properties and tree growth in a Neotropical savanna. Plant and Soil 346: 385–397.
48. Dzwonko Z, Gawronski S (2002) Effect of litter removal on species richness and acidification of a mixed oak-pine woodland. Biological Conservation 106: 389–398.
49. Mo JM, Brown S, Lenart M, Kong GH (1995) Nutrient Dynamics of a Human-Impacted Pine Forest in a Mab Reserve of Subtropical China. Biotropica 27: 290–304.
50. Li YQ, Xu M, Sun OJ, Cui WC (2004) Effects of root and litter exclusion on soil CO_2 efflux and microbial biomass in wet tropical forests. Soil Biology & Biochemistry 36: 2111–2114.
51. Fahey TJ, Battles JJ, Wilson GF (1998) Responses of early successional northern hardwood forests to changes in nutrient availability. Ecological Monographs 68: 183–212.
52. Fisk MC, Fahey TJ (2001) Microbial biomass and nitrogen cycling responses to fertilization and litter removal in young northern hardwood forests. Biogeochemistry 53: 201–223.
53. Bingeman CW, Varner JE, Martin WP (1953) The Effect of the Addition of Organic Materials on the Decomposition of an Organic Soil. Soil Science Society of America Proceedings 17: 34–38.
54. Kuzyakov Y, Friedel JK, Stahr K (2000) Review of mechanisms and quantification of primingeffects. Soil Biology & Biochemistry 32: 1485–1498.

How Ants Drop Out: Ant Abundance on Tropical Mountains

John T. Longino[1]*, Michael G. Branstetter[2], Robert K. Colwell[3]

1 Department of Biology, The University of Utah, Salt Lake City, Utah, United States of America, **2** Department of Entomology, National Museum of Natural History, Smithsonian Institution, Washington, D.C., United States of America, **3** Department of Ecology and Evolutionary Biology, University of Connecticut, Storrs, Connecticut, United States of America, and University of Colorado Museum of Natural History, Boulder, Colorado, United States of America

Abstract

In tropical wet forests, ants are a large proportion of the animal biomass, but the factors determining abundance are not well understood. We characterized ant abundance in the litter layer of 41 mature wet forest sites spread throughout Central America (Chiapas, Guatemala, Honduras, Nicaragua, and Costa Rica) and examined the impact of elevation (as a proxy for temperature) and community species richness. Sites were intentionally chosen to minimize variation in precipitation and seasonality. From sea level to 1500 m ant abundance very gradually declined, community richness declined more rapidly than abundance, and the local frequency of the locally most common species increased. These results suggest that within this elevational zone, density compensation is acting, maintaining high ant abundance as richness declines. In contrast, in sites above 1500 m, ant abundance dropped abruptly to much lower levels. Among these high montane sites, community richness explained much more of the variation in abundance than elevation, and there was no evidence of density compensation. The relative stability of abundance below 1500 m may be caused by opposing effects of temperature on productivity and metabolism. Lower temperatures may decrease productivity and thus the amount of food available for consumers, but slower metabolisms of consumers may allow maintenance of higher biomass at lower resource supply rates. Ant communities at these lower elevations may be highly interactive, the result of continuous habitat presence over geological time. High montane sites may be ephemeral in geological time, resulting in non-interactive communities dominated by historical and stochastic processes. Abundance in these sites may be determined by the number of species that manage to colonize and/or avoid extinction on mountaintops.

Editor: M. Alex Smith, University of Guelph, Canada

Funding: This work was supported by National Science Foundation grants DEB-0072702 (Project ALAS), DEB-0640015/DEB-1157383 (Project LLAMA), DEB-1354739 (Project ADMAC), National Geographic Society grants 7331-02 and 7751-04, and Conservation International. MB was supported by a Buck Postdoctoral Fellowship from the Smithsonian Institution. The funders had no role in study design, data collection and analysis, decision to publish, or preparation of the manuscript.

Competing Interests: The authors have declared that no competing interests exist.

* Email: jacklongino@gmail.com

Introduction

What determines the number of ants in 1 m^2 of forest floor leaf litter? This disarmingly simple question is not easy to answer, because many processes potentially influence ant density: temperature effects on food availability [1–3], temperature effects on metabolism and foraging [3–5], litter depth effects on food availability and habitat space [6–9], moisture effects on food availability and foraging efficiency [10–13], seasonality effects on food and foraging [10] [1], species composition effects through species packing and niche partitioning [14], and availability of species in a regional species pool [15]. Ecologists have a long-standing interest in explaining ecological patterns at multiple spatial scales [4] [5] [15] [16], including the smallest scale relevant to particular groups of organisms. Small-scale abundance patterns in land plants have been extensively studied [17], but less attention has been paid to similar patterns in consumers. This study examines ant abundance at the 1 m^2 scale in 41 Central American forests, across a broad range of elevations. We focus on temperature and species richness as primary explanatory variables.

Longino and Colwell [18] documented ant abundance on the Barva Transect, a continuously-forested elevational gradient in Costa Rica. Ant abundance was curvilinear as a function of elevation, changing relatively little from sea-level to 1500 m and then declining abruptly to near zero at 2000 m. In addition, the study yielded evidence of density compensation: at higher elevations there were fewer species, but more individuals per species, resulting in relatively stable total ant density even as diversity declined. However, the Barva Transect study covered a limited geographic scope, with just seven discrete sampling elevations on a single mountainside. Here, we aim to examine patterns of ant abundance with respect to elevation based on a much larger dataset (34 additional sites), over a larger spatial extent (Costa Rica to southern Mexico), and with a larger number of high-elevation sites. This more comprehensive dataset has revealed broadly similar patterns across all of Central America and documents previously unappreciated patterns of ant decline above 1500 m.

Methods

Research sites were a combination of private land and state protected lands. Research and collecting permits were obtained from El Colegio de la Frontera Sur in Chiapas, Mexico; Universidad del Valle, Guatemala; Consejo Nacional de Areas Protegidas, Guatemala; Escuela Agrícola Panamericana Zamorano, Honduras; Instituto Nacional de Conservación y Desarrollo Forestal, Areas Protegidas y Vida Silvestre, Honduras; Fundación para La Autonomía y Desarrollo de la Costa Atlántica de Nicaragua; Ministerio del Ambiente y los Recursos Naturales, Nicaragua; Jesus Mountain Coffee Group, Nicaragua; The Organization for Tropical Studies, Costa Rica; and the Ministerio de Ambiente y Energía, Costa Rica. No protected species were sampled.

Like the original Barva Transect data, this larger dataset limited the number of contributing variables by sampling exclusively in closed-canopy evergreen wet forest, thus removing or at least greatly reducing effects of aridity, seasonality, and major habitat differences on ant abundance and diversity. We chose widely dispersed sites in Central America, aiming for a compromise between sites that were far enough apart to reduce effects of spatial autocorrelation, yet close enough to reduce effects of dispersal limitation. To focus on temperature and species richness as primary explanatory variables, we selected sites across a range of elevations from 0–2600 m. Because none of our sites were ever exposed to freezing temperatures, adaptations for frost-tolerance were not relevant.

Field sampling was carried out at 41 sites from Chiapas, Mexico, to Costa Rica (Fig. 1, Table 1). We used a subset (7 sites) of the data reported in Longino and Colwell [18], sampled from 2001 to 2007 as part of the Arthropods of La Selva project (ALAS, http://purl.oclc.org/ALAS) and Conservation International's Tropical Ecology Assessment and Monitoring project (TEAM, http://www.teamnetwork.org/en/). The remaining 34 sites were distributed in wet forest throughout Nicaragua, Honduras, Guatemala, and the Mexican state of Chiapas. These sites were part of the Leaf Litter Arthropods of MesoAmerica (LLAMA) project (https://sites.google.com/site/longinollama/) and were sampled from 2008 to 2011, following the same protocols used by the ALAS and TEAM projects for the Costa Rican sites.

Sampling was carried out during the dry season for the ALAS samples, throughout the year for TEAM samples, and during the dry to wet season transition for LLAMA samples. Site histories and disturbance levels varied among sites, but most sampling was performed in relatively mature forest patches. The highlands of Honduras, Guatemala, and Chiapas often have island-like areas of diverse mesophyl cloud forest surrounded by pine or pine/oak forests; LLAMA sampling in these areas was always in the mesophyl cloud forest. Thus, other than elevational differences, the sites were substantially uniform with respect to physiognomy and forest structure.

The sampling unit was a "miniWinkler" sample of arthropods extracted from a 1 m^2 forest floor quadrat, following the methods of Fisher [19]. Litter in a quadrat (including any vegetation and suspended organic matter immediately above the quadrat) was chopped with a machete, gathered into a sifter, and shaken vigorously. Sifting continued until all litter in the plot was sifted or a maximum of 6 L of siftate was obtained. In the latter case, material from different parts of the plot was subsampled. Although not quantified, the proportion of quadrats that required subsampling was low and did not appear to vary with elevation. Siftate was transferred to cloth sacks and moved to a laboratory or shelter, where each sample was suspended in an individual Winkler extractor for three days. Falling arthropods were collected into 95% ethanol. For the three ALAS sites on the Barva Transect and the 34 LLAMA sites, miniWinkler samples were distributed in two haphazardly-placed transects of 50 samples, with 5 m spacing between samples. For the four TEAM sites on the Barva Transect, samples were distributed in 10 transects of 10 samples each, with 10 m spacing between samples. Total sample size was 4099 miniWinkler samples (one sample was lost).

All ants were separated from samples by project staff. The total number of workers was recorded for each sample, as a measure of overall ant abundance. Only a subset of the ants in each sample was identified to species; certain taxa were excluded due to taxonomic impediments (genera that are very difficult to sort to species), habitat edge effects (arboreal species that occasionally occur in Winkler samples), and pest ants that were likely contaminants. Genera excluded due to taxonomic difficulty were *Brachymyrmex* (except the distinctive litter species *B. cavernicola*), *Hypoponera* (except the distinctive species *H. nitidula* and *H. parva*), *Nylanderia*, *Solenopsis* (except *S. geminata*), and *Tapinoma*. Excluded arboreal taxa were *Azteca*, *Camponotus*, *Cephalotes*, *Dolichoderus*, *Myrmelachista*, *Procryptocerus*, *Pseudomyrmex*, and *Xenomyrmex*. Excluded pest ants were *Cardiocondyla*, *Monomorium*, *Paratrechina longicornis*, and *Tetramorium bicarinatum*. Identifications were carried out by Branstetter (*Stenamma*) and Longino (remaining taxa). Voucher specimens were deposited in regional collections in Costa Rica, Honduras, Guatemala, and Chiapas; in the Longino research collection at The University of Utah; and in the Branstetter research collection. The personal research collections of Longino and Branstetter will ultimately be deposited in major institutional collections. All specimen data are posted on AntWeb (www.AntWeb.org).

Bioclimatic variables were extracted for all 41 sites from WorldClim global climate data ([20]; http://www.worldclim.org, Version 1.4, release 3; accessed 11 July 2012) for the latitude and longitude coordinates in Table 1, using "current values" (averages from 1950–2000). Four variables were examined: mean annual temperature, minimum temperature of the coldest month, annual precipitation, and precipitation seasonality (coefficient of variation of monthly averages) (WorldClim variables BIO1, BIO6, BIO12, and BIO15, respectively). Minimum temperature of the coldest month and precipitation seasonality were highly collinear with mean annual temperature and annual precipitation, respectively, and are not considered further. Annual precipitation varied from 1230–4400 mm across sites, but explained negligible variance in ant abundance, suggesting that our selection of wet evergreen sites successfully removed moisture as a contributing variable. Mean annual temperature and elevation were highly collinear ($r^2 = 0.96$). Often relationships were "cleaner" (higher r^2) when regressed on elevation rather than on WorldClim temperature. Elevation may be a better proxy for average temperature at a site. The adiabatic lapse rate insures a constancy of elevation - temperature relationship within a region [21] that may match or exceed the accuracy of WorldClim estimates. Thus, elevation was used as a proxy for temperature in this study.

Within sites, number of workers was approximately normal following log(n+1) transformation. Average worker abundance, the main dependent variable in this study, was the mean of the 100 log(n+1)-transformed values for a site.

Species richness was the number of species in the 100 miniWinkler samples, using only the subset of identified taxa (a "taxonomic sufficiency" approach; see [22]). Observed species richness was used instead of rarefaction to a common sample coverage (*sensu* [23]) because sample coverage was above 0.93 in all cases and the observed richnesses were highly collinear with

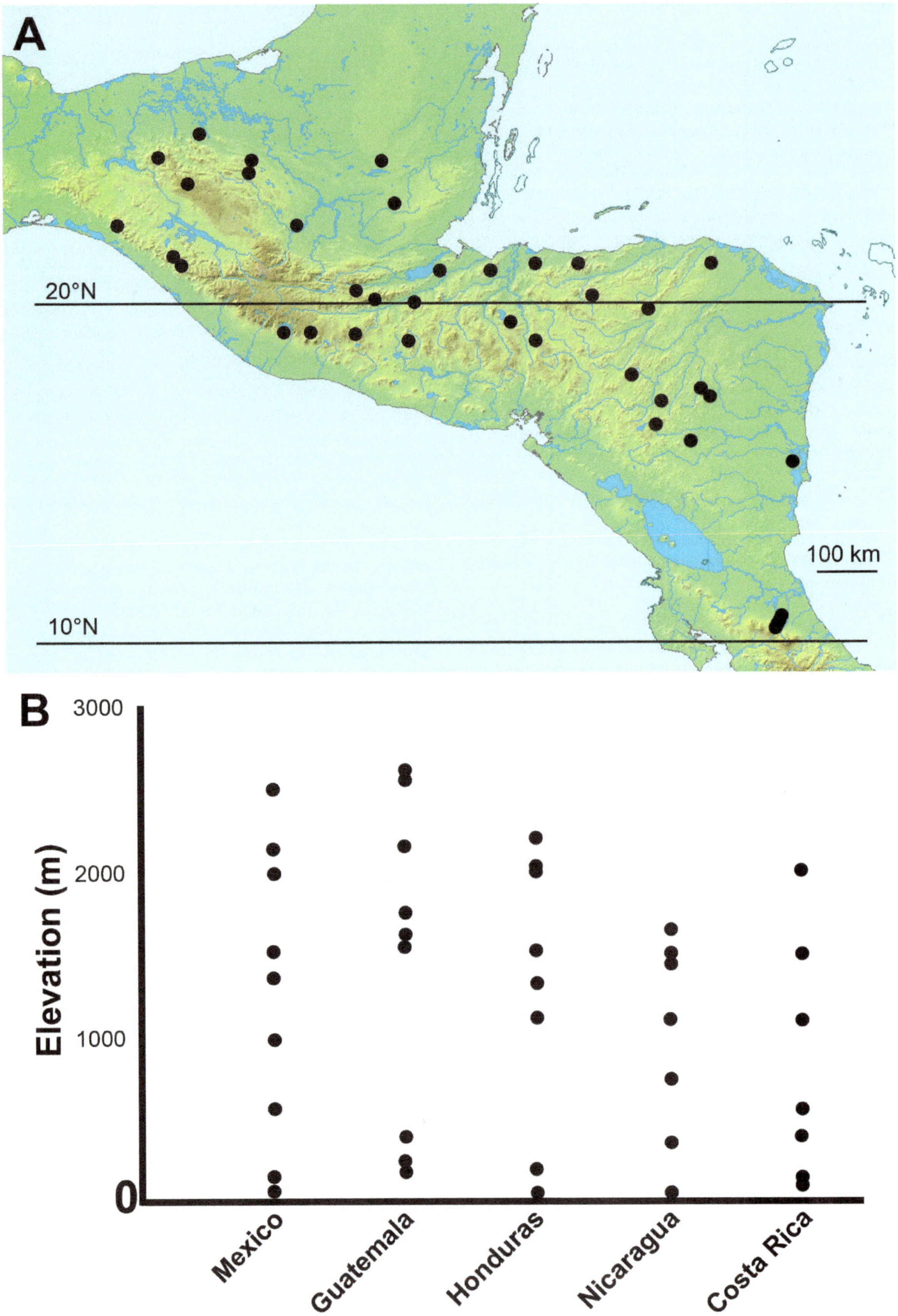

Figure 1. Geographic (A) and elevational (B) distribution of study sites.

Table 1. Sample sites for leaf litter ants in Central America.

Country	Locality	Latitude	Longitude	Elevation (m)
Mexico	La Sepultura	16.160	−93.605	1360
Mexico	El Triunfo low	15.721	−92.951	1520
Mexico	El Triunfo high	15.710	−92.929	2140
Mexico	Coapilla	17.176	−93.132	1990
Mexico	Huitepec	16.747	−92.490	2500
Mexico	Metzabok	17.127	−91.630	570
Mexico	Nahá	16.964	−91.593	985
Mexico	Salto de Agua	17.515	−92.296	70
Mexico	Playón de la Gloria	16.160	−90.901	160
Guatemala	Sierra de las Minas	15.084	−89.946	2560
Guatemala	Biotopo el Quetzal	15.212	−90.215	1750
Guatemala	La Unión	14.947	−89.276	1550
Guatemala	Montaña Chiclera	15.511	−88.861	195
Guatemala	Tikal	17.002	−89.717	250
Guatemala	Machaquilá	16.446	−89.550	400
Guatemala	Cerro Santiago	14.530	−90.149	2600
Guatemala	Cerro Carmona	14.536	−90.694	2150
Guatemala	Volcán Atitlán	14.549	−91.191	1625
Honduras	La Muralla	15.099	−86.741	1530
Honduras	Sierra de Agalta	14.949	−85.915	2020
Honduras	Cerro Comayagua	14.460	−87.545	2000
Honduras	Azul Meambar	14.871	−87.899	1120
Honduras	Guisayote	14.456	−89.069	2200
Honduras	Cusuco	15.487	−88.234	1330
Honduras	Río Platano	15.664	−84.858	60
Honduras	Lancetilla	15.764	−87.457	30
Honduras	Pico Bonito	15.694	−86.863	200
Nicaragua	Musún	12.961	−85.232	750
Nicaragua	Saslaya low	13.769	−84.984	360
Nicaragua	Saslaya high	13.772	−85.012	1110
Nicaragua	Datanlí El Diablo	13.110	−85.868	1440
Nicaragua	Kilambé	13.570	−85.697	1500
Nicaragua	Cerro Jesús	13.982	−86.189	1650
Nicaragua	Wawashang	12.672	−83.716	30
Costa Rica	Barva Transect	10.416	−84.020	120
Costa Rica	Barva Transect	10.404	−84.039	150
Costa Rica	Barva Transect	10.345	−84.058	400
Costa Rica	Barva Transect	10.317	−84.049	570
Costa Rica	Barva Transect	10.267	−84.083	1100
Costa Rica	Barva Transect	10.236	−84.118	1500
Costa Rica	Barva Transect	10.183	−84.117	2000

rarefied richness. Important to our analyses was the assumption that the relationship between included and excluded taxa was unbiased with respect to elevation. Examination of the seven Barva Transect sites, for which all ant species were sorted to species [18], allowed us to evaluate the impact of using a subset of taxa. Subset species richness varied from 61–89% of total species richness across the seven sites, but the proportion was unrelated to elevation ($r^2 = 0.1$).

Average worker abundance was examined as a function of elevation (a proxy for temperature), species richness, and their interaction using linear models ("lm" and "AIC" functions) in R v. 2.15.1 [24]. We report AICc, which corrects AIC for small sample size effects. At each site, we determined which one of the local species was the most abundant at that site, and computed its local frequency of occurrence in the 100 miniWinkler samples. This "frequency of the most common species" was examined as a

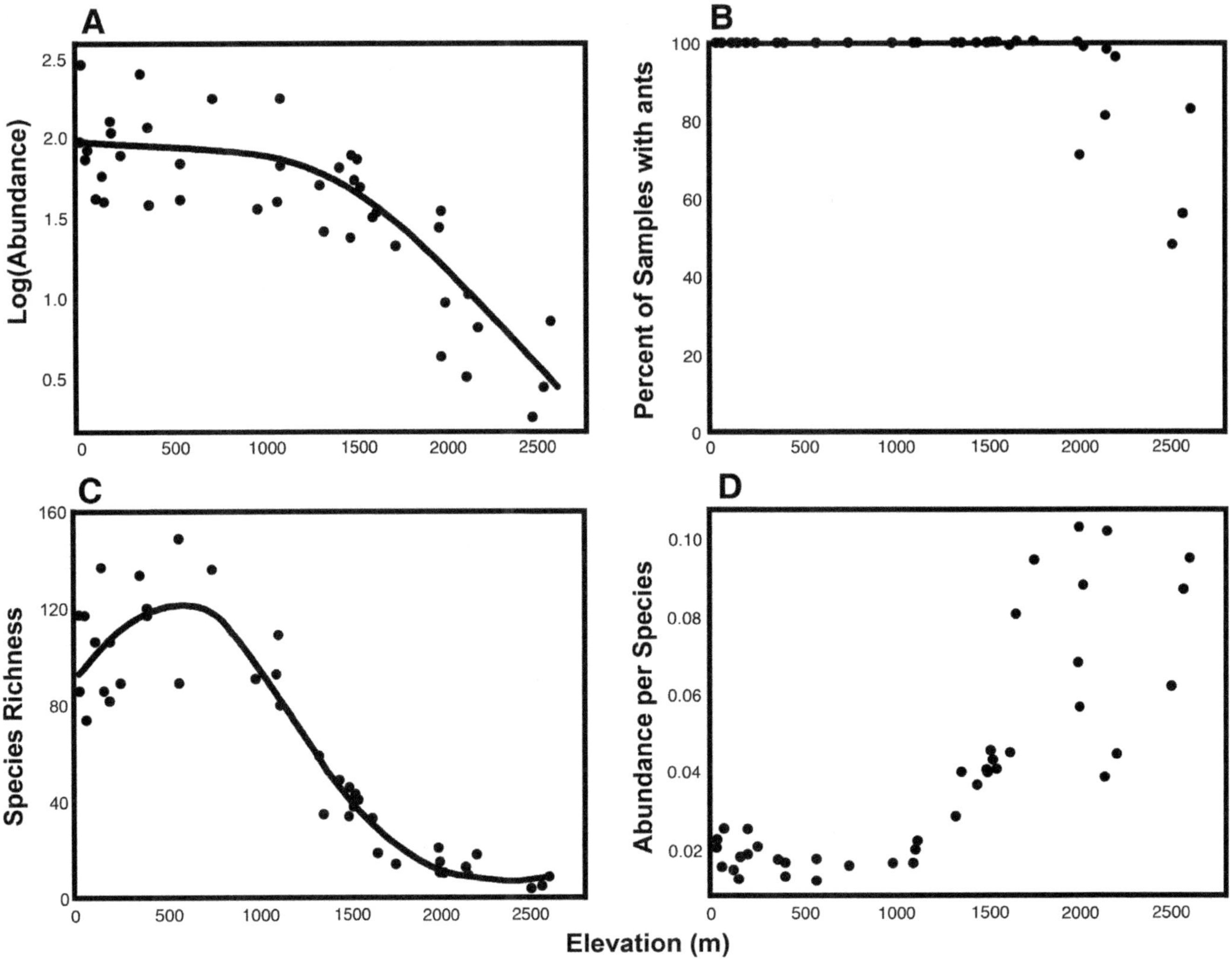

Figure 2. Ant abundance and richness variables as a function of elevation among 41 Central American wet forest sites. A. Log(Abundance) = average of 100 log[n+1] values where n = number of workers in 1 m^2. B. Percent of samples containing ants. C. Species Richness = number of species in 100 samples (taxon subset; see text). D. Abundance per species = Abundance as defined in panel A/Richness as defined in panel C. LOESS curves are included for A and C, using default parameters in R 3.0.2.

function of species richness. An index of abundance per species was calculated as the average worker abundance divided by the total number of species in the 100 miniWinkler samples. It is an index because worker abundance was based on all ants and the species count was based on a subset of ant species (excluding taxonomically problematic genera). Density compensation was inferred by (1) a weak relationship between abundance and species richness, (2) a negative relationship between species richness and the frequency of the most common species, and (3) a positive relationship between abundance per species and elevation.

Results

Average worker abundance very gradually declined from sea level to 1500 m and very steeply declined above 1500 m, with ants becoming extremely rare at the highest sampled sites, 2500–2600 m (Fig. 2A). In all lowland and mid-elevation sites, all or nearly all samples contained ants; whereas in high montane sites, >1500 m, samples showed variable occupancy, with up to 50% lacking ants (Fig. 2B). Species richness showed high values from 0–1000 m, with a peak around 500 m, a sharp drop from 1000–2000 m, and a more gradual decline above 2000 m (Fig. 2C). Abundance per species was flat to 1000 m elevation, rose linearly between 1000 and 1500 m, but then "exploded" above 1500 m, becoming highly variable and unrelated to elevation (Fig. 2D).

The effect of species richness on abundance varied with elevation, as shown by a strong interaction between elevation and richness in linear models (Table 2). Abundance showed a weak positive relationship to species richness in lowland rainforests and mid-elevation cloud forests, and a very strong positive relationship among the high montane forest sites (Fig. 3A). When the ten sites above 1950 m elevation were analyzed separately, the only significant explanatory variable was species richness (Table 2).

The percent of samples containing the single most abundant species at a site varied with species richness (Fig. 3B). A linear model of percent of samples as a function of number of species was not significant, whereas a quadratic model was significant ($p < 0.05$, $r^2 = 0.10$) and with lower AIC value. The same analysis performed on arcsin(squareroot)-transformed frequencies was non-

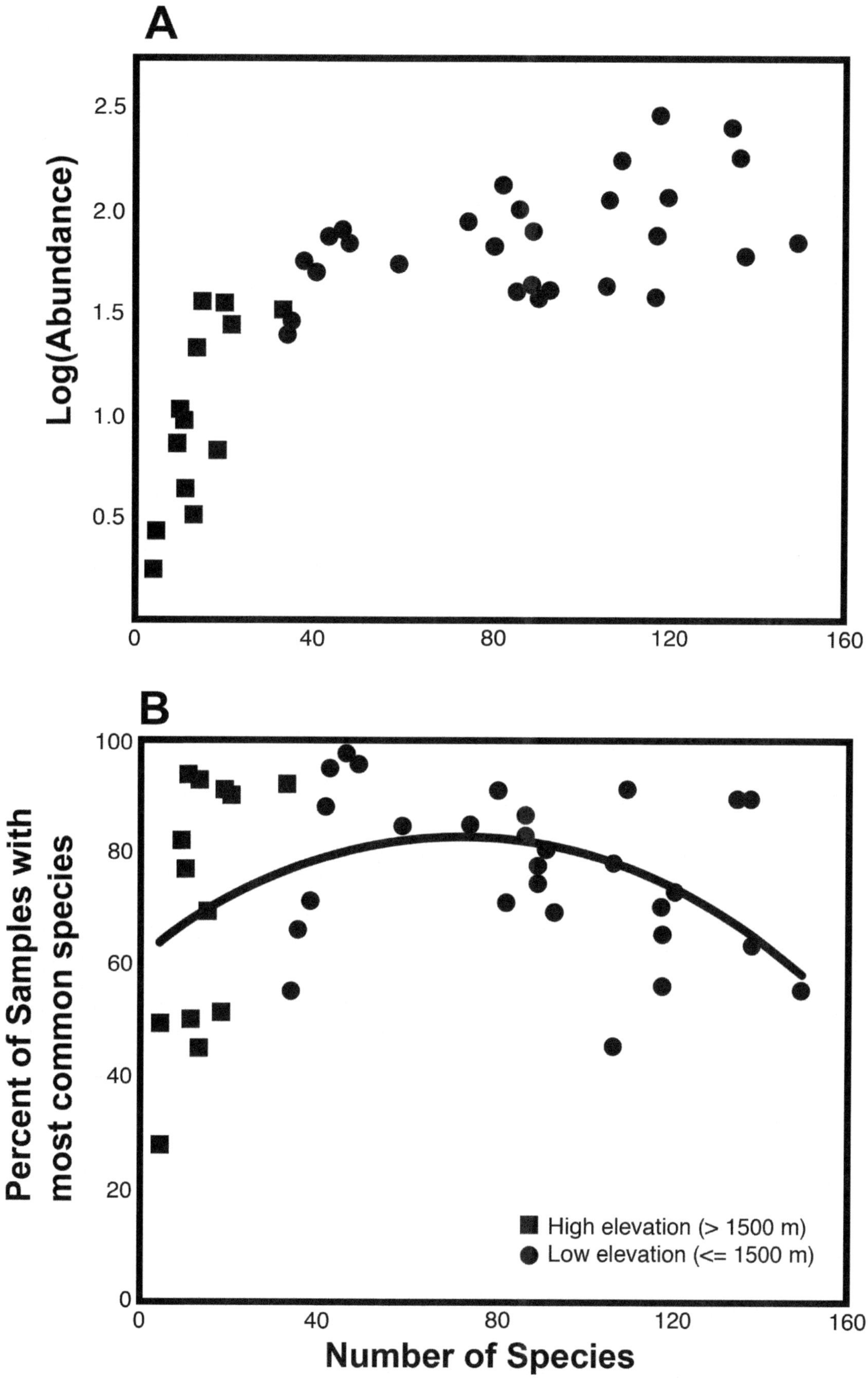
A
Log(Abundance)
2.5
2.0
1.5
1.0
0.5
0
40
80
120
160
B
Percent of Samples with most common species
100
80
60
40
20
0
High elevation (> 1500 m)
Low elevation (<= 1500 m)
Number of Species

Figure 3. Ant abundance variables as a function of species richness among 41 Central American wet forest sites. Number of Species = number in 100 samples (taxon subset; see text). Symbols differentiate low and high elevation sites (elevation rounded to nearest 100 m). A. Log(Abundance) = average of 100 log[*n*+1] values where *n* = number of workers in 1 m^2. B. Percent of samples containing the single most abundant species at each site. The curve shows the fitted quadratic model.

significant ($p<0.065$ for quadratic model) but retained the same trend. The results suggest that among high diversity sites (lowland rainforest and cloud forest), the common species become more common if overall species diversity is lower. In contrast, among high montane sites, the common species decrease in abundance as overall diversity declines.

Discussion

A robust pattern of ant abundance in evergreen wet forests prevails throughout Central America. The average 1 m^2 patch of leaf litter and dead wood on the forest floor is filled with a similar number of ant workers, from steamy lowland rainforest at sea level to cool cloud forest at 1500 m. Above 1500 m there is an abrupt change, and abundance declines to near 0 above 2500 m. This pattern was revealed on one mountainside in Costa Rica [18] and is now shown to be remarkably consistent on similar mountains throughout Central America. The pattern persists whether in continuously forested elevational gradients like the Barva Transect in Costa Rica or in small forest patches surrounded by pine forest, coffee farms, or pastures in Nicaragua, Honduras, Guatemala, and southern Mexico. What factors might explain this pervasive pattern?

The energy limitation hypothesis posits that, for organisms of similar size, abundance of consumers is controlled by available energy, and as a result, abundance will be positively correlated with measures of productivity [25] [16] [2]. Ants provide empirical support for this relationship at a continental scale [1] [2] [14] but not always at local scales [4].

The lack of a productivity - abundance relationship at small scales may be explained by the fact that both producers and consumers are affected by temperature. Brown et al. [3] outlined a metabolic theory of ecology, describing the pervasive effects of metabolism on many traits of ecological communities. They predicted that population density or carrying capacity, K, would follow the relationship:

$$K \propto [R] M^{-3/4} e^{E/kT}$$

where R is a resource supply rate, M is body mass, and the last term is the Boltzmann factor (E is the activation energy of metabolic reactions, k is Boltzmann's constant, and T is absolute temperature in degrees Kelvin). Thus carrying capacity K should increase with supply rate R (the energy limitation hypothesis), decrease with M, and, somewhat paradoxically, decrease with T (at least for plants and ectotherms, where body temperature tracks environmental temperature; [26]). As T increases, all else being equal, metabolism will increase, as will energetic turnover rate, resulting in lower standing biomass. For consumers, the resource supply rate is NPP, which is affected by many factors including stoichiometry, nutrient availability, water, light, and temperature [27]. But NPP is often mainly temperature-limited in mesic sites [27] [1] such as those studied here. In the wet tropics, resource supply rates of all kinds (solar radiation, chemical nutrient availability, water) may be similar across elevations, with only T varying. If plant metabolism is limited by the same Boltzmann factor as consumer metabolism, NPP and temperature effects will act in opposition, so that K may become relatively insensitive to NPP and T. NPP in wet forests is predicted to decline at colder temperatures (higher elevations), and thus the amount of food available for consumers like ants will also decline. However, the ants at those cooler temperatures may lead lives at a slower tempo, keeping the same biomass as lowland ants, but with a lower rate of energy use (see also [28]).

Density compensation patterns similar to those observed on the Barva Transect [18] occurred throughout Central America. Species richness is lower in cloud forests than in lowland rainforests, but ants nevertheless fill up the environment. Average abundance remains high and the most abundant species achieve higher frequencies than they do in lowland rainforests.

An alternative hypothesis we have not fully assessed is that the relative stability of ant abundance is an artifact of an increase in litter depth with elevation. Litter depth and organic matter content increase with elevation in tropical forests [29], and positive correlations of litter depth and ant abundance have been observed in other studies [7] [9]. We may be seeing more ants at higher elevations because we are sampling a greater volume of habitat, whereas ant density per volume is actually decreasing. We cannot fully rule out a litter volume effect, but our sampling method limited the potential impact by imposing a maximum sample volume of 6 L (Methods). When litter volumes were high, only a portion of the litter within a 1 m^2 plot was sifted.

Starting above 1500 m, very different processes are occurring. In high montane forests, abundance sharply declines and is closely related to species richness. There is no evidence of density compensation. A partial explanation for the shift above 1500 m may be the appearance, beginning at about that elevation, of areas unsuitable for any ant species. Within a site, shading causes spatial variation in the thermal environment. Individual points on the forest floor often vary dramatically in the amount of direct solar radiation they receive. One possibility is that at 1500 m and below, all sites are inhabitable by at least one species of ant in the habitat, while above this elevation an increasing proportion of the forest floor is too cold for any ant species to survive. Metabolic theory predicts a stable abundance within habitable space. The abrupt shift to declining worker abundance above 1500 m may simply reflect an increasing proportion of ant-free space in the area sampled.

However, this mechanism by itself does not explain the relationship of abundance to species richness in high montane forests. If habitable space were only a function of temperature, independent of the regional species pool, one would expect abundance to be more closely related to elevation than to species richness. The regional species pool of highland specialists may exhibit a range of cold tolerances, affecting the amount of habitable space. Historical effects of vicariance, dispersal limitation, and age of high mountains may determine the species pool, which in turn determines the amount of habitable space. Hypotheses regarding the effects of ant-free space and regional species pools could be tested with fine-scale microhabitat measurements in montane sites, to see if ant species are tracking particular thermal environments and if montane species vary in cold tolerances.

Table 2. Linear models for effects of elevation (as a proxy for temperature) and species richness on average ant worker abundance at 41 Central American forest sites.

Model	I	R	E	R*E	*p*	r^2	AICc
All sites							
Richness	1.01	0.0089			***	0.56	34.42
Elevation	2.16		−0.0005		***	0.59	31.81
Richness + Elevation	1.68	0.0039	−0.0003		***	0.61	31.53
Richness + Elevation + (Richness*Elevation)	2.24	−0.0032	−0.0007	7.76E-6	***	0.72	18.57
Ten highest sites							
Richness	0.19	0.0563			*	0.45	12.90
Elevation					n.s.	0.28	15.47
Richness + Elevation					n.s.	0.39	18.44

Species richness is the number of species in 100 samples, using a taxon subset of identified ants (see text). Coefficient estimates have column headings I, R, E, R*E for intercept, richness, elevation, and interaction, respectively.

The patterns observed here are emergent properties that arise despite shifting community composition. Longino and Colwell [18] showed sharp transitions in community composition among lowland, cloud forest, and high montane forests. Although not analyzed in detail here, these patterns hold throughout Central America. The lowland rainforest communities show high species and genus-level diversity, with particularly high numbers of species in the hyperdiverse genera *Pheidole* and *Strumigenys*. Common species in the high montane sites are all montane specialists that do not occur in the lowlands. They include a few species in the large genera *Pheidole* and *Strumigenys*, but Central American high montane sites are often dominated by species in the genera *Adelomyrmex* and *Stenamma*. These two genera have clearly diversified and become abundant in the mountains of Central America and are less diverse and less abundant in the lowlands [30] [31]. Mid-elevation cloud forests are intermediate with respect to representation of lowland and highland lineages, but at the species level are distinct from both. There is also strong species turnover latitudinally, with little overlap in species composition between Costa Rica and southern Mexico (Longino et al., unpublished data). More quantitative investigation of phylogenetic structure across these assemblages is clearly needed (e.g., [32]).

Long-term equilibrial processes may be shaping the lowland and cloud forest communities on these gradients. Even with Pleistocene climate fluctuations and large elevational shifts of climate zones, lowland and cloud forest habitats probably remained broadly available and within easy dispersal distance on tropical mountain slopes, allowing community structure to be shaped by competitive interactions over long periods of contact. Is the lower diversity we found in cloud forest habitats driven by equilibrial processes? In ecological time, by the paradoxical prediction of the metabolic theory, does the lower energy availability result in higher minimum population sizes, thereby limiting diversity? In evolutionary time, does lower energy availability increase extinction rates relative to speciation rates? These hypotheses predict stable abundance and diversity relationships over long periods of time. Alternatively, are the patterns stable in ecological time but not in evolutionary time, such that the difference between lowland and cloud forest diversity is due to a lowland head-start during the Tertiary? Are cloud forest faunas younger, the result of planetary cooling during the Pleistocene and the creation of extensive cloud forest habitat? These questions so far remain unanswered.

In contrast to lowland and cloud forest communities, non-equilibrial processes may dominate the high montane communities. A small number of species occur at the highest elevations, and they are constrained to those elevations. Abundance relationships are highly variable and there is no evidence of density compensation at these high elevations. The strongest predictor of ant abundance is the number of species present. Variation in the number of species in these high montane sites is not related to elevation, and thus may be more influenced by stochastic processes of extinction and colonization. High montane sites may be evolutionarily more ephemeral and more island-like, routinely appearing and disappearing through the vagaries of orogeny and, on a shorter time scale, by Pleistocene glacial cycles. In these highest sites, age of habitat and dispersal limitation may be the dominant factors influencing diversity and abundance relationships.

Acknowledgments

We thank the ALAS staff Danilo Brenes, Flor Cascante, Maylin Paniagua, and Ronald Vargas, and TEAM staff Joel Alvarado, Deimer Alvarez, Felix Corrales, Humberto Garcia, Gilbert Hurtado, Johanna Hurtado, Marcos

Molina, and Karol Mora. We thank Project LLAMA co-PI Bob Anderson, staff Laura Sáenz and Jesús Luna Cozar, the 31 student participants, and the many agency staff, reserve managers, and community supporters at our field sites. University of Utah undergraduate Jamie Oswald counted a lot of ants.

Author Contributions

Conceived and designed the experiments: JTL MGB RKC. Performed the experiments: JTL MGB RKC. Analyzed the data: JTL MGB RKC. Contributed reagents/materials/analysis tools: JTL MGB RKC. Wrote the paper: JTL MGB RKC.

References

1. Kaspari M, Alonso L, O'Donnell S (2000) Three energy variables predict ant abundance at a geographical scale. Proceedings of the Royal Society of London Series B-Biological Sciences 267: 485–489.
2. Kaspari M, O'Donnell S, Kercher J (2000) Energy, density, and constraints to species richness: ant assemblages along a productivity gradient. American Naturalist 155: 280–293.
3. Brown JH, Gillooly JF, Allen AP, Savage VM, West GB (2004) Toward a metabolic theory of ecology. Ecology 85: 1771–1789.
4. Sanders NJ, Lessard J-P, Fitzpatrick MC, Dunn RR (2007) Temperature, but not productivity or geometry, predicts elevational diversity gradients in ants across spatial grains. Global Ecology and Biogeography 16: 640–649.
5. Kaspari M, Weiser MD (2012) Energy, taxonomic aggregation, and the geography of ant abundance. Ecography 35: 65–72.
6. Kaspari M (1996) Testing resource-based models of patchiness in four Neotropical litter ant assemblages. Oikos 76: 443–454.
7. Kaspari M, Yanoviak SP (2009) Biogeochemistry and the structure of tropical brown food webs. Ecology 90: 3342–3351.
8. McGlynn TP, Fawcett RM, Clark DA (2009) Litter biomass and nutrient determinants of ant density, nest size, and growth in a Costa Rican tropical wet forest. Biotropica 41: 234–240.
9. Donoso DA, Johnston MK, Clay NA, Kaspari ME (2013) Trees as templates for trophic structure of tropical litter arthropod fauna. Soil Biology and Biochemistry 61: 45–51.
10. Levings SC (1983) Seasonal, annual and among-site variation in the ground ant community of a deciduous tropical forest: some causes of patchy species distributions. Ecological Monographs 53: 435–455.
11. Levings SC, Windsor DM (1984) Litter moisture content as a determinant of litter arthropod distribution and abundance during the dry season on Barro Colorado Island, Panama. Biotropica 16: 125–131.
12. McCain CM (2007) Could temperature and water availability drive elevational species richness patterns? A global case study for bats. Global Ecology and Biogeography 16: 1–13.
13. McCain CM (2010) Global analysis of reptile elevational diversity. Global Ecology and Biogeography 19: 541–553.
14. Kaspari M (2001) Taxonomic level, trophic biology and the regulation of local abundance. Global Ecology and Biogeography 10: 229–244.
15. Kaspari M, Yuan M, Alonso L (2003) Spatial grain and the causes of regional diversity gradients in ants. American Naturalist 161: 459–477.
16. Rosenzweig M (1995) Species Diversity in Space and Time. Cambridge, UK: Cambridge University Press.
17. Grace JB (1999) The factors controlling species density in herbaceous plant communities: an assessment. Perspectives in Plant Ecology, Evolution and Systematics 2: 1–28.
18. Longino JT, Colwell RK (2011) Density compensation, species composition, and richness of ants on a neotropical elevational gradient. Ecosphere 2: art29.
19. Fisher BL (1999) Improving inventory efficiency: A case study of leaf-litter ant diversity in Madagascar. Ecological Applications 9: 714–731.
20. Hijmans RJ, Cameron SE, Parra JL, Jones PG, Jarvis A (2005) Very high resolution interpolated climate surfaces for global land areas. International Journal of Climatology 25: 1965–1978.
21. Fan Y, van den Dool H (2008) A global monthly land surface air temperature analysis for 1948–present. Journal of Geophysical Research: Atmospheres 113: D01103.
22. Groc S, Delabie JHC, Longino JT, Orivel J, Majer JD, et al. (2010) A new method based on taxonomic sufficiency to simplify studies on Neotropical ant assemblages. Biological Conservation 143: 2832–2839.
23. Chao A, Jost L (2012) Coverage-based rarefaction and extrapolation: standardizing samples by completeness rather than size. Ecology 93: 2533–2547.
24. R Core Development Team (2012) R: A language and environment for statistical computing: R Foundation for Statistical Computing, Vienna, Austria.
25. Connell JH, Orias E (1964) The ecological regulation of species diversity. The American Naturalist 98: 399–414.
26. Allen AP, Brown JH, Gillooly JF (2002) Global biodiversity, biochemical kinetics, and the energetic-equivalence rule. Science 297: 1545–1548.
27. Lieth H (1975) Modeling the primary productivity of the world. In: Lieth H, Whittaker RH, editors. Primary Productivity of the Biosphere. New York: Springer. pp. 237–263.
28. Hou C, Kaspari M, Vander Zanden HB, Gillooly JF (2010) Energetic basis of colonial living in social insects. Proceedings of the National Academy of Sciences 107: 3634–3638.
29. Jordan CF (1985) Nutrient Cycling in Tropical Forest Ecosystems. New York, USA: John Wiley & Sons.
30. Branstetter M (2013) Revision of the Middle American clade of the ant genus Stenamma Westwood (Hymenoptera, Formicidae, Myrmicinae). ZooKeys 295: 1–277.
31. Longino JT (2012) A review of the ant genus *Adelomyrmex* Emery 1897 (Hymenoptera, Formicidae) in Central America. Zootaxa 3456: 1–35.
32. Machac A, Janda M, Dunn RR, Sanders NJ (2011) Elevational gradients in phylogenetic structure of ant communities reveal the interplay of biotic and abiotic constraints on diversity. Ecography 34: 364–371.

8

Pulse Increase of Soil N_2O Emission in Response to N Addition in a Temperate Forest on Mt Changbai, Northeast China

Edith Bai[1]*, Wei Li[1,2], Shanlong Li[1,2], Jianfei Sun[1,2], Bo Peng[1,2], Weiwei Dai[1], Ping Jiang[1], Shijie Han[1]

1 State Key Laboratory of Forest and Soil Ecology, Institute of Applied Ecology, Chinese Academy of Sciences, Shenyang, China, **2** College of Resources and Environment, University of Chinese Academy of Sciences, Beijing, China

Abstract

Nitrogen (N) deposition has increased significantly globally since the industrial revolution. Previous studies on the response of gaseous emissions to N deposition have shown controversial results, pointing to the system-specific effect of N addition. Here we conducted an N addition experiment in a temperate natural forest in northeastern China to test how potential changes in N deposition alter soil N_2O emission and its sources from nitrification and denitrification. Soil N_2O emission was measured using closed chamber method and a separate incubation experiment using acetylene inhibition method was carried out to determine denitrification fluxes and the contribution of nitrification and denitrification to N_2O emissions between Jul. and Oct. 2012. An NH_4NO_3 addition of 50 kg N/ha/yr significantly increased N_2O and N_2 emissions, but their "pulse emission" induced by N addition only lasted for two weeks. Mean nitrification-derived N_2O to denitrification-derived N_2O ratio was 0.56 in control plots, indicating higher contribution of denitrification to N_2O emissions in the study area, and this ratio was not influenced by N addition. The N_2O to (N_2+N_2O) ratio was 0.41–0.55 in control plots and was reduced by N addition at one sampling time point. Based on this short term experiment, we propose that N_2O and denitrification rate might increase with increasing N deposition at least by the same fold in the future, which would deteriorate global warming problems.

Editor: Dafeng Hui, Tennessee State University, United States of America

Funding: This work was supported by the Major State Basic Research Development Program of China (973 Program) (2014CB954400 and 2011CB403202) and National Science Foundation of China (31100326). The funders had no role in study design, data collection and analysis, decision to publish, or preparation of the manuscript.

Competing Interests: The authors have declared that no competing interests exist.

* Email: baie@iae.ac.cn

Introduction

Anthropogenic nitrogen (N) has approximately doubled the input of reactive N to the Earth's land surface since the industrial revolution [1,2]. There are three potential fates of anthropogenic N entering terrestrial ecosystem: loss via nitrification and denitrification to the atmosphere as gases, loss via leaching into the aquatic ecosystems, and storage in plants and soils [3]. The gaseous products include NO_x, N_2O, N_2 and NH_3, among which N_2O is a potent greenhouse gas [4] and ozone-depleting agent [5] and has been a major concern for environmental scientists. Although previous studies suggested that many ecosystems have high N retention capacity without evidence of adverse effects of increasing N deposition [6], some other studies found a big increase of N_2O following increase of N inputs [7–9]. Therefore, the response of gaseous emissions to N deposition may be system-specific and depends on the N status [10] and environmental factors of the ecosystem [11].

Soil N_2O flux has been found to be affected by ammonium and nitrate availability [12], water filled pore space (WFPS) [13], soil temperature [14], soil moisture[15], soil pH [16], and carbon (C) availability [17]. When increasing N deposition increases inorganic N availability for nitrification and denitrification, it is expected to increase N_2O emission [18]. However, inorganic N availability may not increase in response to increasing N deposition due to plant uptake and microbial immobilization [6]. In addition, soil pH has been found to decrease in response to increasing N deposition, which may lower N_2O emissions [19]. Decomposition of light fractions of soil C with low density (<1.7 g cm^{-3}) has been found to accelerate under long term N addition [20], which may affect C availability for nitrification and denitrification and thereby N_2O emission.

N_2O is derived from both nitrification and denitrification and it is important to distinguish these processes in order to better understand factors controlling the emission of N_2O [21]. The contribution of the two processes is mainly affected by oxygen availability and forms and availability of inorganic N [22,23]. Therefore, increasing N deposition may impact the ratio of nitrification derived to denitrification derived N_2O. However, very limited data exist on this ratio or the denitrification rate in temperate forest ecosystems [3,24,25].

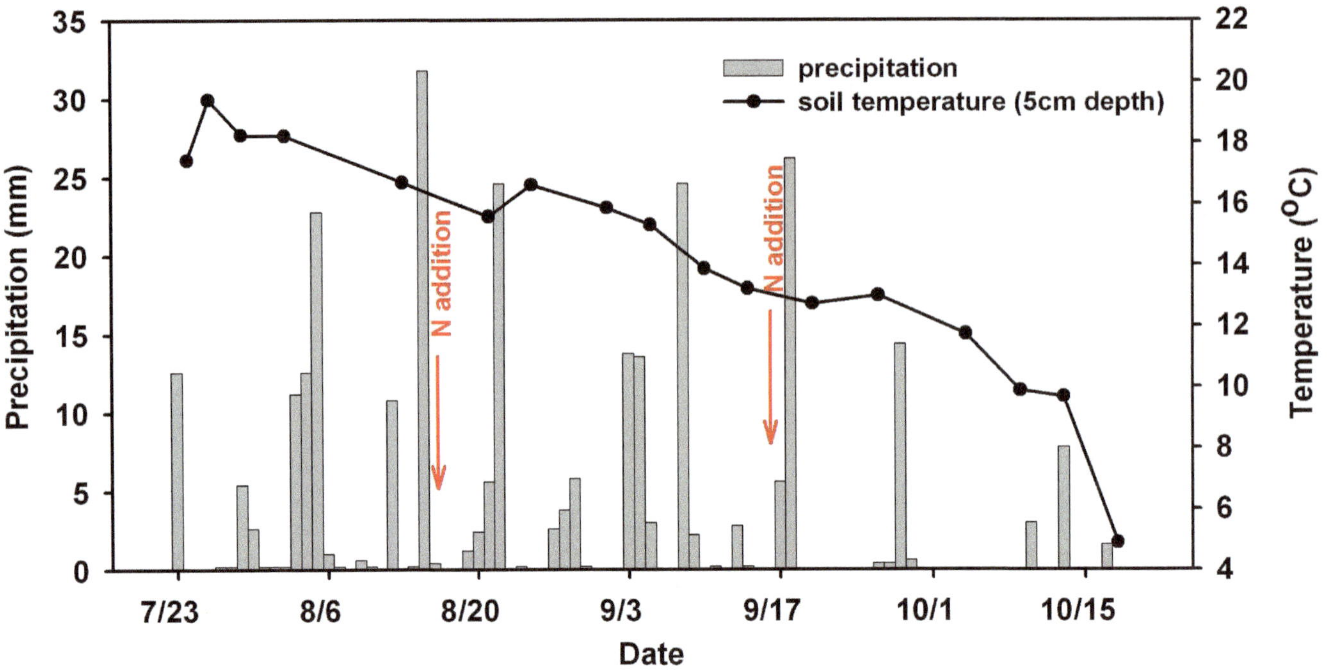

Figure 1. Variations of precipitation and soil temperature (5 cm depth) in the study area during the sampling periods.

This study was carried out in a broad-leaved Korean pine (*Pinus koraiensis* Siebold & Zucc) mixed forest, which is the dominant natural forest type in northeastern China. The system was previously thought to be N-limited but may be experiencing N saturation currently [26]. Hence, it is important to assess how increasing N deposition may affect the N cycling of this forest type. The objectives of our study were: 1) to measure soil N_2O fluxes, ratio of nitrification derived to denitrification derived N_2O, and denitrification rate, for the first time in the study area; 2) to examine the treatment effect of N addition on these fluxes; 3) to investigate the influences of environmental factors on these variables. The hypotheses of this study are: 1) N addition would increase soil N_2O emissions and denitrification rate; 2) soil temperature, moisture, WFPS, pH, and soil texture play an important role in soil N_2O emissions and denitrification.

Materials and Methods

The study was carried out within the research forest of Changbai Forest Ecosystem Research Station (CBFERS) established in 1979 (127°38' E, 41°42' N). No specific permits were required for the described field studies. Research activities within that research forest do not need any specific permissions from any government levels, but need to inform Professor Han (Director CBFERS, co-author of the present paper). The field studies did not involve endangered or protected species.

Study area

The study region is characterized by a typical temperate climate, with long and cold winters, and warm summers. Mean annual temperature is 3.6 °C, with the highest temperature in July, and the lowest in January. Mean annual precipitation is 745 mm, mainly falling between May and September. The growing season is from Jun. to Sep., with mean temperature at 16.7 °C. Natural Korean pine and broad-leaved mixed forest is distributed from approximately 750 to 1100 m asl. The soil is a dark brown soil developed from volcanic ash (Albic Luvisol).

Experimental design

Present N deposition rate in China is 12.89–63.53 kg N ha^{-1} yr^{-1}, with higher values in eastern China [27]. It is predicted that in 2050, many areas in eastern China will have a deposition rate of

Table 1. Soil characteristics of treatment plots.

	pH	Clay (%)	Bulk Density (g/cm^3)	WFPS (%)	Soil T (°C)	Soil moisture (kg H_2O/kg dry soil)	NH_4^+ (mg N/kg soil)	NO_3^- (mg N/kg soil)	Inorganic N (mg N/kg soil)
Control	5.48±0.33[a]	31.67±4.40[a]	0.581±0.013[a]	70.48±15.88[a]	14.25±3.67[a]	94.81±21.99[a]	12.50±9.79[a]	19.11±23.87[a]	31.62±22.39[a]
N-treated	5.47±0.09[a]	30.98±1.11[a]	0.587±0.113[a]	76.07±20.15[a]	14.25±3.67[a]	101.90±26.94[a]	10.40±8.93[a]	22.56±26.26[b]	32.96±22.47[b]

Soils were collected 17 times from Jul. to Oct. 2012. Values are reported as mean ± standard deviation. WFPS: water-filled pore space. T: temperature. Sampling time was used as the within subject factor and N treatment was used as the between subject factor in repeated measures ANOVA. Different letter represents significant difference (p = 0.05) caused N treatment.

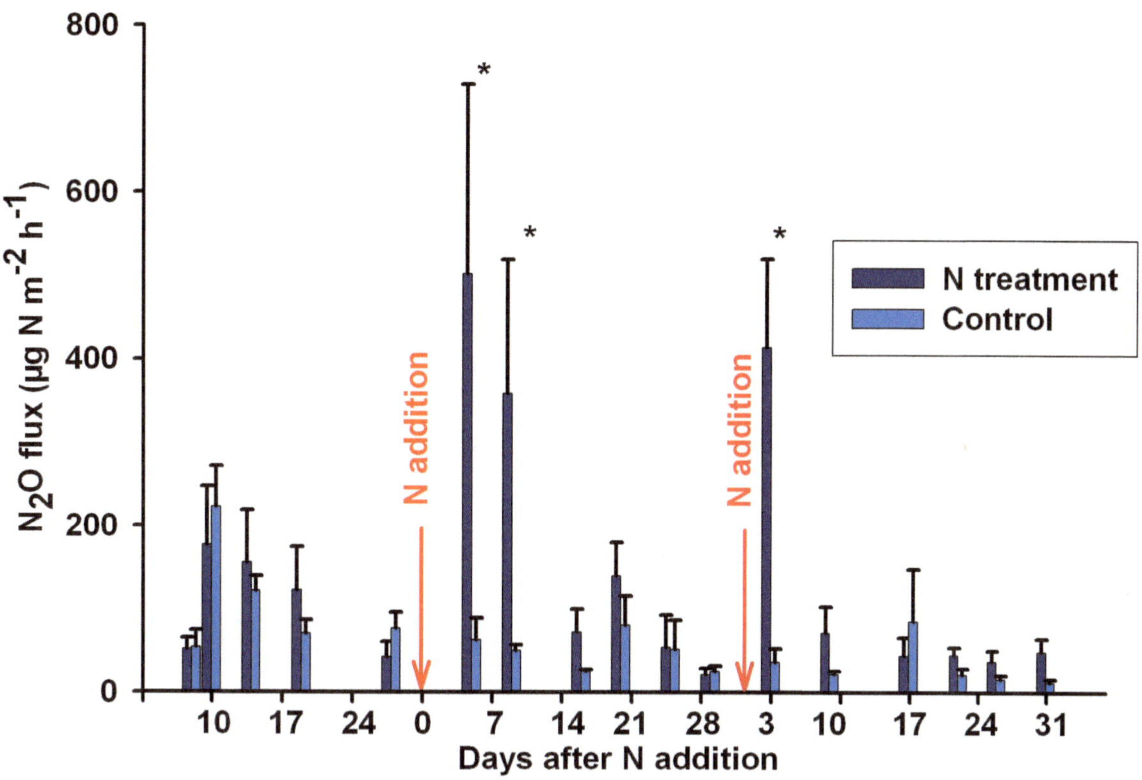

Figure 2. Effects of N addition on N_2O fluxes (n = 3). Arrows represent N addition on that date. Error bars are standard errors. * represents significant effect at p = 0.05.

50 kg N ha^{-1} yr^{-1} [1,28]. Current N deposition rate in the study area is 23 kg N ha^{-1} yr^{-1} [26], and we chose a N addition rate of 50 kg N ha^{-1} yr^{-1} in the treatment as a reasonable estimate of future change. Six sampling plots (25 m×50 m) were established: three replicates for both control and N addition treatment. Plots were separated by 10 m buffer strips. Vegetation was homogeneous based on visual observation.

Starting in 2008, NH_4NO_3 solution has been evenly sprayed for N addition treatment monthly from May to September at a rate of 10 kg N ha^{-1} (starting on May 17th), giving an annual N addition rate of 50 kg N ha^{-1} yr^{-1}. The same volume (40 L) of water was sprayed on the control plots.

Gas sampling and analysis

Soil N_2O emission was measured using closed chamber method [29]. A steel-made chamber (0.5 m×0.5 m×0.5 m) with a septum for gas sampling was placed on the base every time gases were sampled. The bases were installed into the soil one month before initial gas sampling to avoid disturbance on soils. Water was used to seal the connection between the chamber and the base. The chamber was covered with thermal insulation cotton to avoid temperature increase in the chamber. Fans were equipped in the chamber to increase air circulation. Gas sample was collected from each chamber every 30 mins for 2 hours and stored in air bags for measurement.

We sampled for three N addition periods in year 2012, 5–6 times for each period. Samples were taken once every 5–6 days. Given that NH_4NO_3 was sprayed on the day 17th each month, gas samples were collected in three periods: Period one (Jul. 17th–Aug. 16th): 7, 9, 13, 18, and 27 days after N addition; Period two (Aug. 17th–Sep. 16th): 4, 8, 15, 19, 24, and 28 days after N addition; Period three (Sep. 17th to Oct. 18th): 3, 9, 17, 22, 26, and 31 days after N addition. Samples were taken between 9:00 and 11:00 hours and between 15:00 and 17:00 hours because during these two time slots the flux can represent daily average [30]. N_2O concentration was measured by gas chromatography (HP 5890-II). The temperature of the chromatographic column was 55°C and the carrying gas was high purity N_2.

Acetylene inhibition method [31,32] was used to determine denitrification rates and to differentiate sources of N_2O based on the following principle: high concentration of acetylene (10 kPa) could inhibit both nitrification and the reduction of N_2O to N_2 by denitrifiers, so that the otherwise emitted N_2 remained as N_2O. When the acetylene concentration was low (10–100Pa), only nitrification was inhibited [33]. Emitted N_2O could be from three sources: N_2O_{den} (N_2O produced by denitrification); N_2O_{N2}(additional N_2O produced by denitrification, instead of normal N_2 production); N_2O_{nit} (N_2O produced by nitrification). In each sampling plot, soils were incubated *in situ* in three tightly sealed bottles injected with different concentrations of C_2H_2 (purified by H_2SO_4 and distilled water): 0 (nothing inhibited), 60 Pa (nitrification inhibited), and 10 kPa (nitrification and the reduction of N_2O to N_2 inhibited). Correspondingly, measured N_2O of the three incubation bottles were N_2O_{nit}+N_2O_{den}, N_2O_{den}, and N_2O_{den}+N_2O_{N2}, respectively. Therefore, nitrification-derived N_2O, denitrification-derived N_2O, and denitrification-derived N_2

Table 2. Pearson's correlation coefficients between N_2O fluxes and soil characteristics.

Parameter	N_2O µg N/m^2/h	Soil M (kg H_2O/kg dry soil)	WFPS (%)	NH_4^+ (mg N/kg soil)	NO_3^- (mg N/kg soil)	Inorganic N (mg N/kg soil)	pH	Clay (%)	Soil T (°C)
N_2O	1	−0.108	−0.028	−0.103	0.225*	0.210*	0.050	−0.072	0.253*
Soil M		1	0.797*	−0.023	0.079	0.079	0.099	0.169	−0.113
WFPS			1	−0.021	0.144	0.153	0.085	0.023	−0.109
NH_4^+				1	−0.475*	−0.116	−0.015	−0.027	0.016
NO_3^-					1	0.929*	−0.042	−0.014	0.460*
Inorganic N						1	−0.054	−0.027	0.526*
pH							1	0.727*	0.000
Clay								1	0.000

WFPS: water-filled pore space. T: temperature. M: moisture. * represents significant correlation at p = 0.05.

can be calculated from these three measured N_2O fluxes. Denitrification rate is equal to N_2O_{den}+N_2O_{N2}.

60 Pa C_2H_2 was used for low concentration treatment as recommended in previous experiment in the study area [34]. This concentration has also been used in other experiments [35], although 10 Pa is more widely used.

Although the acetylene inhibition technique has not always been satisfactory due to incomplete or non-specific inhibition [36], at present it is still a widely-used method for obtaining a reasonable estimate of denitrification rate and ratio of nitrification to denitrification derived N_2O [3,37–40]. We used an *in situ* incubation method to reduce some potential bias [41].

Three sampling points were selected within each sampling plot. Nine soil cores (5 cm diameter × 10 cm depth) were sampled by PVC tubes at each point. There were 25 holes (0.6 cm diameter) on each PVC tube to ensure acetylene circulation. At each sampling point, every three soil cores were assigned to one of the three incubation bottles. Therefore, we had three soil cores in each incubation bottle at each plot. Before the incubation, the headspace air was replaced by purified C_2H_2. The bottles were then buried *in situ* and 60 ml headspace samples were taken after 24 hours incubation and analyzed for N_2O concentration by gas chromatography (HP 5890-II). Gas samples were collected on Aug. 18th, Aug. 30th, Sep. 10th, Sep. 16th, Sep. 23rd, Sep. 30th, Oct. 7th, and Oct. 15th, 2012.

Soil sampling and analysis

Air temperature and soil temperature at 5 cm depth were recorded *in situ* at each sampling time of both closed chamber experiment and acetylene inhibition experiment using TP3001 thermometer (Boyang, China) (Fig. 1). Precipitation was recorded at the local weather station in the Changbai Mountain Nature Reserve (Fig. 1). Bulk density (BD) in each sampling plot (n = 3) was determined by the core method [42].

One soil sample (0–10 cm) in each plot was collected at each sampling time to determine pH, soil moisture content, and soil NH_4^+ and NO_3^-. Standard oven-drying method was used to measure soil gravimetric water content, which was expressed as kg H_2O/kg dry soil. WFPS was calculated based on Eqn.1:

$$WFPS = \text{soil gravimetric water content} \times BD/PS \qquad (Eqn.1)$$

where PS is soil pore space and is calculated from Eqn.2 [43]:

$$PS = (1\text{-}BD/2.65) \times 100 \qquad (Eqn.2)$$

20 g fresh soil was passed through 2 mm screen and extracted by 50 mL 2 M KCl. The solution was shaken for 1 hour and filtered through rapid filtering paper. Soil inorganic N in extracts was determined by Futura continuous flow analysis system (Alliance, France).

Soil pH was measured on a 1:2.5 soil solution in de-ionized water using a pH meter. Soil texture was determined by the pipette-sedimentation method [44].

Statistical analysis

Repeated measures ANOVA was used to examine the effects of N addition using N treatment method as the between subject factor and sampling time as the within subject factor. All analyses were carried out in SPSS [45].

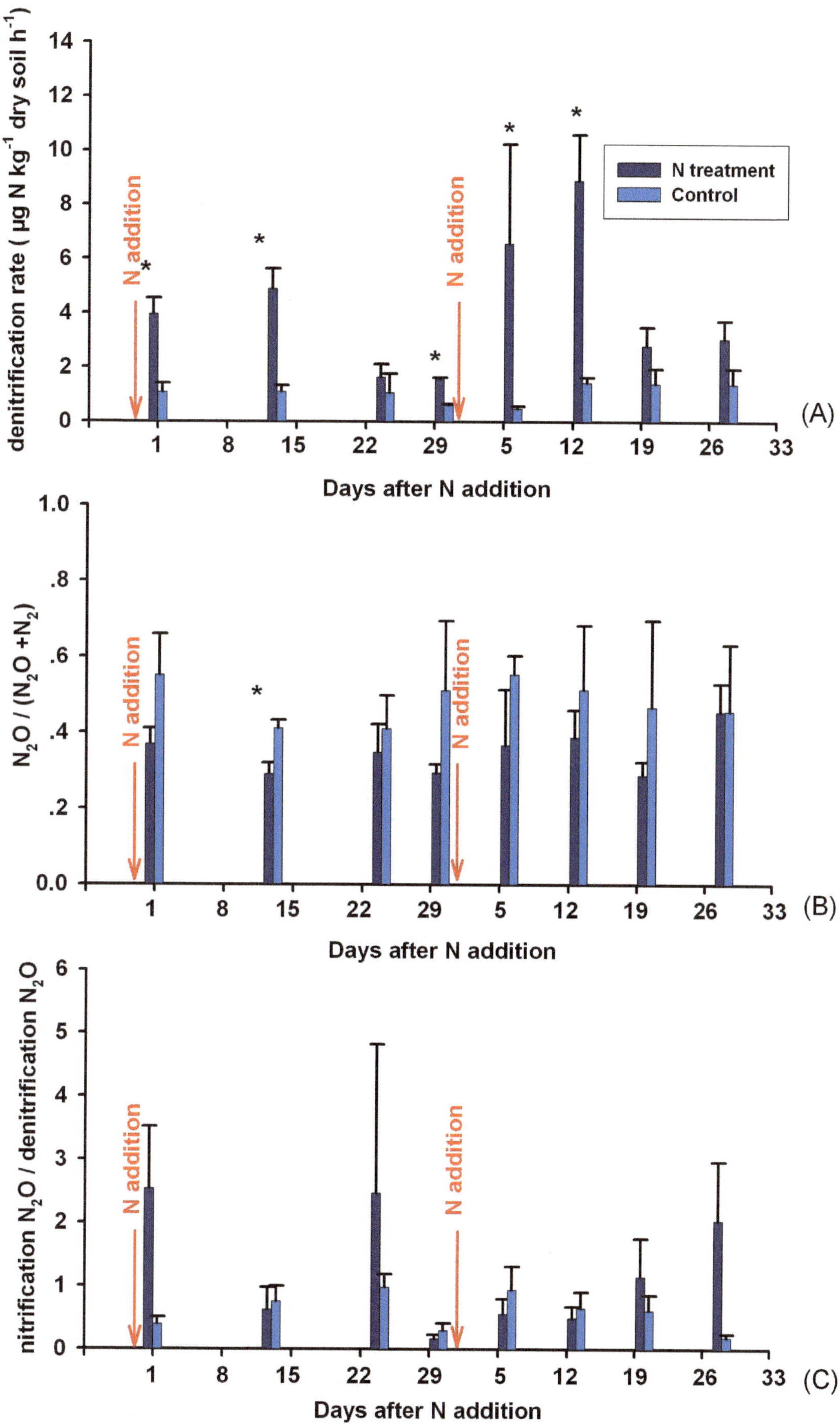
denitrification rate (μg N kg-1 dry soil h-1)
14
12
10
8
6
4
2
0
N addition
N addition
*
*
*
*
*
N treatment
Control
1
8
15
22
29
5
12
19
26
33
(A)
Days after N addition
N2O / (N2O +N2)
1.0
.8
.6
.4
.2
0.0
N addition
N addition
*
1
8
15
22
29
5
12
19
26
33
(B)
Days after N addition
nitrification N2O / denitrification N2O
6
5
4
3
2
1
0
N addition
N addition
1
8
15
22
29
5
12
19
26
33
(C)
Days after N addition

Figure 3. Effect of N addition on denitrification rate (A), N_2O: (N_2O+N_2) ratio (B), and nitrification-derived N_2O: denitrification-derived N_2O ratio (N_2O_{nit}:N_2O_{den}) (C) (n = 3). Arrows represent N addition on that date. * represents significant effect at p = 0.05.

Results

N_2O fluxes

Soil characteristics of the sampling plots were tested 17 times from Jul. to Oct. 2012 and mean results were reported in Table 1 (values were reported as mean ± standard deviation). Mean soil pH was 5.48±0.33 for control plot and 5.47±0.09 for N-treated plot. Soil clay content was 31.67±4.40% and 30.98±1.11% for control and N-treated plot, respectively. Soil bulk density was also similar between control and N-treated plots (0.581±0.013 g cm^{-3} and 0.587±0.113 g cm^{-3} respectively). Soil moisture and WFPS were not significantly different between N-treated plots (101.90±26.94 kg H_2O kg^{-1} dry soil and 76.07±20.15% respectively) and control plots (94.81±21.99 kg H_2O kg^{-1} dry soil and 70.48±15.88% respectively). Mean soil NO_3^- concentration was significantly increased in N-treated plots (22.56±26.26 mg N kg^{-1} soil) compared to control plots (19.11±23.87 mg N kg^{-1} soil) based on repeated measures ANOVA, but mean soil NH_4^+ concentration was not changed (Table 1).

N-treated plots had significantly higher N_2O fluxes right after N addition for period 2 and 3 (Fig. 2). For example, four days after N addition in period 2, mean N_2O emission in N-treated plots reached 501.74 μg N m^{-2} h^{-1}, which was significantly higher than that in control plots (63.31 μg N m^{-2} h^{-1}). This difference between N-treated and control plots declined with time and became minimal on the 15th day after N addition. Similarly, for the next sampling period, mean N_2O emission in N-treated plots was 414.07 μg N m^{-2} h^{-1} three days after N addition, compared with 36.88 μg N m^{-2} h^{-1} in control plots. Only 17 days later, N_2O flux was similar in N-treated (44.13 μg N m^{-2} h^{-1}) and control (85.38 μg N m^{-2} h^{-1}) plots.

Pearson's correlation analysis showed that N_2O emission was positively correlated with soil NO_3^- (r = 0.225), soil total inorganic N (r = 0.210), and soil temperature (r = 0.253) (Table 2). N_2O emission first increased with WFPS then decreased, and the highest N_2O flux rate was observed at 80% WFPS. Soil pH was positively correlated with soil clay content (r = 0.727) (Table 2).

Sources of N_2O and denitrification flux

One day after N addition in August, denitrification rate was significantly higher in N-treated plots (3.95 μg N kg $soil^{-1}$ h^{-1}) than in control plots (0.37 μg N kg $soil^{-1}$ h^{-1}) (Fig. 3a). The effect lasted till the end of the treatment cycle. The same phenomenon was observed for the next sampling period, except that the effect of N addition was not statistically significant on the last two sampling dates (Fig. 3a). The N_2O to (N_2+N_2O) ratios in N-treated plots (0.29–0.45) were all lower than those in control plots (0.41–0.55), although the effect was not statistically significant (Fig. 3b). N-additions on Aug. 17th and Sep. 17th did not affect this pattern (Fig. 3b). We did not find a general pattern for the temporal variation of the N_2O_{nit}: N_2O_{den} ratio or a significant effect of N addition on the ratio (Fig. 3c). The N_2O_{nit}: N_2O_{den} ratio in N-treated plots had a wider range (0.17–4.47) than that in control plots (0.19–0.94).

Correlation analysis indicated both denitrification rate and N_2 flux had a significantly positive correlation with WFPS (r = 0.344 and 0.353 respectively). Neither the N_2O to (N_2+N_2O) ratio nor the N_2O_{nit}: N_2O_{den} ratio correlated with WFPS or any other measured soil variables (Table 3).

Discussion

N_2O emission and its sources

N_2O emission fluxes observed in the control plots were between 0.77 and 382.75 μg N m^{-2} h^{-1}, which mostly is within the range of soil N_2O emission in a 40-year-old pine forest ecosystems in the USA (3–424 μg N m^{-2} h^{-1}) [46]. Our results suggested N addition (50 kg N ha^{-1} yr^{-1}) could increase N_2O fluxes to up to 501 μg N m^{-2} h^{-1} and the monthly sum of N_2O emission in N-treated plots were five-times higher than that in control plots (Fig. 2). The positive correlation between N input rate and the magnitude of N_2O emissions has been found in previous studies [9,11,47]. Previously, Bowden et al. [48] reported N addition at 50–150 kg N/ha/yr only increased N_2O emission from 0.2–1.1 to 0.7–5.2 μg N m^{-2} h^{-1} in the pine forest plantation of Harvard Forest, US, which is much lower than our results. However, our results are consistent with those observations that showed a big increase of N_2O following increase of N inputs [7–9]. Our speculation on the discrepancy is that Harvard Forest in 1990s was N-limited and thereby could retain added N efficiently, while our forest in Changbai Mountain is experiencing "kinetic N saturation" currently [49], with lower N sinks than the N input rate.

N addition effect on N_2O emissions diminished quickly with time after N addition, because added NH_4^+ and NO_3^- were also subject to plant uptake, microbial immobilization, and leaching (Fig. 2). We found the increase of N_2O emission after N addition only lasted for two weeks. Scheer et al. [7] also observed similar "N_2O emission pulses" which lasted for 7 days after N-fertilizer application in combination with irrigation events in a cotton field. This temporal variation suggests more frequent sampling is necessary to capture the dynamics of the emission pulse following N addition.

Precipitation and soil texture, which determines WFPS, may play an important role in soil N_2O emission. Positive correlations between WFPS and N_2O were usually observed because the contribution of denitrification-derived N_2O increases with WFPS [50,51]. Other potential reasons include: microbes were water-limited [52]; rain may disrupt physical aggregate and increase soil organic matter exposure for mineralization and denitrification [53]; and water may alleviate diffusional constraints [54,55]. However, when WFPS is very high (>80%), those factors were not limiting and N_2O fluxes may decline due to their further reduction to N_2 [56,57]. Because our mean WFPS mainly varied between 55% and 100% (Table 1), N_2O emissions increased with WFPS first and then decreased, and the highest N_2O flux rate was observed at WFPS = 80%. We speculate this was because N_2O from both nitrification and denitrification was high at this water and oxygen level.

We found N_2O fluxes were positively correlated with soil temperature (Table 3), which is consistent with many previous findings [58–60]. Both nitrification-derived N_2O [61] and denitrification-derived N_2O [62,63] have been found to be higher at lower pH in acid forest soils. However, nitrification-derived N_2O has more often been found to increase with pH [19]. Our N addition treatment did not affect soil pH (Table 1) and N_2O fluxes were not related to soil pH in our study (Table 2).

Table 3. Pearson's correlation coefficients between soil characteristics and denitrification rate, N_2 emission rate, $N_2O/(N_2O+N_2)$, and nitrification-derived N_2O: denitrification-derived N_2O (N_2O_{nit}:N_2O_{den}) ratios.

	Denitrification µg N/kg soil/h	N_2 µg N/kg soil/h	$N_2O/(N_2O+N_2)$	N_2O_{nit}:N_2O_{den}	WFPS (%)	NO_3^- (mg N/kg soil)	Inorganic N (mg N/kg soil)	pH	Clay (%)	Soil T (°C)	Soil M (kg H_2O/kg dry soil)
Denitrification	1	0.965*	−0.306*	−0.035	0.344*	0.202	−0.001	0.018	−0.092	0.030	0.273
N_2		1	−0.379*	−0.013	0.353*	0.191	0.006	0.054	−0.098	0.049	0.144
$N_2O/(N_2O+N_2)$			1	−0.173	−0.054	−0.131	−0.088	−0.107	−0.050	−0.045	−0.122
N_2O_{nit}:N_2O_{den}				1	−0.054	−0.061	0.045	0.008	0.094	0.029	0.196
WFPS					1	0.249	−0.028	0.131	0.013	−0.237	0.672*
NO_3^-						1	0.659*	0.019	−0.115	0.136	0.137
Inorganic N							1	0.072	−0.017	0.523*	0.022
pH								1	0.727*	0.000	0.105
Clay									1	0.000	0.177
Soil T										1	−0.225

WFPS: water-filled pore space. T: temperature. M: moisture. * represents significant correlation at p = 0.05.

The N_2O_{nit}: N_2O_{den} ratio was not affected by N addition (Table 3). The N_2O_{nit}: N_2O_{den} ratio in N-treated plots had a larger variation (0.17–4.47) than that in control plots (0.19–0.94). Mean N_2O_{nit}: N_2O_{den} ratio was 0.56 and 1.46 in control and N-treated plots respectively, which is lower than what Papen & Butterback-Bahl [47] found (2.33) in spruce and beech forest ecosystem in Germany. The WFPS in our area was high (Table 1), therefore more N_2O was emitted from denitrification. The N_2O_{nit}: N_2O_{den} ratio was not significantly correlated with any soil factors we examined (Table 3), which suggested that the effects of soil moisture, temperature, WFPS, and texture on N_2O_{nit}: N_2O_{den} ratio may be non-linear, or negligible.

Denitrification fluxes and the N_2O to (N_2+N_2O) ratio

Denitrification in the control plots were between 0.126 and 5.460 µg N kg^{-1} soil h^{-1} with a mean at 1.581 µg N kg^{-1} soil h^{-1} (Fig. 3). Bulk density of the study site was between 0.53 and 0.67 g cm^{-3} (Table 1) with a mean at 0.60 g cm^{-3}. Assuming denitrification keeps at the high flux rate during the growing season (May-September) and there is no denitrification in the other months [64], denitrification in the region was roughly estimated to be 0.346 g N m^{-2} yr^{-1}. This is close to the upper end of the modelled average value (0.156–0.368 g N m^{-2} yr^{-1}) in global temperate and boreal forest ecosystems by Bai et al. [65]. *In situ* denitrification data from temperate forest are very limited. Dannenmann et al. [66] reported a denitrification rate of 0.185–0.622 g N m^{-2} yr^{-1} in a mountainous beech forest using seasonally weighted means, although their direct up-scaling of plot means got much higher values (1.4–9.37 g N m^{-2} yr^{-1}). Wolf & Brumme [62] got 0.088–0.351 g N m^{-2} yr^{-1} *in situ* denitrification rates from beech forest floor and mineral soils. Our estimation is comparable to these above reported values.

N addition significantly increased denitrification rates to a mean of 5.539 µg N kg^{-1} soil h^{-1} (Table 3, Fig. 3a). With the current N deposition rate at 23 kg N ha^{-1} yr^{-1} [26], an N addition of 50 kg N ha^{-1} yr^{-1} (2.2 fold) has caused approximately a 2.5 fold increase in denitrification. Therefore, we predict that with the increase of N deposition, denitrification will increase correspondingly, at least by the same fold.

For the two gaseous products of denitrification, more N_2 is emitted after N addition compared with N_2O because the N_2O to (N_2+N_2O) ratio was lower in N-treated plots (Fig. 3b). The N_2O to (N_2+N_2O) ratios were between 0.41–0.55 and 0.29–0.45 in control and N-treated plots respectively. The N_2O to (N_2+N_2O) ratio has been used to estimate denitrification rate because N_2O was easier to measure [3,67]. Our data in control plots were close to the mean value in natural soils (0.43–0.56) compiled by Schlesinger [3]. Increased NO_3^- has been found to increase the N_2O to (N_2+N_2O) ratio because of the greater affinity of NO_3^- as terminal electron acceptor [68–70]. It is not clear why N addition slightly reduced the N_2O to (N_2+N_2O) ratio in our study, and further exploration is needed.

WFPS was positively correlated with N_2 emission, but not with N_2O fluxes (Table 2, Table 3). This is because the relationship between WFPS and N_2 fluxes is unidirectional, while bell-shaped relationship between WFPS and N_2O is more common [13].

In conclusion, N addition did not change soil pH or soil bulk density in the study area. Mainly due to the increase of substrate availability, N addition significantly increased N_2O and denitrification fluxes. However, the effect only lasted for two weeks. Based on this short term experiment, we propose that denitrification rate might increase at least by the same fold as the increase of N deposition in the future. N_2O may also increase with increasing N deposition in our studied forest, deteriorating global warming problems. Future studies on the underlying mechanism [71] behind increasing N_2O emissions following N addition are needed to better understand the processes contributing to increasing N_2O emissions.

Author Contributions

Performed the experiments: WL SL JS BP WD PJ. Analyzed the data: WL. Contributed reagents/materials/analysis tools: EB. Contributed to the writing of the manuscript: EB WL WD. Designed the experiment: EB SH.

References

1. Galloway JN, Dentener FJ, Capone DG, Boyer EW, Howarth RW, et al. (2004) Nitrogen cycles: past, present, and future. Biogeochemistry 70: 153–226.
2. Gruber N, Galloway JN (2008) An Earth-system perspective of the global nitrogen cycle. Nature 451: 293–296.
3. Schlesinger WH (2009) On the fate of anthropogenic nitrogen. Proc Natl Acad Sci USA 106: 203–208.
4. Ehhalt D, Prather M, Dentener F, Derwent R, Dlugokencky E, et al. (2001) Atmospheric chemistry and greenhouse gases. In: J. T . Houghton, Y . Ding, D. J . Griggs, M . Noguer, P. J. v. d . Linden, X . Dai, K . Maskell and C. A. Johnson, editors. IPCC Report 2001. Cambridge, UK: Cambridge University Press. pp. 241–280.
5. Ravishankara AR, Daniel JS, Portmann RW (2009) Nitrous oxide (N_2O): the dominant ozone-depleting substance emitted in the 21st century. Science 326: 123–125.
6. Bernal S, Hedin LO, Likens GE, Gerber S, Buso DC (2012) Complex response of the forest nitrogen cycle to climate change. Proc Natl Acad Sci USA 109: 3406–3411.
7. Scheer C, Wassmann R, Kienzler K, Ibragimov N, Eschanov R (2008) Nitrous oxide emissions from fertilized, irrigated cotton (Gossypium hirsutum L.) in the Aral Sea Basin, Uzbekistan: Influence of nitrogen applications and irrigation practices. Soil Biol Biochem 40: 290–301.
8. Zhang L, Song C, Wang D, Wang Y (2007) Effects of exogenous nitrogen on freshwater marsh plant growth and fluxes in Sanjiang Plain, Northeast China. Atmos Environ 41: 1080–1090.
9. Butterbach-Bahl K, Breuer L, Gasche R, Willibald G, Papen H (2002) Exchange of trace gases between soils and the atmosphere in Scots pine forest ecosystems of the northeastern German lowlands 1. Fluxes of N_2O, NO/NO_2 and CH_4 at forest sites with different N-deposition. Forest Ecol Manag 167: 123–134.
10. Hall SJ, Matson PA (2003) Nutrient status of tropical rain forests influences soil N dynamics after N additions. Ecol Monogr 73: 107–129.
11. Templer PH, Pinder RW, Goodale CL (2012) Effects of nitrogen deposition on greenhouse-gas fluxes for forests and grasslands of North America. Front Ecol Environ 10: 547–553.
12. Groffman PM, Tiedje JM (1989) Denitrification in north temperate forest soils—spatial and temporal patterns at the landscape and seasonal scales. Soil Biol Biochem 21: 613–620.
13. Davidson EA, Keller M, Erickson HE, Verchot LV, Veldkamp E (2000) Testing a conceptual model of soil emissions of nitrous and nitric Oxides. Bioscience 50: 667–680.
14. Heinen M (2006) Simplified denitrification models: Overview and properties. Geoderma 133: 444–463.
15. Schindlbacher A, Zechmeister-Boltenstern S, Butterbach-Bahl K (2004) Effects of soil moisture and temperature on NO, NO_2, and N_2O emissions from European forest soils. J Geophys Res - Atmos 109: D17302.
16. Stevens RJ, Laughlin RJ, Malone JP (1998) Soil pH affects the processes reducing nitrate to nitrous oxide and di-nitrogen. Soil Biol Biochem 30: 1119–1126.
17. Gregorich EG, Rochette P, Hopkins DW, McKim UF, St-Georges P (2006) Tillage-induced environmental conditions in soil and substrate limitation determine biogenic gas production. Soil Biol Biochem 38: 2614–2628.
18. Morse JL, Ardón M, Bernhardt ES (2012) Using environmental variables and soil processes to forecast denitrification potential and nitrous oxide fluxes in coastal plain wetlands across different land uses. J Geophys Res - Biogeo 117: G02023.
19. Baggs EM, Smales CL, Bateman EJ (2010) Changing pH shifts the microbial sourceas well as the magnitude of N_2O emission from soil. Biol Fert Soils 46: 793–805.
20. Neff JC, Townsend AR, Gleixner G, Lehman SJ, Turnbull J, et al. (2002) Variable effects of nitrogen additions on the stability and turnover of soil carbon. Nature 419: 915–917.
21. Kiese R, Butterbach-Bahl K (2002) N_2O and CO_2 emissions from three different tropical forest sites in the wet tropics of Queensland, Australia. Soil Biol Biochem 34: 975–987.
22. Bollmann A, Conrad R (1998) Influence of O_2 availability on NO and N_2O release by nitrification and denitrification in soils. Global Change Biol 4: 387–396.
23. Davidson EA, Matson PA, Vitousek PM, Riley R, Dunkin K, et al. (1993) Processes regulating soil emissions of NO and N_2O in a seasonally dry tropical forest. Ecology 74: 130–139.
24. Ambus P (1998) Nitrous oxide production by denitrification and nitrification in temperate forest, grassland and agricultural soils. Eur J Soil Sci 49: 495–502.
25. Wolf I, Brumme R (2002) Contribution of nitrification and denitrification sources for seasonal N_2O emissions in an acid German forest soil. Soil Biol Biochem 34: 741–744.
26. Wang C, Han S, Zhou Y, Yan C, Cheng X, et al. (2012) Responses of fine roots and soil N availability to short-term nitrogen fertilization in a broad-leaved Korean pine mixed forest in Northeastern China. PLoS ONE 7: e31042.
27. Lü C, Tian H (2007) Spatial and temporal patterns of nitrogen deposition in China: Synthesis of observational data. J Geophys Res - Atmos 112: D22S05.
28. Liu X, Zhang Y, Han W, Tang A, Shen J, et al. (2013) Enhanced nitrogen deposition over China. Nature 494: 459–462.
29. Hutchinson G, Mosier A (1981) Improved soil cover method for field measurement of nitrous oxide fluxes. Soil Sci Soc Am J 45: 311–316.
30. Alves BJR, Smith KA, Flores RA, Cardoso AS, Oliveira WRD, et al. (2012) Selection of the most suitable sampling time for static chambers for the estimation of daily mean N_2O flux from soils. Soil Biol Biochem 46: 129–135.
31. Tiedje J, Simkins S, Groffman P (1989) Perspectives on measurement of denitrification in the field including recommended protocols for acetylene based methods. Plant Soil 115: 261–284.
32. Ryden JC, Skinner JH, Nixon DJ (1987) Soil core incubation system for the field measurement of denitrification using acetylene-inhibition. Soil Biol Biochem 19: 753–757.
33. Klemedtsson L, Svensson B, Rosswall T (1987) Dinitrogen and nitrous oxide produced by denitrification and nitrification in soil with and without barley plants. Plant Soil 99: 303–319.
34. Xu H, Chen G-x, Li A-n, Han S-j, Huang G-h (2000) Nitrification and denitrification as sources of gaseous nitrogen emission from different forest soils in Changbai Mountain. J Forest Res 11: 177–182.
35. Deboer W, Tietema A, Gunnewiek P, Laanbroek HJ (1992) The chemolithotrophic ammonium-oxidizing community in a nitrogen-saturated acid forest soil in relation to pH-dependent nitrifying activity. Soil Biol Biochem 24: 229–234.
36. Felber R, Conen F, Flechard CR, Neftel A (2012) Theoretical and practical limitations of the acetylene inhibition technique to determine total denitrification losses. Biogeosciences 9: 4125–4138.
37. Yamamoto A, Akiyama H, Naokawa T, Miyazaki Y, Honda Y, et al. (2014) Lime-nitrogen application affects nitrification, denitrification, and N_2O emission in an acidic tea soil. Biol Fert Soils 50: 53–62.
38. Zhong L, Du R, Ding K, Kang X, Li FY, et al. (2014) Effects of grazing on N_2O production potential and abundance of nitrifying and denitrifying microbial communities in meadow-steppe grassland in northern China. Soil Biol Biochem 69: 1–10.
39. Boulêtreau S, Salvo E, Lyautey E, Mastrorillo S, Garabetian F (2012) Temperature dependence of denitrification in phototrophic river biofilms. Sci Total Environ 416: 323–328.
40. Stief P, Poulsen M, Nielsen LP, Brix H, Schramm A (2009) Nitrous oxide emission by aquatic macrofauna. Proc Natl Acad Sci USA 106: 4296–4300.
41. Hatch DJ, Jarvis SC, Philipps L (1990) Field measurement of nitrogen mineralization using soil core incubation and acetylene inhibition of nitrification. Plant Soil 124: 97–107.
42. Burke W, Gabriels D, Bouma J (1986) Soil Structure Assessment. Rotterdam, Netherlands: A. A. Balkema Publishers. 92 p.
43. Parton WJ, Holland EA, Del Grosso SJ, Hartman MD, Martin RE, et al. (2001) Generalized model for NO_x and N_2O emissions from soils. J Geophys Res - Atmos 106: 17403–17419.
44. Gee GW, Bauder JW (1986) Particle-size analysis. In: A . Klute, editor editors. Methods of soil analysis, part I: physical and mineralogical methods. Madison, Wisconsin, USA: American Society of Agronomy, Inc. and Soil Science Society of America, Inc. pp. 383–411.
45. SPSS Inc. (Chicago, IL) SPSS for Windows, Version 11.5.
46. Goodroad LL, Keeney DR (1984) Nitrous oxide emission from forest, marsh, and prairie ecosystems. J Environ Qual 13: 448–452.
47. Papen H, Butterbach-Bahl K (1999) A 3-year continuous record of nitrogen trace gas fluxes from untreated and limed soil of a N-saturated spruce and beech forest ecosystem in Germany: 1. N2O emissions. J Geophys Res - Atmos 104: 18487–18503.
48. Bowden RD, Melillo JM, Steudler PA, Aber JD (1991) Effects of nitrogen additions on annual nitrous-oxide fluxes from temperate forest soils in the northeastern United-States. J Geophys Res 96: 9321–9328.
49. Lovett G, Goodale C (2011) A new conceptual model of nitrogen saturation based on experimental nitrogen addition to an oak forest. Ecosystems 14: 615–631.
50. Ruser R, Flessa H, Russow R, Schmidt G, Buegger F, et al. (2006) Emission of N_2O, N_2 and CO_2 from soil fertilized with nitrate: effect of compaction, soil moisture and rewetting. Soil Biol Biochem 38: 263–274.
51. Kachenchart B, Jones DL, Gajaseni N, Edwards-Jones G, Limsakul A (2012) Seasonal nitrous oxide emissions from different land uses and their controlling factors in a tropical riparian ecosystem. Agric, Ecosyst Environ 158: 15–30.
52. Fierer N, Schimel JP, Holden PA (2003) Influence of drying–rewetting frequency on soil bacterial community structure. Microbial Ecol 45: 63–71.

53. Goebel M-O, Bachmann J, Woche SK, Fischer WR (2005) Soil wettability, aggregate stability, and the decomposition of soil organic matter. Geoderma 128: 80–93.
54. Schjønning P, Thomsen IK, Moldrup P, Christensen BT (2003) Linking soil microbial activity to water- and air-phase contents and diffusivities. Soil Sci Soc Am J 67: 156–165.
55. Petersen SO, Schjønning P, Thomsen IK, Christensen BT (2008) Nitrous oxide evolution from structurally intact soil as influenced by tillage and soil water content. Soil Biol Biochem 40: 967–977.
56. Ciarlo E, Conti M, Bartoloni N, Rubio G (2007) The effect of moisture on nitrous oxide emissions from soil and the $N_2O/(N_2O+N_2)$ ratio under laboratory conditions. Biol Fert Soils 43: 675–681.
57. Parton WJ, Mosier AR, Ojima DS, Valentine DW, Schimel DS, et al. (1996) Generalized model for N_2 and N_2O production from nitrification and denitrification. Global Biogeochem Cy 10: 401–412.
58. Li K, Gong Y, Song W, Lv J, Chang Y, et al. (2012) No significant nitrous oxide emissions during spring thaw under grazing and nitrogen addition in an alpine grassland. Global Change Biol 18: 2546–2554.
59. Hénault C, Bizouard F, Laville P, Gabrielle B, Nicoullaud B, et al. (2005) Predicting *in situ* soil N_2O emission using NOE algorithm and soil database. Global Change Biol 11: 115–127.
60. Del Grosso SJ, Parton WJ (2012) Climate change increases soil nitrous oxide emissions. New Phytol 196: 327–328.
61. Martikainen PJ, de Boer W (1993) Nitrous oxide production and nitrification in acidic soil from a dutch coniferous forest. Soil Biol Biochem 25: 343–347.
62. Wolf I, Brumme R (2003) Dinitrogen and nitrous oxide formation in beech forest floor and mineral soils. Soil Sci Soc Am J 67: 1862–1868.
63. Weslien P, Klemedtsson AK, Borjesson G, Klemedtsson L (2009) Strong pH influence on N_2O and CH_4 fluxes from forested organic soils. Eur J Soil Sci 60: 311–320.
64. Mogge B, Kaiser E-A, Munch J-C (1999) Nitrous oxide emissions and denitrification N-losses from agricultural soils in the Bornhöved Lake region: influence of organic fertilizers and land-use. Soil Biol Biochem 31: 1245–1252.
65. Bai E, Houlton BZ, Wang YP (2012) Isotopic identification of nitrogen hotspots across natural terrestrial ecosystems. Biogeosciences 9: 3287–3304.
66. Dannenmann M, Butterbach-Bahl K, Gasche R, Willibald G, Papen H (2008) Dinitrogen emissions and the N_2:N_2O emission ratio of a Rendzic Leptosol as influenced by pH and forest thinning. Soil Biol Biochem 40: 2317–2323.
67. Seitzinger SP (1988) Denitrification in freshwater and coastal marine ecosystems: ecological and geochemical significance. Limnol Oceanogr 33: 702–724.
68. Weier KL, Doran JW, Power JF, Walters DT (1993) Denitrification and the dinitrogen/nitrous oxide ratio as affected by soil water, available carbon, and nitrate. Soil Sci Soc Am J 57: 66–72.
69. Cho CM, Burton DL, Chang C (1997) Kinetic formulation of oxygen consumption and denitrification processes in soil. Can J Soil Sci 77: 253–260.
70. Gillam KM, Zebarth BJ, Burton DL (2008) Nitrous oxide emissions from denitrification and the partitioning of gaseous losses as affected by nitrate and carbon addition and soil aeration. Can J Soil Sci 88: 133–143.
71. Butterbach-Bahl K, Baggs EM, Dannenmann M, Kiese R, Zechmeister-Boltenstern S (2013) Nitrous oxide emissions from soils: how well do we understand the processes and their controls? Philos T Roy Soc B 368: 20130122.

9

Differential Responses of Herbivores and Herbivory to Management in Temperate European Beech

Martin M. Gossner[1,2]*, Esther Pašalić[1], Markus Lange[1,3], Patricia Lange[1], Steffen Boch[4], Dominik Hessenmöller[3], Jörg Müller[5], Stephanie A. Socher[4], Markus Fischer[4], Ernst-Detlef Schulze[3], Wolfgang W. Weisser[1,2]

1 Friedrich-Schiller-University of Jena, Institute of Ecology, Jena, Germany, **2** Technische Universität München, Terrestrial Ecology Research Group, Department of Ecology and Ecosystem Management, Center for Food and Life Sciences Weihenstephan, Freising-Weihenstephan, Germany, **3** Max Planck Institute for Biogeochemistry, Jena, Germany, **4** University of Bern, Institute of Plant Sciences, Bern, Switzerland, **5** University of Potsdam, Institute of Biochemistry and Biology, Potsdam, Germany

Abstract

Forest management not only affects biodiversity but also might alter ecosystem processes mediated by the organisms, i.e. herbivory the removal of plant biomass by plant-eating insects and other arthropod groups. Aiming at revealing general relationships between forest management and herbivory we investigated aboveground arthropod herbivory in 105 plots dominated by European beech in three different regions in Germany in the sun-exposed canopy of mature beech trees and on beech saplings in the understorey. We separately assessed damage by different guilds of herbivores, i.e. chewing, sucking and scraping herbivores, gall-forming insects and mites, and leaf-mining insects. We asked whether herbivory differs among different forest management regimes (unmanaged, uneven-aged managed, even-aged managed) and among age-classes within even-aged forests. We further tested for consistency of relationships between regions, strata and herbivore guilds. On average, almost 80% of beech leaves showed herbivory damage, and about 6% of leaf area was consumed. Chewing damage was most common, whereas leaf sucking and scraping damage were very rare. Damage was generally greater in the canopy than in the understorey, in particular for chewing and scraping damage, and the occurrence of mines. There was little difference in herbivory among differently managed forests and the effects of management on damage differed among regions, strata and damage types. Covariates such as wood volume, tree density and plant diversity weakly influenced herbivory, and effects differed between herbivory types. We conclude that despite of the relatively low number of species attacking beech; arthropod herbivory on beech is generally high. We further conclude that responses of herbivory to forest management are multifaceted and environmental factors such as forest structure variables affecting in particular microclimatic conditions are more likely to explain the variability in herbivory among beech forest plots.

Editor: Hans Henrik Bruun, University Copenhagen, Denmark

Funding: The work has been funded by the DFG Priority Program 1374 "Infrastructure-Biodiversity-Exploratories" (WE 3081/21-1)(www.exploratories.de). In addition, it was supported by the German Research Foundation (DFG; www.dfg.de) and the Technische Universität München within the funding programme Open Access Publishing (http://www.ub.tum.de/open-access-publikationsfonds-der-tum). The funders had no role in study design, data collection and analysis, decision to publish, or preparation of the manuscript.

Competing Interests: The authors have declared that no competing interests exist.

* Email: martin.gossner@tum.de

Introduction

Land use is a main driver of global biodiversity change [1]. Because ecosystem processes are mediated by interactions between organisms the ecosystem harbours, any change in biodiversity is likely to also affect the processes mediated by these organisms [1]. Thus, understanding the links between land use, biodiversity and ecosystem processes is currently a major field of research in ecology [2]. Arthropods make up most of the metazoan biodiversity in terrestrial ecosystems [3] and play an important role for ecosystem functioning [4]. One such process is arthropod herbivory. A plant-eating life-style is particularly prevalent among insects worldwide, with about 400,000 species of herbivores [5], but it is also common among other arthropod taxa such as mites (e.g. gall-forming mites). Arthropod herbivory may affect a number of other ecosystem processes such as the movement of water from soil to the atmosphere [6] and nutrient dynamics [7,8]. While arthropod herbivory is considered to be generally low in temperate forest tree canopies, i.e. $<5\%$ of leaf area eaten, except in outbreak situations [9], it may have significant effects on processes even at endemic levels [10] such as affecting plant community composition through affecting competitive interactions among plants [11]. Furthermore, arthropod herbivores also have much potential to mediate effects that cascade up and down trophic chains in ecosystems [12].

In forests, there are various ways in which forest management may affect the composition of the arthropod community and consequently also processes mediated by arthropods such as herbivory. Forestry influences tree species composition, which is a main factor for the composition of the herbivore community [13–15] and may also affect herbivory [16–18]. This has already been shown for temperate European forests (e.g., [16]), but forest management effects within forests that are dominated by the same tree species but are managed differently are less clear. Such management effects might be mainly mediated by changes in forest structure which is a generally strong driver of herbivory in forest ecosystems [19]. Forestry, also affects tree age distribution and thus horizontal and vertical structuring of the forest stand and this has been shown to influence the arthropod community through the provision of different niches [20] and herbivore resource selection [19]. Because habitat requirements of arthropods often include particular abiotic factors [21,22], forestry may also alter the arthropod community and herbivory through changes in e.g. light regime, precipitation or temperature. Stand and tree age might additionally affect herbivory by age-related changes in phenology and chemistry [23,24]. It has also been shown that most herbivorous species peak during particular host developmental stages [19]. Finally, any change in the biotic composition of a forest ecosystem may affect the complex species interactions (e.g., [25]), with consequences for the herbivore community and therefore herbivory. Predator populations might, for example, more strongly suffer from reduced structural complexity in managed forests and top-down control might thus be less effective (enemy hypothesis, [26]).

In Central Europe, European Beech *Fagus sylvatica* L. is the dominant deciduous tree species that still covers ca. 14–15 Mio ha, despite heavy logging since the Roman era and afforestation with species such as spruce [27]. Most of these beech forests are, however, managed and even beech forests nowadays classified as unmanaged have been disturbed by humans at some time in the past [28]. In high forests (that is, forests regenerated from seedlings and managed mainly for timber production) forest managers generally distinguish between two silvicultural practices: uneven-aged and even-aged forest management. The two practices result in forest stands differing in vertical and horizontal spatial structure. Uneven-aged forest management (also termed selection system/ cutting) aims at continuously providing timber at the stand level. Therefore it comprises different forest developmental stages at very high spatial grain through selective harvesting as well as selective thinning of tree individuals. Characteristics of selection system forests are a non-closed canopy in the uppermost layer, a high degree of canopy roughness through layering and a high variability in the age/size distribution of trees, with young/small trees dominating in abundance and old/mature trees dominating in biomass. In contrast, even-aged forest management (age-class forestry) aims at continuously providing timber at the forest enterprise level (that is on landscape level). The trees of age-classed forest stands belong to more or less confined age cohorts. Vertical structure of even-aged stands is thus much lower than in uneven-aged stands and tree density and canopy openness highly depends on the developmental stage. In contrast to these two general forest management approaches, unmanaged primeval beech forests are characterised by small-scale disturbances [29,30], which result in a fine-grained mosaic of developmental stages [31]. As the shade tolerant beech is able to efficiently occupy small gaps, natural beech forests tend to build up pure forests with less than 5% mixed species [32,33]. However, the unmanaged forests considered in this study not yet reached the structure of primeval beech forests.

In the present study we aimed to test for general effects of beech forest management on arthropod herbivory caused by a number of particular species, by contrasting unmanaged, selection cutting, and age-class forest stands of different developmental stages. We selected three regions in Germany characterised by different geology, climate and forest history which might lead to diverging management effects on herbivory. Moreover we assessed herbivory in two different strata, i.e. in the sun-exposed beech canopy and on understorey beech saplings. Beech shows leaf-dimorphism between sun and shade leaves [34] with known consequences for particular herbivore species [35]. Because different herbivorous species themselves might differ in their response to forest management and forest management might cause differential responses of leaf traits in both strata, different strata need to be assessed to test for general management effects.

First, we asked whether herbivory differs between managed and unmanaged forests. We predicted that herbivory is generally higher in the less-complex managed forests. We further predict that this pattern is independent of region and forest stratum investigated due to similar underlying mechanisms. The strength of the relationship is, however, expected to vary, e.g. because abiotic conditions in the sun-exposed canopy change less conspicuously in response to forest management than in the understorey, and because some herbivores such as less mobile mites which live in open galls might be more susceptible to a change in light and hence moisture regime than other herbivores that either develop in more strongly protected galls and mines or are mobile and thus can escape unfavourable conditions [36]. Moreover free-living herbivores such as chewers and suckers might more strongly depend on non-food resources such as dead wood etc. as protection against predators and climatic extremes than concealed living herbivores such as miners and gall-inducers [37]. As these resources are expected to be more abundant in unmanaged forests, stronger management effects on free-living herbivores might be expected.

Second, we asked whether selection cutting forestry might decrease the level of herbivory compared to age-class forestry because it is assumed those uneven-aged forests have a higher structural complexity than even-aged forests. We predict an intermediate herbivory level between age-class and unmanaged forests in the understorey as well as in the canopy.

Third, we asked whether within age-class forests herbivory changes across age-classes. We predict significant difference among beech developmental stages. These differences are expected to differ among herbivore species due to adaptations to leaf chemistry [19] and this should be more pronounced in specialists such as gall-inducers. We generally predict a decreasing herbivory with age because fewer resources are invested in growth and thus more can be allocated to defensive functions [38]. Moreover, stand density decreases with age and this might affect herbivores by, first, increasing exposure of herbivores to antagonists while dispersing from one tree to another, resulting in decreasing herbivory with age in all herbivore groups, second, decreasing herbivore survival due to increased exposure to climatic extremes and changing plant traits, resulting in decreasing herbivory with age, in particular in herbivore groups that are more exposed but less mobile such as gall mites, and third changing resource allocation and thus changing investment in defence due to two contrasting mechanisms; decreased competition with an expected decrease in herbivory with age and increasing drought stress with an expected increase in herbivory with age (e.g. [9,38]).

Materials and Methods

Ethics statements

Field work permits were issued by the responsible state environmental offices of Baden-Württemberg, Thüringen, and Brandenburg (according to § 72 BbgNatSchG). The study sites comprise state forests and protected areas such as the National Park Hainich and some nature reserves within the biosphere reserves Schwäbische Alb and Schorfheide-Chorin, as well as in the forest of Keula, Hainich-Dün. During this study no species that are protected by European or national laws were sampled.

Study area and research plots

The study was conducted within the framework of the Biodiversity Exploratory project in three regions of Germany (www.biodiversity-exploratories.de, [2]); Schwäbische Alb (460–860 m a.s.l.) in the South-West (09°10′49″–09°35′54″ E/48°20′28″–48°32′02″ N), Hainich-Dün (285–550 m a.s.l.) in the Central part (10°10′24″–10°46′45″ E/50°56′14″–51°22′43″ N) and Schorfheide-Chorin (3–140 m a.s.l.) in the North-East (13°23′27″–14°08′53″ E/52°47′25″–53°13′26″ N) (detailed information, e.g. on geology and soil types: www.biodiversity-exploratories.de and [2]). The three study regions differ in climatic conditions; in the Schwäbische Alb mean annual temperature is 6.0–7.0°C with a mean annual precipitation of 700–1,000 mm, in Hainich-Dün 6.5–8.0°C (500–800 mm) and in Schorfheide-Chorin 8.0–8.5°C (500–600 mm). The Schwäbische Alb is a highly fragmented, mixed forest landscape dominated by European beech (*Fagus sylvatica*; 46%) and Norway spruce (*Picea abies* (l.) Karst.; 24%). The Hainich-Dün region contains the largest unfragmented forest area in Germany dominated by broad-leaved trees, conifers comprise only 12%. The largest part of the Schorfheide-Chorin is covered by forests of Scots pine (*Pinus sylvestris* L.; 39%), beech (*F. sylvatica*; 12%) and Sessile oak (*Quercus petraea* (Matt.) Liebl.; 9%).

Forest plots (representing different forest stands and thus different management units) in the three regions were selected using a stratified random sampling design [2]. About 500 candidate sites representing major forest types were selected for each region. Surveys of soils, vegetation, and management were conducted in all sites. From the candidate points, 50 one-hectare forest plots were selected across the whole range of forest management intensities on the typical soils in the respective region (Schwäbische Alb: Cambisol/Leptosol; Hainch-Dün: Luvisol/Stagnosol; Schorfheide-Chorin: Cambisol). A number of additional criteria were employed for plot selection, e.g., a distance of at least 200 m between the borders of each plot and replication within one management unit was not tolerated [2]. The stratified random selection of plots was also used to reduce spatial autocorrelation problems.

In this study only plots where beech (*Fagus sylvatica*) occurred in the upper tree layer and comprised >70% cover of the tree layer were chosen for herbivory assessment: 38 plots in the Schwäbische Alb, 46 plots in the Hainich-Dün, and 21 plots in the Schorfheide-Chorin (Table 1, Figure S1). Research plots represented different management intensities. We distinguished between unmanaged and managed forests even though in Central Europe no virgin forests exist and as unmanaged classified forests have been disturbed by humans. Unmanaged stands in the Schwäbische Alb have been unmanaged for several decades, except for small interventions such as removal of spruce trees in order to achieve the aims of protecting the old beech trees. In Hainich-Dün half of the unmanaged plots are located in the core area of the Hainich National Park. The forest is a former coppice with standards forest which remained unmanaged for 50 years. The other unmanaged plots are located in the surrounding area and are unmanaged for 20 years. In Schorfheide-Chorin plots are part of a former royal hunting area and were taken out of management about 30 years ago.

Managed forests were classified according to harvesting strategies: 1) selection cutting forests where only selected tree individuals are harvested resulting in forests with an uneven-age structure (only in Hainich-Dün). To achieve the required heterogeneity not only sawn timber (60–65 cm diameter at breast height (dbh)), but also trees with low diameter are continuously harvested. 2) age-class forests where trees are from one age-cohort (see introduction). Within age-class forests four different developmental stages were considered (Table 1). Thickets are characterised by trees of dbh<7 cm which corresponds to a tree age of < 30 years, pole wood (7 to 15 cm dbh; 30–50 years), and timber (> 15 cm dbh; >90 years). Rotation time for age-class beech forests is around 160 (±30) years. Because of a change in management practices, the younger age-class forests did not originate from clear-cuts with subsequent planting, but included both natural seedling recruitment and the preservation of some mature trees (120–180 years) that were cut only once the young cohort reached an age of 10–20 years. We named this intermediated stage between old timber and thickets 'timber with regeneration'. Some studied thickets of the Schwäbische Alb still contained single mature trees, leading to a slightly higher mean diameter compared to the other two regions (Table 1).

Herbivory assessment

Leaf herbivory was estimated separately in the canopy and in saplings (henceforth 'understorey') of European beech to test for differences between the strata from June to August 2009. To collect leaves from the canopy five trees were selected in each plot by walking a transect northwest-southeast through the plot area and choosing the first five trees in the upper canopy layer high enough to have sun-exposed leaves. Due to different developmental stages studied, tree size varied between stands from approximately eight meters in thickets to maximum 40 meters in old timber stages. Of each tree one small branch with at least 50 leaves was harvested in the southern, sun-exposed part of the upper canopy with the help of a crossbow. A bolt fixed to a fishing line was shot over the selected branch. Then a thicker rope was fixed to the fishing line and pulled over the branch. By pulling on both ends of the rope the branch was broken off so that it fell to the ground. It was important to harvest the branch as high as possible, to ensure that leaves were sun-exposed. The method revealed to be very effective (on average every second shot was successful) and is recommended for future studies when no permanent canopy access technique such as canopy cranes or walkways are available. In total 50 leaves starting from the tip of each branch were assessed for herbivory, resulting in a total of 50 leaves×5 trees = 250 leaves from each forest stand.

In the understorey saplings less than 30 cm and with at least two fully developed leaves were investigated. All leaves represent shade leaves in this stratum. Herbivory was examined in all saplings within two circles with a radius of 1 m in each plot. If necessary the radius of the circles was enlarged (maximum 1 ha plot size) until five individuals were enclosed by the virtual circle. All leaves of each sapling were examined. In some plots, mainly in the thicket and pole wood stage, only few saplings were found. Only plots with at least five saplings were used for analysis (missing plots see Table 1). On average herbivory was assessed on 96 leaves (minimum: 10, maximum: 243) of on average 18 saplings (5, 30) per plot. For analysis, herbivory was assessed at the plot level, by summing all leaves with and without damage per tree and treating tree as random effect in the statistical models (see section 'Data analysis').

Damage types

Each harvested leaf was immediately examined for herbivory damage. We distinguished five main different damage types. For each leaf the occurrence of every damage type was noted. Overall damage was defined as number of leaves with damage. Addition-

Table 1. Number of study plots in which we assessed herbivory of canopy trees and saplings of European beech (*Fagus sylvatica*) and the number of plots included in models 1 to 3 (model 1: 'managed-unmanaged comparison', model 2: 'Hainich-Dün', model 3: 'age-class comparison'; for details see material and methods).

Region	Management type	Development stage	Stand height [m]	Canopy only	Sapling only	Canopy and sapling	Plot set model 1	Plot set model 2	Plot set model 3
Schwäbische Alb	managed (age-class)	thicket	16 (7,38)	6		2	8		8
		pole wood	16 (10,26)	5		2	7		7
		timber	31 (26,40)	1	1*	10	12		11
		timber with regeneration	35 (33,37)			6	6		6
	unmanaged		29 (21,35)			5	5		
				12	**1**	**25**	**38**		**32**
Hainich-Dün	managed (age-class)	thicket	8 (7,10)	5		0	5	5	5
		pole wood	14 (10,19)	3		1	4	4	4
		timber	35 (31,40)			8	8	8	8
		timber with regeneration	35 (31,38)			3	3	3	3
	managed (selection cutting)		34 (29,39)			13		13	
	unmanaged		31 (24,36)			13	13	13	
				8		**38**	**33**	**46**	**20**
Schorfheide-Chorin	managed (age-class)	timber	37 (32,41)	1		6	7		
		timber with regeneration	37 (30, 43)			7	7		
	unmanaged		38 (28,42)	3		4	7		
				4		**17**	**21**		
Sum				**24**	**1**	**80**	**92**	**46**	**52**

*harvesting leaves in the canopy was not allowed in this plot.
As in thickets and pole woods only a few seedlings occurred, in some plots seedling herbivory could not be assessed. Additional mean, minimum and maximum stand height is given. Please note that thickets in the Schwäbische Alb contain a few large trees resulting in a relatively high mean value.

ally the percentage of leaf area removed was assessed. This was estimated for each leaf in eight categories (0%, 0–1%, 1–5%, 6–10%, 11–25%, 26–50%, 51–75%, >75%) by eye using a series of leaf templates that showed examples of damage representing the critical percentages of leaf area removed that distinguished between the different categories. The mean of these categories (0%, 0.5%, 3%, 8%, 18%, 38%, 63%, 88%) were used to illustrate the overall strength of herbivore, but because of the large span of herbivory within especially the last classes we do not statistically analyse these percentage values. Instead, we focus on the number of leaves damaged. Figure S2 illustrates the different damage types.

a) Chewing damage. We found the typical circular chewing damage caused by adult beetles of *Orchestes fagi* (Linnaeus 1758) in spring, and also chewing damage that could be not specified. For analysis, we counted the number of leaves with overall chewing damage and number of leaves with chewing damage caused by *O. fagi*.

b) Scraping damage. Scraping damage (of the epidermis) on deciduous trees is mainly caused by Lepidoptera, Symphyta and Coleoptera larvae. We counted the number of leaves with scraping damage.

c) Damage by gall mites and gall midges. Beech trees were attacked by a number of different gall-inducing species. We distinguished between three different eriophyid gall mites (order Prostigmata) and two gall midges (order Diptera). Gall mites were *Aceria nervisequa* (Canestrini 1891), *Aceria nervisequa faginea* (Nalepa 1920) and *Acalitus stenaspis* (Nalepa 1891). *A. nervisequa* forms haired, often red pigmented galls along the leave veins on the upper side of the leaf. *A. nervisequa faginea* forms white, pannose spots between the leave veins on the undersides of leaves. *A. stenapis* forms a rolled-up leave edge.

Gall midges were *Mikiola fagi* (Hartig 1839) and *Hartigiola annulipes* (Hartig 1839). Galls of *M. fagi* are green and later red, ovate, occur on the upper side of the leaf, and are 4–12 mm long. Galls of *H. annulipes* are cylindrical, occur on the upper side of the leaf, are about 4 mm long, and often haired brownish. In the case that galls were broken or died in an earlier stage, they were still counted, because they could be clearly identified. We counted the number of leaves infected by each species and summed the number of leaves attacked by gall mites or gall midges.

d) Mining damage. Mining arthropods on beech occur in the orders Coleoptera and Lepidoptera. Larvae of the beetle *Orchestes fagi* form a very characteristic mine. The first part of the mine is a gradually widening serpentine tunnel. Later - most often at the tip of the leave - it becomes a large irregular blotch mine. Mines from the leaf-mining moths *Phyllonorycter maestingella* (Müller 1764) and *Parornix fagivora* (Frey, 1861) are similar, they have a long blotch mine usually between two veins from the rachis reaching almost to the leave edge. A clear differentiation between these two species based on mine morphology only is rather difficult. Most observed mines were probably caused by *P. maestingella*, and we refer to the mines caused by either of these two moths as '*Phyllonorycter*-group'. In addition, various species of the moth genus *Stigmella* live on beech where they create an S-shaped mine, usually within the confines of two veins. Mines of the different species cannot be distinguished and were pooled as '*Stigmella* spp.' mines. We counted the number of infested leaves per herbivore taxon and the total number of leaves with mines.

e) Sucking. On each leaf we recorded the presence or absence of sucking damage induced by insects of Hemiptera (e.g. aphids, cicadas and true bugs), recognised by small whitish dots. In addition, the occurrence of the woolly beech aphid *Phyllaphis fagi* (Linnaeus 1767) was noted, identified by the presence of woolly wax on the lower side of the leaves.

Covariates

To obtain more mechanistic insight into relationship between herbivory and forest management, we used a number of explanatory variables of forest structure (assessed by forest inventory) and plant community (vegetation relevés) as covariates in our analyses.

a) Forest inventory. In a core area of each plot a forest inventory was performed [39]. Diameter at breast height (dbh) and tree height were measured for each tree in concentric circles with the radius 12.62 m (for trees >29.9 cm dbh; 500 m^2 when projected to the ground), 7.98 m (for trees >19.9 cm & <30.0 cm dbh; 200 m^2) and 5.64 m (for trees 7.0–19.9 cm dbh; 100 m^2). Within each circle the positions and species identity of all trees (above 7 cm dbh) were recorded. For height measurements the Vertex III-system (Haglof Company Group, Sweden) was used. Dead wood was measured separately for coarse woody debris (CWD, length>50 cm and diameter≥20 cm, m^3/ha) and fine woody debris (FWD, diameter≥5 cm, m^3/ha).

From these parameters, we calculated the variables basal area (m^2/ha), number of trees (dbh>7 cm), the 90th quantile of tree height (m) (henceforth 'stand height') and solid volume (wood/timber greater than 7 cm in diameter, m^3/ha; henceforth 'wood volume') which were used along with CWD and FWD as descriptors of forest structure in the analyses.

b) Vegetation relevés. We sampled vascular plants in spring and late summer of the same year in a 20 m×20 m subplot concentric with the forest inventory circle. We identified all vascular plant species including shrubs and trees and estimated the percentage cover per species separately [for details see 40]. The number of vascular plant species and the cover of tree species> 5 m (henceforth 'cover tree layer') were used for the analyses. In addition, we calculated the Shannon diversity of vascular plants by summing cover data from spring and summer.

Data analysis

All analyses were conducted in R 3.0.2 (www.R-project.org).

a) General procedure. In order to analyse the land-use effects on herbivory, we carried out three different analyses to account for the unbalanced number of different forest management types in the three regions. Depending on the model, we tested for differences among regions, among forest management types and/or developmental stages, and between canopy and sapling herbivory (factor *stratum*). In addition, the interactions between factors were considered. These were treated as fixed effects. Additionally covariates describing forest structure and plant diversity (see paragraphs "Forest inventory" and "Vegetation relevés") were added to the model after testing for correlations (see below). Forest stand was used as random factor in all models to account for pseudoreplication within management units.

We performed generalised linear mixed effects model fit by Laplace approximation for binomial errors, using the function 'glmer' in the lme4 package. Tukey contrasts were used for post-hoc comparisons with the function 'glht' within the multcomp package.

We used 'cbind' function to combine the number of leaves with and without damage per tree for all damage types and thus account for differences in sample size. In this case R adds the two columns together to produce the correct binomial denominator.

b) Selection of covariates. In order to test for the independence of covariates but also to reduce the number of covariates a Principal Component Analysis (PCA) was carried out using the prcomp-function to calculate a singular value decomposition of the centred data matrix, not by using Eigenvalues on the covariance matrix. Because multicollinearity highly influences the outcome of multiple regressions, we preferred an a-priori variable selection based on a PCA on all potential covariates over a step-wise selection procedure in which all variables are used in the initial model. This approach is a commonly used method in ecology (e.g., [41,42]). For each of the three models a separate PCA was computed, using the subset of plot sets selected for the respective model (Table 1). We considered all axes needed to explain at least 70% of total variability among plots. For all models these were the first three axes (for details see File S1).

Wood volume and basal area were highly correlated with axis 1 in all three plot sets (Table S1 in File S1). We choose wood volume instead of basal area as the first covariate for all three models as the most intuitive descriptor of forest biomass.

Axis two was in plot sets one (model 1) and three (model 3) highly correlated to tree number and the cover of the tree layer. We choose tree number as the second covariate for the 'managed-unmanaged-comparison' and 'age-class-comparison'. In plot set two (model 2), axis two was highly correlated with the 90th quantile of tree height, with plant diversity and with the cover of tree layer (Table S1 in File S1). We choose plant diversity as the second covariate for model 2.

PCA axis three was in plots set one highly correlated with plant diversity, and in plot sets two and three with CWD. Thus, we choose plant diversity as third covariate for 'managed-unmanaged-comparison' (model 1) and CWD as third covariate for 'Hainich-Dün' (model 2) and 'age-class-comparison' (model 3).

c) Model details. In the first analysis (**model 1** *'managed-unmanaged comparison'*) we compared unmanaged and managed forests across all regions. Therefore a total of 92 plots in the three regions were considered, 25 unmanaged plots and 67 managed plots (only even-aged age-class forests, Table 1). Uneven-aged selection cutting forests were not included in the analysis because they only occurred in the Hainich-Dün region. The model had the following specification:

$$glmer(y \sim wood\ volume + tree\ number + plant\ diversity + region \times management \times stratum + (1|Plot), family = binomial)$$

In the second analysis (**model 2** *'Hainich-Dün'*) we compared unmanaged, selection cutting and age-class forests in the Hainich-Dün region. A total of 46 plots were included (Table 1). The model had the following specification:

$$glmer(y \sim wood\ volume + plant\ diversity + CWD + management \times stratum + (1|Plot), family = binomial)$$

In the third analysis (**model 3** *'age-class comparison'*) we compared the different developmental stages within the age-class forests. 52 plots of the Schwäbische Alb and the Hainich-Dün were considered (Table 1). We used only data from the canopy herbivory assessment, because many plots of the young developmental stages had no saplings (Table 1). Plots of the Schorfheide-Chorin were excluded, because of the lack of young developmental stages in this region. The model had the following specification:

$$glmer(y \sim wood\ volume + tree\ number + CWD + region \times developmental\ stage + (1|Plot), family = binomial)$$

Each model was used for the following response variables: a) overall damage, i.e. for each leaf we assessed presence of herbivory, b) particular herbivory damage, i.e. separate models were carried out for chewing damage, scraping damage, gall mites, gall midges, mines and sucking damage, and c) the occurrence of single species or species groups (chewing damage: *Orchestes fagi*, gall mites: *Aceria nervisequa, Aceria nervisequa faginea* and *Acalitus stenapis*, gall midges: *Mikiola fagi* and *Hartigiola annulipes*, and mines: *O. fagi, Phyllonorycter*-group and *Stigmella* sp. and occurrence of *Phyllaphis fagi*.

Results

Overall occurrence of herbivory and herbivores

Altogether 33,760 leaves of *F. sylvatica* were examined, 26,000 in the canopy (250 from the 5 branches per plot) and 7,760 in the understorey (95.8±4.5 SE per plot) from an average of 17.6±0.6 SE saplings in this stratum. Overall 25,902 leaves were damaged (77%), 20,463 leaves (79%) in the canopy and 5,439 leaves (70%) in the understorey. The most frequent damage in both strata was chewing damage (Figure 1). In the canopy it was followed by the occurrence of gall mites, mines, sucking damage, scraping damage, gall midges and *Phyllaphis fagi*. In the understorey the second most frequent damage was sucking damage followed by the occurrence of gall mites, *P. fagi*, mines and gall midges. Scraping damage occurred only rarely in the understorey. The percentage of leaf area loss per tree ranged from 0.05–45% (mean 6.7%) in the canopy and from 0–79% (mean 5.8%) in the understorey (Figure 1).

Effects of region, stratum and management on different damage types (models 1, 2)

While overall damage of leaves was high, i.e. most leaves were damaged by at least one type of herbivore, there were marked differences between damage types between regions (model 1), between strata (model 1 and 2), and also with respect to the effect of management (model 1 and 2) (Figure 2 and 3, Table 2; for details see File S2). Overall damage was significantly higher in the Schwäbische Alb (mean 80%) and Hainich–Dün (73%) than in the Schorfheide-Chorin (71%), but different damage types responded differently to region; total chewing damage, *Orchestes fagi* (adult chewing and larval mining), mines of *Phyllonorycter*-group and total sucking damage were highest in the Schwäbische Alb, occurrence of *Hartigiola annulipes* galls and *Phyllaphis fagi* in the Schorfheide-Chorin, and gall mites, *Stigmella* spp. mines, and *Mikiola fagi* galls in the Hainch-Dün (see Table 2 and File S2).

Herbivory differed greatly between the canopy and the understorey (model 1 and 2), but the direction and strength depended on damage type (Figure 2, Table 2). Mines (canopy: 16%, understorey: 1% of leaves damaged) and gall midges (4%, 0.5%) as well as scraping damage (5%, 0%) were observed on a significantly higher proportion of leaves in the canopy than in the understorey, although in mines and midges this effect was not significant in all regions (significant interaction region×stratum; Table 2 and File S2). In contrast, *Phyllaphis fagi* was much more abundant in the understorey (0.4%, 12%). Difference between strata regarding the proportion of leaves damaged by sucking and gall mites depended strongly on region and management

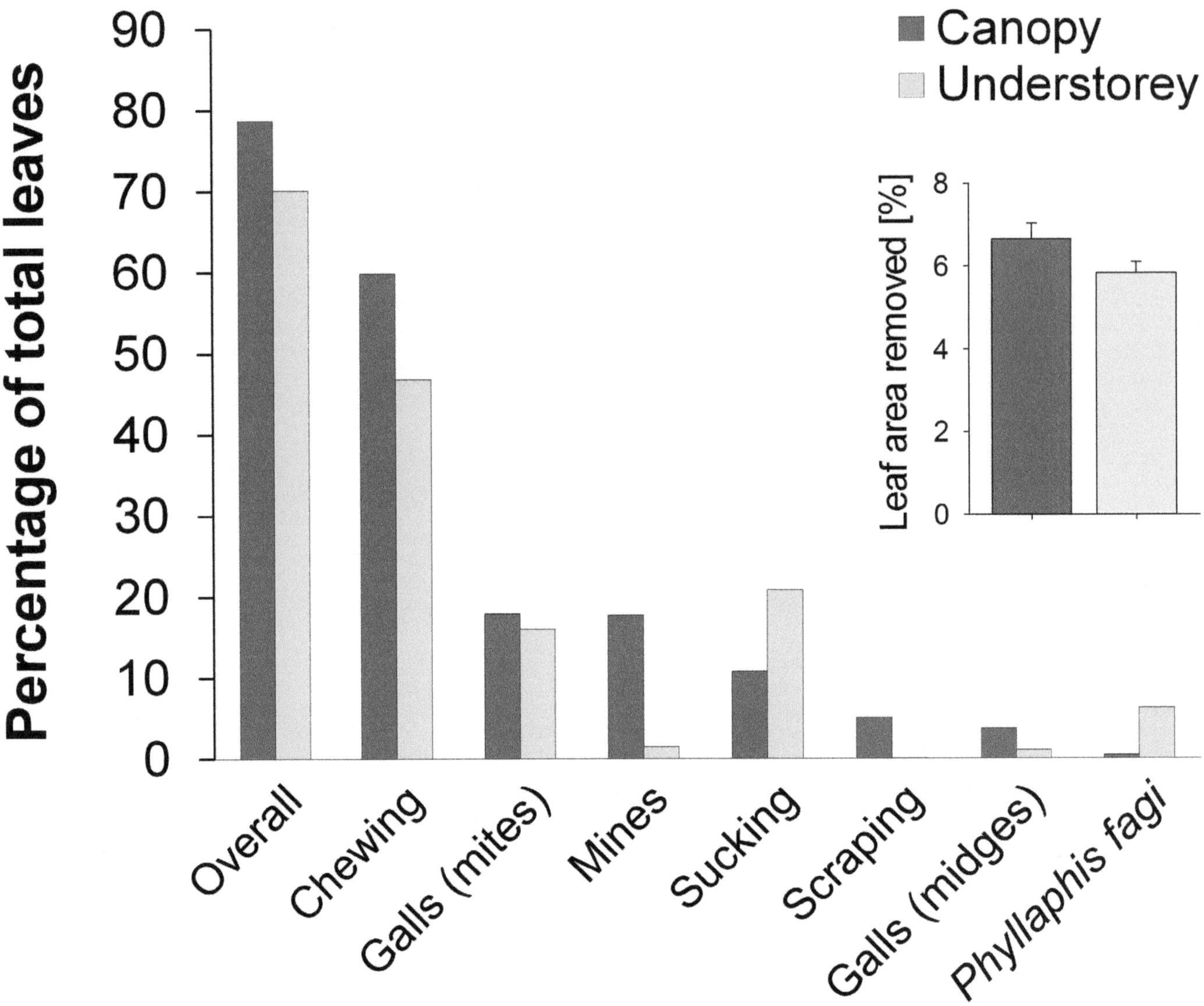

Figure 1. Herbivory of different damage types in the canopy and understory. Percentage of leaves damaged in the canopy and understory of European beech (*Fagus sylvatica*) separated by different damage types and occurrence of *Phyllaphis fagi* (canopy: N = 26,000 leaves, understorey: N = 7,760). The inset shows the average percentage of leaf area removed per tree (± standard error).

(significant region×stratum and region×management interaction; Figure 2, Table 2). Sucking damage, for example, was higher in the understorey in the Schwäbische Alb and Hanich-Dün, but higher in the canopy in the Schorfheide-Chorin. The occurrence of gall mites in the Hainich-Dün highly depended on management, with lower proportion of leaves damaged in the canopy of unmanaged, but higher in managed forests (Figure 2).

The effect of management (model 1) was generally smaller than the effect of region and particularly of the stratum (Figure 2, Table 2). Overall, gall mites were significantly more abundant in unmanaged than in managed forests (Table 2) and this was in particular pronounced in the Hainich-Dün with 20% (understorey) and 8% (canopy) higher number of leaves damaged in unmanaged forests (Figure 3). This difference was mainly caused by the two species *Aceria nervisequa faginea* and *Acalitus stenaspis*. In contrast, a higher percentage of leaves were damaged by sucking in managed forests, with up to 28% higher number of leaves damaged in the understorey (Schwäbische Alb) and 8% in the canopy (Schorfheide-Chorin) of managed compared to unmanaged forests (Figure 3). In the Hainich-Dün no effect of management on sucking damage was found (significant region× management interaction; Table 2, Figure 2). Responses of other damage types to management depended on region and stratum (significant region×management and management×stratum interactions; Table 2, Figure 3). The two gall midges showed contrasting responses to management with *Mikiola fagi* being more abundant in managed and *Hartigiola annulipes* in unmanaged forests (Table 2). Effects, however, were weak with differences in the proportion of damaged leaves between managed and unmanaged forests of less than 2%.

Effects of selection cutting forestry in the Hainich-Dün (model 2) on herbivory were observed in only a few cases. We found lower overall and *Aceria nervisequa faginea* damage in uneven-aged selection cutting compared to even-aged age-class forests (Tab. 2).

The covariates used in our study, in particular measures of forest structure and plant diversity, had generally not a strong

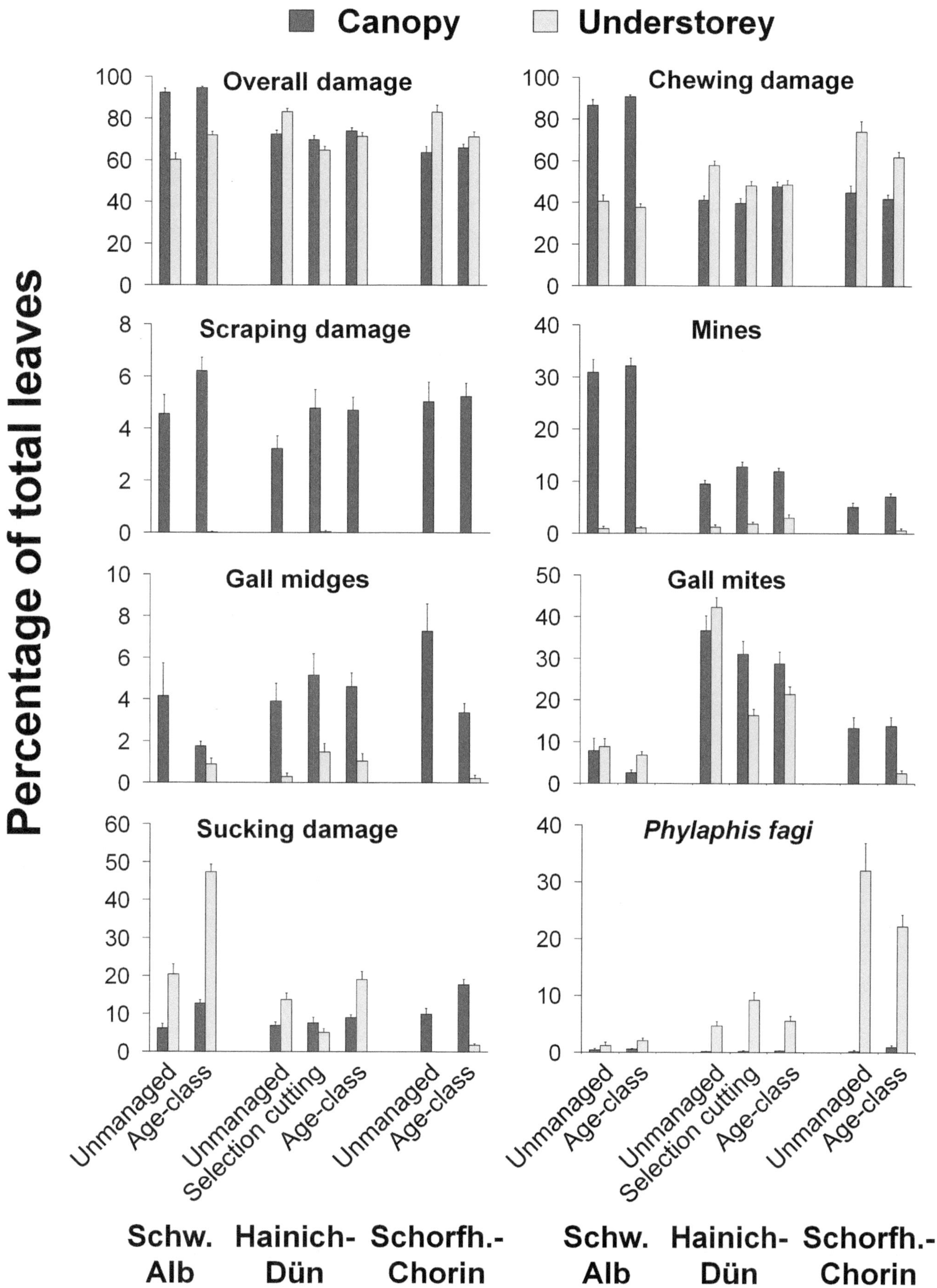
Canopy
Understorey
Percentage of total leaves
Overall damage
Chewing damage
Scraping damage
Mines
Gall midges
Gall mites
Sucking damage
Phylaphis fagi
Unmanaged
Age-class
Unmanaged
Selection cutting
Age-class
Unmanaged
Age-class
Schw. Alb
Hainich-Dün
Schorfh.-Chorin

Figure 2. Herbivory of different damage types in different regions and forest types. Percentage of leaves damaged in the canopy and understory of European beech (*Fagus sylvatica*) in stands of different management intensity (canopy: N = 26,000 leaves, understorey: N = 7,760). For significance of differences see Table 2 and Tables S1 to S3 in File S2. Due to the design of the experiment, means were compared in two models (Table 1), one including the region and omitting the selection cutting forests in Hainich-Dün (model 1, Table 2), and one restricted to Hainich-Dün including the selection cutting forests (model 2, Table S1 in File S2). Please see 'Materials and Methods' for detailed explanations.

effect on the proportion of leaves with damage (Table 3, File S2). Stand wood volume and tree number were often significant, but had inconsistent effects. Increasing wood volume positively affected gall mites, mines of *Stigmella* spp., and the occurrence of *Phyllaphis fagi*, but negatively affected chewing damage, galls of *Mikiola fagi*, mines of *Orchestes fagi* and sucking damage. Increasing tree number caused higher overall and chewing damage and a higher number of gall midges, but decreased gall

Table 2. Significant results of all models (model 1 'managed-unmanaged-comparison'; model 2 'Hainich-Dün'; model 3 'age-class comparison').

Damage type	Region	Stratum	Management	Interactions
Overall	ALB, HAI>SCH$_1$ ALB>HAI$_3$	canopy> understorey$_{1,2}$	age-class> selection cutting$_2$	region×stratum$_1$ region×age-classes$_3$ management×stratum$_2$ region×stratum×management$_1$
Chewing total	ALB>SCH>HAI$_1$ ALB>HAI$_3$	canopy>understorey$_1$ understorey>canopy$_2$		region×stratum$_1$ management×stratum$_{1,2}$
Orchestes fagi	ALB>HAI, SCH$_1$	canopy>understorey$_{1,2}$		region×stratum$_1$
	ALB>HAI$_3$			
Scraping total		canopy>understorey$_1$		
Gall mites total	HAI>ALB, SCH$_1$ HAI>ALB$_3$	understorey>canopy$_1$ canopy>understorey$_2$	unmanaged> managed$_1$	region×management$_1$ region×stratum$_1$ management×stratum$_{1,2}$ region×stratum×management$_1$
Aceria nervisequa	HAI>ALB, SCH$_1$ HAI>ALB$_3$	canopy>understorey$_{1,2}$		region×stratum$_1$ management×stratum$_2$
Aceria Nervisequa faginea	HAI>ALB, SCH$_1$ HAI>ALB$_3$	understorey>canopy$_1$ canopy>understorey$_2$	unmanaged> managed$_1$ age-class> selection cutting$_2$	region×management$_1$ region×stratum$_1$ management×stratum$_{1,2}$ region×stratum×management$_1$
Acalitus stenaspis	HAI>ALB, SCH$_1$ HAI>ALB$_3$	understorey>canopy$_1$ canopy>understorey$_2$	unmanaged> managed$_1$	region×management$_1$ region×stratum$_1$ management×stratum$_2$ region×stratum×management$_1$
Gall midges total	HAI, SCH>ALB$_1$	canopy>understorey$_2$	managed> unmanaged$_1$	region×management$_1$ region×stratum$_1$ management×stratum$_2$
Mikiola fagi	HAI>ALB> SCH$_1$	canopy>understorey$_2$	managed> unmanaged$_1$	region×management$_1$ region×stratum$_1$
Hartigiola annulipes	SCH>HAI>ALB$_1$	canopy>understorey$_2$	unmanaged> managed$_1$	
Mines	ALB>HAI> SCH$_1$ ALB> HAI$_3$	canopy>understorey$_{1,2}$		region×stratum$_1$ region×age-classes$_3$ management×stratum$_2$
Orchestes fagi	ALB>HAI> SCH$_1$	canopy>understorey$_1$	pole wood>thicket> timber>timber with regeneration$_3$	region×stratum$_1$
	ALB>HAI$_3$			
Stigmella spp.	HAI>ALB> SCH$_1$	canopy>understorey$_{1,2}$		region×stratum$_1$ management×stratum$_2$ region×stratum×management$_1$
Phyllonorycter-group	ALB>HAI> SCH$_1$	canopy>understorey$_2$		region×age-classes$_3$
	ALB>HAI$_3$			
Sucking total	ALB>HAI> SCH$_1$	understorey>canopy$_{1,2}$	managed> unmanaged$_1$	region×stratum$_1$ region×age-classes$_3$ management×stratum$_{1,2}$
Phyllaphis fagi	SCH>HAI> ALB$_1$	understorey>canopy$_{1,2}$		region×stratum$_1$

Models which showed significant effects are indicated by subscripted numbers. Generalized linear mixed effects models fit by Laplace approximation (lmer) were applied followed by a post-hoc comparison using Tukey contrasts. Please note that model 3 includes only herbivory in the sun-exposed canopy. For details see File S2.

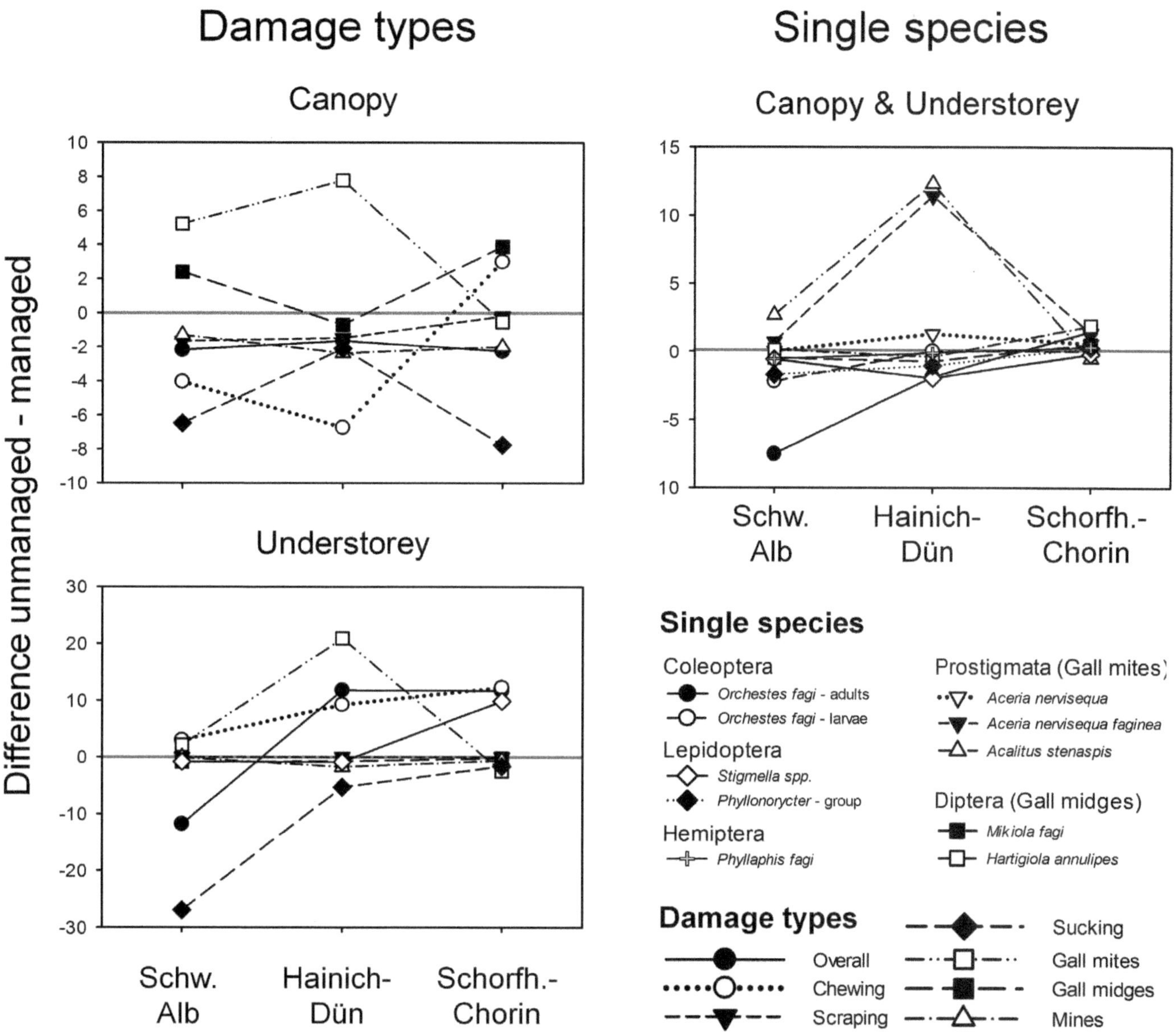

Figure 3. Differences in herbivory between managed and unmanaged stands. Difference in mean damage, measured as percentage of leaves damaged, between managed and unmanaged forests, separated by stratum and damage type (left) and single species (top right). Symbols above zero-line indicate higher damage in unmanaged and symbols below higher damage in managed forests. Results are based on model 1. See Table 2 for statistical details.

mites and mines of *Stigmella* spp. Increasing plant diversity positively influenced the occurrence of gall mites, mines of *O. fagi* and *Phyllonorycter*-group, and *P. fagi*, and affected chewing damage negatively (Table 3).

Effects of stand age (model 3)

In the age-class comparison (model 3, Table 2 and Table S3 in File S2) effects of region were broadly consistent to model 1. Significant effects of developmental stages of age-classes across regions were found for mines of *Orchestes fagi* only. Proportion of leaves with mines decrease from pole wood stands (mean 16%) to thicket (11%) to timber (9%) to timber with regeneration (4%; Table 2). In mines of the *Phyllonorycter*-group and in sucking damage effects depended on region. Mines of the *Phyllonorycter*-group were more frequent in thickets than other developmental stages, but only in the Hainich-Dün. Sucking damage was highest in thickets of the Schwäbische Alb, but pole woods in the Hainich-Dün.

Discussion

In this study, we tested how forest management affects arthropod herbivory of European beech in a large scale approach including different regions, two strata, and several damage types caused by different taxa and guilds. Our results show that while overall levels of herbivory were high, with almost 80% of all leaves in a stand showing damage, the percentage of leave area removed by arthropod herbivores in non-outbreak situations was low. The average herbivory rates found in this study, 6%, were in the range of average percentage damage per year mentioned by other studies in the canopy and understorey of temperate broad-leaved

Table 3. Significant effects (↑ positive, ↓ negative) of covariates regarding all models (model 1 'managed-unmanaged comparison'; model 2 'Hainich-Dün'; model 3 'age-class comparison'.

Covariates	Overall damage	Chewing damage	*Orchestes fagi*	Scraping damage	Gall mites	*Aceria nervisequa*	*Aceria nervisequa faginea*	*Acalitus stenaspis*	Gall midges	*Mikiola fagi*	*Hartigiola annulipes*	Mines	*Orchestes fagi*	*Stigmella* spp.	*Phyllonorycter*-group	Sucking damage	*Phyllaphis fagi*
Solid volume (models 1, 2, 3)		$\downarrow_{1,2}$			$\uparrow_{1,2}$	$\uparrow_{1,2,3}$	$\uparrow_{1,2}$	$\uparrow_{1,2}$		$\downarrow_{1}$			$\downarrow_{1,2}$	$\uparrow_{3}$		$\downarrow_{1}$	$\uparrow_{3}$
Tree number (model 1, 3)	$\uparrow_{1}$	$\uparrow_{1}$			$\downarrow_{1}$	$\downarrow_{1,3}$	$\downarrow_{1}$	$\downarrow_{1}$	$\uparrow_{1}$	$\uparrow_{1}$				$\downarrow_{1}$			
Plant diversity (model 1, 2)		$\downarrow_{2}$			$\uparrow_{1,2}$	$\uparrow_{1,2}$	$\uparrow_{1,2}$	$\uparrow_{2}$				$\uparrow_{2}$	$\uparrow_{1}$		$\uparrow_{2}$		$\uparrow_{1}$
CWD (model 2, 3)												$\downarrow_{2}$			$\downarrow_{2}$	$\uparrow_{3}$	

The number next to the arrows indicate respective model with significant results.

forests [9,16,43,44]. Differences in the proportion of leaves with herbivore damage between the differently managed forest types were generally small and not generally higher in less complex managed forests, as hypothesised. There was substantial variability in the effects of management on the different damages types, and the effects often differed between regions or between canopy leaves and leaves of saplings in the understorey. For example, damage by gall mites (understorey and canopy) and gall midges (canopy) were higher in unmanaged than in managed forest, while sucking damage was more common in managed forests (Figure 3). This is despite the fact that there was significant variation in attack rates among the forest stands. Our first conclusion is therefore that forest management, defined by management types, is not a very good predictor for the ecosystem process of arthropod herbivory, but can explain the occurrence of single herbivore species and damage types. We also included a number of covariates describing forest structure and plant diversity in our statistical models, but these covariates were rarely significant and effect directions differed among damage types, and are thus also not clear predictors of overall herbivore damage to leaves. In the following, we will describe possible underlying mechanisms.

Effects of forest management

In a recent meta-analysis of species richness in European forests among arthropods a higher species richness of Carabids, which are mainly predators in forests, were observed in unmanaged compared to managed forests [28]. Based on this finding, one could hypothesise that herbivory damage should be lower in unmanaged stands compared to managed stands due to a more effective control by predators (enemies hypothesis, [26]). However, our data do not support this assumption; management was often not a significant factor and if yes, effect sizes were often small. One possible reason is that in fact herbivore abundances and diversities are not consistently lower in one management type, i.e. unmanaged forests. Unpublished data from the same study show no significant difference in abundance and species richness of chewers and suckers between unmanaged and managed beech forests across regions. Another explanation is that lower herbivore abundances must not translate accordingly into lower rates of herbivory. While we did not sample herbivores and predators in the same year, a year before (2008) we found no differences in abundance of the important ground-dwelling predator beetles Carabidae and Staphylinidae between managed and unmanaged beech forests of the same regions, while species richness was even lower in unmanaged forests [45]. In fact, previous studies have also found several exceptions to the general trend of higher abundance and diversity of predators and lower abundance of herbivores in unmanaged forests. Chumak et al. [46] showed for European beech forests in the Carpathians and Switzerland, which both show similar plant diversity [29], that the abundance and species numbers of herbivores as well as predators were higher in managed forests than in unmanaged counterparts. On the other hand, Summerville and Crist [47] found no differences in species number or abundance of moths sampled from managed and unmanaged forest stands. They emphasised the importance of the surrounding landscape for the community composition within and among forest stands, which was not tested in our study. In contrast, Savilaakso et al. [48] showed that herbivory and species richness were significantly lower in recently logged compartments than in forests 40 years after selective logging. In addition they found significant differences in community composition between logged compartments and natural forests. Thus, general effects of management revealed to be not independent of region and stratum as predicted, but varied in our study among species and arthropod groups [28,49]. Also in contrast to our prediction, no clear differentiation in responses between free-living herbivores such as chewers and suckers and concealed living herbivores such as miners and gall-inducers was observe, although former are predicted to depend more strongly on non-food resources such as dead wood for protection [37] which is expected to be higher in unmanaged forests [50].

We predicted that the strength of the relationship with management is different in the studied strata, e.g. because abiotic conditions in the sun-exposed canopy change less strongly in response to forest management than in the understorey, and because some herbivore species may be more susceptible to the increase in temperature and possibly decrease in moisture in the sun-exposed crown. Management effects in our study indeed often depended on the stratum investigated, i.e. herbivory in the canopy vs. understorey responded differently to the management regimes, but the effects were complex and depended on the species investigated. For example, in the understorey chewing damage tended to be generally higher in the unmanaged forests, whereas in the canopy it was higher in managed forest (except Schorfheide-Chorin; Figure 3). For gall mites in the canopy, occurrence increased from age-class to selection cutting to unmanaged forests, while in the understorey it was lowest in selection cutting forests, followed by age-class and unmanaged forests. This indicates that management may have different effects on understorey and canopy conditions [51,52] or on different ontogenetic stages (saplings vs. mature trees). Most likely changing abiotic conditions following harvesting activities, which are expected to be more pronounced in the understorey, change the suitability of leaves for herbivores due to changing physical and chemical leaf traits as well as suitability of microclimatic conditions for herbivore species. For example, the lowest occurrence of gall mites in more open selection cutting forests might be explained by a higher risk of desiccation because of their less mobile and exposed lifestyle (they build open galls [36]).

In contrast to our prediction the uneven-aged selection cutting forests of the Hainich-Dün did not generally show an intermediate herbivory rate between unmanaged and age-class forests. The overall damage was lowest, but herbivory of *M. fagi* was highest compared to other forest types. Moreover, effects clearly depended on the stratum; while in the canopy herbivory rate were indeed often intermediate, in the understorey they mostly showed extremes, for example significantly lower herbivory rates of gall mites compared to other forest types (see above). According to this, other studies also found different effects of selection cutting forest on herbivores, ranging from no effect [53] to a positive effect [54]. Selection cutting may resemble the disturbance caused in natural forests of selective tree death, except for the removal of the dead wood and the higher frequency and spatially more homogenous tree felling events yet it strongly depends on the frequency of selective harvest, thinning and other forestry measures how intensive this type of forestry is.

Within the age-class forests, there were significant differences among the developmental stages only for mines and sucking damage. In contrast to our prediction, gall-inducers with highest degree of specialisation did not respond to stand age. For mines of *Orchestes fagi* and *Phyllonorycter*-group (only Hainich-Dün) and sucking damage occurrence of damage decreased from young (thicket and pole wood) to old (timber and timber with regeneration) forest stands. We proposed several mechanisms related to resource allocation and tree density underlying changes in herbivory during tree ontogeny. Resistance to herbivores is likely to change during beech development and might decrease rather than increase with age from seedlings to mature trees [38].

Thus, an increase in damage by less specialised species with stand age would have been expected. However, strong competition in thicket and pole wood stage might result in less defence investment, higher nitrogen content and thus higher herbivory rates in young stages [55]. Throop and Lerdau [56] confirmed the assumption of a positive effect of high nitrogen content on the individual performance of sap-sucking insects and their population levels and Gossner et al. [57] proposed similar mechanisms for the higher percentage of sap-sucking Heteroptera in young oak stands on former agricultural fields compared to forest sites due to differences in plant stress. Maleque et al. [58] found a general decrease of understorey insect herbivores with forest age which was attributed to the general decrease in forest floor vegetation with stand age. In our study, plant diversity positively affected most herbivores, among these miners, and thus plant diversity might partly explain observed patterns (Table 3, see also below).

One likely explanation for the small differences among management types despite the high variability among individual stands is that management types as used here are only broad descriptors of forest management. This is in particular true for the managed-unmanaged comparison as a management period of 40 years and less may not be sufficient to erase any effects of past management on biodiversity and ecosystem processes such as herbivory [59]. Similarly, the age-class forests differed in age and probably also in the precise way in which thinning and other management actions were carried out, causing heterogeneity between stands within the same class. Thus, while our results suggests that there is little effect of forest management on arthropod herbivory within the context of our managed vs. unmanaged comparison, possibly more detailed descriptors of human action in each stand, e.g. the absolute or relative amount of wood taken out of the forest, have a better explanatory power for different damage types. This will be the task of future work.

Differences in herbivory among regions

The three study regions represent different climate conditions ranging from warm dry conditions in the Schorfheide-Chorin, to rather cold and wet conditions in the Schwäbische Alb. Arthropod species richness in the studied forests was generally highest in the Schorfheide-Chorin (1,521 species), followed by the Schwäbische Alb (1,018) and lowest in the Hainich-Dün (943) [60]. Because the composition of canopy arthropod assemblages varies with latitude, precipitation [61] and elevation [62], these differences are expected to affect herbivory and this is why we treated region as a factor in all analyses. Because the differences in latitude affect the dates of leaf shoot, we carried out our herbivory assessment in mid-summer, when there were no systematic differences in phenology between the regions. Moreover landscape structure might have additionally contributed to observed differences among regions. The strong regional differences in the occurrence of some of the herbivores have, to our knowledge, not been reported before. For example, in the Schwäbische Alb, mines, chewing, and sucking damage were more frequent than in the other regions. In the Hainich-Dün gall mites and the gall midge *Mikiola fagi* peaked and in the Schorfheide-Chorin the occurrences of *Hartigiola annulipes* and *Phyllaphis fagi* was most frequent. Although regional differences might be overestimated due to changing weather conditions between North and South Germany among years, latitudinal gradients in feeding type-specific herbivory have already been observed in other parts of the world [63].

Management effects also varied between regions. For example, occurrence of galls was higher in managed than in unmanaged forests in the Hainich-Dün, while in the Schwäbische Alb and Schorfheide-Chorin we found the opposite result. Particularly gall induction depends on specific physiological adaptations of the gall inducer, entailing strong host plant specificity of most gall-inducing species [64]. The observed higher genetic differentiation of beech trees in unmanaged and managed forests of the Schwäbische Alb and the Schorfheide-Chorin compared to the Hainich-Dün [65] might therefore serve as possible explanation. For other tree species an effect of plant genetic differences on herbivore communities has already been reported [66].

Herbivory in the canopy and the understorey

In most cases we found a higher proportion of damaged leaves in the canopy than in the understorey, more precisely in leaves of small saplings. For sucking damage and occurrence of *Phyllaphis fagi* we found the opposite and for gall mites preferences depended on region and management. It has frequently been shown that arthropod community compositions differ between the canopy and the understorey, for a variety of herbivores [67,68], spiders [69], for Lepidoptera but not Coleoptera [70] [for a review see 71]. Sun leaves that more frequently occur in the canopy are smaller, thicker and tougher, while shaded leaves are larger and thinner which can affect herbivory [72,73]. Leaf chemistry also affects the attraction of leaves as resource [23,74]. In addition, the microclimate in the understorey is moister and shadier, in particular in dense forests while the canopy is more often exposed to wind and temperature changes. Some species such as endophyllously living herbivores might prefer or tolerate that [e.g. 35], other free-living species might not. Moreover, ontogenetic stage of trees needs to be considered [74]. Our study emphasises that it is important to standardise where assessments of herbivory and herbivores are made [cf. 75].

Effect of covariates

In addition to management types, we added a number of covariates to our analysis, i.e. stand variables that are affected by management and that could mediate effects on arthropod damage. Forest standing biomass, measured as wood volume and tree number affected a number of damage types, such as chewing damage and occurrence of gall mites. Increasing standing biomass increases the resources available for herbivores and can positively affect herbivore abundances. If the increase in abundance is stronger than the increase in available leaf biomass then herbivory is expected to increase. While gall mites increased in abundance with increasing biomass and decreasing tree number, chewers and gall midges showed the opposite. This suggests diverging preferences with gall mites preferring older forests with shady conditions to prevent desiccation (see above) and chewers and gall midges preferring canopy areas with higher solar radiation [35].

Coarse woody debris (CWD) describes the amounts of dead wood in forests. While we did not investigate insects feeding on dead wood, CWD nevertheless affected mines of *Phyllonorycter*-group (−) and sucking damage (+). Many organisms indirectly depend on dead wood, e.g. as shelter or overwintering habitat [76]. This might either be beneficial when herbivores use dead wood structure themselves or disadvantageous when antagonists (e.g. parasitoids) benefit from increased dead wood supply. It can be assumed that free-living herbivores such as suckers use dead wood structures as shelter and overwintering habitat (own observations) and their abundance is thus supported by an increased availability. Additionally parasitoids of herbivores might be supported by these structures and thus parasitisation rates of concealed living herbivores might be increased.

An important consequence of forest management is plant diversity, which differs between the studied forests, but showed no significant difference between studied managed and unmanaged beech forests [40]. In our study, we included stands with more than 70% European beech, based on basal stem area. As a consequence, overall vascular plant diversity in studied beech forests was dominated by the diversity of the understorey [40]. Several damage types were affected by plant diversity, but effects were weak and direction differed among damage types. Varying effects among damage types have also been found in other studies [16,62,77]. While chewing damage on beech decreased with plant diversity in our study, gall mites, mines and *Phyllaphis fagi* increased in abundance. Antagonists of chewers might be supported by higher plant diversity supressing herbivore populations [78], but the mechanism underlying the positive relationship between plant diversity and the occurrence of gall mites and mines which are specialised on beech is less clear.

Conclusions

Our study shows that arthropod herbivory on European beech leaves is common across different forest management regimes in Central Europe. Because beech hosts far fewer specified insects species than e.g. oak [79] and does not suffer from outbreaks of bark beetle or other major forest pests such as the winter moth, beech herbivory has generally received relatively little attention from entomologists and foresters, except for the beech weevil *O. fagi*. Our study suggests that herbivory may well be a significant factor for the fitness of beech. Although only 6% leave area was removed on average, this is substantially higher than e.g. in the grasslands of the same regions [1% on average; 80]. Moreover, herbivory of sap-sucking insects is likely to be underestimated because it is less conspicuous and difficult to quantify. It has, for example, been shown that phloem feeders can remove as much plant biomass as chewers [81]. Forest management affected the different types of herbivore damage in a complex way, depending on the region considered, the stratum, or the developmental stage in the case of beech age-class forests. We found no clear overall difference in herbivory between unmanaged and managed forests, suggesting that effects of management on herbivory are not per se related to management categories. Rather, the variability in responses between forest stands within a particular category suggests that details of the management are important. Future studies may include additional components of forest management into the analyses, such as the thinning regime, the age when trees are harvested, the use of machinery etc., all of which may affect the relationship between trees and insects.

Acknowledgments

We thank the managers of the three exploratories, Swen Renner, Sonja Gockel, Kerstin Wiesner, and Martin Gorke for their work in maintaining the plot and project infrastructure; Simone Pfeiffer and Christiane Fischer giving support through the central office, Michael Owonibi for managing the central data base, and Markus Fischer, Eduard Linsenmair, Dominik Hessenmöller, Jens Nieschulze, Daniel Prati, Ingo Schöning, François Buscot, Ernst-Detlef Schulze, Wolfgang W. Weisser and the late Elisabeth Kalko for their role in setting up the Biodiversity Exploratories project. We are grateful to Peter Schall and Sebastian T. Meyer for fruitful discussions and Claudia Seilwinder for providing site maps. Field work permits were issued by the responsible state environmental offices of Baden-Württemberg, Thüringen, and Brandenburg (according to § 72 BbgNatSchG).

Author Contributions

Conceived and designed the experiments: MMG MF EDS WWW. Performed the experiments: EP ML PL MMG SB DH JM SAS. Analyzed the data: EP MMG ML. Contributed reagents/materials/analysis tools: MMG EP ML PL SB DH JM SAS EDS MF WWW. Contributed to the writing of the manuscript: MMG EP WWW. Thoroughly revised the manuscript: ML PL SB DH JM SAS EDS MF.

References

1. Sala OE, Chapin FS, Armesto JJ, Berlow E, Bloomfield J, et al. (2000) Global biodiversity scenarios for the year 2100. Science 287: 1770–1774.
2. Fischer M, Bossdorf O, Gockel S, Hansel F, Hemp A, et al. (2010) Implementing large-scale and long-term functional biodiversity research: The Biodiversity Exploratories. Basic and Applied Ecology 11: 473–485.
3. Southwood TRE (1978) The components of diversity. In: Mound L.A WN, editor. Diversity of insect faunas. London: Blackwell Scientific Publications. 19–40.
4. Weisser WW, Siemann E, editors (2004) Insects and ecosystem function. Berlin Heidelberg: Springer-Verlag.
5. Schoonhoven LM, Loon JJAV, Dicke M (2005) Insect-plant biology. Oxford: Oxford University Press. 440 p.
6. Cunningham SA, Pullen KR, Colloff MJ (2009) Whole-tree sap flow is substantially diminished by leaf herbivory. Oecologia 158: 633–640.
7. Fonte SJ, Schowalter TD (2005) The influence of a neotropical herbivore (*Lamponius portoricensis*) on nutrient cycling and soil processes. Oecologia 146: 423–431.
8. Loreau M (1995) Consumers as maximizers of matter and energy flow in ecosystems. American Naturalist: 22–42.
9. Franklin RT (1970) Insect influences on the forest canopy. In: Reichle DE, editor. Analysis of temperate forest ecosystems. Berlin: Springer. 86–99.
10. Stadler B, Solinger S, Michalzik B (2001) Insect herbivores and the nutrient flow from the canopy to the soil in coniferous and deciduous forests. Oecologia 126: 104–113.
11. Frost CJ, Hunter MD (2008) Insect herbivores and their frass affect *Quercus rubra* leaf quality and initial stages of subsequent litter decomposition. Oikos 117: 13–22.
12. Schmitz OJ (2008) Herbivory from individuals to ecosystems. Annual Review of Ecology Evolution and Systematics 39: 133–152.
13. Sobek S, Gossner MM, Scherber C, Steffan-Dewenter I, Tscharntke T (2009) Tree diversity drives abundance and spatiotemporal beta-diversity of true bugs (Heteroptera). Ecological Entomology 34: 772–782.
14. Sobek S, Steffan-Dewenter I, Scherber C, Tscharntke T (2009) Spatiotemporal changes of beetle communities across a tree diversity gradient. Diversity and Distributions 15: 660–670.
15. Boncina A (2011) Conceptual approaches to integrate nature conservation into forest management: a Central European perspective. International Forestry Review 13: 13–22.
16. Sobek S, Scherber C, Steffan-Dewenter I, Tscharntke T (2009) Sapling herbivory, invertebrate herbivores and predators across a natural tree diversity gradient in Germany's largest connected deciduous forest. Oecologia 160: 279–288.
17. Jactel H, Brockerhoff EG (2007) Tree diversity reduces herbivory by forest insects. Ecology Letters 10: 835–848.
18. Vehvilainen H, Koricheva J, Ruohomaki K (2007) Tree species diversity influences herbivore abundance and damage: meta-analysis of long-term forest experiments. Oecologia 152: 287–298.
19. Schowalter TD, Hargrove WW, Crossley DA (1986) Herbivory in forested ecosystems. Annual Review of Entomology 31: 177–196.
20. Schowalter TD (1995) Canopy arthropod communities in relation to forest age and alternative harvest practices in western Oregon. Forest Ecology and Management 78: 115–125.
21. De Cauwer B, Reheul D, De Laethauwer S, Nijs I, Milbau A (2006) Effect of light an botanical species richness on insect diversity. Agronomy for Sustainable Development 26: 35–43.

22. Abrahamczyk S, Gottleuber P, Matauschek C, Kessler M (2011) Diversity and community compostion of euglossine bee assamblages (Hymenoptera: Apidae) in western Amazonia. Biodiversity and Conservation 20: 2981–3001.
23. Donaldson JR, Stevens MT, Barnhill HR, Lindroth RL (2006) Age-related shifts in leaf chemistry of clonal aspen (*Populus tremuloides*). Journal of Chemical Ecology 32: 1415–1429.
24. Coley PD (1980) Effects of leaf age and plant life-history patterns on herbivory. Nature 284: 545–546.
25. Bailey JK, Whitham TG (2003) Interactions among elk, aspen, galling sawflies and insectivorous birds. Oikos 101: 127–134.
26. Root RB (1973) The organisation of a plant-arthropod association in simple and diverse habitats: the fauna of collards, *Brassica olacea*. Ecological Monographs 43: 95–120.
27. Brunet J, Fritz Ö, Richnau G (2010) Biodiversity in European beech forests – a review with recommendations for sustainable forest management. Ecological Bulletins 53: 77–94.
28. Paillet Y, Berges L, Hjalten J, Odor P, Avon C, et al. (2010) Biodiversity differences between managed and unmanaged forests: Meta-analysis of species richness in Europe. Conservation Biology 24: 101–112.
29. Commarmot B, Bachofen H, Bundziak Y, Bürgi A, Ramp B, et al. (2005) Structures of virgin and managed beech forests in Uholka (Ukraine) and Sihlwald (Switzerland): a comparative study. Forest Snow and Landscape Research 79: 45–56.
30. Meyer P, Tabaku V, Lüpke Bv (2003) Die Struktur albanischer Rotbuchen-Urwälder–Ableitungen für eine naturnahe Buchenwirtschaft. Forstwissenschaftliches Centralblatt vereinigt mit Tharandter forstliches Jahrbuch 122: 47–58.
31. Bengtsson J, Nilsson SG, Franc A, Menozzi P (2000) Biodiversity, disturbances, ecosystem function and management of European forests. Forest Ecology and Management 132: 39–50.
32. Commarmot B, Brändli U-B, Hamor F, Lavnyy V (2013) Inventory of the largest primeval beech forest in Europe. A Swiss-Ukrainian scientific adventure. Birmensdorf: Swiss Federal Research Institute WSL, L'viv, Ukrainian National Forestry University, Rakhiv, Carpathian Biosphere Reserve. 69 p.
33. Trotsiuk V, Hobi ML, Commarmot B (2012) Age structure and disturbance dynamics of the relic virgin beech forest Uholka (Ukrainian Carpathians). Forest Ecology and Management 265: 181–190.
34. Nielsen BO, Ejlersen A (1977) The distribution pattern of herbivory in a beech canopy. Ecological Entomology 2: 293–300.
35. Kampichler C, Teschner M (2002) The spatial distribution of leaf galls of *Mikiola fagi* (Diptera: Cecidomyiidae) and *Neuroterus quercusbaccarum* (Hymenoptera: Cynipidae) in the canopy of a Central European mixed forest. European Journal of Entomology 99: 79–84.
36. Kula E, Buchta I, Hadaš P (2008) Effects of site conditions on the occurrence of *Acalitus rudis* (Canestrini) and *Eriophyes leionotus* (Nalepa) (Acari: Eriophyidae) in Czech Republic. International Journal of Acarology 34: 155–166.
37. Campbell RW, Sloan RJ (1977) Release of gypsy moth populations from innocuous levels. Environmental Entomology 6: 323–330.
38. Boege K, Marquis RJ (2005) Facing herbivory as you grow up: the ontogeny of resistance in plants. Trends in Ecology & Evolution 20: 441–448.
39. Hessenmöller D, Nieschulze J, von Lüpke N, Schulze ED (2011) Identification of forest management types from ground-based and remotely sensed variables and the effects of forest management on forest structure and composition. Forstarchiv 82: 171–183.
40. Boch S, Prati D, Müller J, Socher S, Baumbach H, et al. (2013) High plant species richness indicates management-related disturbances rather than the conservation status of forests. Basic and Applied Ecology 14: 496–505.
41. Kahmen A, Perner J, Audorff V, Weisser W, Buchmann N (2005) Effects of plant diversity, community composition and environmental parameters on productivity in montane European grasslands. Oecologia 142: 606–615.
42. Unsicker SB, Franzke A, Specht J, Köhler G, Linz J, et al. (2010) Plant species richness in montane grasslands affects the fitness of a generalist grasshopper species. Ecology 91: 1083–1091.
43. Rinker HB, Lowman MD (2004) Insect herbivory in tropical forests. Forest canopies 2nd edition Elsevier/Academic Press, San Diego, CA: 359–386.
44. Coley PD, Barone JA (1996) Herbivory and plant defenses in tropical forests. Annual Review of Ecology and Systematics 27: 305–335.
45. Lange M, Türke M, Pašalić E, Boch S, Hessenmöller D, et al. (2014) Effects of forest management on ground-dwelling beetles (Coleoptera; Carabidae, Staphylinidae) in Central Europe are mainly mediated by changes in forest structure. Forest Ecology and Management 329: 166–176.
46. Chumak V, Duelli P, Rizun V, Obrist MK, Wirz P (2005) Arthropod biodiversity in virgin and managed forests in Central Europe. Forest Snow and Landscape Research 79: 101–109.
47. Summerville KS, Crist TO (2002) Effects of timber harvest on forest Lepidoptera: Community, guild, and species responses. Ecological Applications 12: 820–835.
48. Savilaakso S, Koivisto J, Veteli TO, Pusenius J, Roininen H (2009) Long lasting impact of forest harvesting on the diversity of herbivorous insects. Biodiversity and Conservation 18: 3931–3948.
49. Taki H, Inoue T, Tanaka H, Makihara H, Sueyoshi M, et al. (2010) Responses of community structure, diversity, and abundance of understorey plants and insect assemblages to thinning in plantations. Forest Ecology and Management 259: 607–613.
50. Meyer P, Schmidt M (2011) Accumulation of dead wood in abandoned beech (*Fagus sylvatica* L.) forests in northwestern Germany. Forest Ecology and Management 261: 342–352.
51. Gossner MM (2009) Light intensity affects spatial distribution of Heteroptera in deciduous forests. European Journal of Entomology 106: 241–252.
52. Gossner MM, Floren A, Weisser WW, Linsenmair KE (2013) Effect of dead wood enrichment in the canopy and on the forest floor on beetle guild composition. Forest Ecology and Management 302: 404–413.
53. Atlegrim O, Sjöberg K (1996) Effects of clear-cutting and single-tree selection harvests on herbivorous insect larvae feeding on bilberry (*Vaccicium myrtillus*) in uneven-aged boreal *Picea abies* forests. Forest Ecology and Management 87: 139–148.
54. Nol E, Douglas H, Crins WJ (2006) Responses of Syrphids, Elaterids and Bees to Single-tree selection harvesting in Algonquin Provincial Park, Ontario. Canadian Field Naturalist 120: 15–21.
55. Thomas AT, Hodkinson ID (1991) Nitrogen, water stress and the feeding efficiency of lepidopteran herbivores. Journal of Applied Ecology 28: 703–720.
56. Throop HL, Lerdau MT (2004) Effects of nitrogen deposition on insect herbivory: Implications for community and ecosystem processes. Ecosystems 7: 109–133.
57. Gossner M, Engel K, Jessel B (2008) Plant and arthropod communities in young oak stands: are they determined by site history? Biodiversity and Conservation 17: 3165–3180.
58. Maleque MA, Maeto K, Makino S, Goto H, Tanaka H, et al. (2010) A chronosequence of understorey parasitic wasp assemblages in secondary broad-leaved forests in a Japanese 'satoyama' landscape. Insect Conservation and Diversity 3: 143–151.
59. Lindenmayer DB, Laurance WF (2012) A history of hubris – Cautionary lessons in ecologically sustainable forest management. Biological Conservation 151: 11–16.
60. Gossner MM, Fonseca CR, Pašalić E, Türke M, Lange M, et al. (2014) Limitations to the use of arthropods as temperate forests indicators. Biodiversity and Conservation 23: 945–962.
61. Progar RA, Schowalter TD (2002) Canopy arthropod assemblages along a precipitation and latitudinal gradient among Douglas-fir *Pseudotsuga menziesii* forests in the Pacific Northwest of the United States. Ecography 25: 129–138.
62. Ober HK, Hayes JP (2010) Determinants of nocturnal Lepidopteran diversity and community structure in a conifer-dominated forest. Biodiversity and Conservation 19: 761–774.
63. Hiura T, Nakamura M (2013) Different mechanisms explain feeding type-specific patterns of latitudinal variation in herbivore damage among diverse feeding types of herbivorous insects. Basic and Applied Ecology 14: 480–488.
64. Raman A, Schaefer CW, Withers TM (2005) Biology, ecology, and evolution of gall-inducing arthropods. Enfield: Science Publishers.
65. KC R (2011) Spatial dynamics of intraspecific genetic variation in European beech (*Fagus sylvatica* L.) [Dissertation]. Göttingen: Georg-August University of Göttingen. 125 p.
66. Hochwender CG, Fritz RS (2004) Plant genetic differences influence herbivore community structure: evidence from a hybrid willow system. Oecologia 138: 547–557.
67. Le Corff J, Marquis RJ (1999) Differences between understorey and canopy in herbivore community composition and leaf quality for two oak species in Missouri. Ecological Entomology 24: 46–58.
68. Basset Y (2001) Communities of insect herbivores foraging on saplings versus mature trees of Pourouma bicolor (Cecropiaceae) in Panama. Oecologia 129: 253–260.
69. Larrivee M, Buddle CM (2009) Diversity of canopy and understorey spiders in north-temperate hardwood forests. Agricultural and Forest Entomology 11: 225–237.
70. Hirao T, Murakami M, Kashizaki A (2009) Importance of the understory stratum to entomofaunal diversity in a temperate deciduous forest. Ecological Research 24: 263–272.
71. Ulyshen MD (2011) Arthropod vertical stratification in temperate deciduous forests: Implications for conservation-oriented management. Forest Ecology and Management 261: 1479–1489.
72. Abrams MD, Kubiske ME (1990) Leaf structural characteristics of 31 hardwood and conifer tree species in central Wisconsin: influence of light regime and shade-tolerance rank. Forest Ecology and Management 31: 245–253.
73. James SA, Bell DT (2000) Influence of light availability on leaf structure and growth of two Eucalyptus globulus ssp. globulus provenances. Tree Physiology 20: 1007–1018.
74. Boege K (2005) Herbivore attack in *Casearia nitida* influenced by plant ontogenetic variation in foliage quality and plant architecture. Oecologia 143: 117–125.
75. Kowalski E, Gossner MM, Türke M, Lange M, Veddeler D, et al. (2011) The use of forest inventory data for placing flight-interception traps in the forest canopy. Entomologia Experimentalis et Applicata 104: 35–44.
76. Siitonen J, Jonsson BG (2012) Other associations with dead wood material. In: Stokland JN, Siitonen J, Jonsson BG, editors. Biodiversity in Dead Wood. Cambridge: Cambridge University Press. 58–81.
77. Schuldt A, Baruffol M, Bohnke M, Bruelheide H, Hardtle W, et al. (2010) Tree diversity promotes insect herbivory in subtropical forests of south-east China. Journal of Ecology 98: 917–926.

78. Fabian Y, Sandau N, Bruggisser OT, Aebi A, Kehrli P, et al. (2014) Plant diversity in a nutshell: testing for small-scale effects on trap nesting wild bees and wasps. Ecosphere 5.
79. Brändle M, Brandl R (2001) Species richness of insects and mites on trees: expanding Southwood. Journal of Animal Ecology 70: 491–504.
80. Gossner MM, Weisser WW, Meyer ST (2014) Invertebrate herbivory decreases along a gradient of increasing land-use intensity in German grasslands. Basic and Applied Ecology 15: 347–352.
81. Leigh Jr EG (1999) Tropical forest ecology: a view from Barro Colorado Island. New York, USA: Oxford University Press.

10

Effects of the Interception of Litterfall by the Understory on Carbon Cycling in Eucalyptus Plantations of South China

Long Yang[1,2*,9], Jun Wang[2,9], Yuhui Huang[3], Dafeng Hui[4], Meili Wen[1]

1 Centre of Resource and Environment, Guangzhou Institute of Geography, Guangzhou, China, **2** Key Laboratory of Vegetation Restoration and Management of Degraded Ecosystems, South China Botanical Garden, Chinese Academy of Sciences, Guangzhou, China, **3** Guangdong Academy of Forestry, Guangzhou, China, **4** Department of Biological Sciences, Tennessee State University, Nashville, Tennessee, United States of America

Abstract

For the purposes of forest restoration, carbon (C) fixation, and economic improvement, eucalyptus (*Eucalyptus urophylla*) has been widely planted in South China. The understory of eucalyptus plantations is often occupied by a dense community of the fern *Dicranopteris dichotoma*, which intercepts tree canopy leaf litter before it reaches the ground. To understand the effects of this interception of litterfall on C cycling in eucalyptus plantations, we quantified the mass of intercepted litter and the influences of litterfall interception on litter decomposition and soil respiration. The total mass of *E. urophylla* litterfall collected on the understory was similar to that collected by the traditional litter trap method. All of the eucalyptus litterfall is intercepted by the *D. dichotoma* canopy. Of the litterfall that was intercepted by *D. dichotoma*, 20–40% and 60–80% was intercepted by the top (50–100 cm) and bottom (0–50 cm) of the understory canopy, respectively. Intercepted litterfall decomposed faster at the bottom of understory canopy (at the base of the plants) than at the top, and decomposition was slower on the soil surface in the absence of understory than on any location in the understory canopy. Soil respiration was highest when both the understory and litter were present and was lowest when both the understory and litter were absent. These results indicate that litterfall interception changed carbon flow between aboveground and belowground through litter decomposition and soil respiration, which changed carbon cycling in eucalyptus plantations. The effects of the understory on litter decomposition and soil respiration should be considered in ecosystem carbon models.

Editor: Ben Bond-Lamberty, DOE Pacific Northwest National Laboratory, United States of America

Funding: This research was funded by the National Natural Science Foundation of China (No. 31000212), the Special Foundation for Young Scientists of Guangdong Academy of Sciences (2009-46), and the Key Laboratory of Vegetation Restoration and Management of Degraded Ecosystems, Chinese Academy of Sciences. The funders had no role in study design, data collection and analysis, decision to publish, or preparation of the manuscript.

* Email: yanglong@scib.ac.cn

9 These authors contributed equally to this work.

Introduction

Forests play an important role in terrestrial carbon (C) cycling [1,2,3]. Because of deforestation, the forest C stock in China had decreased by 0.62 Pg by the 1970s [4]. Since then, long-term forest restoration has increased C sequestration in China [5,6], and the establishment of plantations has increased the forest C stock by 0.45 Pg C and at a rate of 0.021 Pg C yr^{-1} [4]. China has more artificial plantations than any other country [7]. Presently, artificial plantations occupy 3379 ha and represent 32% of the forested area in China. In Guangdong Province in South China, for example, the forest cover increased from 26% in 1979 to 50% in 1998 because of artificial vegetation restoration. These plantations are considered to be important C pools that may influence climate change at the regional scale [8].

The eucalyptus *Eucalyptus urophylla*, which is native to Australia, has become a widely planted and economically important tree in South China because it grows rapidly and has other desirable characteristics. By 2010, eucalyptus plantations in South China occupied approximately 2.6 million ha [9] and were considered to be an important C sink [10]. In eucalyptus plantations, the understory vegetation is often dominated by the fern *Dicranopteris dichotoma*. Although the understory vegetation accounts for only a small portion of the total plantation biomass, it plays an important role in nutrient cycling and total production of forest ecosystems. Because of its higher nutrient content and faster biomass turnover [11], understory vegetation is considered to be a driver in forest ecosystems [12]; the understory affects many processes, such as tree seedling establishment [13,14], litter decomposition [15,16], and soil respiration [17,18].

D. dichotoma (Gleicheniaceae) is a heliophyte fern that usually forms a dense single-species layer under the canopy of the trees in eucalyptus plantations because of its vigorous clonal growth [19] and allelopathic effects [20,21,22]. In eucalyptus plantations, most litter falling from the canopy is intercepted by the *D. dichotoma* foliage before it reaches the ground. Litter is regarded as the most important C source of forest soils. Litterfall interception may change the spatial distribution of litterfall, which would affect litter decomposition and soil respiration and thus C cycling in forest ecosystems. However, most previous studies have assumed that canopy-derived litter falls directly to the forest floor. The effect of litterfall interception by the understory C carbon cycling has not been carefully investigated.

To better understand the effects of litterfall interception by the understory vegetation, we conducted three experiments in three 6-year-old experimental eucalyptus plantations in South China. In these experiments, we determined: (1) the total mass and proportion of eucalyptus litterfall intercepted by the understory fern *D. dichotoma*; (2) the effect of litterfall interception on the rate at which the litterfall decomposes; and (3) the effect of litterfall interception on soil respiration.

Materials and Methods

Site description

The study was conducted at the Heshan National Field Research Station of Forest Ecosystem (112 °50'E, 22 °34'N), which is located near Heshan, Guangdong, China. All necessary permits for the field experiments were obtained from the South China Botanical Garden, Chinese Academy of Sciences. This station is one of the core stations of the Chinese Ecological Research Network (CERN) and occupies 40 ha. The location is characterized by south subtropical monsoon climate, and the soil is laterite. There is a distinctive wet season (from April to September) and dry season (from October to March). The mean annual temperature is 22.6°C, and the mean annual precipitation is 1700 mm. The annual solar radiation is 4350.5 MJ m^{-2}.

As a result of long-term and severe human disturbance, the soil in the area of the station had been eroded, and the original vegetation had almost disappeared before the station was established in 1983. With the establishment of the station, the degraded land was planted with many native and exotic species as part of restoration research. The native tree species included *Castanopsis hystrix*, *Liquidambar formosana*, and *Michelia macclurei*, and the exotic tree species include *E. urophylla* and *Acacia crassicarpa*. The climax plant community in this region is subtropical monsoon evergreen broad-leaved forest and includes members of the *Lauraceae*, *Euphorbiaceae*, and *Fagaceae*.

Mass and proportion of litterfall interception (experiment 1)

In this study, three 6-year-old forest sites were selected in which *E. urophylla* was the dominant tree and a dense community of *D. dichotoma* dominated the understory (Fig. 1A). Each site occupied ca. 1 ha, and adjacent sites were separated by ca. 100 m. The diameter at breast height of *E. urophylla* was 10.11±1.54 cm, and the height of *D. dichotoma* was 108±13 cm in 2011. The intercepted litterfall was quantified by establishing nine plots (1 m×1 m each) with three treatments and three replicates at each of the three field sites. The three treatments are referred to as litter trap (LT), open *D. dichotoma* (OD), and litter baffle (LB). The LT treatment used the traditional litter trap method. In the LT treatment, all understory *D. dichotoma* was cleared, and litter was collected using a nylon net (1 m×1 m), which was held horizontally at 100 cm above the soil surface (the average height of *D. dichotoma*) by four 1-m-high PVC tubes (4 cm diameter). The litter collected in the LT treatment can be used to estimate the total litterfall without interception by the understory. In the OD treatment, litter was collected by hand from the foliage of the *D. dichotoma* in a 1 m×1 m area. Litter in the OD treatment could move horizontally and vertically. In the LB or litter baffle treatment, litter was collected from the foliage of the *D. dichotoma* in a 1 m×1 m area as with the OD treatment, but the area was enclosed with PVC boards, which extended from the soil surface to 1 m above the soil surface. The boards or baffles prevented horizontal movement of the litter. The litter in all plots was collected every 2 months from May 2011 to May 2012. At the time of collecting, the litter in the OD and LB plots was separated into two groups according to height in the understory (0–50 cm and 50–100 cm, as measured from the ground). All litter was weighed after it had been oven-dried at 65°C for 72 h.

Litter decomposition (experiment 2)

Litter decomposition was quantified in three plots (1 m×1 m each) at each field site. Fresh litter of *E. urophylla* was collected with litter traps, air-dried for 15 days, and then added to 10 cm×10 cm litter bags (10 g of air-dried litter per bag). The mesh size on the upper side of the nylon litter bags was 0.8 mm to access the soil organisms and the mesh size on the bottom side was 0.25 mm to prevent litter loss. Thirty litter bags were placed in each plot at different heights: 10 were placed on the top of the *D. dichotoma* canopy (ca. 100 cm above the forest floor); 10 were placed at mid-height of the *D. dichotoma* canopy (ca. 50 cm above the forest floor); and 10 were placed on the *D. dichotoma* roots that grew on the surface of the forest floor. Another 10 litterbags were placed directly on the ground in an area where all of the *D. dichotoma* had been removed. One litterbag per treatment per plot was retrieved every 2 months from April 2011 to June 2012. After recovery, the litterbags were opened, and the decomposing litter was carefully removed and cleaned of adhering soil particles. The litter was then oven-dried at 65°C to constant mass and weighed. The *E. urophylla*

Figure 1. A *Eucalyptus urophylla* plantation with a dense understory of the fern *Dicranopteris dichotoma* in South China (A) and litterfall interception by *D. dichotoma* (B).

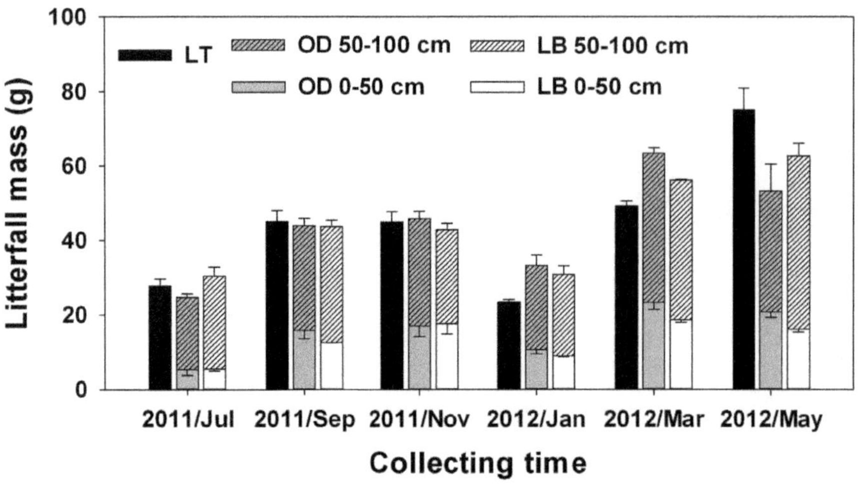

Figure 2. Mass of litterfall intercepted (means ± SE) by traditional litter traps (LT), the natural *Dicranopteris dichotoma* understory (OD), and the *D. dichotoma* understory with bordering baffles to prevent horizontal movement of litter (LB) (experiment 1). The mass of litter captured in LT can be considered equivalent to the total mass of tree litter that can potentially fall to the forest floor.

litter decomposition rate (decay constant, k) was determined with a negative exponential decay model [23,24]: $y = e^{-kt}$, where y is the proportion of the initial biomass remaining at time t (yr^{-1}), and k is the litter decay constant. The model was fitted by nonlinear regression in Sigmaplot 12.0 (Systat Software, San Jose, CA).

Soil respiration (experiment 3)

This experiment tested the effects of litterfall and the understory fern on soil respiration. In each of the three field sites, three blocks were established. Each block (2 m×2 m) included four 1 m×1 m plots to accommodate four treatments (two levels of each of two factors, i.e., ± understory fern and ± litter). The spatial separation of blocks and plots were about 10 m and 20 cm, respectively. The four treatments were: 1) minus understory fern and minus litter (-D-E); 2) minus understory fern and plus litter (-D+E); 3) plus understory fern and minus litter (+D-E); and 4) plus understory fern and plus litter (+D+E). The plus treatments involved complete retention of the fern or litter, and the minus treatments involved complete removal of the fern or litter (removal was done each week to maintain the treatments). The ferns were uprooted by hand, and the disturbance from this treatment was greatly reduced, for the root of the fern was shallow and the experiment started after 2 months of treatment. In the +E treatments, the litter was placed on the surface roots at the base of the fern. In each plot, one soil collar (PVC pipe, 10 cm in diameter and 5 cm in height) was set up in the soil (2 cm deep) for soil respiration measurement. In total, there were 12 collars (3 plots×4 treatments) at each site. Soil respiration rate was measured twice each month from April 2011 to March 2012 with a soil chamber (Li-6400-09) connected to a Li-6400 portable gas exchange analyzer (Li-Cor, Biosciences, Lincoln, Nebraska, USA). All measurements of soil respiration were conducted between 09:00 and 14:00 each time with no precipitation, and three measurements were taken at each sampling point. Soil temperature was recorded simultaneity by soil temperature probe that connected to the Li-6400. Soil humidity was measured by a time domain reflectometry (TDR, TRIME-FM, IMKO, Ettlingen, Germany). The regression relationship between soil respiration and soil temperature were fitted by exponential growth model: $R_s = a{\cdot}e^{b{\cdot}T_s}$, where R_s is the soil respiration rate; T_s is soil temperature at 5°C depth; a and b are fitted constants. Q_{10} was calculated as $Q_{10} = e^{10b}$. The regression relationship between soil respiration and soil humidity were fitted by linear model: $R_s = a + bM_s$, where M_s is soil humidity at 5°C depth; a and b are fitted constants.

Table 1. Results of ANOVA for the effects of time (date of litter collection), site, and treatment (LT, OD, or LB) on the total mass of intercepted litterfall (experiment 1).

Source	*df*	F	*P*
Time	5	35.488	**<0.001**
Site	2	1.819	0.174
Treatment	2	0.013	0.987
Time×Site	10	0.133	0.999
Time×Treatment	10	0.502	0.879
Site×Treatment	4	0.298	0.878
Time×Site×Treatment	20	0.064	1.000

Note: LT refers to litter collected on nylon sheets (litter traps); OD refers to litter collected on natural *D. dichotoma* foliage; LB refers to litter collected on *D. dichotoma* foliage with baffles on the plot borders that prevented the horizontal movement of the litter.

Table 2. Results of ANOVAs for the effects of time (date of litter collection), site, and height (position in the understory canopy: from 0–50 vs. 50–100 cm from the ground) on the total mass of litter intercepted by the canopy in the OD treatment (litter collected on natural *D. dichotoma* foliage) and the LB treatment (litter collected on *D. dichotoma* foliage with baffles on the plot borders that prevented the horizontal movement of litter) (experiment 1).

Source	*df*	OD		LB	
		F	*p*	F	*p*
Time	5	18.999	**<0.001**	11.030	**<0.001**
Site	2	6.185	**0.006**	0.483	0.622
Height	1	104.106	**<0.001**	122.837	**<0.001**
Time×Site	10	0.111	1.000	0.053	1.000
Time×Height	5	0.054	0.998	0.588	0.709
Site×Height	2	0.130	0.878	0.054	0.947
Time×Site×Height	10	0.143	0.999	0.045	1.000

Statistical analysis

For experiment 1, a three-way analysis of variance (ANOVA) was used to determine the effects of time, site, treatment, and their interactions on litterfall interception by the understory fern *D. dichotoma*; in addition, a three-way ANOVA was used to determine the effects of time, site, height in the *D. dichotoma* understory, and their interactions on litterfall interception in the OD and LB treatments. For experiment 2, one-way ANOVAs were used to determine the effect of height in the understory and absence of understory on mass remaining in litter bags at the end of the experiment and on the decomposition decay constant. For experiment 3, four-way ANOVAs were used to test the effects of season (wet or dry), site, litter (±), understory (±), and their interactions on soil respiration. When an effect was significant at $\alpha = 0.05$, means were compared by least significant difference (LSD) tests with SYSTAT 11.0 (Systat Software Inc., USA). Graphs were made with Sigmaplot 9.0 (Systat Software, Inc., Point Richmond, CA, USA).

Results

Mass and proportion of litterfall intercepted by the understory (experiment 1)

The mass of litterfall was unaffected by treatment, site, or the interactions but was affected by time (Table 1). The largest amount of leaf litter was collected in May 2012, and the smallest amount was collected in January 2012 (Fig. 2). For OD and LB treatments, litter interception was significantly affected by height in the understory (0–50 cm or bottom vs. 50–100 cm or top) and time (Table 2). For the OD treatment, more leaf litter was collected from the bottom (61–80% of the total) than from the top (22–39%) of the *D. dichotoma* patch (Fig. 2). For the LB treatment, more leaf litter was collected from the bottom (59–81%) than from the top (19–41%) of the *D. dichotoma* patch.

Litter decomposition (experiment 2)

The litter mass remaining in all treatments followed a typical litter decomposition pattern and was always greater for litter bags kept in the *D. dichotoma* canopy (at 0, 50, or 100 cm) than on the surface of the soil in an area without understory (control) (Fig. 3A). Litter decomposition was the highest during the first 2 months of the experiment. At the end of the experiment, the litter mass remaining differed among the treatments ($F = 13.513$, $p = 0.006$). At the end of the experiment, litter mass remaining was 41.8, 54.2, and 62.7% at 0 cm, 50 cm, 100 cm height, respectively, in the *D. dichotoma* canopy and was 69.1% on the bare ground without *D. dichotoma.* The litter decomposition rate (decay constant, *k*) declined with distance from the ground to canopy ($p = 0.003$, $F = 17.803$, $df = 2$) and was lowest on the bare ground without *D. dichotoma* (Fig. 3B).

Soil respiration (experiment 3)

Season (wet vs. dry), litter (±), understory (±), and the interaction between season and litter significantly affected soil respiration (Table 3). Regardless of litter or understory treatment, the soil respiration rate was much higher in the wet season (from April to September) than in the dry season (from October to March) (Fig. 4). In the dry season, soil respiration was higher in the +D+E treatment than in the other three treatments (Fig. 4). In the wet season and over the entire experiment, the treatments were in the following order with respect to soil respiration rates: +D+E>-D+E>+D-E>-D-E (Fig. 4). The relationship between soil respiration rate and soil temperature was well fit by an exponential growth regression model ($p<0.0001$). Soil temperature explained 40%, 48%, 38% and 27% of the variations in +D+E, +D-E, -D+E and-D-E treatments, respectively (Fig. 5, Table 4). Q_{10} was ranged from 2.45 to 4.38, and the highest value appeared in +D-E treatment. Soil humidity have no linear relationship with soil respiration ($p>$ 0.05 in all treatments).

Discussion

Mass of litterfall

Litterfall interception is greatly influenced by plant architecture (arrangement of branches and leaves) and morphology (stalks, spines, and leaf size) [25]. *D. dichotoma* propagates with spores or clones and always forms a dense understory [20]. The above-ground biomass of *D. dichotoma* ranges from 1360 to 3411 kg ha^{-1}, and the understory canopy cover can be as high as 100% [26]. *D. dichotoma* intercepts litterfall from tree species in its interlaced fronds and stalks. Because *E. urophylla* leaves are relatively large [27] and light [28], they are readily intercepted and supported by the *D. dichotoma* canopy (Fig. 1B). The results of our first experiment indicated that virtually all of the eucalyptus litterfall is intercepted by the *D. dichotoma* canopy because the quantity of litterfall collected from the understory canopy was similar to that collected by traditional traps, which presumably capture nearly 100% of the litterfall. The quantity of intercepted litterfall was also

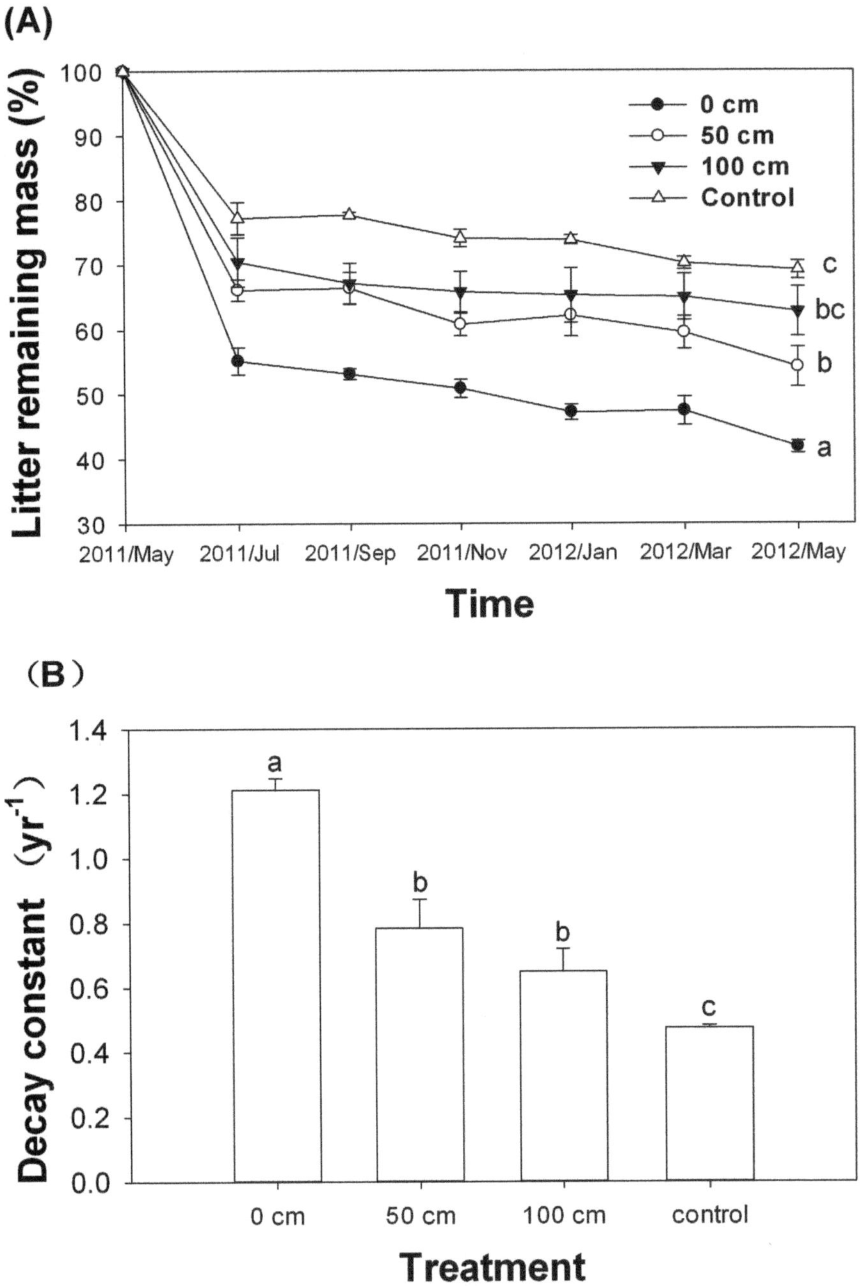

Figure 3. Litter mass remaining (A) and litter decomposition rate (*k*) (B) at different heights (0, 50, and 100 cm) in the *Dicranopteris dichotoma* understory or on the bare ground without understory (control) (experiment 2). Error bars represent standard errors, and different letters indicate significant differences at $\alpha = 0.05$.

reported to be substantial in other ecosystems [25,31], and we quantified the mass of litterfall interception by the fern and compared litterfall collected using different methods. Our results indicated that the OD and LB methods provided reliable estimation of litterfall. In the current study, the annual litterfall in the eucalyptus forests was about 2.657 Mg ha^{-1}. This value was comparable to those in previous studies in eucalyptus forests [29,30].

Litterfall interception in our study also varied in time and space. With regard to time, litterfall interception (and total litterfall) was highest from March to May (i.e., between the dry and wet season) and was lowest in the wet season (from May to July). With regard to space, more litterfall was intercepted at the bottom (0–50 cm height) than at the top of the understory (50–100 cm height).

Effects of litterfall interception on litter decomposition

Litter decomposition is influenced by the physicochemical environment, litter quality, and the composition of the decomposer community [31]. Litter quality was excluded as a variable in our decomposition experiment current study because the litter that was placed in the litter bags was identical. Decomposition, however, can be influenced by the understory vegetation because the understory can affect the location of litter [25] and therefore

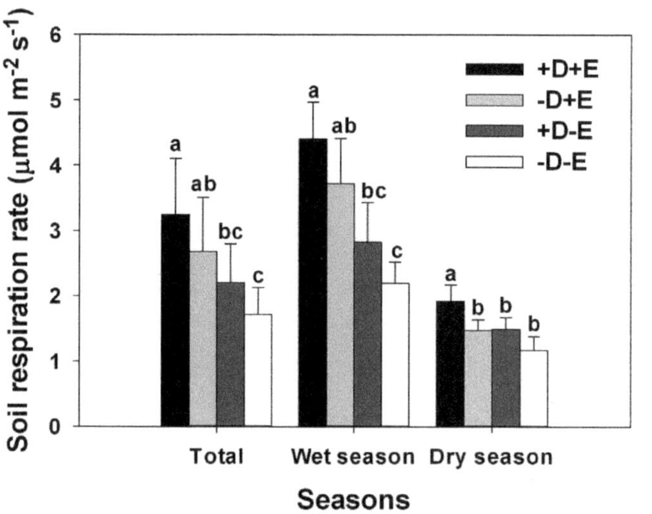

Figure 4. Soil respiration rate as affected by treatment (-D-E, -D+E, +D-E, and +D+E) and season (wet and dry season) in *Eucalyptus urophylla* plantations (experiment 3). D refers to the *Dicranopteris dichotoma* understory, which was present or absent (±). E refers to tree litter, which was present (at the base of *D. dichotoma* or on bare ground) or absent (±). Different letters above bars indicate significant differences (at $\alpha = 0.05$) among the treatments in the same season. Error bars represent SEs.

affect the decomposition microenvironment. Litter decay was reduced when litterfall was intercepted by the *D. dichotoma* canopy, probably because of spatial heterogeneity of physicochemical environment and the composition of the decomposer community in the understory fern. Humidity may be a more important factor which can significantly affects microbial composition and activity and litter decay rate [32,33]. For example, in *Cinnamomum camphora* plantation forest in South China, litterfall interception of understory vegetation delayed the litter decomposition, for water content on the crown may inhibited microbial activities [34]. Although light could also change microbial community characteristics and affect litter decomposition, it may not be an important factor in this study, as litter decomposition controlled by photodegradation happened in arid or semi-arid ecosystem [35], not in subtropical ecosystem. The composition of the decomposer community could change the litter decay rate, but we did not determine the composition of the decomposer community.

We also found that the litter decayed more quickly in the *D. dichotoma* canopy than on bare soil without canopy. This result was consistent with Dearden and Wardle [25], who reported that litter of canopy species in the crowns of the understory ferns decayed faster than that on the ground. In another study, O'Connell [16] found that the understory legume *Acacia pulchella* could increase the decomposition rate of *E. marginata* canopy litter. The results of this study also indicate that the overall litter decomposition rate of tree leaf litter in forests with understory may be underestimated when litter bags are places on bare ground in the absence of understory.

Effects of litterfall interception on soil respiration

Soil respiration includes autotrophic respiration from root and heterotrophic respiration from microbe and soil fauna. Soil respiration showed obvious seasonal variation, higher in wetter season and lower in drier season, which affected by south subtropical monsoon climate. Warmer temperature and abundant water content may increase the activity of microbe and fauna. In our study, both key factors (understory vegetation and litter) may influence soil respiration. Our results concerning the effects of the understory on soil respiration are consistent with several previous studies. In the same eucalyptus forests of South China, for example, removal of understory vegetation reduced annual soil respiration by 6% [35]. In a deciduous broad-leaved forest, removal of understory dwarf bamboo (*Sasa senanensis*) also reduced soil respiration [36]. In the current study of eucalyptus forests, removal of litter reduced soil respiration by 36%. The reported contributions of litter to soil respiration in artificial and natural forests in China and other countries have ranged from 12 to 41% [17,35,37,38,39,40,41,42]. Soil respiration rate was higher in the +D-E treatment than in the -D-E treatment, indicating that the

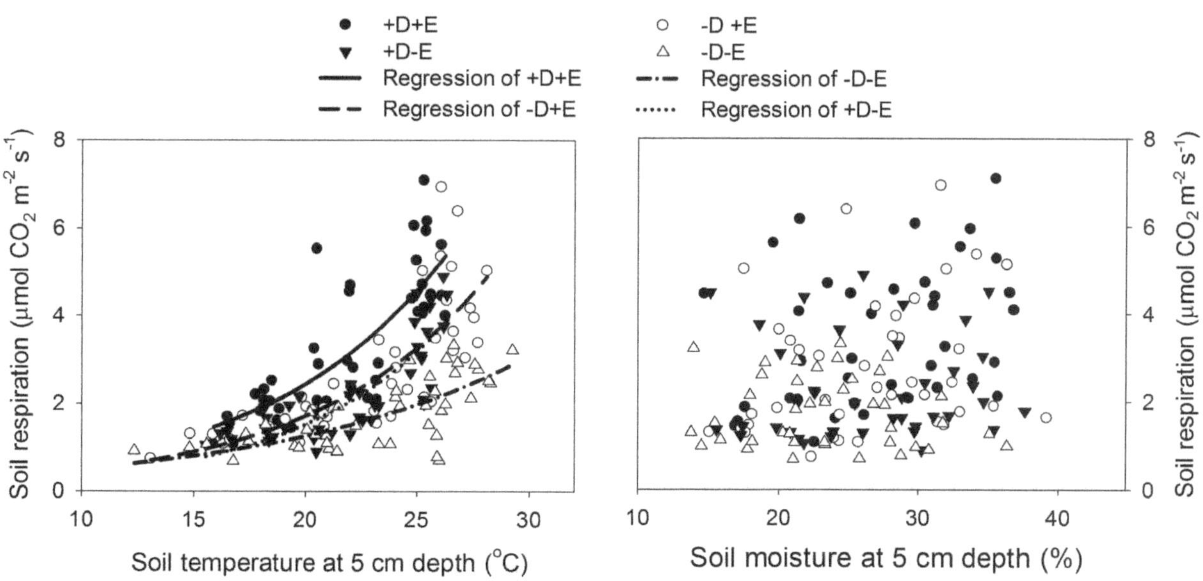

Figure 5. Relationship between soil respiration rate and soil temperature (A) or soil humidity (B) under all treatments: -D-E, -D+E, +D-E, and +D+E.

Table 3. Results of ANOVA for the effects of season (wet vs. dry), site, litter (±), understory (±), and their interactions on the soil respiration rate.

Source	*df*	F	*p*
Season	1	126.216	**0.000**
Site	2	0.810	0.447
Litter	1	40.766	**0.000**
Understory	1	11.838	**0.001**
Season×Site	2	1.055	0.351
Season×Litter	1	12.002	**0.001**
Site×Litter	2	0.294	0.746
Season×Site×Litter	2	0.064	0.938
Season×Understory	1	0.565	0.453
Site×Understory	2	0.235	0.791
Season×Site×Understory	2	0.003	0.997
Litter×Understory	1	0.159	0.691
Season×Litter×Understory	1	0.029	0.866
Site×Litter×Understory	2	0.840	0.434
Season×Site×Litter×Understory	2	0.390	0.678

understory substantially affected soil respiration even in the absence of litter. This can be attributed to respiration by *D. dichotoma* roots, which are shallow and abundant; root biomass is usually higher than aboveground biomass for *D. dichotoma* [26].

Soil temperature and soil humidity are the most important environmental factors, which influence the soil respiration. However, the relationship varied along with spatio-temportal difference. Some study found soil respiration was just related with soil temperature or soil humidity [43], or soil respiration was related with soil temperature and soil humidity [44,45]. Our results, that soil respiration was just related with soil temperature and had no relationship with soil humidity, were different from previous studies in South China [35]. Lower soil porosity may impede CO_2 release when soil humidity increase. In climax vegetation of South China, subtropical monsoon evergreen broad-leaved forest, both of soil temperature and soil humidity affected soil respiration, and soil humidity had greater influence [46]. While in artificial plantations of monocultures, lower canopy cover, higher sun radiation of understory layer and weaker buffer of air temperature led to spatial heterogeneity of soil temperature and microbial activity, which may be the reason why soil respiration was sensitive to soil temperature. Under global warming condition, the temperature sensitivity of soil respiration has become the focus topic and Q_{10} was widely used [47,48]. Most studies found that Q_{10} was ranged from 1.3 to 3.3, and decreased with latitude reducing [49]. Higher Q_{10} in Eucalyptus plantations (>2.4), means that more CO_2 will be released from soil respiration in the future, especially when litterfall were intercepted by understory fern (+D-E treatment, $Q_{10}=4.38$).

Effects of litterfall interception on carbon cycling

Litterfall interception by the understory is a ubiquitous phenomenon in subtropical and tropical forests because the understory in such forests is always dense. In this study, which was

Table 4. Effects of soil temperature (T_s) and soil moisture (M_s) on the variation of soil respiration rate (R_s) under all treatments (-D-E, -D+E, +D-E, and +D+E).

Regression	Treatment	*p*	R^2	*a*	*b*	Q_{10}
R_s & T_s	-D-E	<0.0001	0.2661	0.2091	0.0898	2.45
	-D+E	<0.0001	0.3777	0.1295	0.1292	3.64
	+D-E	<0.0001	0.4787	0.0791	0.1479	4.38
	+D+E	<0.0001	0.3988	0.1995	0.1254	3.50
R_s & M_s	-D-E	0.474	0.0002	−0.0166	2.1563	-
	-D+E	0.107	0.0047	0.0676	1.0099	-
	+D-E	0.779	0.0000	0.0088	2.0629	-
	+D+E	0.053	0.0095	0.0801	1.3011	-

The regression relationship between R_s and T_s (R_s & T_s) was fitted by exponential growth model $R_s = a{\cdot}e^{b{\cdot}T_s}$, $Q_{10}=e^{10{\cdot}b}$; The regression relationship between R_s and M_s (R_s & M_s) was fitted by linear model $R_s = a + bM_s$.

conducted in eucalyptus plantations in South China, we quantified the vertical distribution of litterfall in the understory, which was dense and dominated by the fern *D. dichotoma*, and tested the effects of litterfall interception on litter decomposition and soil respiration. Although heterogeneous soil structure appeared, the results from smaller plots measured in our study for even plantation landscape indicate that litter interception by the understory fern *D. dichotoma* changes two important processes of carbon flow between aboveground and belowground, litter decomposition and soil respiration, which changes carbon cycling.

As part of vegetation restoration efforts, plantations now occupy large areas worldwide. Such plantations are considered to be important carbon pools that might mitigate climate change at a regional scale. Although understory vegetation acts as a transitional layer that links aboveground and belowground processes, the ecosystem functions of understory vegetation in plantations have generally been ignored because the understory biomass is much less than the tree biomass. - However, understory vegetation is an important driving force of nutrient cycling and productivity in forest ecosystems [12]. In addition to accumulating biomass, understory plants reduce litter decomposition rates and soil respiration rates and therefore enhance carbon sequestration as a consequence of litterfall interception. Because litterfall interception will affect carbon budgets and cycling, the effects of litterfall interception by understory vegetation should be considered in biogeochemical models of forest ecosystems.

Author Contributions

Conceived and designed the experiments: LY YH JW. Performed the experiments: LY YH JW. Analyzed the data: LY YH JW. Contributed reagents/materials/analysis tools: LY YH JW DH MW. Wrote the paper: LY YH JW.

References

1. Bonan GB (2008) Forests and climate change: forcings, feedbacks, and the climate benefits of forests. Science 320: 1444–1449.
2. Cadisch G, Giller KE (1996) Driven by nature: plant litter quality and decomposition. London: CAB International.
3. Kauffman JB, Hughes RF, Heider C (2009) Carbon pool and biomass dynamics associatd with deforestation, land use, and agricultural abandonment in the neotropics. Ecological Applications 19: 1211–1222.
4. Fang JY, Chen AP (2001) Dynamic forest biomass carbon pools in China and their significance. Acta Botanica Sinica 43: 967–973.
5. Fang JY, Chen AP, Peng CH, Zhao SQ, Ci LJ (2001) Changes in forest biomass carbon storage in China between 1949 and 1998. Science 292: 2320–2322.
6. Piao SL, Fang JY, Zhu B, Tan K (2005) Forest biomass carbon stocks in China over the past 2 decades: estimation based on integrated inventory and satellite data. Journal of Geophysical research- Biogeoscience 110, G01006.
7. China State Forestry Bureau (2005) The sixth National Forest Inventory (1999–2003). Beijing: China Forestry Publishing House.
8. Peng SL, Hou YP, Chen BM (2009) Vegetation restoration and its effects on carbon balance in Guangdong Province, China. Restoration Ecology 17: 487–494.
9. Wu JP, Liu ZF, Wang XL, Sun YX, Zhou LX, et al. (2011) Effects of understory removal and tree girdling on soil microbial community composition and litter decomposition in two Eucalyptus plantations in South China. Functional Ecology 25: 921–931.
10. Chen DM, Zhang Y, Lin YB, Chen H, Fu SL (2009) Stand level estimation of root respiration for two subtropical plantations based on in situ measurement of specific root respiration. Forest Ecology Management 257: 2088–2097.
11. Yarie J (1980) The role of understory vegetation in the nutrient cycle of forested ecosystems in the mountain hemlock biogeoclimatic zone. Ecology 6: 1498–1514.
12. Nilsson M, Wardle DA (2005) Understory vegetation as a forest ecosystem driver: evidence from the northern Swedish boreal forest. Frontiers in Ecology and Environment 3: 421–428.
13. George LO, Bazzaz FA (1999) The fern understory as an ecological filter: emergence and establishment of canopy-tree seedlings. Ecology 80: 833–845.
14. Wang J, Ren H, Yang L, Duan WJ (2009) Establishment and early growth of introduced indigenous tree species in typical plantations and shrubland in South China. Forest Ecology and Management 258: 1293–1300.
15. Ehrenfeld JG, Kourtev P, Huang W (2001) Changes in soil functions following invasions of exotic understory plants in deciduous forests. Ecological Applications 11: 1287–1300.
16. O'Connell AM (1986) Effect of legume understory on decomposition and nutrient content of eucalypt forest litter. Plant and Soil 92: 235–248.
17. Li YQ, Xu M, Sun OJ, Cui WC (2004) Effects of root and litter exclusion on soil CO_2 efflux and microbial biomass in wet tropical forests. Soil Biology and Biochemistry 36: 2111–2114.
18. Wu J, Liu Z, Chen D, Huang G, Zhou L, et al. (2011) Understory plants can make substantial contributions to soil respiration: Evidence from two subtropical plantations. Soil Biology and Biochemistry 43: 2355–2357.
19. Flora of China Editorial committee (1999) Flora of China. Beijing: Science Press.
20. Lin XX (2004) The regulation of growth and development of *Dicranopteris dichotoma* and its propagative technology. Fujian Soil and Water Conservation 16: 60–62.
21. Ye JX, Hong RC (1987) The effect of the maceration extract of *Dicranopteris dichotoma* on the growth of several plant species. Acta Phytoeclogica et Geobotanica Sinica 11: 203–211.
22. Yuan YR, Li XY (2007) Preliminary study on the mechanism of allelopathy in *Dicranopteris pedata*. Journal of Anhui Agricultural Science 35: 5047–5048.
23. Hui D, Jackson RB (2009) Assessing interactive responses in litter decomposition in mixed species litter. Plant Soil 314: 263–271.
24. Swift MJ, Heal OW, Anderson JM (1979) Decomposition in terrestrial ecosystems. Oxford: Blackwell Scientific Publications.
25. Dearden FM, Wardle DA (2008) The potential for forest canopy litterfall interception by a dense fern understorey, and the consequences for litter decomposition. Oiks 117: 83–92.
26. Liu YC, Liu QJ, Wang HQ, Ma ZQ, Xu WJ (2008) Characteristics of biomass allocation of *Dicranopteris dichotoma*. Chinese Journal of Ecology 27: 705–711.
27. Zhu HG, Zhao JL, Wen YG, Hou RH (2010) Leaf area calculation model of *E. urophylla* and *E. grandis×E. urophylla*. Journal of Anhui Agricultural Science. 38.
28. Zhang JG, Fu SL, Wen DZ, Zhang LL (2009) Relationships of key leaf traits of 16 woody plant species in low subtropical China. Journal of Tropical and Subtropical Botany 17: 395–400.
29. Liu WF, Fan HB, Gao CF, Huang RZ, Su BQ (2009) Litter production and its nutrient fluxes in an age sequence of Eucalyptus plantations. Chinese Journal of Ecology 28: 1928–1934.
30. Xue L, He YJ, Qu M, Wu M, Xu Y (2005) Water holding characteristics of litter in plantations in South China. Acta Phytoeologica Sinica 29: 415–421.
31. Alvarez-Sanchez J, Guevara S (1999) Litter interception on *Astrocaryum mexicanum* Liebm. (Palmae) in a tropical rain forest. Biotropica 31: 89–92.
32. Dirks I, Navon Y, Kanas D, Dumbur R, Grünzweig JM (2010) Atmospheric water vapor as driver of litter decomposition in Mediterranean shrubland and grassland during rainless seasons. Global Change Biology 16: 2799–2812.
33. Schimel JP, Gulledge JM, Clein-Curley JS, Lindstrom JE, Braddock JF (1999) Moisture effects on microbial activity and community structure in decomposing birch litter in the Alaskan taiga. Soil Biology and Biochemistry 31: 831–838.
34. He X, Lin Y, Han G, Ma T (2013) Litterfall interception by understory vegetation delayed litter decomposition in *Cinnamomum camphora* plantation forest. Plant and Soil 372: 207–219.
35. Wang XL, Zhao J, Wu JP, Chen H, Lin YB, et al. (2010) Impacts of understory species removal and/or addition on soil respiration in a mixed forest plantation with native species in Southern China. Forest Ecology and Management 261: 1053–1060.
36. Yashiro Y, Shizu Y, Adachi T, Ohtsuka T, Lee NY, et al. (2012) The effect of dense understory dwarf bamboo (*Sasa senanensis*) on soil respiration before and after clearcutting of cool temperate deciduous broad-leaved forest. Ecological Research 27: 577–586.
37. Deng Q, Liu SZ, Liu JX, Meng Z, Zhang DQ (2007) Contribution of litter-fall to soil respiration and its affecting factors in southern subtropical forests of China. Advance in Earth Sciences 22: 976–986.
38. Wang GJ, Tian DL, Yan WD, Zhu F, Xiang WH, et al. (2009) Effects of aboveground litter exclusion and addition on soil respiration in a *Cunninghamia lanceolata* plantation in China. Chinese Journal of Plant Ecology 33: 739–747.
39. Wang LL, Song CC, Guo YD, Liu DY, Yang GS (2009) Contribution of litter to soil respiration under different land-use types in Sanjiang Plain. Environmental Science 30: 3130–3135.
40. Wang YD, Wang HM, Ma ZQ, Wen XF, Li QK, et al. (2009) Contribution of aboveground litter decomposition to soil respiration in a subtropical coniferous plantation in Southern China. Asia-Pacific Journal of Atmospheric Sciences 45: 137–147.
41. Wu CS, Sha LQ, Zhang YP (2012) Effect of litter on soil respiration and its temperature sensitivity in a montane evergreen broad-leaved forest in Ailao Mountains, Yunnan. Journal of North-East Forestry University 40: 37–40.
42. Zimmermann M, Meir P, Bird M, Malhi Y, Ccahuana A (2009) Litter contribution to diurnal and annual soil respiration in a tropical montane cloud forest. Soil Biology and Biochemistry 41: 1338–1340.

43. Gough C, Seiler J (2004) The influence of environmental, soil carbon, root, and stand characteristics on soil CO_2 efflux in loblolly pine (*Pinus taeda* L.) plantations located on the South Carolina Coastal Plain. Forest Ecology and Management 191: 353–363.
44. Tang X, Liu S, Zhou G, Zhang D, Zhou C (2006) Soil-atmospheric exchange of CO_2, CH_4, and N_2O in three subtropical forest ecosystems in southern China. Global Change Biology 12: 546–560.
45. Xu M, Qi Y (2001) Soil-surface CO_2 efflux and its spatial and temporal variations in a young ponderosa pine plantation in northern California. Global Change Biology 7: 667–677.
46. Huang Y, Zhou G, Tang X, Jiang H, Zhang D, et al. (2011) Estimated soil respiration rates decreased with long-term soil microclimate changes in successional forests in Southern China. Environmental Management 48: 1189–1197.
47. Fang C, Smith P, Moncrieff J, Smith J (2005) Similar response of labile and resistant soil organic matter pools to changes in temperature. Nature 433: 57–59.
48. Luo YQ, Wan S, Hui D, Wallace LL (2001) Acclimatization of soil respiration to warming in a tall grass prairie. Nature 413: 622–625.
49. Raich JW, Schlesinger WH (1992) The global carbon dioxide flux in soil respiration and its relationship to vegetation and climate. Tellus 44: 81–99.

11

Conserving Tropical Tree Diversity and Forest Structure: The Value of Small Rainforest Patches in Moderately-Managed Landscapes

Manuel A. Hernández-Ruedas[1], Víctor Arroyo-Rodríguez[1]*, Jorge A. Meave[2], Miguel Martínez-Ramos[1], Guillermo Ibarra-Manríquez[1], Esteban Martínez[3], Gilberto Jamangapé[4], Felipe P. L. Melo[5], Bráulio A. Santos[6]

1 Centro de Investigaciones en Ecosistemas, Universidad Nacional Autónoma de México, Morelia, Michoacán, Mexico, **2** Departamento de Ecología y Recursos Naturales, Facultad de Ciencias, Universidad Nacional Autónoma de México, Mexico City, Mexico, **3** Departamento de Botánica, Instituto de Biología, Universidad Nacional Autónoma de México, Mexico City, Mexico, **4** Ejido Loma Bonita, Ocosingo, Chiapas, Mexico, **5** Departamento de Botânica, Universidade Federal de Pernambuco, Recife, Pernambuco, Brazil, **6** Departamento de Sistemática e Ecologia, Universidade Federal da Paraíba, João Pessoa, Paraíba, Brazil

Abstract

Rainforests are undergoing severe deforestation and fragmentation worldwide. A huge amount of small forest patches are being created, but their value in conserving biodiversity and forest structure is still controversial. Here, we demonstrate that in a species-rich and moderately-managed Mexican tropical landscape small rainforest patches (<100 ha) can be highly valuable for the conservation of tree diversity and forest structure. These patches showed diverse communities of native plants, including endangered species, and a new record for the country. Although the number of logged trees increased in smaller patches, patch size was a poor indicator of basal area, stem density, number of species, genera and families, and community evenness. Cumulative species-area curves indicated that all patches had a similar contribution to the regional species diversity. This idea also was supported by the fact that patches strongly differed in floristic composition (high β-diversity), independently of patch size. Thus, in agreement with the land-sharing approach, our findings support that small forest patches in moderately-managed landscapes should be included in conservation initiatives to maintain landscape heterogeneity, species diversity, and ecosystem services.

Editor: Norman W.H. Mason, Landcare Research, New Zealand

Funding: This research was funded by the Programa de Apoyo a Proyectos de Investigación e Innovación Tecnológica (PAPIIT), DGAPA-UNAM (Projects IA-203111, IB-200812 and RR-280812). MAHR obtained a scholarship from the CONACyT, Mexico. The Centro de Investigaciones en Ecosistemas (UNAM) provided logical support. H. Ferreira and A. Palencia provided technical support, and A. Navarrete provided the SPOT images. C. Dobler digitized the maps and S. Nicasio-Arzeta elaborated the Figure 1. A. Garmendia, M. San-José and M. Solà assisted the field work. MAHR is grateful to M.G.F. Mendes, K.M. Demetrio, U. González-Gordillo, and their families for their support. The authors are indebted to the local people of the Marqués de Comillas región. Part of this paper was written while MAHR was on a research stay at the Departamento de Botânica, Universidade Federal de Pernambuco, and Departamento de Sistemática e Ecologia, Universidade Federal da Paraíba funded by the Programa de Apoyo a los Estudios de Posgrado (PAEP), UNAM, and the Programa de Becas Mixtas para Becarios CONACyT nacionales, CONACyT. The funders had no role in study design, data collection and analysis, decision to publish, or preparation of the manuscript.

Competing Interests: The authors have declared that no competing interests exist.

* E-mail: victorarroyo_rodriguez@hotmail.com

Introduction

Approximately 83% of earth's land surface has been altered by human action [1], leading to the rapid destruction and fragmentation of terrestrial ecosystems. In the tropics, habitat loss and degradation are main drivers of biodiversity loss [2], but the effectiveness of forest patches to retain large subsets of species diversity and forest structure is still controversial [3,4], particularly when considering tropical plants [5,6]. For example, whereas some studies argue that highly fragmented tropical landscapes can maintain high levels of the original diversity [7,8], others demonstrate that species diversity in forest patches declines rapidly after forest fragmentation [9,10].

In particular, increasing tree mortality rates have been reported near forest edges [9], particularly among emergent species [11,12]. This process may lead to declines in tree species richness in smaller forest patches [10,13], particularly near forest edges [11,12]. Logging and tree mortality can also alter forest structure, as they usually provoke the collapse of tree biomass [9]. However, under certain circumstances (e.g., in recently fragmented landscapes with lower deforestation levels), small and large forest patches can maintain similar tree species diversities [7,8]. Thus, to correctly assess the conservation value of small forest patches it is necessary to conduct more studies, encompassing landscapes and regions with contrasting management histories and intensities of land use. Such studies are particularly valuable if performed within tropical biodiversity hotspots [6,7,14].

The Mesoamerican region is among the most important tropical biodiversity hotspots because of their high number of species and high rates of deforestation [15]. The Lacandon rainforest contributes greatly to the floristic diversity of Mesoamerica, especially the Mexican portion of this forest, where ca. 3000 vascular plant species have been recorded [16]. Unfortunately, the rapid loss and fragmentation of this forest are seriously threatening this extraordinary biodiversity. Several studies have described the composition and forest structure of the Mexican Lacandon rainforest [e.g., 16–18]; however, most of them have been carried out in regenerating forest patches or within the Chajul Biological Station; an area of well-preserved continuous forest. Here we assess, for the first time, the value of small old-growth forest patches (<100 ha) for the maintenance of tree diversity and forest structure in the region. Specifically, we evaluate whether tree species diversity and forest structure are related to patch size, and assess the relative contribution of small forest patches to regional diversity.

Evidence indicates that the value of small patches for biodiversity conservation in human-modified landscapes is higher in recently fragmented landscapes [6], with higher remaining forest cover [7,19,20], and embedded in a heterogeneous landscape matrix [21,22]. Biodiversity maintenance is also dependent on the 'health' of food webs [4]. Deforestation in our study area is relatively recent (<40 years ago), the remaining forest cover is considerably high (ca. 40%; Figure 1), the matrix surrounding forest patches is very heterogeneous (see Methods), and the region still maintains their original fauna [23]. Within this landscape context, we hypothesized that small rainforest patches would be highly valuable for the conservation of tree diversity and forest structure. This hypothesis implies weak correlations between patch size and vegetation attributes, and a high contribution of small forest patches to regional diversity.

Materials and Methods

Ethics Statement

All necessary permits were obtained for this study. It adhered to the laws of the Mexican Government (SEMARNAT, Secretaría de Medio Ambiente y Recursos Naturales) to sample and collect animals and plants in Lacandona (permit no. SGPA/DGVS/09606). Although our institution, Universidad Nacional Autónoma de Mexico (UNAM), does not yet have an Institutional Review Board (IRB) or a similar governing body of ethics, this project was approved by the institutional authorities from UNAM and the Programa de Apoyo a Proyectos de Investigación e Innovación Tecnológica (PAPIIT), DGAPA-UNAM (Projects IA-203111, IB-200812 and RR-280812). The owners of the forest patches gave us the permission to perform the research in the study sites.

Study Area

The Lacandon rainforest comprises parts of Belize, Guatemala, and Mexico, and is one of the largest areas of tropical rainforest (ca. 800,000 ha) in Mesoamerica. The study was carried out in the Mexican portion of the Lacandon rainforest, Chiapas, Mexico (16°04′–16°21′N, 90°40′–91°06′W; 100–400 m a.s.l.; Figure 1). The climate is warm and humid, with a mean annual temperature of 24.1°C and average annual rainfall of 2875 mm.

In 1978, the Montes Azules Biosphere Reserve (MABR) was created, encompassing 331,200 ha of continuous forest. Outside this reserve, the Marqués de Comillas region (MCR) encompasses 203,999 ha of disturbed forests and human settlements. The dominant vegetation types in both regions are semi-deciduous tropical rainforest and lowland tropical rainforest [16,24]. During the 1970s, the MCS suffered a colonization process by a multicultural population [25], and it has been largely deforested since, especially between 1984 and 1986 [25], owing to national policies that promoted agroforestry, agriculture, and cattle ranching. The main cause of the massive deforestation in the MCS was the so-called 'Plan Piloto Forestal de Marqués de Comillas', resulting in tree down-cutting in 30% of the area during 1998 and 1999 [24]. Nowadays, approximately 40% of old-growth forest cover remains as remnant patches of variable sizes embedded in a heterogeneous matrix of secondary forests, human settlements, cattle pastures, and crop fields.

The study was conducted in two adjacent areas separated by the Lacantún River: the continuous forest of MABR and the fragmented forest of MCR (Figure 1). In the MCR, we selected 26 isolated old-growth forest patches (ranging from 2 to 92 ha). The forest patches were selected according to the following criteria: (i) they represented discrete forest masses separated by cattle pastures and crop fields, (ii) they were separated from each other as much as possible to increase independence among them, and (iii) they were all located between 100 and 250 m a.s.l. to avoid the effects of altitudinal (and soil) gradients. The size of all forest patches was calculated using recent SPOT-5 satellite images (March 2011) and Quantum GIS 1.7.4. (Quantum GIS Development Team, 2012). In addition, as reference sites (100% of forest cover), we selected four buffer areas of 100 ha within the continuous forest of MABR, which were spaced 4 km apart (on average) and at least 1 km from the nearest edge of the Lacantún River (Figure 1).

Vegetation Sampling

We used Gentry's [26] protocol to sample vegetation in each site, but recording only tree species (including palms) with a diameter at breast height (dbh) ≥2.5 cm in ten 50×2-m transects randomly located at each site (0.1 ha per site). Unidentified specimens were collected for their identification at the National Herbarium of Mexico (MEXU, Mexico City). Plant names followed the Missouri Botanical Garden electronic database (Tropicos) available at http://www.tropicos.org. To assess differences among sites in tree mortality and logging, we also recorded the number of dead and logged trees within transects.

Data Analyses

To avoid pseudoreplication problems, we summed up the data of the 10 transects from each site and considered the sites as independent samples for the following analyses. In particular, we recorded the number of families, genera, species, stems, and dead and logged trees within each site. We also estimated the total basal area by summing the basal area of all stems (assuming the stem cross-section area as a circle). To assess the inventory completeness within each site, we employed the coverage estimator suggested by Chao and Jost [27], which is a less biased estimator of sample completeness:

$$Cn = 1 - \frac{f1}{n}\left[\frac{(n-1)f1}{(n-1)f1 + 2f2}\right].$$

where f_1 and f_2 are the number of singletons and doubletons in the sample, respectively, and n is the number of individuals. Because sample coverage varied among sites (Table 1), our estimates of the number of species, genera and families could be biased by differences in sample completeness [27]. Thus, following Chao & Jost [27] and Chao et al. [28], we estimated the number of species, genera and families in each site using coverage-based rarefaction (interpolation) with the iNEXT software [29]. In particular, we considered the minimum completeness for all sites (0.82 in species, 0.86 in genera and 0.93 in families) to have reliable and comparable estimates of the number of species, genera and families based on samples of equal completeness (equal coverage) [27,28].

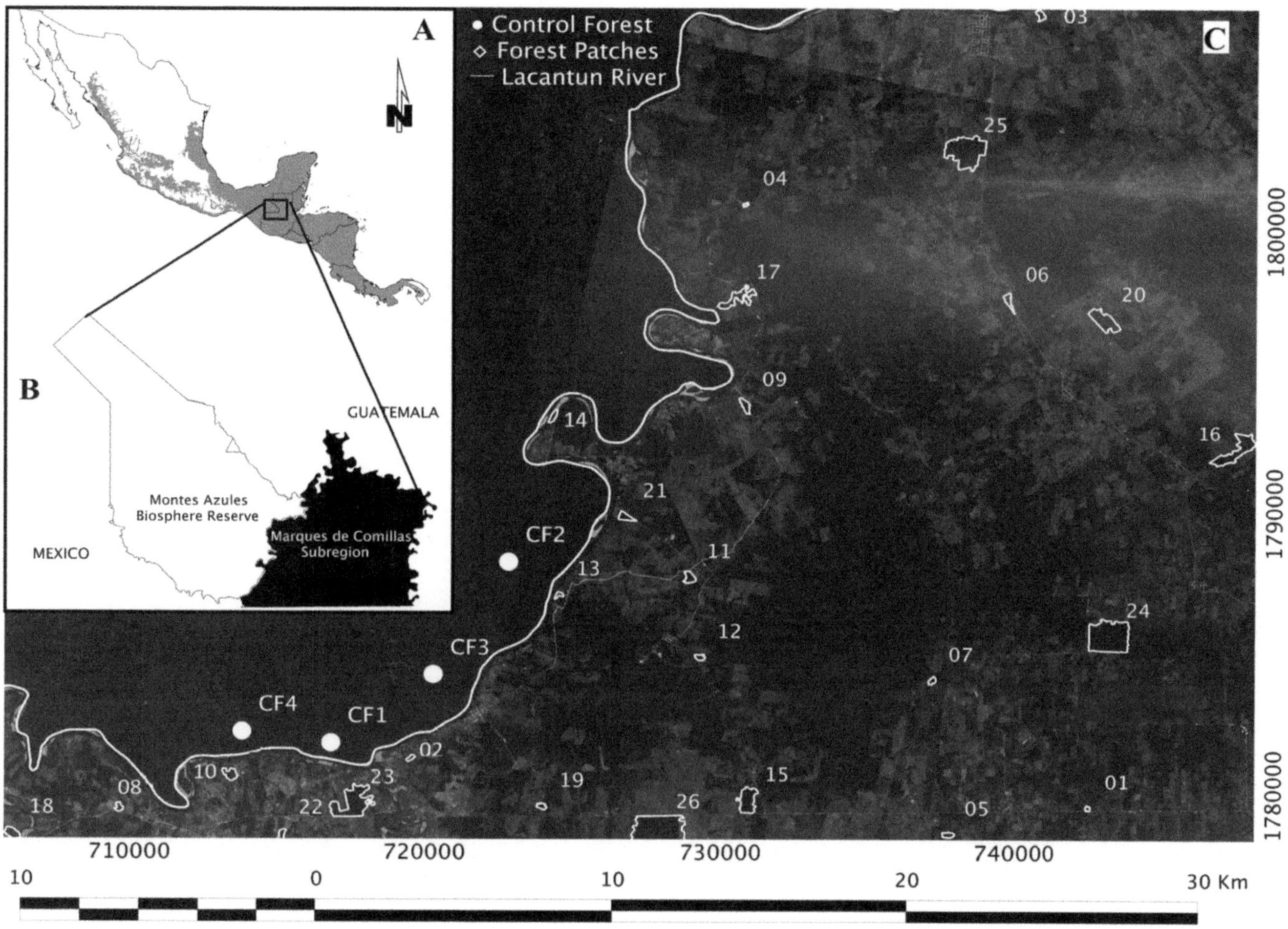

Figure 1. Location of the Mesoamerican biodiversity hotspot (a), and the study area in the Lacandon rainforest, Mexico (b). The 26 forest patches located in the Marqués de Comillas region and the four reference areas studied within the continuous forest of the Montes Azules Biosphere Reserve are also indicated (c).

We also assessed changes in community evenness among sites using the evenness factor (*EF*) proposed by Jost [30]: $EF={}^{2}D/S$, where ${}^{2}D$ is the inverse Simpson concentration and S is the total number of species in the sample. This index ranges between 1 (all species are equally common) and nearly $1/S$ (the community is dominated by one species [30]). Roughly, *EF* is interpreted as the proportion of dominant species in the community [30].

To test whether tree species diversity and forest structure were related to patch size, we performed simple linear regression models between patch size ($\log_{10}$-transformed), taken as the independent variable, and all vegetation attributes (i.e., rarefied number of species, genera, families, density of stems and dead and logged trees, total basal area, and *EF* per site), taken as response variables. Reference sites were included in the models assuming that they covered 100 ha. To evaluate the changes in species composition across sites, we estimated two measures of compositional similarity, the Jaccard index based on the presence and absence of each species, and the abundance-based Bray-Curtis index. Then, we performed a non-metric multidimensional scaling (NMDS) using the Bray-Curtis index to assess the overall differences in species composition among sites. The sites were classified based on different size classes (<10 ha, 10–30 ha, >30–50 ha, >50–100 ha, and continuous forest sites), and a multivariate analysis of variance (MANOVA) was performed to test for significant differences among size categories over the ordination dimensions generated by the NMDS analysis. We also used Mantel tests to assess whether variations in floristic composition between sites (Bray-Curtis index) were related to patch location (i.e., log-transformed inter-site isolation distances) and/or to differences between sites in size. *P*-values were calculated using the distribution of the *R* coefficients obtained from 10,000 permutations.

Finally, to assess the contribution of small forest patches to regional diversity, we used cumulative species-area curves [7,31]. The sites were ordered based on their area from small to large, and then from large to small [32]. We plotted the cumulative number of species versus the number of sites added to compare the cumulative observed number of species in the small to large and large to small curves. To evaluate the effect of habitat subdivision on tree species diversity, we calculated the Saturation Index (SI) of Quinn and Harrison [32] using the procedure used by Cook [33]. This index ranges from -1 to $+1$; based on the same sampling effort, positive values indicate that collections of small patches contain more species than fewer larger patches, while negative values indicate the opposite. To assess the relative importance of the landscape spatial context on these results, we compared these species-area curves with those found in three neighboring fragmented tropical landscapes located in the Los Tuxtlas

Table 1. Number of plant species, genera and families in 26 forest patches (FP) and 4 reference sites within a continuous forest (CF) sampled in the fragmented Lacandon rainforest, Chiapas, Mexico.

Sites	Size (ha)	Species			Genera			Families		
		Obs	$\hat{C}_n$	Exp	Obs	$\hat{C}_n$	Exp	Obs	$\hat{C}_n$	Exp
FP1	2.4	65	0.92	50	56	0.94	45	34	0.97	29
FP2	2.8	38	0.91	27	33	0.93	24	18	0.96	14
FP3	4.8	36	0.94	22	32	0.95	23	26	0.96	22
FP4	5.7	58	0.87	48	51	0.89	45	26	0.97	22
FP5	6	67	0.88	54	55	0.91	46	33	0.97	29
FP6	12.7	58	0.89	49	49	0.93	43	27	0.98	25
FP7	12.7	72	0.86	68	59	0.90	56	29	0.96	26
FP8	14.5	58	0.92	45	51	0.93	43	27	0.99	23
FP9	18	43	0.85	39	36	0.89	34	22	0.96	20
FP10	20.3	66	0.89	52	50	0.93	39	31	0.97	27
FP11	20.6	68	0.89	58	57	0.91	50	35	0.96	32
FP12	22.4	62	0.85	55	54	0.89	49	32	0.93	32
FP13	28.1	68	0.84	62	59	0.86	59	35	0.94	32
FP14	33	31	0.91	14	26	0.93	15	19	0.97	16
FP15	33.2	76	0.88	64	61	0.92	53	34	0.97	30
FP16	33.2	55	0.90	44	46	0.93	39	29	0.97	26
FP17	33.4	59	0.83	57	47	0.88	45	26	0.95	25
FP18	37.8	59	0.87	55	51	0.90	49	28	0.99	27
FP19	37.8	75	0.82	75	64	0.86	64	35	0.97	31
FP20	42.3	69	0.86	62	57	0.90	51	29	0.97	26
FP21	51.3	47	0.92	39	43	0.93	38	22	0.99	19
FP22	62.9	64	0.88	51	49	0.92	39	33	0.96	30
FP23	65.5	63	0.89	50	53	0.92	44	25	0.97	19
FP24	72.1	60	0.89	43	48	0.93	36	29	0.97	24
FP25	75.9	58	0.90	41	51	0.92	40	28	0.97	24
FP26	91.9	52	0.92	30	45	0.94	31	24	0.99	20
CF1	100	74	0.82	74	61	0.90	58	30	0.98	27
CF2	100	51	0.89	35	38	0.93	27	25	0.97	20
CF3	100	70	0.85	64	57	0.86	57	33	0.95	30
CF4	100	55	0.88	47	46	0.92	41	24	0.98	21

The observed values (Obs) and the expected (Exp) after a coverage-based rarefaction (interpolation) are indicated for each site. The coverage estimator suggested by Chao and Jost [27] is also indicated for each case ($\hat{C}_n$, see Methods).

Table 2. Vegetation structure in 26 forest patches (FP) and 4 reference sites within a continuous forest (CF) sampled in the fragmented Lacandon rainforest, Chiapas, Mexico.

Sites	Vegetation characteristics				
	# stems	# dead trees	# logged trees	BA (m^2)	*EF*
FP1	260	18	15	3.00	0.39
FP2	168	13	4	3.67	0.35
FP3	209	16	0	3.93	0.37
FP4	197	12	4	2.57	0.42
FP5	228	21	3	3.40	0.47
FP6	185	13	10	2.79	0.50
FP7	193	17	5	4.35	0.54
FP8	214	22	2	3.84	0.56
FP9	140	7	7	3.36	0.17
FP10	231	9	5	2.70	0.50
FP11	216	28	7	2.67	0.55
FP12	192	19	3	4.31	0.39
FP13	199	12	5	4.15	0.52
FP14	174	10	2	4.84	0.29
FP15	237	24	0	1.98	0.54
FP16	206	9	3	2.65	0.33
FP17	165	13	0	3.57	0.41
FP18	151	10	4	2.95	0.63
FP19	239	18	0	2.48	0.27
FP20	209	11	6	3.54	0.47
FP21	154	8	7	2.57	0.59
FP22	229	28	10	3.85	0.41
FP23	238	10	1	5.11	0.36
FP24	255	8	1	3.32	0.23
FP25	233	17	1	3.02	0.44
FP26	279	14	0	2.29	0.13
CF1	198	11	1	2.14	0.39
CF2	228	11	1	3.40	0.18
CF3	212	17	0	5.54	0.50
CF4	183	8	0	3.36	0.25

The number of stems, dead trees, and logged trees are indicated, as well as basal area (BA, m^2) and the evenness factor (*EF*) estimated for ten 50×2-m transects (0.1 ha) per site. Sites are arranged in order of increasing size (see site sizes in Table 1). *EF* represents the proportion of dominant species in the community.

rainforest, Mexico, but with higher deforestation levels (24%, 11% and 4% of remaining forest cover). The species-area curves in this case were generated using the same protocol described above, based on a vegetation data set that was collected by VAR using the same sampling methodology and sampling effort as in the Lacandona rainforest [7,8]. The only difference was that Arroyo-Rodríguez et al. [7,8] included not only trees and palms, but also lianas. Thus, we excluded lianas from the original data set of VAR to make the results comparable with the present study.

Results

We recorded 6222 trees from 55 families, 144 genera, and 228 species (including 24 morphospecies) in a total sampled area of 3 ha (Table S1). On average (± SD), we recorded 28±5 families, 50±9 genera, and 59±12 species per site (Table 1). Stem density averaged 207±34 stems/0.1 ha (Table 2). We also recorded 15±6 dead trees, and 4±4 logged trees per site. Basal area averaged 3.4±0.87 m^2 per site. Finally, the evenness factor averaged 0.41±0.13 (Table 2).

The four largest families in the region were Fabaceae (27 species), Rubiaceae (13), Malvaceae (11), and Moraceae (10), which combined accounted for 30% of all species recorded (Table S1). In contrast, 25 families (46%) were represented by only one species. Acanthaceae, Asteraceae, Cannabaceae, Malpighiaceae, and Nyctaginaceae (9% of all families) were recorded in only one forest patch, and Passifloraceae (2%) was exclusive to a single continuous forest area. The most diverse genus was *Piper* (Piperaceae) with six species, whereas 106 genera (75%) were represented by a single species (Table S1).

All species were native to the region. Eleven out of the 204 species (5%) are classified within a risk category by the Mexican government: *Bactris major*, listed as species subjected to special protection; *Astronium graveolens*, *Bravaisia integerrima*, *Calophyllum*

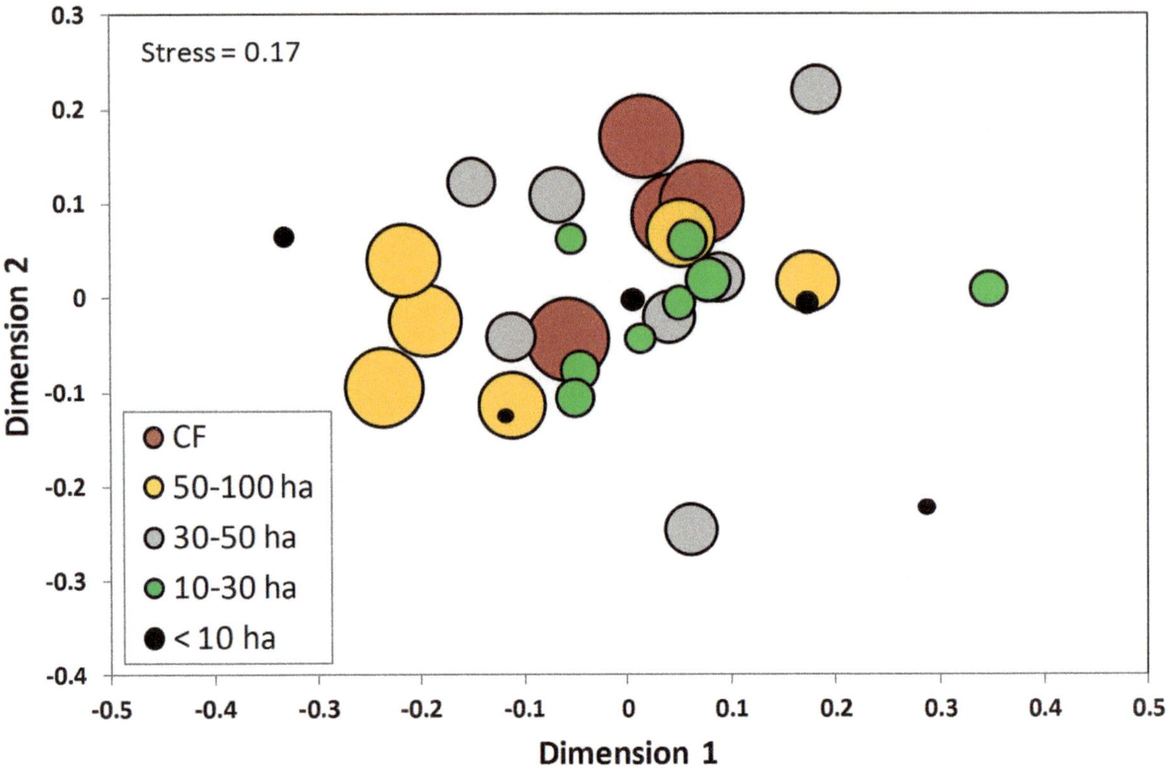

Figure 2. Non-metric multidimensional scaling ordination of tree species across each site sampled in the fragmented Lacandon rainforest, Mexico. Dot size is directly proportional to site size. CF = continuous forest.

brasiliense, *Cryosophila stauracantha*, *Geonoma interrupta*, *Guatteria anomala*, *Magnolia mexicana*, and *Spondias radlkoferi*, listed as threatened species; and *Ormosia isthmensis* and *Vatairea lundellii*, listed as endangered species (Table S1). All these species were recorded in forest patches (Table S1). Furthermore, we reported for the first time the occurrence of *Ouratea crassinervia* Engl. (Ochnaceae) for the Mexican flora (see online Appendix S1). This species was recorded in two areas within the continuous forest and in two forest patches (Table S1).

Analyzing the full set of sites (n = 30), there were no significant relationships between patch size and vegetation attributes: rarefied number of families ($R^2 = 0.002$, $P = 0.80$), genera ($R^2 = 0.022$, $P = 0.43$), and species ($R^2 = 0.034$, $P = 0.33$), density of stems ($R^2 = 0.016$, $P = 0.50$) and dead trees ($R^2 = 0.041$, $P = 0.28$), basal area ($R^2 = 0.001$, $P = 0.91$) and evenness factor ($R^2 = 0.045$, $P = 0.26$). Only the density of logged trees was negatively correlated with patch size ($R^2 = 0.207$, $P = 0.01$). Based on the presence and absence of species within the sites (Jaccard index), we found that on average they shared 27±9% species, indicating a high species turnover among sites. Considering the abundance of individuals, the NMDS ordination showed a strong variation in species composition among sites, and this variation was unrelated to patch size (Figure 2), as MANOVA did not detect differences among size classes in any of the ordination dimensions generated by the NMDS analysis ($F_{8,48} = 0.896$, $P = 0.53$). Furthermore, the Mantel tests revealed a non-significant correlation between floristic similarity and differences in size among sites ($r = -0.03$, $P = 0.5$), and a significant but very weak correlation between floristic similarity and geographic distances among sites ($r = 0.15$, $P = 0.004$).

The cumulative species-area curves displayed similar patterns of species accumulation in both treatments, either from small to large and large to small (Figure 3). The saturation index (SI = −0.003) indicated that the resulting species-area curves were indistinguishable, regardless of whether small or large patches were added first. This pattern was almost identical to that observed in the landscape with lower deforestation level (24% of forest cover) at Los Tuxtlas, Mexico; however, in this region, the landscape with highest deforestation level (4% of forest cover) showed that collections of small patches contained a lower number of species than fewer larger patches (SI = −0.14; Figure 4), indicating that the relative contribution of small forest patches to regional diversity declines in landscapes with lower forest cover.

Discussion

It is clear from our study that small forest patches in the fragmented Lacandon rainforest can still harbor original levels of tree species diversity and forest structure, as if they were part of a continuous forest. These patches maintained diverse communities of native tree species, including endangered taxa (see Table S1) and a new record for the Mexican flora (see additional and recent new records for the region in [34–36]). Surprisingly, patch size was a poor indicator of tree species diversity and forest structure, indicating that even the smallest forest patches were highly similar to continuous forest sites in terms of total basal area, community evenness, stem density and number of species, genera, and families. Also, the proportion of pioneer species (i.e., early-colonizer species) within each site – an indicative of forest disturbance [10–13] – was independent of site size (see online Figure S1 in Appendix S2). Finally, each patch conserved a particular floristic composition (i.e., high β-diversity among sites),

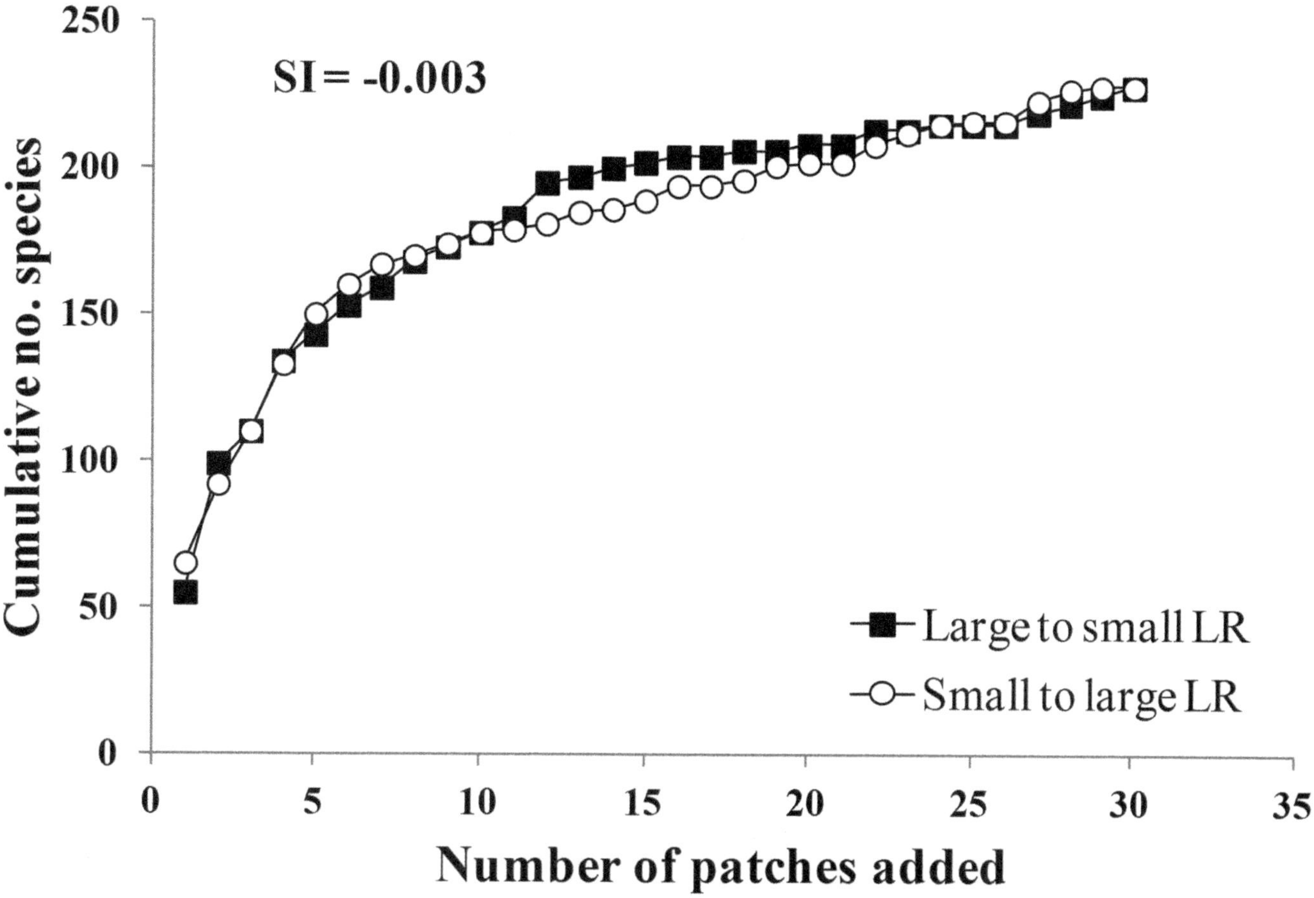

Figure 3. Cumulative species-area curves for Lacandon rainforest, Mexico. Cumulative number of species sampled in 0.1 per patch versus number of patches added. Patches were added from small to large or large to small, respectively, and then corresponding cumulative species counts were obtained. The value of Saturation Index is indicated.

which was determined neither by differences in patch size nor by geographical distance. Thus, the preservation of both small and large patches is necessary for conserving regional plant species diversity.

Our findings contrast with those found in other fragmented tropical landscapes with higher deforestation levels ([7]; Figure 4) and longer history of contemporary human activity (e.g., [10]). For example, several studies from South American fragmented rainforests indicate that many tree species can disappear from small forest patches surrounded by pastures or sugarcane fields [10–13]. This local species extirpation has been mainly related to edge effects, i.e., to altered biotic and abiotic conditions near forest edges [10–13]. Several studies have reported that a number of disturbance-adapted native species (e.g., short- and long-lived light-demanding species) can dominate plant communities in fragmented forests with pronounced edge effects [11,37]. This process can reduce community evenness and increase floristic homogenization (i.e., reduced β-diversity) across multiple spatial scales [13,38].

These contrasting findings may be related to the fact that species maintenance in forest patches largely depends on the landscape context [6]. For example, several studies have demonstrated that species extirpation in smaller forest patches is only evident below a given threshold of landscape forest cover (e.g., <11% [7]; <10–30% [39]; <20% [40]), and our study landscape has a notably greater remaining forest cover (ca. 40%). Additionally, human population density in the region is amongst the lowest in Mexico (4.8 people/km^2 [41]), and regional land use is dominated by familiar cattle ranches and *milpas*, a sort of small-scale agriculture for subsistence and local commerce that is known to have a lower impact if compared with agribusiness activities typical from other tropical countries, such as Brazil [2,6]. This reinforces the context-dependency in the determination of how much biodiversity a human-modified landscape can harbour [6] and claims for the attention to the socioecological context where tropical landscapes rely.

The ability of small forest patches to maintain biodiversity also depends on the landscape matrix [6,22,42], as it determines the severity of edge effects [43–45]. Edge effects, or the environmental changes that occur near forest edges, can increase tree mortality rates in forest patches, particularly when the landscape is dominated by agricultural matrices of intensive uses (e.g., large pastures and monocultures [9,43]). However, in fragmented landscapes with a relatively high forest cover and a heterogeneous matrix, such as the Lacandon region, we may expect such environmental changes to be attenuated [43,44]. Thus, these processes, in conjunction with the fact that many tropical trees are long-lived [46], can contribute to prevent tree mortality and maintain the overall patterns of forest structure even in the smallest patches, at least in the short-term. Certainly, further studies are needed to monitor tree mortality and assess the long-term maintenance of species diversity within these patches.

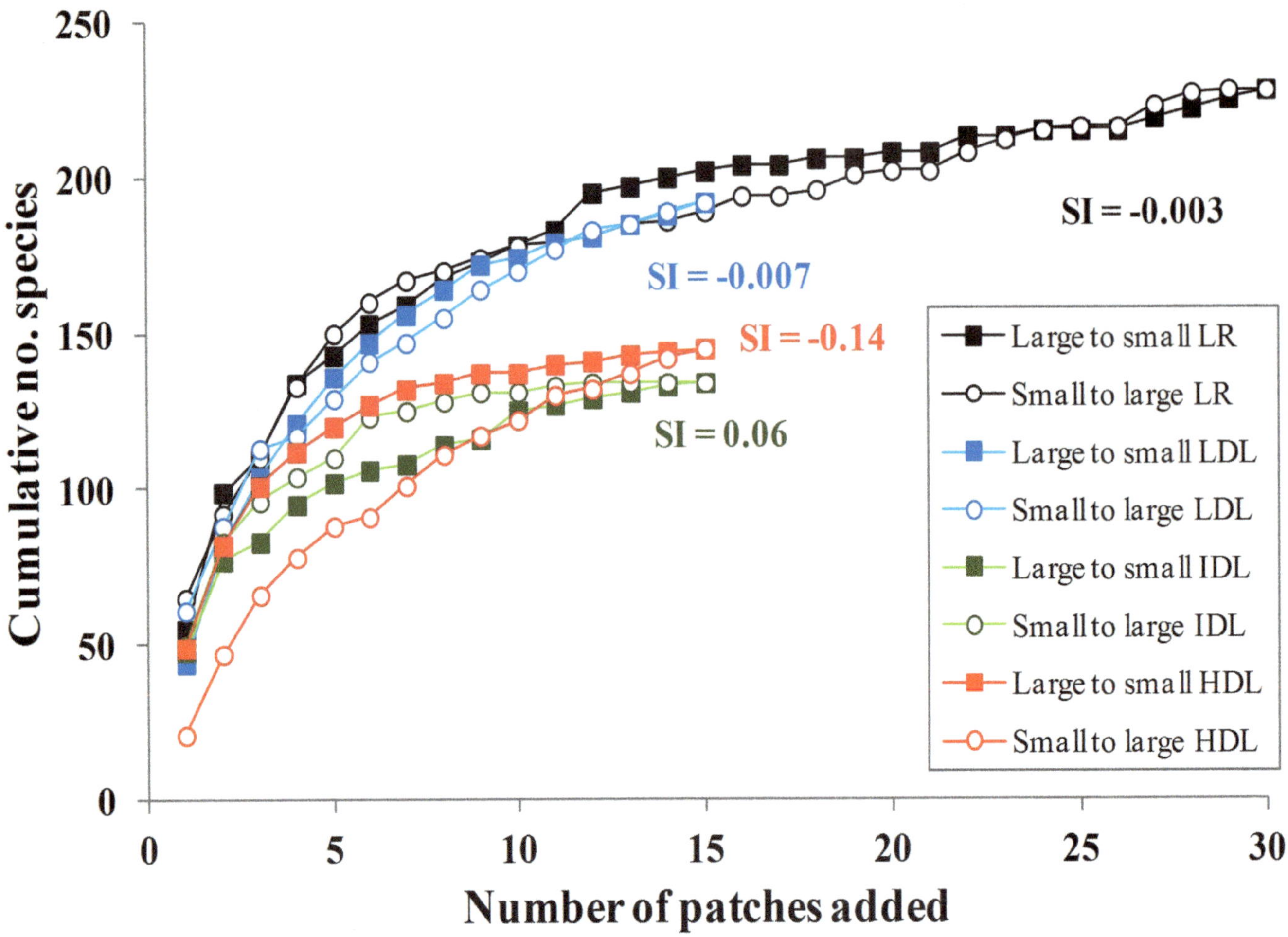

Figure 4. Cumulative number of tree species sampled in 0.1 ha per patch versus the number of patches added in four landscapes with different degrees of deforestation. In the Lacandon rainforest (LR; with approximately 40% of remaining forest cover) we sampled 30 forest patches (present study). Using the same sampling methodology and sampling effort as in LR, in the Los Tuxtlas rainforest, Mexico we sampled 15 patches in three landscapes (see Arroyo-Rodríguez et al. [7]): LDL (lowest deforestation, 24%); IDL (intermediate deforestation, 11%); and HDL (highest deforestation, 4%). In all cases, patches were added from large to small (squares) or small to large (circles), and then, corresponding cumulative species counts were obtained. The Saturation Index (SI) indicated that only in the landscape with highest deforestation level (HDL) larger patches accumulated more species than smaller ones with the same sample intensity.

Management and Conservation Implications

Our findings indicate that a half century of human occupancy in the region has not resulted in significant changes in tree assemblages within small forest patches. This is crucial for providing suitable habitat required by many closed-canopy forest animal groups, and the preservation of the ecological processes in which they are involved (e.g., seed dispersal, seed predation, and pollination), and that are necessary for forest regeneration. The importance of small patches goes beyond their values as species reservoirs, as they can also act as buffer zones in which some management practices can be allowed, reducing the pressure over the largest patches. Even the smallest patches can contribute to landscape connectivity, because a set of several small patches distributed across the landscape may serve as stepping stones, increasing the connectivity among patches and enhancing inter-patch animal and plant dispersal [7,21,31].

Unfortunately, the long-term persistence of these patches is uncertain, as the focus of national and international conservation efforts has been to safeguard the largest forest remnants. In particular, economic mechanisms and incentives such as the payments for ecosystem services (PES) have been increasingly implemented as a promissory conservation approach in this and many other Mexican forests [47]. Yet in the Lacandon region, the recently created 'Programa Especial Selva Lacandona 2013' [48] only offers economic benefits (ca. 77 US$/ha/year) to landowners and local communities that are preserving >25 ha of forest. Below this area, local people cannot apply to this program, limiting the economic and ecological value of small forest patches. This situation will obviously increase the anthropogenic pressures on these small, but highly valuable forest patches.

Nonetheless, we are not arguing that funding and other conservation efforts of larger forest patches should be re-directed towards smaller ones. The great value of large blocks of original ecosystem for conservation is incontrovertible, but what is questionable is the exclusion of smaller forest patches from conservation initiatives. In the tropical rainforest of Mexico, either in the Gulf of Mexico [7,8] or in the Lacandon region (present study), not protecting smaller forest patches is a waste of biodiversity. We stress that, supporting a land-sharing approach, future conservation and management initiatives should consider the importance of protecting both small- and large-sized forest patches.

Acknowledgments

The Centro de Investigaciones en Ecosistemas (UNAM) provided logical support. H. Ferreira, A. Valencia and A. López provided technical support, and A. Navarrete provided the SPOT images. C. Dobler digitized the maps and S. Nicasio-Arzeta elaborated the Figure 1. A. Garmendia, M. San-José and M. Solà assisted the field work. MAHR is grateful to M.G.F. Mendes, K.M. Demetrio, U. González-Gordillo, and their families for their support. We are indebted to the local people of the Marqués de Comillas región. Part of this paper was written while MAHR was on a research stay at the Departamento de Botânica, Universidade Federal de Pernambuco, and Departamento de Sistemática e Ecologia, Universidade Federal da Paraíba.

Author Contributions

Conceived and designed the experiments: VAR MAHR JAM MMR. Performed the experiments: MAHR GJ GIM EM. Analyzed the data: MAHR VAR. Contributed reagents/materials/analysis tools: VAR MAHR JAM MMR EM FPLM BAS. Wrote the paper: MAHR VAR JAM MMR EM BAS FPLM GIM. Plant species identification: MAHR GIM EM.

References

1. Sanderson EW, Jaiteh M, Levy MA, Redford KH, Wannebo AV, et al. (2002) The human footprint and the last of the wild. Bioscience 52: 891–904.
2. Laurance WF, Sayer J, Cassman KG (2014) Agricultural expansion and its impacts on tropical nature. Trends Ecol Evol 29: 107–116.
3. Lindenmayer D, Hobbs RJ, Montague-Drake R, Alexandra J, Bennett A, et al. (2008) A checklist for ecological management of landscapes for conservation. Ecol Lett 11: 78–91.
4. Gardner TA, Barlow J, Chazdon R, Ewers RM, Harvey CA, et al. (2009) Prospects for tropical forest biodiversity in a human-modified world. Ecol Lett 12: 561–582.
5. Chazdon RL, Harvey CA, Komar O, Griffith DM, Ferguson BG, et al. (2009) Beyond reserves: a research agenda for conserving biodiversity in human-modified tropical landscapes. Biotropica 41: 142–153.
6. Melo FPL, Arroyo-Rodríguez V, Fahrig L, Martínez-Ramos M, Tabarelli M (2013) On the hope for biodiversity-friendly tropical landscapes. Trends Ecol Evol 28: 461–468.
7. Arroyo-Rodríguez V, Pineda E, Escobar F, Benítez-Malvido J (2009) Value of small patches in the conservation of plant-species diversity in highly fragmented rainforest. Conserv Biol 23: 729–739.
8. Arroyo-Rodríguez V, Cavender-Bares J, Escobar F, Melo FPL, Tabarelli M, et al. (2012) Maintenance of tree phylogenetic diversity in a highly fragmented rain forest. J Ecol 100: 702–711.
9. Laurance WF, Delamônica P, Laurance SG, Vasconcelos HL, Lovejoy TE (2000) Rainforest fragmentation kills big trees. Nature 404: 836.
10. Silva JMC, Tabarelli M (2000) Tree species impoverishment and the future flora of the Atlantic forest of northeast Brazil. Nature 404: 72–74.
11. Laurance WF, Nascimento HEM, Laurance SG, Andrade A, Ribeiro JELS, et al. (2006) Rapid decay of tree-community composition in Amazonian forest fragments. Proc Natl Acad Sci USA 103: 19010–19014.
12. Santos BA, Peres CA, Oliveira MA, Grillo A, Alves-Costa CP, et al. (2008) Drastic erosion in functional attributes of tree assemblages in Atlantic forest fragments of northeastern Brazil. Biol Conserv 141: 249–260.
13. Tabarelli M, Peres CA, Melo FPL (2012) The 'few winners and many losers' paradigm revisited: emerging prospects for tropical forest biodiversity. Biol Conserv 155: 136–140.
14. Turner IM, Corlett RT (1996) The conservation value of small, isolated fragments of lowland tropical rain forest. Trends Ecol Evol 11: 330–333.
15. Myers N, Mittermeier RA, Mittermeier CG, da Fonseca GAB, Kent J (2000) Biodiversity hotspots for conservation priorities. Nature 403: 853–858.
16. Martínez E, Ramos CH, Chiang F (1994) Lista florística de la Lacandona, Chiapas. Bol Soc Bot Mex 54: 99–177.
17. Ibarra-Manríquez G, Martínez-Ramos M (2002) Landscape variation of liana communities in a Neotropical rain forest. Plant Ecol 160: 91–112.
18. van Breugel M, Martínez-Ramos M, Bongers F (2006) Community dynamics during early secondary succession in Mexican tropical rain forests. J Trop Ecol 22: 663–674.
19. Fahrig L (2003) Effects of habitat fragmentation on biodiversity. Annu Rev Ecol Evol Syst 34: 487–515.
20. Pardini R, Bueno AA, Gardner TA, Prado PI, Metzger JP (2010) Beyond the fragmentation threshold hypothesis: regime shifts in biodiversity across fragmented landscapes. PLoS ONE 5: e13666.
21. Ewers RM, Didham RK (2006) Confounding factors in the detection of species responses to habitat fragmentation. Biol Rev 81: 117–142.
22. Fahrig L, Baudry J, Brotons L, Burel FG, Crist TO, et al. (2011) Functional landscape heterogeneity and animal biodiversity in agricultural landscapes. Ecol Lett 14: 101–112.
23. Garmendia A, Arroyo-Rodríguez V, Estrada A, Naranjo E, Stoner KE (2013) Landscape and patch attributes impacting medium- and large-sized terrestrial mammals in a fragmented rain forest. J Trop Ecol 29: 331–344.
24. Martínez E (2003) Marquéz de Comillas. In: Lichtinger V, Enkerlin E, Enríquez C, García P, editors. La Deforestación en 24 Regiones PRODERS. Mexico City: Secretaría de Medio Ambiente y Recursos Naturales, 124–131.
25. Aguilar LF, Mora CS (1992) Colonización y deterioro de la selva Lacandona. Rev Geogr 116: 67–84.
26. Gentry AH (1982) Patterns of Neotropical plant species diversity. In: Hecht MK, Wallace B, Prance GT, editors. Evolutionary Biology. New York: Plenum Press, 1–84.
27. Chao A, Jost L (2012) Coverage-based rarefaction and extrapolation: standardizing samples by completeness rather than size. Ecology 93: 2533–2547.
28. Chao A, Gotelli NJ, Hsieh TC, Sander EL, Ma KH, et al. (2014) Rarefaction and extrapolation with Hill numbers: a framework for sampling and estimation in species diversity studies. Ecol Monogr (doi:10.1890/13-0133.1).
29. Hsieh TC, Ma KH, Chao A (2013) iNEXT online: interpolation and extrapolation (Version 1.3.0) Software. Available: http://chao.stat.nthu.edu.tw/blog/software-download/.
30. Jost L (2010) The relation between evenness and diversity. Diversity 2: 207–232.
31. Fischer J, Lindenmayer DB (2002) Small patches can be valuable for biodiversity conservation: two case studies on birds in southeastern Australia. Biol Conserv 106: 129–136.
32. Quinn JF, Harrison SP (1988) Effects of habitat fragmentation and isolation on species richness: evidence from biogeographic patterns. Oecologia 75: 132–140.
33. Cook RR (1995) The relationship between nested subsets, habitat subdivision, and species diversity. Oecologia 101: 204–210.
34. Ibarra-Manríquez G, Cornejo-Tenorio G, González-Castañeda N, Piedra-Malagón EM, Luna A (2012) El género *Ficus* L. (Moraceae) en México. Bot Sci 90: 389–452.
35. Santos BA, Lombera R, Benítez-Malvido J (2009) New records of *Heliconia* (Heliconiaceae) for the region of Chajul, Southern Mexico, and their potential use in biodiversity-friendly cropping systems. Rev Mex Biodivers 80: 857–860.
36. Schutzman B, Vovides AP (1998) A new *Zamia* (Zamiaceae, Cycadales) from Eastern Chiapas, Mexico. Novon 8: 441–446.
37. Santos GGA, Santos BA, Nascimento HEM, Tabarelli M (2012) Contrasting demographic structure of short- and long-lived pioneer tree species on Amazonian forest edges. Biotropica 44: 771–778.
38. Lôbo D, Leão T, Melo FPL, Santos AMM, Tabarelli M (2011) Forest fragmentation drives Atlantic forest of northeastern Brazil to biotic homogenization. Divers Distrib 17: 287–296.
39. Andrén H (1994) Effects of habitat fragmentation on birds and mammals in landscapes with different proportions of suitable habitat: a review. Oikos 71: 355–366.
40. Fahrig L (1998) When does fragmentation of breeding habitat affect population survival? Ecol Model 105: 273–292.
41. SEDESOL (2014) Secretaría de Desarrollo Social. Microrregiones-Catálogo de localidades. Available: http://www.microrregiones.gob.mx/catloc/LocdeMun.aspx?tipo = clave&campo = loc&ent = 07&mun = 116 Accessed 10 January 2014.
42. Perfecto I, Vandermeer J. Wright A (2009) Nature's matrix: linking agriculture, conservation and food sovereignty. London: Earthscan.

43. Mesquita R, Delamonica P, Laurance WF (1999) Effects of surrounding vegetation on edge-related tree mortality in Amazonian forest fragments. Biol Conserv 91: 129–134.
44. Murcia C (1995) Edge effects in fragmented forests: implications for conservation. Trends Ecol Evol 10: 58–62.
45. Pinto SRR, Mendes G, Santos AMM, Dantas M, Tabarelli M, et al. (2010) Landscape attributes drive complex spatial microclimate configuration of Brazilian Atlantic forest fragments. Trop Conserv Sci 3: 399–402.
46. Martínez-Ramos M, Álvarez-Buylla ER (1998) How old are tropical rain forest trees? Trends Plant Sci 3: 400–405.
47. CONAFOR (2010) Coordinación General de Producción y Productividad. Gerencia de Servicios Ambientales del Bosque. Mexico City: Comisión Nacional Forestal.
48. CONAFOR (2013). Programa Especial Selva Lacandona 2013. Coordinación General de Producción y Productividad. Gerencia de Servicios Ambientales del Bosque. Mexico City: Comisión Nacional Forestal.

12

Large Sample Area and Size Are Needed for Forest Soil Seed Bank Studies to Ensure Low Discrepancy with Standing Vegetation

You-xin Shen[1,2]*, Wei-li Liu[3], Yu-hui Li[4], Hui-lin Guan[4]

1 Key Laboratory of Tropical Forest Ecology, Xishuangbanna Tropical Botanical Garden, Chinese Academy of Sciences, Menglun, Yunnan, China, **2** Restoration Ecology Group, Xishuangbanna Tropical Botanical Garden, Chinese Academy of Sciences, Kunming, China, **3** Kunming Metallurgical Research Institute, Kunming, China, **4** Yunnan Normal University, Kunming, China

Abstract

A large number of small-sized samples invariably shows that woody species are absent from forest soil seed banks, leading to a large discrepancy with the seedling bank on the forest floor. We ask: 1) Does this conventional sampling strategy limit the detection of seeds of woody species? 2) Are large sample areas and sample sizes needed for higher recovery of seeds of woody species? We collected 100 samples that were 10 cm (length) ×10 cm (width) ×10 cm (depth), referred to as larger number of small-sized samples (LNSS) in a 1 ha forest plot, and placed them to germinate in a greenhouse, and collected 30 samples that were 1 m×1 m×10 cm, referred to as small number of large-sized samples (SNLS) and placed them (10 each) in a nearby secondary forest, shrub land and grass land. Only 15.7% of woody plant species of the forest stand were detected by the 100 LNSS, contrasting with 22.9%, 37.3% and 20.5% woody plant species being detected by SNLS in the secondary forest, shrub land and grassland, respectively. The increased number of species vs. sampled areas confirmed power-law relationships for forest stand, the LNSS and SNLS at all three recipient sites. Our results, although based on one forest, indicate that conventional LNSS did not yield a high percentage of detection for woody species, but SNLS strategy yielded a higher percentage of detection for woody species in the seed bank if samples were exposed to a better field germination environment. A 4 m^2 minimum sample area derived from power equations is larger than the sampled area in most studies in the literature. Increased sample size also is needed to obtain an increased sample area if the number of samples is to remain relatively low.

Editor: Franco Biondi, University of Nevada, Reno, United States of America

Funding: Funding was provided by grants from the "Light in Western China" Program of Chinese Academy of Sciences and the National Nature Scientific Foundation of China (41262013, 41371514, 41361075). The funders had no role in study design, data collection and analysis, decision to publish, or preparation of the manuscript.

Competing Interests: The authors have declared that no competing interests exist.

* Email: yxshen@xtbg.ac.cn

Introduction

The presence of viable seeds in the soil has been extensively documented [1–3], and seeds in the soil may include those dispersed directly from existing plants at the site, persisting from previous vegetation, and arriving via secondary dispersal from other sites [4–6]. Ecologists, evolutionary biologists and restorationists are interested in the composition and density of seeds in the soil seed bank and thus many studies have been conducted [2,5]. However, the lack of consistent guidelines for sampling seed banks remains as an obstacle in making it possible to compare the results of seed bank studies, especially of forests.

In the great majority of studies, soil samples are taken by excavating soil cores or quadrats of varying depths, spreading the soil in trays under conditions suitable for germination, and identifying and counting the seedlings to calculate soil seed bank density and richness [4,5,7]. To save space and accelerate germination, soil samples can be sieved to reduce soil volume and then seeds are allowed to germinate [8,9]. Seeds have clustered spatial distributions [10], which produce large sampling variances and may lead to imprecise abundance estimates. The precision (narrow confidence interval) of seed-number estimates may be improved by taking a large number of small samples (LNSS) [11]. This LNSS sampling strategy is successful in grasslands and agricultural lands for estimation of both seed density and diversity [11,12]. However, in most forests, the LNSS method does not adequately detect seeds of woody species. Seeds of predominantly climax (woody) species in extant tropical forests often are absent from soil seed banks, but those of weedy and short-lived pioneer species can be abundant [13]. Clark *et al.* [14] reviewed the forest soil seed bank literature (90 studies) from 1969 to 1998 and found seeds of woody plants to be very rare, while those of herbaceous taxa were common. Hopfensperger [15] reviewed 108 articles published between 1945 and 2006 on the similarity between above and belowground species composition and found lower similarities in forests than in grasslands and in wetlands. Another comprehensive review from Europe also showed low similarities between forest-soil seed banks and standing vegetation [16]. However, on the forest floor, many species produce seedlings under a range of environmental conditions. This large discrepancy between lack of seeds in the soil seed bank studies and presence of seedlings limits valid interpretations of the soil seed bank and forest stand relationships, and it could lead to misjudgment of restoration potentials in degraded and disturbed forest sites.

Difficulties in detecting seeds of woody species in soil samples from forests may be attributed to spatially structured deposition patterns [17], short-lived seeds [16], ephemeral soil seed banks [15], and predation [18]. However, sampling methods, which could be part of the problem, are rarely discussed. Thus we need to consider minimum sample areas, the smallest area within which species are adequately represented, and the sample size, which varies widely among areas and research interests [14].

Appropriate sampled soil areas or volumes can be determined in a manner analogous to species-area curves for aboveground plants, and many such studies are found in the literature [19,20]. Forcella [12] found that the combined soil surface areas of replicates should be about 1000 cm^2 to ascertain the number of species with seed reserves in pasture soils in Australia. However, for forest soil seed banks, the species-area relationships suggest that a small sample size and soil surface areas might not yield a high detection percentage of woody species. For example, the largest number of species detected was only 34 in the accumulated species area curve for soil seed bank samples collected in a secondary tropical rain forest in Costa Rica [21]. These authors collected 121 soil samples (4.7 cm in diameter) at the intersection points of a 10×10 m grid in a 1 ha plot with a total area of 893 cm^2. Although they did not separate woody species from herbaceous species, a low detection percentage of woody species could be inferred since this number of species was far lower than the $135.5\pm6.5/0.5$ ha of understory species (not including canopy species) in the site [22]. Thus, we hypothesized that larger soil sample sizes and sampled areas will allow for more accurate determination of the presence and abundance of seeds of woody species in forest soils.

In our study, a large number of small soil sample (LNSS) and a small number of large sample (SNLS) were taken from a diverse karst forest in Yunnan Province of southern China, and the germination of woody species was determined. Our specific objectives were to: 1) Determine the effectiveness of the LNSS vs. the SNLS strategy in detecting woody-plant seeds in forest-soil seed banks; 2) Analyze the relationship between the number of woody species and sample area; 3) Suggest an appropriate sampling area and size for forest soil seed-bank studies.

Material and Methods

Study site

This study was conducted at Shilin Stone Forest Geographical Park (24°38′–24°58′N, 103°11′–103°29′E), Yunnan Province, southwest China. This area is 1950 m a.s.l., and it has a mean annual precipitation of 967.9 mm, 80% of which falls between May and October. Mean annual temperature is 16.2 °C, mean monthly maximum temperature is 20.7°C (July), and mean monthly minimum temperature is 8.2°C (January) [23]. The growing season coincides with the rainy season and with moderate temperature periods from late spring into summer [24]. The karst landscape consists of rock gaps, rock ditches, small rock caves and rock slots, with about 30% of the land surface being covered by soil. The soil is shallow and patchily distributed on or between these various rock surfaces. The primary forest on this karst land is an evergreen broadleaved forest mixed with few deciduous species [25,26]. Most tree species shed their seeds in autumn (September to November) [27]. All necessary permits were obtained for field survey and soil replacement experiment from Shilin Stone Forest

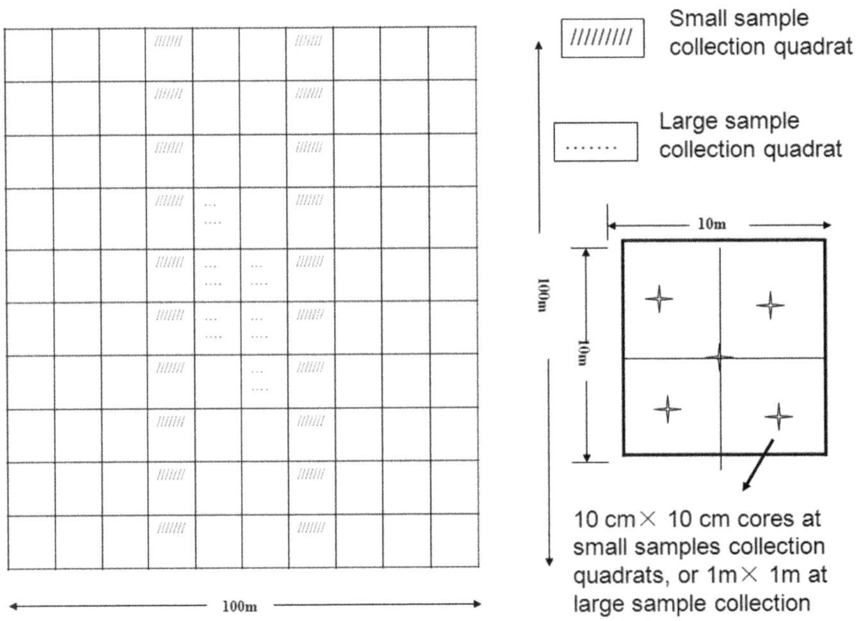

Figure 1. Illustration of sample designs for the vegetation survey and two types of soil sample collection for the soil seed bank study. A 100 m×100 m plot was divided into 10 m grids for the vegetation survey. Twenty quadrats of two vertical transects were selected for collection of a large number of small samples (LNSS) 100 cm×100 cm (slanted lines), and 6 quadrats were selected for collection a small number of large samples (SNLS) 1 m×1 m (dots).

Management Bureau, Shilin County. No endangered species was involved.

A 100 m×100 m plot was established on a hill slope and divided into a 10 m grid (Fig. 1). In each of the quadrats (10 m×10 m), tree individuals with the main shoot ≥3 cm DBH (diameter at breast height) were identified and recorded. At the center of each quadrat, a 2 m×2 m small quadrat was established, and all woody individuals >30 cm height were identified and recorded.

Soil sample collection methods

Two types of soil sample were collected to determine forest soil seed bank: 1) Large number of small samples (LNSS); 2) Small number of large samples (SNLS).

Two 100 m (length) ×10 m (width) vertical transects spaced 20 m apart along the slopes of the vegetation survey plot were selected. The transect was three grids wide (30 m) from one edge of the vegetation plot. All 20 quadrats in the two transects were chosen for LNSS. Six quadrats (10 m×10 m) at the most central part of the plot survey were selected for SNLS (Fig. 1). We further divided the sample quadrats into 5 m×5 m sub-quadrats. Soil samples 10 cm (length) ×10 cm (width) ×10 cm (depth) were collected at the center of the 20 LNSS sample quadrats and at the center of each sub-quadrat, thus 5 small sized soil samples were taken from each LNSS quadrat and 100 small samples were taken in total. Following the same arrangement, five 1 m (length) ×1 m (width) ×10 cm (depth) large sized soil samples were taken from each of SNLS quadrat and 30 larger-sized samples were taken in total.

All soil samples were collected at the end of November, when most of woody plants had shed their seeds. Those samples may contain both transient and persistent soil seeds [27]. Since the soil was patchily distributed, LNSS were taken nearest the marked sampling point to avoid collecting stones. For SNLS, we collected extra soil to compensate for soil-area loses if the 1 m×1 m sample sites were partially occupied by rock outcrops.

Seed germination of LNSS

All LNSS samples were transported to the Xishuangbanna Tropical Botanical Garden, Kunming (80 km away from the Park, 1892 m a.s.l.) after collection. Each LNSS was washed through a 4 mm mesh sieve to eliminate coarse materials, and then through a 0.21 mm mesh sieve to eliminate fine materials [8]. Seeds larger than the 4 mm mesh were carefully collected and returned to the residues. The concentrated residue, which contained all seeds from the sample, was spread out evenly onto a 3 cm layer of perlite in plastic seed trays (approx. 18 cm×10 cm×10 cm deep). Trays were placed in a nonheated greenhouse in the Xishuangbanna Tropical Botanical Garden, Kunming, and seedling emergence was monitored. The sides of the greenhouse were covered by a double layer of fine nylon mesh cloth and the roof with a sheet of transparent polyethylene. Other trays containing sterilized soil were placed randomly within the greenhouse to monitor airborne seed contamination. All trays were monitored and watered, usually twice a day. Seedlings of woody species were counted and discarded as soon as they could be identified, or were transplanted into 15-cm-diameter pots filled with fertile soil and grown until species identification was possible. Soil in the trays was stirred two to three times during the germination monitoring period, normally after a large flush of germination. Recording was ceased at the end of the following November, 1 year after soil sample collections.

Soil replacement and germination monitoring for SNLS

All the soil from the 30 SNLS was hand mixed, weighed and divided evenly into 30 parts (referred to as soil sources) to homogenize replicates.

Similar karst slopes covered with secondary forest, shrub and grass were chosen as recipient sites. There was no forest similar to that of the soil collection site in the areas surrounding the three recipient sites. Ten cm deep 1 m×1 m plots were prepared by removing all plants, soils and stones from each of the three sites. The 30 SNLS soil mixtures were transferred and placed on each of the 30 plots in all three recipient sites. A thin layer of rice straw was placed on the surface of each plot to limit moisture loss. The straw was removed between April 6–10, before the beginning of the rainy season (May). There were 1870–4800 seeds m^{-2} year round in primary forest in this study area [27], most of which were herbaceous. Counting those herbaceous seedlings in the 30 m^2 field plots was highly labor intensive. Further, preventing local herbaceous seed entry to the plot was impossible as many herbaceous seeds were small in size and dispersed through the year. Thus, we excluded herb-seedlings counting and discarded all of them. Seedlings of woody species were identified and counted at half- month intervals from June to September 2009, and then at 1-month intervals until December.

The number of live and dead seedlings of each species in each plot was counted at each observation time (t). After recording, dead seedlings were removed to prevent recounting. We did not remove live seedlings in any plots since they might contribute to the restoration of degraded sites. We denote TSt(x) as the total number of seeds germinated of species x up to time t, NSt(x) as the total seedlings of species x at time t, NAt(x) as live seedlings of species x at time t, and NDt(x) as dead seedlings of species x at time t, so the total number of seedlings of a species at a plot at time t was:

$$NS_{t(x)} = NA_{t(x)} + ND_{t(x)}$$

At counting time t+1, the number of seedlings of species x included the newly emerged ones, the live ones and dead ones of time t. The total number of seeds germinated at time t+1 was:

$TS_{t+1(x)} = TS_{t(x)} + (NS_{t+1(x)} - NA_{t(x)})$ if $NA_{t+1(x)} > NA_{t(x)}$ (if new seedlings emerged)

or

$TS_{t+1(x)} = TS_{t(x)}$ if $NA_{t+1(x)} \leq NA_{t(x)}$ (if no new seedlings emerged).

Thus, for species x in plot i, the total number of seeds germinated was the TS(x) at the end of December. The total number of seeds germinated in a plot was the sum of seeds germinated for each species.

Numerical analyses

The differences between total seed number and total species number were tested by one-way ANOVA among the three recipient sites for SNLS. Where main effects were significant ($P < 0.05$), Duncan's multiple range tests were used to compare differences between means of seed germination between recipient sites.

We defined infrequent species as those with only one individual in the vegetation plot, or only one germinated seed in LNSS or SNLS.

We transferred each set of soil-replacement results and soil seed bank data (Appendix S1) to S (species)-by-H (quadrat) incidence matrices (presence-absence). Sample order was randomized 100

Table 1. Comparisons of species composition in the forest survey and in the soil seed bank using two collection strategies: large number of small samples (LNSS) and small number of large samples (SNLS) to replace soils of secondary forest, shrub and grass land.

	Plant survey	SNLS-Secondary forest	SNLS- Shrub	SNLS-Grass	LNSS
Sample size	10 m×10 m	1 m×1 m	1 m×1 m	1 m×1 m	**10 cm×10 cm**
No. of samples	100	10	10	10	**100**
No. of species per sample (mean ±SD)*	13.2 ±3.04	7.2±1.29^{A}	10.8±1.54^{B}	5.5±1.14^{A}	**0.2±0.24**
No. of seeds/ m^{-2}*		33.4±4.6^{B}	40.5±8.6^{B}	15.9±2.4^{A}	**355±62.9**
Total no. of species	83	19	31	17	**13**
Infrequent species	24	1	4	1	**7**
Similarity**	1	0.313	0.393	0.202	**0.226**
Estimated richness	**85**	**19**	**31**	**17**	**13**

*Different uppercase letters indicate the significant difference between two SNLS sites.
** = 2c/(a+b), where c is the number of common species, a is the number of species in the vegetation and b is the number of species in the soil seed bank.

times to calculate a mean species accumulation curve within each plant survey plot, LNSS, and each of the recipient site for SNLS [28] in "EstimateS 9.10" [29]. Two-parameter power equations [30] often are used for species-area relationships [20,31]. The mean predicted species accumulations (S) were fitted to power equations for the area resampled (A) ($S=cA^{z}$) for vegetation, LNSS and each of the recipient site for SNLS.

Results

Eighty-three woody species from 75 genera in 55 families were identified in the 1 ha vegetation plot (Appendix S1). Twenty-four species had only one individual in the vegetation plot. Seeds of 32 species from 30 genera in 19 families germinated from the three SNLS recipient sites (Appendix S1). The number of species identified at the secondary forest site, shrub site and grass site was

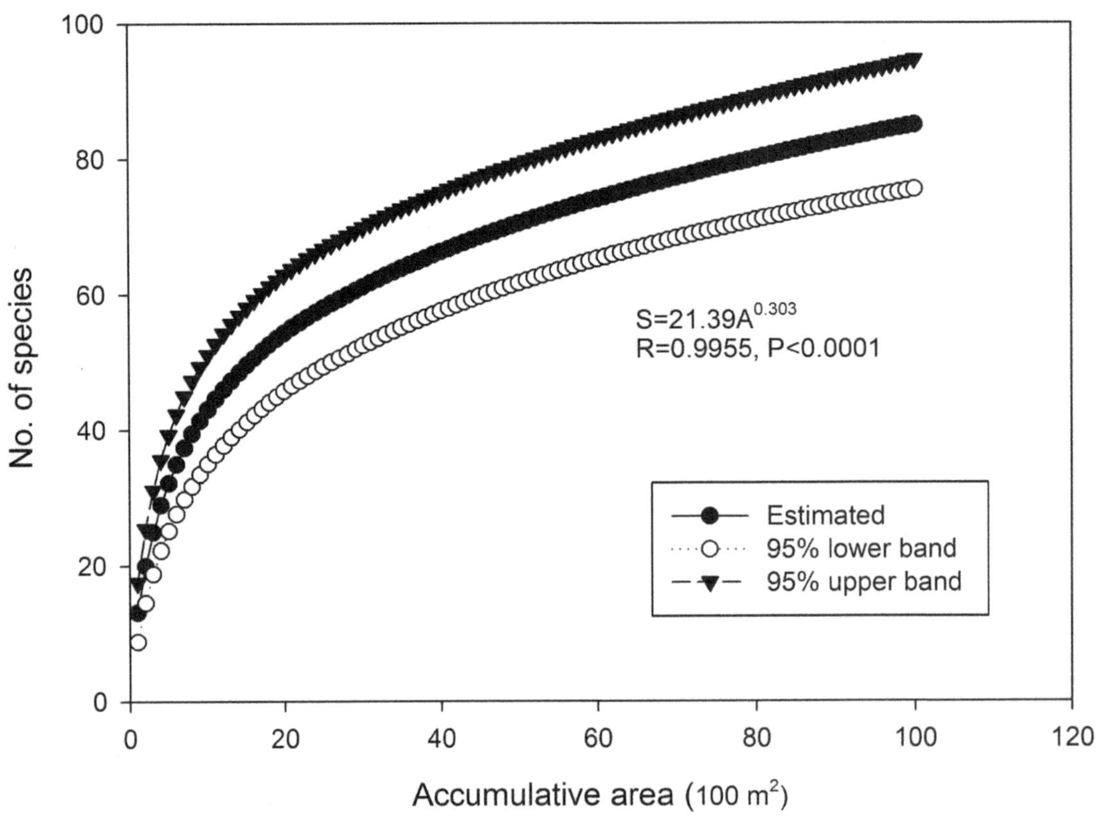

Figure 2. Mean woody species accumulation curve and 95% confidence intervals for karst forest in SW China. Samples were pooled from 100 quadrats of a 100 m×100 m vegetation survey and then randomized 100 times in "EstimateS 9.10" (Colwell 2013).

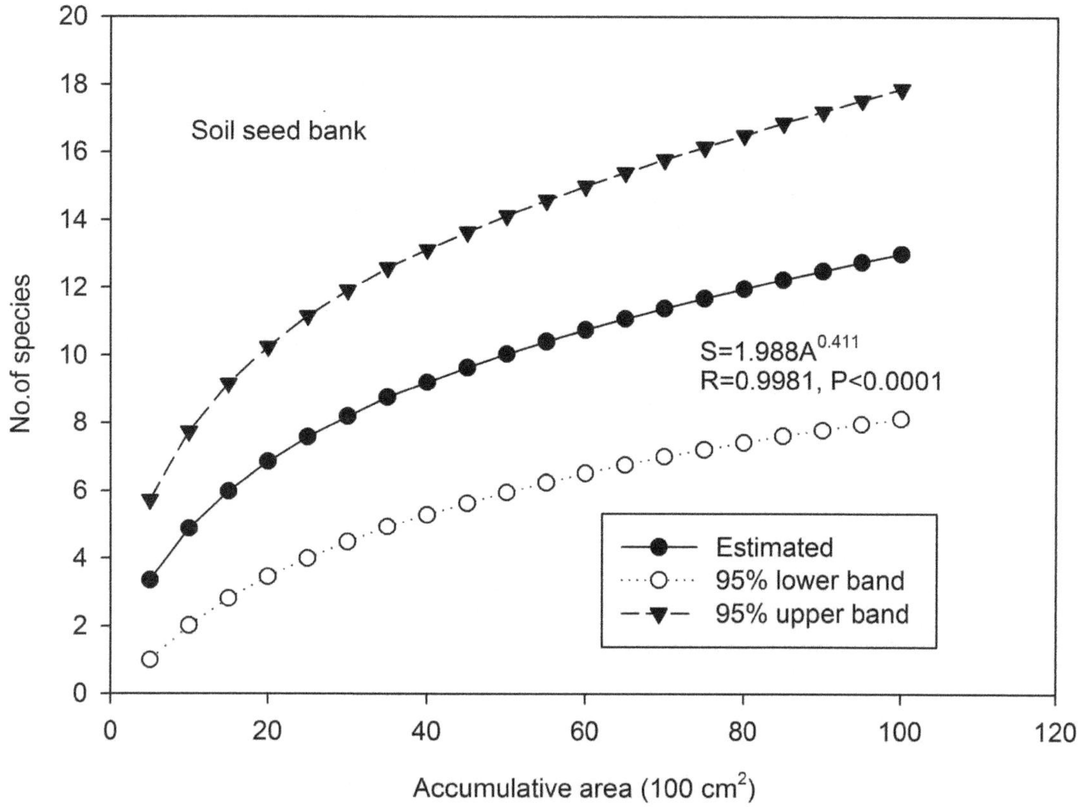

Figure 3. Mean woody species accumulation curve of seeds and its 95% confidence intervals in soil seed bank of a karst forest in SW China. Samples were pooled from 100 quadrats of the soil seed bank by collecting 10 cm×10 cm×10 cm small soil samples and then randomized 100 times in "EstimateS 9.10" (Colwell 2013).

22.9%, 37.3% and 20.5%, respectively, of the total number of woody species in the 1 ha vegetation plot. However, only 13 species from 11 genera of 9 families were identified from LNSS in the greenhouse, which only accounted for 15.7% of woody species in the forest stand (Table 1). Seeds of the most abundant five tree species, *Cyclobalanopsis glaucoides, Olea yunnanensis, Pistacia weinmannifolia, Neolitsea homilantha, Carpinus mobeigiana*, which had >100 individuals in the 1 ha plot, were found in the SNLS (Appendix S1). However, only seeds of two of these species, *N. homilantha* and *C. mobeigiana*, were found in LNSS plots (Appendix S1). Generally, the chances of finding seedlings of abundant trees, shrubs and woody lianas were higher in SNLS than in LNSS (Appendix S1). Many species were infrequently (only germinated in 1 sample) encountered in both LNSS and SNLS plots (Table 1). Two *Smilax* species were present in the forest stand. However, the *Smilax* seedlings could not be identified at the species level. Eight species present in the soil seed bank were absent from the vegetation.

On average, each 10 m×10 m forest stand plot contained 13.2±0.3 (mean ±SE) species, each 1 m×1 m SNLS geminated 5.5±1.14 (grassland) to 10.8±1.54 (shrub land) woody species via soil replacement, and each 10 cm×10 cm LNSS germinated only 0.2±0.24 woody species (Table 1). The number of seeds germinated from 1 m^2 soil in LNSS in greenhouse is far larger than any one of the SNLS in the field.

The number of species ($F = 6.49$, $P = 0.005$) and the number of seeds ($F = 4.76$, $P = 0.017$) that germinated in the three recipient sites differed significantly. Both the highest number of species and number of seeds were found at the shrub site (Table 1).

The number of species initially increases steeply with the sample areas pooled for vegetation plots, SNLS and LNSS, but then approaches an asymptote, in accordance with the power-law equation (Fig. 2, 3, 4). The simulated richness was 102.4% of the number of species observed in the 1 ha vegetation survey plot, and it was equal to the number of species identified in the three recipient sites of SNLS method and of species identified in the SNLS method.

Discussion

Even with 100 small-sized samples only 13 woody species were found by the LNSS method. This number is far lower than the number of woody species identified in the vegetation survey (83 species, Table 1), even excluding infrequent species (24 species that have only one individual each in vegetation plot). However, this result coincides with most other forest soil seed-bank studies, such as in North America [14], Europe [16], tropical wet forests [21], and central African rainforests [13]. On the other hand, 31, 19, and 17 species were identified in a total of 10 SNLS (1 m×1 m size) at the shrub, the secondary forest and grass recipient sites; this was 33.7%, 21.7% and 18.1%, respectively, of the total number of woody species in the 100 m×100 m vegetative plot. In addition, 84.5%, 51.8% and 46.3% of the estimated number of species in the six quadrats, from which SNLS samples were collected (by the power-law equation from the plant survey (Fig. 2), where $A = 6\times100$ m^2, $S = 36.7$). The SNLS method captured seeds of all dominant tree species in the standing vegetation. This means that by collecting 10 samples 1 m×1 m in size and placing them in an environment conducive to germination, such as that of shrub

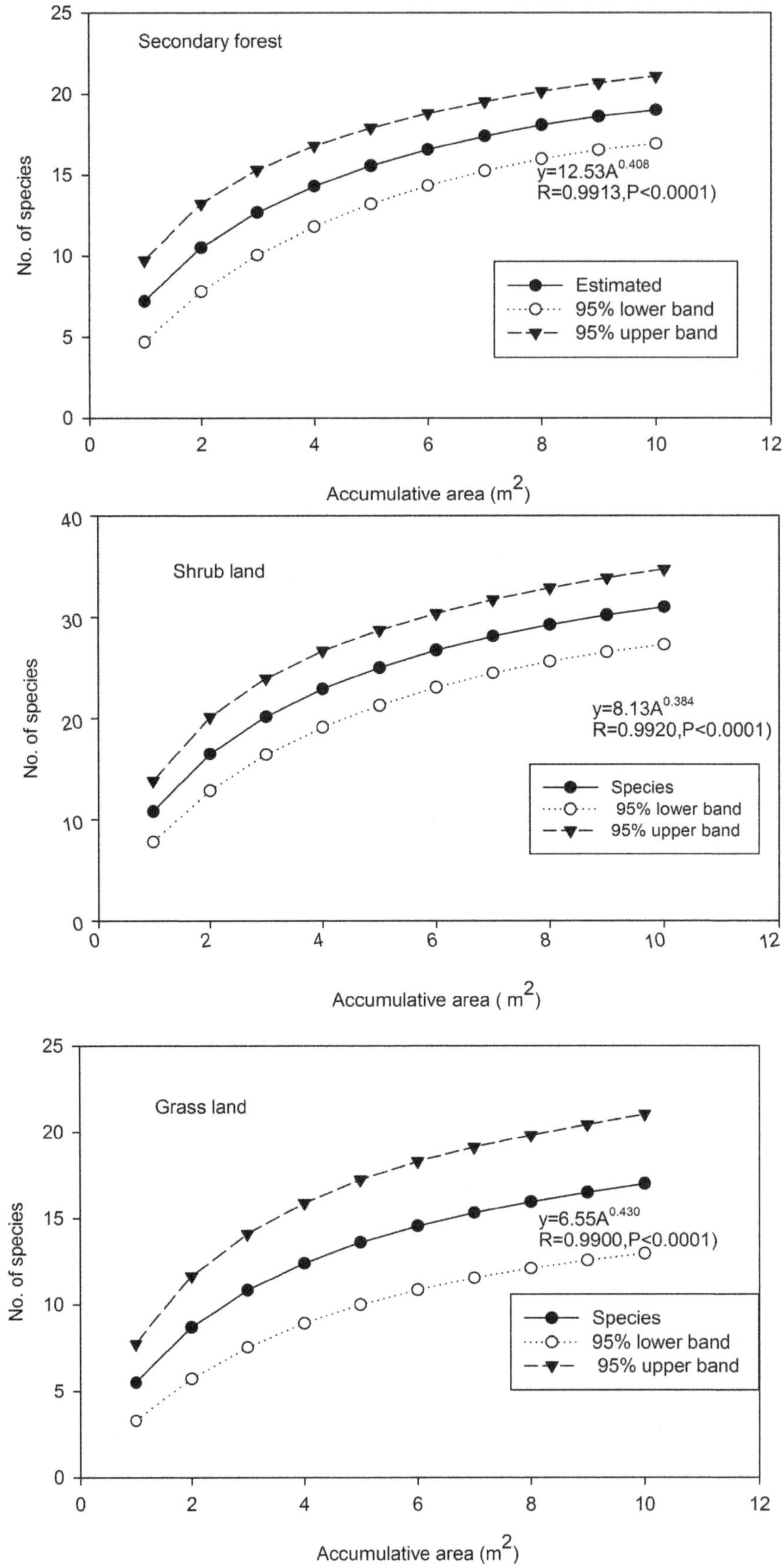

Figure 4. Mean woody species accumulation curve of seeds and 95% confidence intervals in soil seed bank of a karst forest in SW China. Samples were pooled from 10 quadrats of the soil seed bank by collecting ×1 m larger soil samples to replace soil plots of secondary forest, shrub and grass land and then randomized 100 times in "EstimateS 9.10" (Colwell 2013).

site in this study, a woody-plant seed bank that more accurately describes the forest stand was detected.

As germination conditions for LNSS and SNLS samples were not identical, comparing them may present some limitations. The differences among the three recipient sites provide information about the importance of germination environments (Table 1). However, the germination environment of SNLS, even at the most favorable field germination site - the shrub recipient site, may not equal the environment for LNSS in the greenhouse. Thus, empirically, the number of species and seedlings germinated at the three recipient sites were not at the maximum. This means a lower estimation of detection percentages on woody species might be obtained from the three recipients. So, we can confidently say the 10 SNLS (1 m×1 m size) had a higher percentage of detection for woody species than the 100 SNSS (10 cm×10 cm size).

Sample area size (A) is the fundamental element to determine the number of species in soil seed bank studies. The species–area relationship has traditionally been approximated by a power law [20], and it is widely applied to describe the pattern of increasing numbers of plant species with study-area size [20,28,29], as well as in soil seed bank studies [21]. The species-area curves in our study all follow power law pattern ($S = cA^z$) in stand forest, LNSS and SNLS methods (Figs. 2, 3, 4), by which the number of species against sampling area initially increases steeply but then approaches an asymptote.

The z-values in the power-law equation describes the rate of accumulation of species and the c value describes the interception with the increase of area in the logarithmic space ($\log S = z\log A + \log C$). These two parameters are affected by the sample size/total area, number of species per sample, etc. [19]. In our study, lower z-value variations were found among equations for the stand forest (0.303, Fig. 2), LNSS (0.411, Fig. 3), and the three SNLS equations at different recipient sites (0.408, 0.364, 0.430, Fig. 4); these z values were also within the range (0.064 to 1. 312) of those found in an inland species-area data analysis [19]. However, the c value varied greatly among the three equations, possibly affected by sample sizes since the c value was (i) lowest in the LNSS method (1.988, Fig. 3), in which sample size was the smallest (100 cm^2), (ii) highest in the plant survey (21.39, Fig 2) in which sample size was the largest (100 m^2), and (iii) an intermediate value in the SNLS method (12.53, 8.13, 6.55, Fig. 4) in which sample size was intermediate (1 m^2). However, unlike the stand forest, the c value for the soil seed bank also was affected by germination environments, as it varied among the three recipient sites.

Following the methods used in vegetation surveys, a minimum area can be determined from the power-law equations for forest soil seed-bank studies. The 10% ratio, which is "the point along the species determined / sample area size curve at which a 10% increase of the total sampling area yields only 10% more species of the total number recorded," is one criterion to determine suitable minimum areas [32,33]. Using the constructed equations to calculate the number of species at a stimulated sample area, we found that before the LNSS reaches 2000 cm^2, or SNLS reaches 4 m^2, the increase percentages of species determined will be greater than 10%. Correspondingly, predicted S (number of species) was 7 in LNSS, 23 in SNLS at shrub recipient site, 14 in SNLS at secondary forest recipient site, and 12 in SNLS at grass recipient site. This means that the 2000 cm^2 minimum area under LNSS would reach saturation with 8.2% of the species of the 1 ha plot and 27.1%, 16.5% and 14.1% in the 4 m^2 minimum area under SNLS by germination at shrub, secondary forest and grass recipient sites, respectively. Since the SNLS was collected from only six quadrats, the rates of saturation also could be 62.7%, 38.1% and 32.7% at shrub, secondary forest and grass recipient sites (by the power-law equation of plant survey (Fig. 2), when $A = 6 \times 100\ m^2$, $S = 36.7$). Thus, we can take 4 m^2 as the minimum area for forest soil seed bank studies in our study area to fulfill the 10% increase ratio and also to yield a higher saturation ratio (2/3 at the most favorable recipient site) of a stand forest plot. This minimum area is a far larger area than the recommended minimum areas in agricultural land [12]. Unfortunately, most forest soil seed bank studies do not use as large a minimum area as in our study. For instance, Clark *et al.* [14] reviewed 90 forest soil seed-bank studies and found most $A<0.2\ m^2$, which is far smaller than the minimum area in this study. Thus, we recommend a large sample area to ensure a high detection percentage on woody species.

In the power-law equation, the estimated number of species can be expressed as: $S = cA^z = c \times SS^z \times NS^z$. Theoretically, an increase of either SS (sample size) or NS (number of sample) can cause S (number of species) to increase. A large number of small samples (LNSS) presents logistical problems, particularly if seeds are to be counted by germination in a greenhouse. Further, an increased number of samples would be limited since each sample must be bagged, potted, counted, and cared for separately, and this can be very time consuming. For this reason, presumably, many researchers have chosen to use relatively large sample-unit sizes. Resolution could be found by taking smaller samples from large quadrats and pooling them together [11]. Even with this arrangement, greenhouse germination and seedling care usually take a long time, especially for forest soils. A smaller sample size also may reduce the soil depth and expose large seeds, potentially reducing seed germination and seedling survival. On the other hand, collection of a small number of larger samples may damage the forest floor, and uncontrolled field environments would bring larger variation in results (Table 1). Thus, we need to maintain a balance between sample size and number of samples for empirical and practical reasons. Theoretically, if we take samples larger than 10 cm×10 cm, i.e. 20 cm×20 cm, and keep the same collection method as the 10 cm×10 cm, we can get a minimum area $= 4\ m^2$, and $S = 23.3$ following the existing LNSS power-law equation (Fig. 3). This is 28.1% of the total plant species within the 1 ha vegetation plot.

Several studies propose a small sample size but large number of samples to account for clustered spatial distribution of seeds [7,11]. We might have anticipated that the LNSS sampling method would yield higher S than the SNLS method when the same A (sample area) value was considered. The simulation following the power-law equation of LNSS (Fig. 3) and of SNLS for the shrub site (Fig. 4) did not show a sharp difference between these two methods (i.e, when $A = 2\ m^2$, $S = 17.5$ for LNSS and $S = 16.6$ for SNLS shrub site, when $A = 3\ m^2$, $S = 20.7$ for LNSS and $S = 19.6$ for SNLS shrub site). Thus, the sample size in our forest soil seed bank study did not yield the anticipated effect. Probably, mixing SNLS samples before transplantation into recipient plots mitigated this clustering effect, and clustered spatial distributions could be mitigated by increased number of samples and increased sample sizes. On the other hand, a reasonable number of samples could provide enough replication to control for statistical variations. Thus, a minimum number of samples should be maintained.

Our study confirms the shortage of conventional large number of small sized samples (LNSS) in detecting seeds of woody species. This may in part explain low woody-plant seed detection [34] and low similarities between forest stands and soil seed banks [15], since most studies in the literature had a smaller sampling area [14,35] than ours (1 m^2 in total for the 100 LNSS). Consequently, this high discrepancy has limited applications of soil seed-bank results in forest ecology and restoration studies [13,16]. Our study also demonstrates that seeds of species dominant in the forest stand could be detected by increased total sample areas and sizes such as SNLS. This would provide more information to explain regeneration and recruitment dynamics of woody populations, communities, and also to assess the restoration capability of degraded forest sites both. Thus, a large sample area and size are needed for forest soil seed bank studies to detect woody species to ensure low discrepancy with stand forest. In practice, larger sample areas and sizes may be difficult to achieve and might also damage forest soil floors. We can compensate for the area increase by reducing the sample depth in soil volume, for instance, we can reduce the depth to 3 to 5 cm in most forest soils [36,37] since woody seeds are not efficiently transferred to deep soil layers [4,7]. Reducing the amount of soil could also reduce the burden of transporting soil samples to the greenhouse and facilitate germination in greenhouse, thus help mitigate germination variation in noncontrolled natural environments, such as our three recipient sites.

Due to our study's design, the field germination environment of a small number of large sized samples (SNLS) might be less favorable as that of LNSS in the greenhouse, together with variation in the field germination environment at the three recipient sites, we might not get the maximum detection percentage of woody species to construct species-area curves for SNLS sample strategies. On the other hand, we did not consider the fine-scale (i.e. within 1 m×1 m SNLS in our study) spatial seed bank structure, which could be used to design lumped soil cores in reducing soil collection amount of the sample [10]. These would limit the precision in estimating the minimal sample area and sample size. Thus, 4 m^2 as the minimum area fulfilling the 10% increase ratio might not be precise for forest soil seed bank studies in our study area and similar areas. However, 4 m^2 might be the first estimation of minimum area for forest soil seed bank studies in a particular forest. Thus, it should be further tested at other study areas.

Acknowledgments

We thank staff at the Shilin Stone Forest Management Bureau, Shilin County for their support in soil source collection and transfer and in maintaining of all the sites during the experiment.

Author Contributions

Conceived and designed the experiments: YS. Performed the experiments: YS WL. Analyzed the data: YS WL YL HG. Contributed reagents/materials/analysis tools: YS. Contributed to the writing of the manuscript: YS.

References

1. Darwin CR (1859) On the origin of species by means of natural selection, or the preservation of favoured races in the struggle for life. John Murray: Londen.
2. Garwood NC (1989) Tropical soil seed banks: a review. In: Leck MA, Parker VT, Simpson RL, editors.Ecology of soil seed banks.San Diego: Academic Press Inc. pp. 149–210.
3. Gross KL (1990) A comparison of methods for estimating seed numbers in the soil. J Ecol 78: 1079–1093.
4. Baskin CC, Baskin JM (1998) Seeds: ecology, biogeography, and evolution of dormancy and germination. San Diego: Academic Press.
5. Thompson K (2000) The functional ecology of soil seed banks. In: Fenner M, editor.Seeds: the ecology of regeneration in plant communities.Wallingford: CAB International. pp. 215–235.
6. Forey E, Dutoit T (2012) Vegetation, soils and seed banks of limestone grasslands are still impacted by former cultivation one century after abandonment. Community Ecol 13: 194–202.
7. Ferner M, Thompson K (2005) The ecology of seeds. Cambridge, UK: Cambridge University Press.
8. Ter Heerdt GNJ, Verweij GL, Bekker RM, Bakker JP (1996) An improved method for seed bank analysis: seedling emergence after removing the soil by sieving. Funct Ecol 10: 144–151.
9. Bossuyt B, Heyn M, Hermy M (2000) Concentrating samples estimates a larger seed bank density of a forest soil. Funct Ecol 14: 766–767.
10. Plue J, Hermy M (2012) Consistent seed bank spatial structure across semi-natural habitats determines plot sampling. J Veg Sci 23: 505–516.
11. Bigwood DW, Inouye DW (1988) Spatial pattern analysis of seed bank: An improved method and optimized sampling. Ecology 69: 497–507.
12. Forcella F (1984) A Species-area curve for buried viable seeds. Aust J Agr Res 35: 645–652.
13. Dainou K, Bauduin A, Bourland N, Gillet JF, Feteke F, et al. (2011) Soil seed bank characteristics in Cameroonian rainforests and implications for post-logging forest recovery. Ecol Eng 37: 1499–1506.
14. Clark JS, Beckage B, Camill P, Cleveland B, HilleRisLambers J, et al. (1999) Interpreting recruitment limitation in forests. Am J Bot 86: 1–16.
15. Hopfensperger KN (2007) A review of similarity between seed bank and standing vegetation across ecosystems. Oikos 116: 1438–1448.
16. Bossuyt B, Honnay O (2008) Can the seed bank be used for ecological restoration? An overview of seed bank characteristics in European communities. J Veg Sci 19: 875–884.
17. Plue J, Goyens G, Van Meirvenne M, Verheyen K, Hermy M (2010) Small-scale seed-bank patterns in a forest soil. Seed Sci Res 20: 13–22.
18. Garcia-Orth X, Martinez-Ramos M (2008) Seed dynamics of early and late successional tree species in tropical abandoned pastures: Seed burial as a way of evading predation. Restor Ecol 16: 435–443.
19. Triantis KA, Guilhaumon F, Whittaker RJ (2012) The island species-area relationship: biology and statistics. J Biogeogr 39: 215–231.
20. Storch D, Keil P, Jetz W (2012) Universal species-area and endemics-area relationships at continental scales. Nature 488: 78-+.
21. Butler BJ, Chazdon RL (1998) Species richness, spatial variation, and abundance of the soil seed bank of a secondary tropical rain forest. Biotropica 30: 214–222.
22. Denslow JS (1995) Disturbance and diversity in tropical rain forests - the density effect. Ecol Appl 5: 962–968.
23. Zhang FM, Geng H, Li YH, Liang YN, Yang YH, et al. (1997) Study on the Lunan Stone Forest karst, China. Kunming (in Chinese): Yunnan Science and Technology Press.
24. Wu ZY (1980) Vegetation of China. Beijing (in Chinese): Science publisher.
25. Jin ZZ, Peng J (1998) Vegetation of Kunming. Kunming (in Chinese): Yunnan Science and Technology Press.
26. Shen YX, Liu WY, Li YH, Cui JW (2005) Community ecology study on karst semi-humid evergreen broad-leaved forest at the central part of Yunnan. Guihaia 25: 321–326(in Chinese with English abstract).
27. Shen YX, Liu WY, Cao M, Li YH (2007) Seasonal variation in density and species richness of soil seed-banks in karst forests and degraded vegetation in central Yunnan, SW China. Seed Sci Res 17: 99–107.
28. Colwell RK, Mao CX, Chang J (2004) Interpolating, extrapolating, and comparing incidence-based species accumulation curves. Ecology 85: 2717–2727.
29. Colwell RK (2013) EstimateS: Statistical estimation of species richness and shared species from samples, Version 9. Available: http://purl.oclc.org/estimates. Accessed 2013 Jun19.
30. Arrhenius O (1921) Species and area. J Ecol 9: 95–99.
31. He FL, Legendre P (1996) On species-area relations. Am Nat 148: 719–737.
32. Cain SA (1938) The species-area curve. Am Mid Natt 19: 573–586.
33. Cannone N (2004) Minimum area assessment and different sampling approaches for the study of vegetation communities in Antarctica. Antarct Sci 16: 157–164.
34. Esmailzadeh O, Hosseini SM, Tabari M (2011) Relationship between soil seed bank and above-ground vegetation of a mixed-deciduous temperate forest in northern Iran. J Agr Sci Tech-Iran 13: 411–424.
35. Dupuy JM, Chazdon RL (1998) Long-term effects of forest regrowth and selective logging on the seed bank of tropical forests in NE Costa Rica. Biotropica 30: 223–237.
36. Pereira-Diniz SG, Ranal MA (2006) Germinable soil seed bank of a gallery forest in Brazilian Cerrado. Plant Ecol 183: 337–348.
37. Du XJ, Guo QF, Gao XM, Ma KP (2007) Seed rain, soil seed bank, seed loss and regeneration of Castanopsis fargesii (Fagaceae) in a subtropical evergreen broad-leaved forest. For Ecol Manag 238: 212–219.

13

Rates of Litter Decomposition and Soil Respiration in Relation to Soil Temperature and Water in Different-Aged *Pinus massoniana* Forests in the Three Gorges Reservoir Area, China

Wenfa Xiao[1]*, Xiaogai Ge[1,2]*, Lixiong Zeng[1], Zhilin Huang[1], Jingpin Lei[3], Benzhi Zhou[2], Maihe Li[4,5]

1 State Forestry Administration Key Laboratory of Forest Ecology and Environment, Research Institute of Forest Ecology, Environment and Protection, Chinese Academy of Forestry, Beijing, PR China, **2** Research Institute of Subtropical Forestry, Chinese Academy of Forestry, Fuyang, Zhejiang, PR China, **3** State Forestry Administration Key Laboratory of Forest Silviculture, Research Institute of Forestry, Chinese Academy of Forestry, Beijing, PR China, **4** Swiss Federal Research Institute WSL, Birmensdorf, Switzerland, **5** State Key Laboratory of Forest and Soil Ecology, Institute of Applied Ecology, Chinese Academy of Sciences, Shenyang, PR China

Abstract

To better understand the soil carbon dynamics and cycling in terrestrial ecosystems in response to environmental changes, we studied soil respiration, litter decomposition, and their relations to soil temperature and soil water content for 18-months (Aug. 2010–Jan. 2012) in three different-aged *Pinus massoniana* forests in the Three Gorges Reservoir Area, China. Across the experimental period, the mean total soil respiration and litter respiration were 1.94 and 0.81, 2.00 and 0.60, 2.19 and 0.71 μmol CO_2 m^{-2} s^{-1}, and the litter dry mass remaining was 57.6%, 56.2% and 61.3% in the 20-, 30-, and 46-year-old forests, respectively. We found that the temporal variations of soil respiration and litter decomposition rates can be well explained by soil temperature at 5 cm depth. Both the total soil respiration and litter respiration were significantly positively correlated with the litter decomposition rates. The mean contribution of the litter respiration to the total soil respiration was 31.0%–45.9% for the three different-aged forests. The present study found that the total soil respiration was not significantly affected by forest age when *P. masonniana* stands exceed a certain age (e.g. >20 years old), but it increased significantly with increased soil temperature. Hence, forest management strategies need to protect the understory vegetation to limit soil warming, in order to reduce the CO_2 emission under the currently rapid global warming. The contribution of litter decomposition to the total soil respiration varies across spatial and temporal scales. This indicates the need for separate consideration of soil and litter respiration when assessing the climate impacts on forest carbon cycling.

Editor: Dafeng Hui, Tennessee State University, United States of America

Funding: This work was supported by the Special Research Programme In Forestry Industry, State Forestry Administration of China (Project No. 201104008), National 'Twelfth Five-Year" Plan for Science & Technology Project of China (2011BAD38B04), Zigui National Forest Ecological Research Station and CFERN&GENE Award Funds on Ecological Paper. The funders had no role in study design, data collection and analysis, decision to publish, or preparation of the manuscript.

Competing Interests: This study was partly supported by CFERN&GENE Award Funds. There are no patents, products in development or marketed products to declare.

* Email: xiaowenf@caf.ac.cn (WX); gexiaogai2006@163.com (XG)

These authors contributed equally to this work.

Introduction

Soil respiration is a major process controlling carbon (C) loss from terrestrial ecosystems [1]. Globally, soil respiration releases approximately 80 Pg C into the atmosphere per year [2], which is estimated to account for 20–38% of the total annual biogenic CO_2 emissions to the atmosphere [3]. Given the predicted increase in atmospheric CO_2 concentrations, litter as a main C source for the total soil CO_2 efflux will increase [4]. Litterfall represents a major flux of the vegetative C to soil, and hence, changes in litter inputs are likely to have wide-reaching consequences for soil C dynamics [5]. Decomposition plays an integral role in determining soil C entering atmosphere [6,7], and the fate of C contained in litter, therefore, plays an important role in the long-term C sequestration in forest soils [8,9]. The total respiration together with litter inputs data could be used to evaluate C dynamics in soils [10,11]. To better understand and predict the soil C dynamics and cycling in terrestrial ecosystems in a changing world, studies of the relationships between soil respiration and litter decomposition in response to environmental changes are still required.

The aboveground plant litter fluxes have been found to be strongly correlated with soil respiration, and litterfall manipulations have strong effects on soil CO_2 efflux [12]. Several studies have investigated the contribution of the leaf litter layer to soil respiration [4,13,14,15]. Raich [2] showed that soil C flux from respiration was 2.8–3.0 times higher than the C flux from the aboveground litter production. Buchmann [16] found that the annual C loss through soil respiration (710 g C m^{-2} yr^{-1}) was higher than the C input of the annual aboveground litterfall (240 g C m^{-2} yr^{-1}). Prévost-Bouré et al. [4] indicated that addition of fresh litter significantly increased the total soil CO_2 efflux. Sayer et al. [5] indicated that soil respiration was on average 20% lower in the litter removal and 43% higher in the litter addition treatment compared to the controls.

On a global scale, soil respiration is mainly controlled by temperature and precipitation [17]. Many studies have shown that soil temperature is the primary factor determining the rates of soil respiration [18] and litter decomposition, and thus, soil warming increased the decomposition rate of fine woody debris in temperate forests [19]. However, Yuste et al.[17] found that soil respiration was insensitive to temperature in temperate maritime forests during late spring and summer when soil water content was limited. Fierer et al. [20] showed that the temperature sensitivity of litter decomposition varied with litter type and the extent of decomposition. These results indicate that soil respiration and litter decomposition depend on climatic variables such as soil temperature and precipitation [21]. It is thus necessary to see whether the temperature sensitivity of soil respiration and litter decomposition keeps consistent in a changing world.

Pinus massoniana Lamb. is a native species with high adaptability and tolerance to drought and low fertility soil but with low productivity in the Chinese subtropical regions. *P. massoniana* forests cover an area of ~2 million square kilometers in China, and are one of the most important vegetation types with key ecological importance in the Three Gorges Reservoir Area [22]. However, the relationships between soil respiration and litter decomposition in *P. massoniana* forests have not yet been studied. Based on long-term litter decomposition studies, Aerts and de Caluwe [23] concluded that the initial litter respiration rates are reliable indicators for long-term litter decomposability. The present study aimed to investigate the linkage of soil respiration and litter decomposition in different-aged *P. massoniana* forests, to provide basic knowledge for assessing the forest carbon budget and dynamics in those forests. We hypothesized that 1) the dynamics of soil respiration are reliable indicators for litter decomposition rate, 2) the soil respiration and litter decomposition have similar sensitivity to temperature, and 3) the co-variant effects of temperature and soil water contents on both litter decomposition and soil respiration are significant in different-aged *P. massoniana* forests.

Materials and Methods

Ethics Statement

This work was approved by Zigui National Forest Ecological Research Station. Our study related to litter decomposition and soil respiration did not involve endangered or protected species, and did not damage or destroy the vegetation and animals, although the study sites were located within a protected area for wildlife in the Three Gorges Reservoir Area, China.

Study site and forests

The study sites were located in Zigui county (110°00′14″–111°18′41″E, 30°38′14″–31°11′31″N), Hubei province, China. The region has a subtropical monsoon climate with mean annual temperature of 17–19°C and mean annual precipitation of 1000–1250 mm mainly occurring between April and September. The forests studied are *P. massoniana* pure stands on Haplic Luvisol soil [24], with *Camellia oleifera*, *Loropetalum chinensis*, *Cotinus coggygria* and *Echinochloa crusgalli*, *Veronicastrum villosulum* in the understorey. The characteristics of the forests/sites were summarized in Table 1. Table 1 showed that the 30-year-old forests had lower mean values of soil organic matter and total nitrogen contents compared to the other two forests.

The experiment included 20-, 30-, and 46-year-old *P. massoniana* forests each with three spatially separated stands ($n = 3$) at similar elevations (±50 m). Each stand has an area of more than 1 ha. In the center of each stand, a 20 m×30 m plot was established in June 2010. The rates of litter decomposition and soil respiration were determined in each plot.

Soil respiration, soil temperature and soil water content

Six polyvinyl chloride (PVC) collars (19.6 cm in inner diameter and 10 cm in height) were randomly installed to a soil depth of 5 cm for soil respiration sampling within each plot in July 2010. The litters were kept intact in three of these six collars, and were removed in the other three randomly selected collars ($n = 3$). The first measurement was conducted in August 2010, i.e. one month after the collar establishment. Soil respiration was measured using Li-8100 portable soil CO_2 flux system (LI-COR Inc., Lincoln, NE, USA) [25]. All soil respiration measurements were made between 08:00 hours and 11:00 hours (local time), to avoid bias of diurnal changes, according to our pilot tests (data not shown). Each measurement usually takes about 12 min. All the PVC collars were installed permanently throughout the experimental period. The soil respiration was measured monthly from August 2010 to January 2012 (i.e. a total of 18 times of repeated measures).

Soil temperature (T_5) and soil moisture at 5 cm soil depth were recorded adjacent to each soil respiration collar. Soil temperature was measured with a soil temperature probe. Volumetric soil water content (SWC) was measured using a theta probe (EM50, Decagon, USA), which was calibrated to the soil type in the plots following the procedure described by Delta-T (Luan et al. 2011).

Litter decomposition experiment

Leaf litter decomposition was studied using the standard litterbag technique [26]. In each stand, freshly fallen leaf litters were collected by hand and air-dried in May 2010, and then well-mixed for the decomposition experiment. Sub-samples of air-dried leaf litters were oven-dried at 70°C to constant weight for calculating the conversion coefficient between air-dried and oven-dried leaf. The initial litter quality was analyzed and summarized in Table 2. Table 2 showed that the leaf litter of the 30-year-old stands had higher C/N and lignin/N ratios than the leaf litters of the other two stands.

Twenty grams of air-dried leaf litters were placed in a nylon-mesh bag with 20 cm×20 cm in size and 1-mm mesh. We placed 36 litter-filled bags in each plot in June 2010, and a total of 324 decomposition bags (3 forests ×3 stands ×36 bags/stand) were placed. Litterbags lying flat on the litter layer surface were fastened using steel screens. Freshly fallen litters above the litterbags were removed monthly. Six bags in each stand were retrieved at 90 d (2nd November, 2010), 180 d (2nd February, 2011), 270 d (2nd May, 2011), 360 d (2nd August, 2011), 450 d (2nd November, 2011) and 540 d (31st January, 2012) of field decomposition, respectively. Each litterbag was brushed to remove the external soils and litters, and then the remaining litter was placed in a paper bag and oven-dried at 70°C to constant weight. Litter water content (LWC) was also calculated for each sample.

Data analysis

The total soil respiration rate was defined as the sum of the litter respiration plus the litter-free soil respiration. According to Sayer et al. [5], soil and litter respiration was calculated using the equation (1)

$$R_L = R_s - R_{(L-)} \tag{1}$$

where R_L is the litter respiration, $R_{(L-)}$ is the litter-free soil respiration (i.e. litter in collar removed), and Rs is the total soil respiration. Each R was calculated using the equation (2) [3]:

Table 1. Summary of stand and site characteristics measured in summer 2010 (mean ±1 SD; $n=3$).

	Forests		
	20-year-old	30-year-old	46-year-old
Stand characteristics			
Elevation (m)	950	350	990
DBH (cm)	16.1±1.15	25.3±0.93	33.1±1.09
Height (m)	15.6–19.0	17.5–21.6	18.4–24.8
Stand density (ha^{-1})	800	710	575
Litter layer depth (cm)	4.03±1.59	5.77±0.85	6.34±1.99
Litter litterfall ($t \cdot ha^{-1} \cdot a^{-1}$)	3.38±0.07	4.69±0.20	5.60±0.23
Litter standing crop ($t \cdot ha^{-1}$)	9.35±5.14	9.26±2.97	14.05±6.40
Soil characteristics			
Soil depth (cm)	70-100	60-80	70-100
Soil organic matter ($g \cdot kg^{-1}$)	51.90±8.38	19.87±6.78	69.62±11.57
Total nitrogen ($g \cdot kg^{-1}$)	2.36±0.58	1.24±0.18	3.33±0.70
Soil bulk density ($g \cdot cm^{-3}$)	1.13±0.14	1.53±0.08	1.08±0.14
pH value	5.12±0.33	5.43±0.09	4.50±0.15

Note DBH = mean diameter at breast height.

$$R=\alpha e^{\beta T} \quad (2)$$

where R is the soil or litter respiration, T is the temperature at 5 cm soil depth (T_5), α is the respiration rate at $T=0°C$, and β is a fitted temperature-response coefficient, according to Rey [27].

The temperature sensitivity (Q_{10}) of soil respiration (equation 3) [28] and litter mass loss (equation 4) [29] was calculated using the following equations:

$$\text{Soil respiration } Q_{10}=e^{10\beta} \quad (3)$$

$$\text{Litter decomposition } Q_{10}=(k_2/k_1)^{[10/(T2-T1)]} \quad (4)$$

where β is a fitted parameter, k_1 and k_2 are the rate constants for a process of interest at two observed temperatures T_1 and T_2.

The co-variant effects of soil temperature and soil moisture on soil respiration were fitted using equation (5), according to Saiz et al. [5].

$$Rs=(ae^{bT})(cSWC+dSWC^2) \quad (5)$$

where Rs is the total soil respiration, T is the soil temperature at 5 cm depth, SWC is the soil water content at 5 cm soil depth, a, b, c, d are fitted parameters.

The decomposition rates were estimated using a single exponential decay model (6) [30]:

$$M_t=M_0\, e^{kt} \quad (6)$$

where M_t is the litter dry mass at time t, M_0 is the initial litter mass, t is the sampling time interval, and k is the annual decay constant.

Repeated measures ANOVAs were used to test the effects of forest age (between subject), time (within subject), and their interaction on soil respiration, litter respiration, litter-free soil respiration, and litter decomposition rate. One-way and Univariate ANOVAs were applied to compare the litter leaf loss rate followed by Tukey's test. Exponential regression analyses were used to examine the relationships between environmental factors (soil temperature and soil water content) and respirations or litter decomposition rates, as well as the relationships between soil respirations and litter decomposition rates. A log-transformed

Table 2. Initial substrate quality (mean ±1SD, $n=3$) of leaf litter in *Pinus massoniana* forests studied.

Initial litter quality	20-yr-old forests	30-yr-old forests	46-yr-old forests
C %	56.25±0.10a	56.54±0.20a	57.21±2.45a
N %	0.98±0.40a	0.74±0.09a	0.85±0.15a
C/N ratio	62.81±20.56b	76.94±9.56a	68.17±14.45ab
Lignin %	34.27±1.31a	34.31±0.42a	34.24±0.16a
Lignin/N ratio	34.89±3.19b	46.63±5.23a	33.49±9.96b

Note: different letters within a row indicate significant difference ($P<0.05$) among the stands.

Table 3. Effects of forest age and sampling time on the total soil respiration, litter-free soil respiration, and litter respiration in *Pinus massoniana* forests tested using repeated measures ANOVAs.

	Between subject			Within subject					
	Forests			Time			Forest × Time		
	df	F	*P*	df	F	*P*	df	F	*P*
Across the experimental period									
Total soil respiration	2	0.323	0.725	17	88.023	0.00	34	4.2	0.00
Litter-free soil respiration	2	0.782	0.463	17	10.401	0.00	34	2.985	0.00
Litter respiration	2	1.114	0.336	17	120.884	0.00	34	3.873	0.00
April - October									
Total soil respiration	2	1.215	0.312	9	28.301	0.00	18	2.563	0.001
Litter-free soil respiration	2	0.669	0.521	9	42.848	0.00	18	2.719	0.000
Litter respiration	2	5.986	0.007	9	3.826	0.00	18	1.842	0.022
November - March									
Total soil respiration	2	3.355	0.050	7	25.834	0.00	14	6.190	0.00
Litter-free soil respiration	2	7.855	0.003	7	12.238	0.00	14	6.043	0.00
Litter respiration	2	0.541	0.59	7	14.635	0.00	14	7.196	0.00

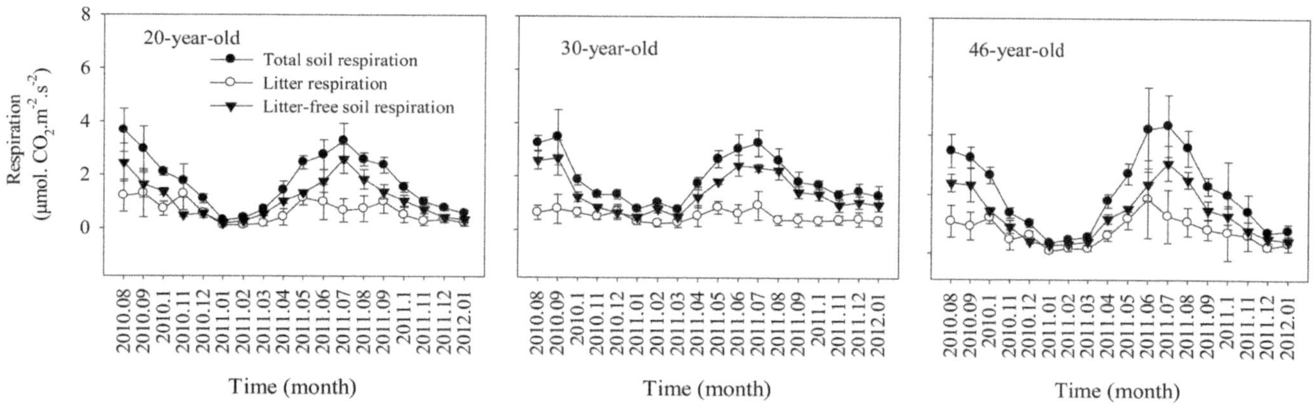

Figure 1. Seasonal patterns and monthly dynamics of the total soil respiration (•), litter respiration (○), and the litter-free soil respiration (▼) in different-aged *Pinus massoniana* forests (mean ±1SD, *n*=9).

model was applied to calculate the annual decay rate constants (*k*) from litter dry mass remaining. All statistical analyses were performed using SPSS 16.0 software package for windows (SPSS Inc., Chicago, USA).

Results

Dynamics of soil and litter respiration

The litter respiration, litter-free soil respiration, and the total soil respiration rates did not vary with forest age across the experimental period (Table 3). Values of all those parameters changed significantly with time, showing obvious seasonality. The forest age effects on litter and soil respiration changed with time, showing significant stand × time interactions (Table 3). The litter respiration had the lowest value in January, and peaked in May/June, whereas the total soil respiration and the litter-free soil respiration had lower rates in winter and higher rates in summer, with the highest rates in August and lowest values in January (Fig. 1). Thus, the highest value of litter respiration occurred nearly one month earlier than the peak values of the total soil respiration and the litter-free soil respiration (Fig. 1).

The mean annual litter-free soil respiration was 1.47±0.92 (mean±1SD), 1.58±0.83, and 1.68±0.99 µmol CO_2 m^{-2} s^{-1} in the 20-, 30 and 46-year-old stands, respectively (Fig. 1). The mean litter respiration was 0.81±0.66, 0.60±0.66, and 0.71±0.77 µmol CO_2 m^{-2} s^{-1} in the 20-, 30-, and 46-year-old stands, respectively (Fig. 1). The litter respiration rates were always lower than the litter-free soil respiration rates, except for some cases in November and December (Fig. 1).

The total soil respiration tended to be somewhat higher ($P>0.05$) in the 46-year-old stands than in the 20- and 30-year-old stands in summer, but it was significantly higher ($P=0.05$) in the 30-year-old stands than in the other two stands in winter (Table 3, Fig. 1). The total soil respiration rates increased from spring to summer and reached the maximum values of 3–4 µmol CO_2 m^{-2} s^{-1} in July/August (Fig. 1). The mean total soil respiration across the experimental period was 1.94±1.28, 2.00±1.0, and 2.19±1.51 µmol CO_2 m^{-2} s^{-1} in the 20-, 30-, and 46-year-old stands, respectively (Fig. 1).

Litter decomposition and its relationship to soil respiration

The leaf litter decomposition rates did not statistically differ among the different-aged stands (data not shown) and followed a similar pattern through the whole decomposition time (Fig. 2). After 540 days' decomposition in the field, the remaining litter ranged from 57.6% in the 20-year-old stands, to 56.1% in the 30-year-old stands, and 61.3% in the 46-year-old stands compared to their initial mass (Fig. 2). The litter decomposition rates were relatively slow and constant during the first six month, and then increased with the onset of the rainy season. The decomposition rates were lower during the dry period from March to June (180–270 d) than those during the hot season in summer.

An exponential model of decomposition was fitted to the experimental data (Fig. 2). The annual decomposition rate coefficients (*k*) were 0.29 (20-year-old stands), 0.33 (30-year-old stands), and 0.30 (46-year-old stands). The predicated remaining litter was 64.9%, 61.1%, and 64.5% in the 20-, 30-, and 46-year-old stands after 540 days of field decomposition, respectively. The predicted data of litter mass remaining slightly overestimated the measured data (Fig. 2). The unexpected higher values of litter mass remaining measured at 270 d in the 30- and 46-year-old stands may be caused by bias with respect to uncontrolled field experimental conditions and to lower decomposition rate during the dry period from March to May (e.g. 180–270 days) (Fig. 2).

The total soil respiration, litter respiration, and litter-free soil respiration in relation to litter decomposition had similar patterns for the three stands, showing a quadratic function ($R^2=0.37$, 0.45, 0.30, and $P=0.03$, 0.01, 0.07, respectively; see Fig. 3). The mean percentage contribution of litter respiration to the total soil respiration was 45.9% (ranging from 19.1 to 72.6% across the experimental period) in the 20-year-old stands, 31.0% (14.3–47.6%) in the 30-year-old stands, and 38.3% (15.6–61.0%) in the 46-year-old stands. The contribution of litter respiration to the total soil respiration was larger in winter than in summer.

Soil respiration and litter decomposition rate in relation to soil temperature and water content

The annual mean soil temperature (T_5) was 12.64°C (ranging from 2.40°C in January to 23.60°C in July) in the 20-year-old stands, 15.08°C (2.80 to 25.80°C) in the 30-year-old stands, and 12.83°C (2.70 to 25.53°C) in the 46-year-old stands (Fig. 4). The annual mean soil water content and litter water content (LWC) were 19.70% and 20.86%, 16.28% and 16.70%, 20.29% and 20.36% in the 20-, 30-, and 46-year-old stands, respectively (Fig. 4).

There were significant exponential relationships between soil temperature (T_5) and the total soil respiration ($R^2=0.80$, $P<0.001$; Fig. 5a), litter respiration ($R^2=0.53$, $P<0.001$; Fig. 5c), or litter-

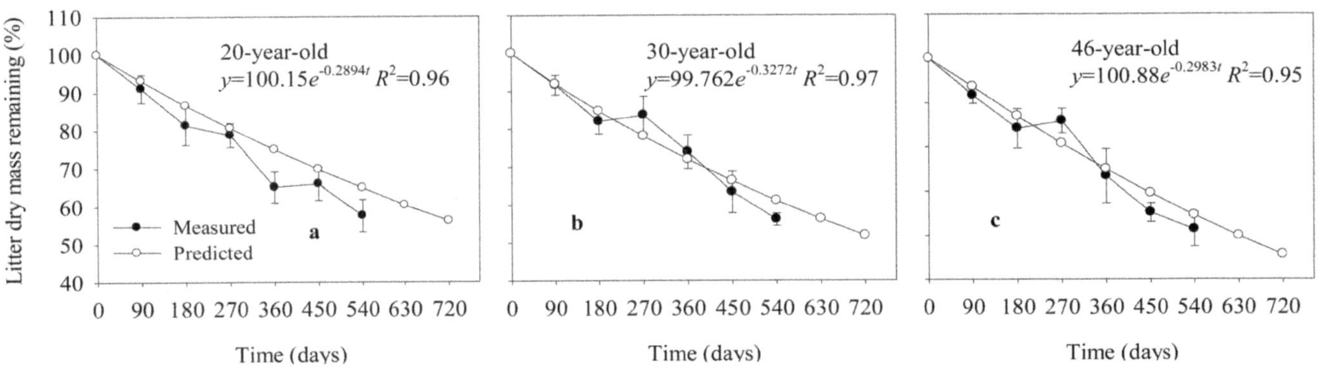

Figure 2. Observed (•) and predicted (○) values of litter dry mass remaining across the experimental period in different-aged *Pinus massoniana* forests (mean ±1SD, *n*=9).

free soil respiration ($R^2=0.83$, $P<0.001$; Fig. 5e). The corresponding Q_{10} was 2.25 for the total soil respiration, 2.10 for the litter respiration, and 2.36 for the litter-free soil respiration. No clear relationships between respirations and soil water contents were found (Fig.5b, d, f).

The co-variant effects of soil temperature and moisture on soil respiration were significant for the three stands ($R^2=0.70$, 0.42, 0.69, all $P<0.001$; Table 4). There were no combined effects of soil temperature and soil or litter water contents on litter respiration (Table 4).

Soil moisture was found to be marginally significantly linearly correlated with litter decomposition ($R^2=0.043$, $P=0.051$; Fig. 6b). The litter decomposition was significantly quadratically correlated with soil temperature ($R^2=0.58$, $P<0.001$; Fig. 6a). The temperature sensitivity (Q_{10} value) of litter mass loss was 1.5.

Discussion

Effects of forest age on soil and litter respiration

Ewel et al. [31] suggested that soil respiration decreased with forest age in temperate forests, while it increased with forest age in tropical and subtropical forests. The present study found that forest age affected the litter respiration during the growing season and influenced the litter-free soil respiration and the total soil respiration during the dormancy period (Table 3). Our results suggest that the effects of forest age on soil respiration may be mainly caused by stand roots in terms of root quantity (roots biomass) and quality (fine roots fraction) which increase with increasing forest age [32]. Previous studies indicated that forest age had a significantly positive influence on soil respiration in *Pinus taeda* plantations [33] and in warm-temperate oak forests [32]. Tang et al. [34] and Bolstad et al. [35] found that soil respiration first increased and then decreased with increasing forest age. Saiz et al. [36] found that soil respiration decreased with forest age during the younger stages of Sitka spruce forests. These results indicate that the effects of forest age on soil respiration differ with tree species and stand structure probably associated with the amount of fine roots and the quality of soil C pools [36].

Our results showed that the forest age effect on litter and soil respiration was nonlinear. For example, the mean litter respiration rate was 0.66, 0.54, and 0.74 μmol CO_2 m^{-2} s^{-1} for the 20-, 30-, and 46-year-old stands, respectively. This lower litter respiration in the 30-year-old stands seemed to be a result of the lower soil organic matter and nitrogen contents (Table 1), which may have led to decreases in soil microbial activities [37]. The lower litter respiration in the 30-year-old stands may also be resulted from the lower initial litter quality compared to that in the other two stands (Table 2) because litter quality directly affects litter decomposition rates [1]. Hättenschwiler and Gasser [38] showed that the decomposition rate was significantly positively correlated with litter N concentration ($R^2=0.75$, $P=0.025$) but negatively correlated with litter C/N ratio ($R^2=0.72$, $P=0.032$) and

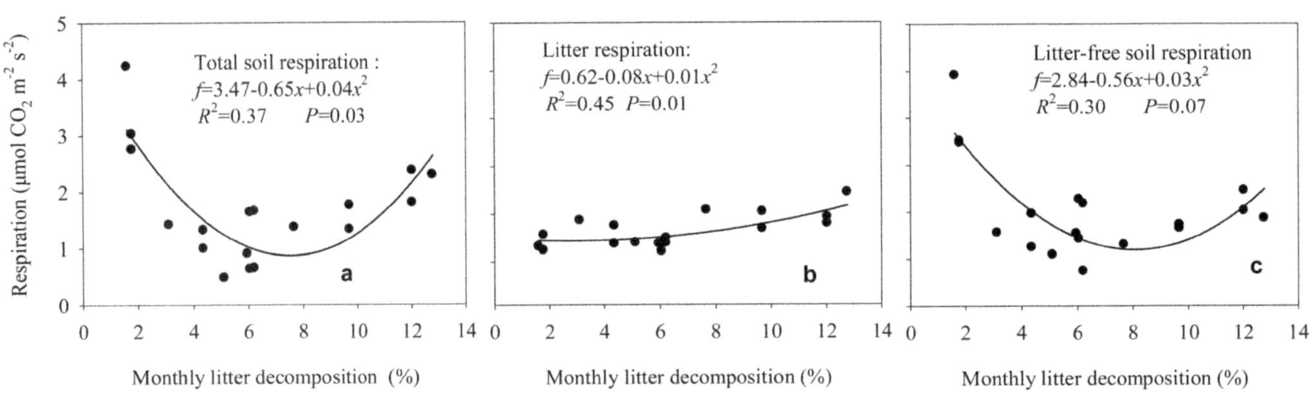

Figure 3. Relationships between mean monthly litter decomposition and the total soil respiration (a), litter respiration (b), and litter-free soil respiration rate (c) for different-aged *Pinus massoniana* forests. Note: respiration rate is the mean value across the litter decomposition period.

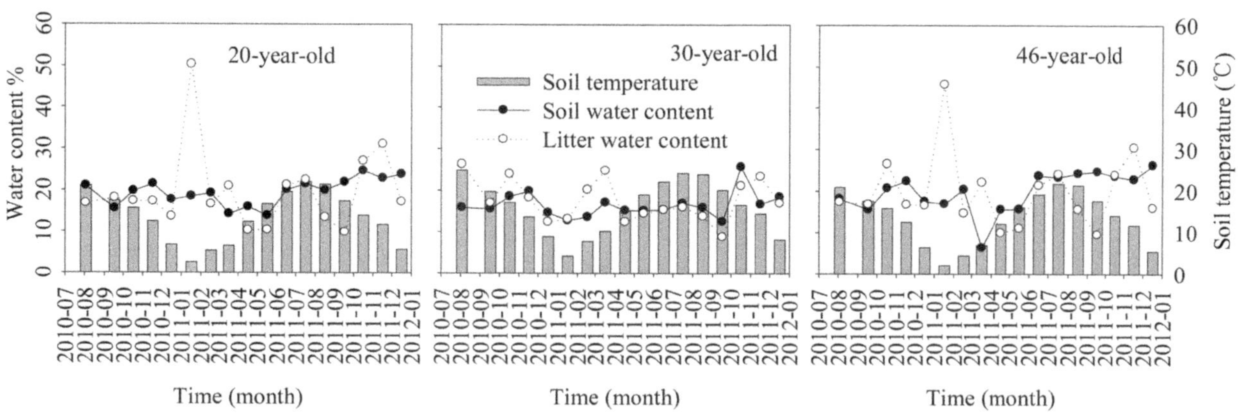

Figure 4. Soil temperature (5 cm soil depth), soil water content (5 cm soil depth) and litter water content in different-aged *Pinus massoniana* forests studied across the experiment period.

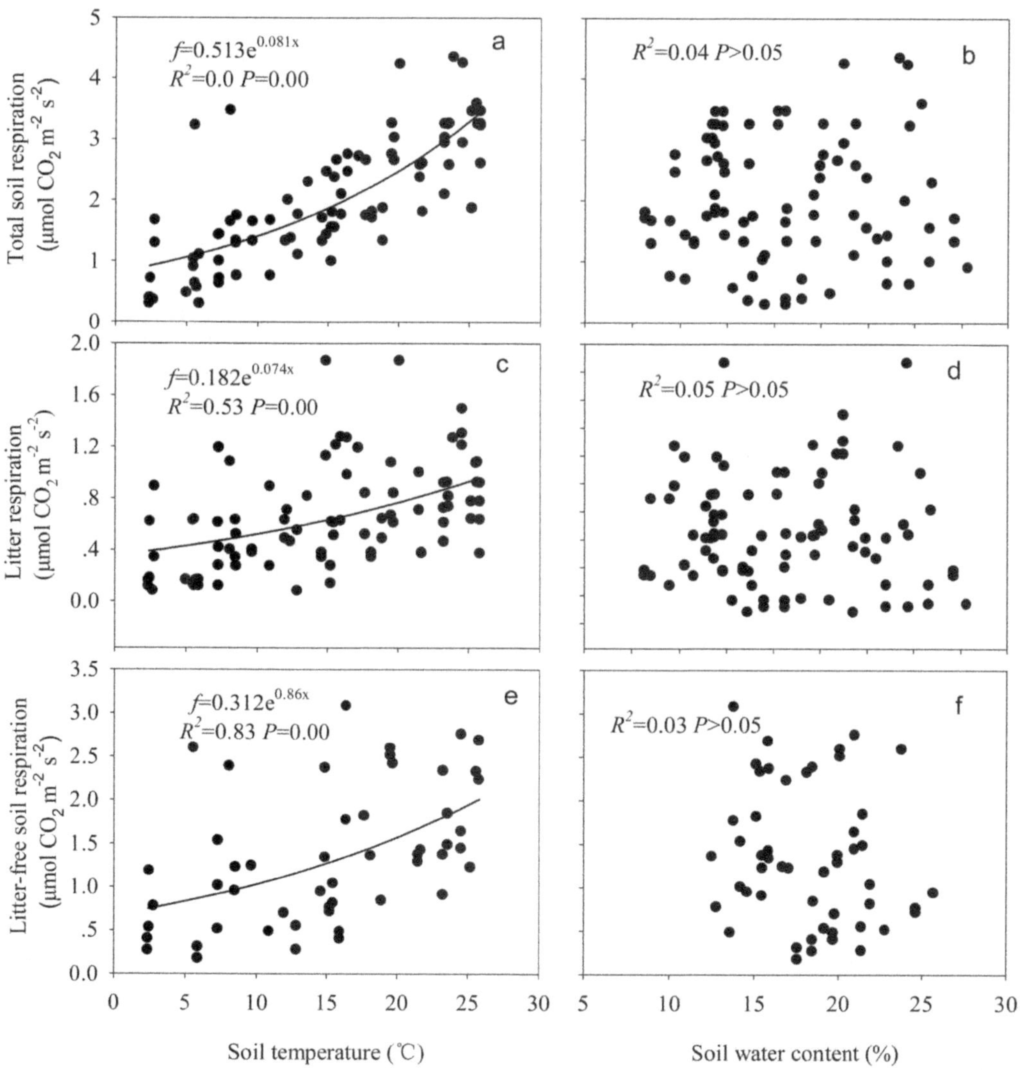

Figure 5. Relationships between the total soil respiration (a, b), litter respiration (c, d), litter-free soil respiration (e, f) and soil temperature at 5 cm depth or soil water content at 5 cm depth in different-aged *Pinus massoniana* forests during the 18-months field observations.

Table 4. The functions and model fit parameters (n, R^2 and P-values) between measured soil respiration with soil temperature (T), soil moisture content (SWC) and litter water content (LWC) in *Pinus massoniana* forests.

Stand	Equations	n	R^2	P
Total soil respiration in relation to T and SWC				
20-year-old stands	$Rs=0.040e^{0.072T}$ $(1.555SWC-0.037SWC^2)$	162	0.695	<0.001
30-year-old stands	$Rs=0.060e^{0.048T}$ $(1.729SWC-0.045SWC^2)$	162	0.420	<0.001
46-year-old stands	$Rs=0.040e^{0.062T}$ $(1.380SWC-0.017SWC^2)$	162	0.693	<0.001
Litter respiration in relation to T and SWC				
20-year-old stands	$Rs=0.038e^{0.051T}$ $(1.158SWC-0.035SWC^2)$	162	0.201	>0.05
30-year-old stands	$Rs=0.042e^{0.002T}$ $(1.618SWC-0.049SWC^2)$	162	0.031	>0.05
46-year-old stands	$Rs=0.020e^{0.047T}$ $(1.106SWC-0.007SWC^2)$	162	0.240	>0.05
Litter respiration in relation to T and LWC				
20-year-old stands	$Rs=0.039e^{0.047T}$ $(0.841LWC-0.019LWC^2)$	162	0.163	>0.05
30-year-old stands	$Rs=0.052e^{0.002T}$ $(0.564LWC-0.017LWC^2)$	162	0.157	>0.05
46-year-old stands	$Rs=0.049e^{0.054T}$ $(0.603LWC-0.011LWC^2)$	162	0.256	>0.05

lignin/N ratio ($R^2=0.75$, $P=0.026$). Bini et al. [39] found that plant litter with a C/N ratio of <25 degraded easily. Fanin et al. [40] reported that litter substrate quality (N, P content) was the most important factor explaining the observed spatial variations in soil respiration, and higher respiration rates were associated with high litter N and P contents.

Saiz et al. [36] indicated that the interactive effects of abiotic and biotic factors may modify the influences of forest age on soil respiration. Previous studies have reported both increased [41] and decreased [42] decomposition rates along a chronosequence of forest stands. Our present study found that litter decomposition rates did not significantly differ with forest age (Table 3), this may be caused by the difference in soil nutrients among the three stands (Table 1). Our previous study with *P. massoniana* stands found that the decomposition rates were faster in soil nutrient-poor than in nutrient-rich stands [26]. Similarly, Barlow et al. [43] found that the secondary species (*Vismia* sp. and *Bellucia* sp.) decomposed slower than the primary forest (*Bertholettia* sp.) along successional stages. Pandey et al. [44] showed that natural Oak (*Quercus* sp.) forests had lower soil nutrient level and greater litterfall with faster decomposition rates compared to Oak plantation forests. These studies suggest that not only forest age but also other factors such as site conditions (e.g. soil nutrients) and litter quality interact to affect litter decomposition [26,43].

Effects of soil temperature and water contents on respiration and litter decomposition

Variation in soil temperature is one of the most important factors determining the seasonal and diurnal variations in soil respiration [17], and soil respiration is highly sensitive to changes in surface temperature [45]. The present study found that the significant effects of soil temperature on respirations can be well-described with simple exponential regression models (Fig5a, c, e), which support the results of Shi et al. [46] and Zimmermann et al. [1]. Fang and Moncrieff [45] showed that the responses of soil respiration to temperature were commonly described using exponential equations, and thus the temperature sensitivity of soil respiration (Q_{10}) reduced at high temperature range. On the other hand, Davidson et al. [28] found that the temperature sensitivity of soil respiration (Q_{10}) decreased with drought or water deficit. Luan et al. [47] suggested that both soil chemical and physical characters contributed to the Q_{10} variations. Hence, our study

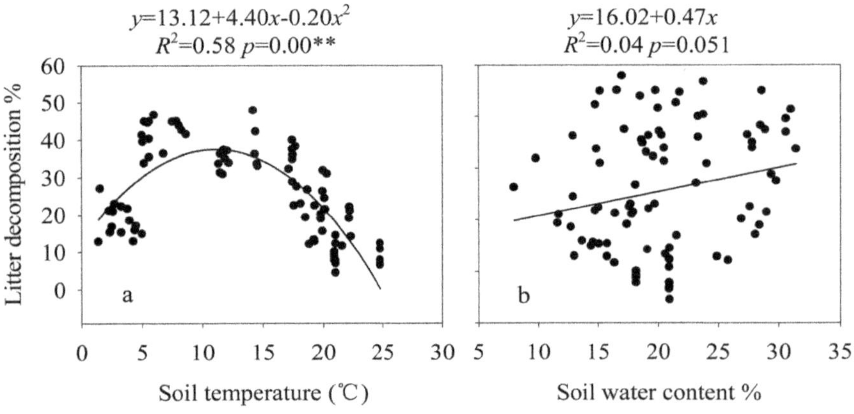

Figure 6. Relationships between litter decomposition (%) and soil temperature (a) or soil water content (b) in different-aged *Pinus massoniana* forests.

found that Q_{10} was 2.25 for the total soil respiration, 2.10 for the litter respiration, and 2.36 for the litter-free soil respiration.

Soil respiration is generally assumed to be strongly controlled by water availability [48]. In our study, we did not find any clear relationships between respirations and soil water contents (Fig. 5b, d and f). Previous studies indicated a wide range of relationships between respiration and moisture including linear [49], quadratic [50], exponential [28], logarithmic [51], and hyperbolic [52] relations, indicating that the physical (e.g.. diffusion), physiological (osmoregulation), and biochemical (enzyme dynamics) factors interact to affect the respiration-moisture relationships [48]. Recently, Moyano et al. [53] stated that further studies should concentrate on reducing uncertainties in the moisture-respiration relationships.

Soil temperature significantly interacted with soil water to affect the soil respiration (Table 4), it is probably because higher temperature may lead to lower soil water content and reverse [28]. Borken et al. [54] showed that the combined effects of temperature and water content on soil respiration were multiple linear regression. Lellei-Kovács et al. [55] showed that the interaction between temperature and moisture on soil respiration was straightforward for a linear model.

Davidson [28] found that soil temperature explained 80% of the variation of soil respiration, and there was no obvious relationship between soil water content and soil respiration. Peng et al. [56] pointed out that the sensitivity of soil respiration to temperature or moisture has not yet been adequately quantified, because most of the published results tend to be site-specific and no models have been widely accepted and commonly used.

Consistent with the results of Fiere et al. [20], we found that the effects of soil temperature on litter decomposition rates were quadratically well-described ($R^2 = 0.58$, $P<0.001$), which indicates that both lower or higher temperature lead to decreased decomposition rates (Fig. 6a). However, our results are inconsistent with the results of Aerts [57] who stated that warming resulted in increased decomposition rate.

Litter decomposition rates were found to positively respond to increased soil moisture [58]. In the present study, higher temperature resulted in lower soil moisture in the 30-year-old stands (Fig. 4). Higher temperature usually stimulates but lower soil moisture decreases the respiration and decomposition rates [55]. Cortze [21] and De Santo et al. [59] reported that soil moisture was most important during the early decomposition stage rather than the late stages. Under field experimental conditions, however, it is difficult to detect the net effects of moisture on respiration or decomposition because all factors interact to affect the litter decomposition and the effects of temperature might become stronger with increasing moisture [19].

Climate factors such as temperature and moisture alone and combined affect litter decomposition [26]. Butenschoen et al. [60] suggested that litter decomposition increased with increasing temperature in the high moisture treatment and decreased with increasing temperature in the low moisture treatment. Many studies failed to find positive responses of litter decomposition to warming when moisture is limited [57]. Cortez [21] reported that the relationships between litter decomposition rate and the ratio of soil humidity to temperature (H/T) showed a polynomial function ($y = ax^2+bx+c$), and soil temperature seemed to be the main determining factor in wet sites, while soil moisture was the most important factor during the early decomposition stages in dry period.

Relationships between soil respiration and litter decomposition

Litter provides the major C source for soil respiration [6]. Subke et al. [7] showed that an increase in litter input promotes rhizosphere respiration and rhizosphere activity, leading to increases in soil respiration. Reynolds and Hunter [6] showed that the soil respiration was significantly reduced by litter removal. Increased litter input stimulated soil microbial activity and soil C loss by microbial respiration [61]. In our study, the total soil respiration (Fig. 3a) was positively quadratically correlated with litter decomposition rate, which indicates that the soil respiration rates may be reliable indicators for long-term litter decomposability and litter carbon dynamics. Similarly, Aerts and de Caluwe [23] reported that the litter respiration rates were positively correlated with litter mass loss rates.

We found that both litter decomposition rate and soil respiration had similar temperature sensitivity and both peaked in summer. These may imply that both soil respiration and litter decomposition in the present study are controlled by similar environmental factors [26]. Moreover, we found that litter respiration peaked nearly one month earlier than the litter-free soil respiration and the total soil respiration (Fig. 1), which may be partly caused by the seasonality of the microbial community composition [46] associated with the environmental conditions, as reported by Berryman et al. [62].

The addition or exclusion of fresh litter was found to significantly increase or decrease the total soil CO_2 efflux, respectively [4,61]. The present study found that the mean contribution of the litter respiration to the total soil respiration ranged from 31.0% (30-year-old stands) to 45.9% (20-year-old stands) (see also [1,27]). Buchmann [16] estimated that soil respiration rates reduced by 10–20% when the litter and semi-decomposed litter layer were removed, and even reduced by up to 30-40% when the humus layer was also additionally removed.

The contribution of the litter respiration to the total soil respiration seems to vary with study area, tree species, and soil fertility [63]. Berger et al. [10] suggested that decomposing litter contributed 22–32% (base-rich sites) and 11–28% (base-poor sites) to the total soil respiration. In temperate coniferous forest ecosystems, the estimated contribution of leaf litter respiration to the total soil respiration exhibit a large seasonal variation from 2% (early spring) to 20% (mid summer) [4]. Cisneros-Dozal et al. [66] also found that the contribution of leaf litter decomposition to the total soil respiration increased from 5±2% (6±3 mg C m^{-2} hr^{-1}) during a transient drought to 37±8% (63±18 mg C m^{-2} hr^{-1}) immediately following water addition, indicating the effects of water availability on the contribution. In our study, the contribution of the leaf litter respiration to the total soil respiration was larger in winter than in summer. A possible explanation is that lower winter temperature limits the root respiration, and the variation of litter respiration is less important to the total soil respiration than the root respiration does [63]. Therefore, the soil respiration is obviously affected by many biotic and abiotic factors including root biomass and root activity across time [64].

In conclusion, the present paper revealed that soil moisture and temperature play complex roles in determining the respiration and decomposition across spatial and temporal scales in the *P. massoniana* forests studied. The apparent temperature sensitivity of soil respiration and litter decomposition is influenced not only by soil water but also by a wide range of factors including soil nutrients and litter quality [55,65]. The contribution of litter decomposition to the total soil respiration varies across spatial and temporal scales. These findings emphasize the need for separate consideration of soil and litter respiration when assessing climate impacts on forest carbon cycling [62].

Acknowledgments

We are grateful to the staff of the Key Laboratory of Forest Ecology and Environment for chemical analysis. We also thank anonymous reviewers for helpful suggestions on the manuscript. The funders had no role in study design, data collection and analysis, decision to publish, or preparation of the manuscript.

Author Contributions

Conceived and designed the experiments: WX XG. Performed the experiments: XG LZ ZH JL. Analyzed the data: XG ML. Contributed reagents/materials/analysis tools: XG WX ML. Wrote the paper: XG ML BZ.

References

1. Zimmermann M, Meir P, Bird M, Malhi Y, Ccahuana A (2009) Litter contribution to diurnal and annual soil respiration in a tropical montane cloud forest. Soil Biology and Biochemistry 41: 1–3.
2. Raich JW (1998) Aboveground productivity and soil respiration in three Hawaiian rain forests. Forest Ecology and Management 107: 309–318.
3. Raich JW, Schlesinger WH (1992) The global carbon dioxide flux in soil respiration and its relationship to vegetation and climate. Tellus 44B: 81–89.
4. Prévost-bouré NC, Soudani K, Damesin C, Berveiller D, Lata JC, et al. (2010) Increase in aboveground fresh litter quantity over-stimulates soil respiration in a temperate deciduous forest. Applied Soil Ecology 46: 26–34.
5. Sayer EJ, Powers JS, Tanner EVJ (2007) Increased litterfall in tropical forests boosts the transfer of soil CO_2 to the atmosphere. PLoS ONE 12: e1299.
6. Reynolds BC, Hunter MD (2001) Responses of soil respiration, soil nutrients, and litter decomposition to inputs from canopy herbivores. Soil Biology and Biochemistry 33: 1641–1652.
7. Subke JA, Hahn V, Battipaglia G, Linder S, Buchmann N, et al. (2004) Feedback interactions between needle litter decomposition and rhizosphere activity. Oecologia 139: 551–559.
8. Smolander A, Levula T, Kitunen V (2008) Response of litter decomposition and soil C and N transformations in a Norway spruce thinning stand to removal of logging residue. Forest Ecology and Management 256: 1080–1086.
9. Fioretto A, Musacchio A, Andolfi G, De Santo AV (1998) Decomposition dynamics of litters of various pine species in a Corsican pine forest. Soil Biology and Biochemistry 30: 721–727.
10. Berger TW, Inselsbacher E, Zechmeister-Boltenstern S (2010) Carbon dioxide emissions of soils under pure and mixed stands of beech and spruce, affected by decomposing foliage litter mixtures. Soil Biology and Biochemistry 42: 986–997.
11. Meier CL, Bowman WD (2008) Links between plant litter chemistry, species diversity, and below-ground ecosystem function. Proceedings of the National Academy of Sciences of the United States of America 105: 19780–19785.
12. Schaefer DA, Feng WT, Zou XM (2009) Plant carbon inputs and environmental factors strongly affect soil respiration in a substropical forest of southwestern China. Soil Biology and Biochemistry 41: 1000–1007.
13. Rubino M, Dungait JAJ, Evershed RP, Bertolini T, Angelis PD, et al. (2010) Carbon input belowground is the major C flux contributing to leaf litter mass loss: Evidences from a ^{13}C labelled-leaf litter experiment. Soil Biology and Biochemistry 42: 1009–1016.
14. Kammer A, Schmidt MWI, Hagedorn F (2012) Decomposition pathway of ^{13}C-depleted leaf litter in forest soils of the Swiss Jura. Biogeochemistry 108: 395–411.
15. Tu TTN, Egasse C, Zeller B, Bardoux G, Biron P, et al. (2011) Early degradation of plant alkanes in soils: a litterbag experiment using C-13-labelled leaves. Soil Biology and Biochemistry 11: 2222–2228.
16. Buchmann N (2000) Biotic and abiotic factors controlling soil respiration rates in *Picea abied* stands. Soil Biology and Biochemistry 32: 1625–1635.
17. Yuste JC, Janssens IA, Carrara A, Meiresonne L, Ceulemans R (2003) Interactive effects of temperature and precipitation on soil respiration in a temperate maritime pine forest. Tree Physiology 23: 1263–1270.
18. Bond-Lamberty B, Thomson A (2010) Temperature-associated increased in the global soil respiration record. Nature 464: 579–583.
19. Berbeco MR, Melillo JM, Orians CM (2012) Soil warming accelerates decomposition of fine woody debris. Plant and soil 344: 10–1007.
20. Fiere N, Craine JM, McLauchlan K, Schiml JP (2005) Litter quality and the temperature sensitivity of decomposition. Ecology 86: 320–326.
21. Cortez J (1998) Field decomposition of leaf litters: relationship between decomposition rates and soil moisture, soil temperature and earthworm activity. Soil Biology and Biochemistry 30: 783–793.
22. Xiao WF, Li JW, Yu CQ, Ma J, Cheng RM, et al. (2000) Terrestrial animal and plant ecology of the Three Gorges of Yangtze River. Chongqing: Southwest China Normal University Press. 320–321.
23. Aerts R, Caluwe H (1997) Initial litter respiration as indicator for long-term leaf litter decomposition of *Carex* species. Oikos 80: 353–361.
24. Gong ZT (2003) Chinese soil taxonomy (Revised Proposal). Beijing: China Science Press.
25. Yan LM, Chen SP, Huang JH, Lin GH (2010) Differential responses of auto- and heterotropic soil respiration to water and nitrogen addition in a semiard temperature steppe. Global Change Biology 16: 2345–2357.
26. Aerts R (1997) Climate, leaf litter chemistry and leaf litter decomposition in terrestrial ecosystems: a triangular relationship. Oikos 79: 439–449.
27. Rey A, Pegoraro E, Tedeschi V, De Parri I, Jarvis PG, et al. (2002) Annual variation in soil respiration and its components in a coppice oak forest in Central Italy. Global Change Biology 8: 851–866.
28. Davidson EA, Belk E, Boone RD (1998) Soil water content and temperature as independent or confounded factors controlling soil respiration in a temperate mixed hardwood forest. Global Change Biology 4: 217–227.
29. Kirschbaum MUF (1995) The temperature dependence of soil organic matter decomposition, and the effect of global warming on soil organic C storage. Soil Biology and Biochemistry 27: 753–760.
30. Olson JS (1963) Energy storage and the balance of producers and decomposers in ecological systems. Ecology 44: 322–331.
31. Ewel KC, Cropper WP, Glolz HL (1986) Soil CO_2 evolution in Florida slash pine plantains.I. Changes through time. Canadian Journal of Forest Research 17: 325–329.
32. Luan JW, Liu SR, Wang JX, Zhu XL, Shi ZM (2011) Rhizospheric and heterotrophic respiration of a warm-temperate oak. Soil Biology and Biochemistry 43: 503–512.
33. Wiseman PE, Seiler JR (2004) Soil CO_2 efflux across four age classed of plantation loblolly pine (*Pinus taeda* L.) on the Virginia Piedmont. Forest Ecology and Management 192: 297–311.
34. Tang JW, Bolstad PV, Martin JG (2009) soil carbon fluxes and stocks in a Great Lakes forest chronosequence. Global Change Biology 15: 145–155.
35. Bolstad PV, Davis KJ, Martin J, Cook BD, Wang W (2004) Component and whole-system respiration fluxes in northern deciduous forests. Tree Physiology 24: 493–504.
36. Saiz G, Byrne KA, Butterbach-Bahl K, Kiese R, Blujden V, et al. (2006) Forest age-related effects on soil respiration in a first rotation Sitka spruce chronosequence in central Ireland. Global Change Biology 12: 1007–1020.
37. Li FL, Liu M, Li ZP, Jiang CY, Han FX, et al. (2013) Changes in soil microbial biomass and functional diversity with a nitrogen gradient in soil columns. Applied Soil Ecology 64: 1–6.
38. Hättenschwiler S, Gasser P (2005) Soil animals alter plant litter diversity effects on decomposition. Proceedings of the National Academy of Sciences of the United States of America 102: 1519–1524.
39. Bini D, Dos Santos CA, Carmom KBD, Kishino N, Andrade G, et al. (2013) Effects of land use on soil organic carbon and microbial processes associated with soil health in southern Brazil. European Journal of Soil Biology 55: 117–123.
40. Fanin N, Hättenschwiler S, Barantal S, Schimann H, Fromin N (2011) Does variability in litter quality determine soil microbial respiration in an Amazonian rainforest? Soil Biology and Biochemistry 43: 1014–1022.
41. Xuluc-Tolosa FJ, Vester HFM, Ramírez-Marcial N, Castellanos-Albores J, Lawrence D (2003) Leaf litter decomposition of tree specied in three successional phases of Tropical dry secondary forest in Campeche, Mexico. Forest Ecology and Management 174: 401–412.
42. Vasconcelos HL, Laurance WF (2005) Influence of habitat, litter type, and soil invertebrates on leaf-litter decomposition in a fragmented Amazonian landscape. Oecologia 144: 456–462.
43. Barlow J, Gardner TA, Ferreira LV, Peres CA (2007) Litter fall and decomposition in primary, secondary and plantation forests in the Brazilian Amazon. Forest Ecology and Management 247: 91–97.
44. Pandey RR, Sharma G, Tripathi SK, Singh AK (2007) Litterfall, litter decomposition and nutrient dynamics in a subtropical natural oak forest and managed plantation in northeastern India. Forest Ecology and Management 240: 96–104.
45. Fang C, Moncrieff JB (2001) The dependence of soil CO_2 efflux on temperature. Soil Biology and Biochemistry 33: 155–165.
46. Shi WY, Zhang JG, Yan MJ, Yamanaka N, Du S (2012) Seasonal and diurnal dynamic s of soil respiration fluxes in two typical forests on the semiarid Loess Plateau of China: temperature sensitivities of autotrophs and heterotrophs and analyses of integrated driving factors. Soil Biology and Biochemistry 52: 99–107.
47. Luan JW, Liu SR, Wang JX, Zhu XL (2013) Factors affecting spatial variation of annual apparent Q_{10} of soil respiration in two warm temperature forests. PLoS ONE 8: e64167.
48. Moyano FE, Manzoni S, Chenu C (2013) Responses of soil heterotrophic respiration to moisture availability: an exploration of processes and models. Soil Biology and Biochemistry 59: 72–85.
49. Epron D, Farque L, Lucot É, Badot PM (1998) Soil CO_2 efflux in a beech forest: dependence on soil temperature and soil water content. Annual of Forest Science 56: 221–226.

50. Tang JW, Baldocchi DD (2005) Spatial-temporal variation in soil respiration in an oak-grass savanna ecosystem in California and its partitioning into autotrophic and heterotrophic component. Biogeochemistry 73: 183–207.
51. Raich JW, Potter CS, Bhagawati D (2002) Interannual variability in global soil respiration, 1980-94. Global Change Biology 8: 800–812.
52. Hanson PJ, Wullschleger SD, Bohlman SA, Todd DE (1993) Seasonal and topographic patterns of forest floor CO_2 efflux from an upland oak forest. Tree Physiology 13: 1–15.
53. Moyano FE, Vasilyeva N, Bouckaert L, Cook F, Craine J, et al. (2012) The moisture response of soil heterotrophic respiration: interaction with soil properties. Biogeochemistry 9: 1173–1182.
54. Borken W, Davidson EA, Savage K, Gaudinski J, Trumbore SE (2003) Drying and wetting effects on carbon dioxide release from organic Horizons. Soil Science Society of America Journal 67: 1888–1896.
55. Lellei-Kovács E, Kovács-Láng E, Botta-Dukát Z, Kalapos T, Emmett B, et al. (2011) Thresholds and interactive effects of soil moisture on the temperature response of soil respiration. European Journal of Soil Biology 47: 247–255.
56. Peng SS, Piao SL, Wang T, Sun JY, Shen ZH (2009) Temperature sensitivity of soil respiration in different ecosystems in China. Soil Biology and Biochemistry 41: 1008–1014.
57. Aerts R (2006) The freezer defrosting: global warming and litter decomposition rates in cold biomes. Journal of Ecology 94: 713–724.
58. Lee H, Fitzgerald J, Hewins DB, McCulley RL, Archer SR, et al. (2014) Soil moisture and soil-litter mixing effects on surface litter decomposition: A controlled environment assessment. Soil Biology and Biochemistry 72: 123–132.
59. De Santo AV, Berg B, Rutigliano FA, Alfani A, Floretto A (1993) Factors regulating early-stage decomposition of needle litters in five different coniferous forests. Soil Biology and Biochemistry 25: 1423–1433.
60. Butenschoen O, Scheu S, Eisenhauer N (2011) Interactive effects of warming, soil humidity and plant diversity on litter decomposition and microbial activity. Soil Biology and Biochemistry 43: 1902–1907.
61. Liu LL, King JS, Booker FL, Giardina CP, Allen HL, et al. (2009) Enhanced litter input rather than changes in litter chemistry drive soil carbon and nitrogen cycles under elevated CO_2: a microcosm study. Global Change Biology 15: 441–453.
62. Berryman EM, Marshall JD, Kavanagh K (2014) Decoupling litter respiration from whole-soil soil respiration along an elevation gradient in a Rochy Mountain mixed conifer forest. Canadian Journal of Forest Research 44. Doi.org/10.1139/cjfr-2013-0334.
63. Wang CK, Yang JY, Zhang QZ (2006) Soil respiration in six temperate forests in China. Global Change Biology 12: 2103–2114.
64. Bowden RD, Nadelhoffer KJ, Boone RD, Melillo JM, Garrison JB (1993) Contribution of aboveground litter, belowground litter, and root respiration to total soil respiration in a temperate mixed hardwood forest. Canadian Journal of Forest Research 23: 1402–1407.
65. Kirschbaum MUF (2004) Soil respiration under prolonged soil warming: are rate reductions caused by acclimation or substate loss? Global Change Biology 10: 1870–1877.
66. Cisneros-Dozal LM, Trumbore SE, Hanson PJ (2007) Effect of moisture on leaf litter decomposition and its contribution to soil respiration in a temperature forest. Journal of Geophysical Research 112: G1013.

Numerical Responses of Saproxylic Beetles to Rapid Increases in Dead Wood Availability following Geometrid Moth Outbreaks in Sub-Arctic Mountain Birch Forest

Ole Petter Laksforsmo Vindstad[1]*, Sabrina Schultze[2], Jane Uhd Jepsen[3], Martin Biuw[3], Lauri Kapari[1], Anne Sverdrup-Thygeson[4], Rolf Anker Ims[1]

1 Department of Arctic and Marine Biology, University of Tromsø, Tromsø, Norway, **2** Department of Ecology, Leuphana University Lüneburg, Lüneburg, Germany, **3** Norwegian Institute for Nature Research, Fram Centre, Tromsø, Norway, **4** Norwegian University of Life Sciences, Department of Ecology and Natural Resource Management, Aas, Norway

Abstract

Saproxylic insects play an important part in decomposing dead wood in healthy forest ecosystems, but little is known about their role in the aftermath of large-scale forest mortality caused by pest insect outbreaks. We used window traps to study short-term changes in the abundance and community structure of saproxylic beetles following extensive mortality of mountain birch in sub-arctic northern Norway caused by an outbreak of geometrid moths. Three to five years after the outbreak, the proportion of obligate saproxylic individuals in the beetle community was roughly 10% higher in forest damaged by the outbreak than in undamaged forest. This was mainly due to two early-successional saproxylic beetle species. Facultative saproxylic beetles showed no consistent differences between damaged and undamaged forest. These findings would suggest a weak numerical response of the saproxylic beetle community to the dead wood left by the outbreak. We suggest that species-specific preferences for certain wood decay stages may limit the number of saproxylic species that respond numerically to an outbreak at a particular time, and that increases in responding species may be constrained by limitations to the amount of dead wood that can be exploited within a given timeframe (i.e. satiation effects). Low diversity of beetle species or slow development of larvae in our cold sub-arctic study region may also limit numerical responses. Our study suggests that saproxylic beetles, owing to weak numerical responses, may so far have played a minor role in decomposing the vast quantities of dead wood left by the moth outbreak.

Editor: Judi Hewitt, University of Waikato (National Institute of Water and Atmospheric Research), New Zealand

Funding: This study was funded by the University of Tromsø and the research council of Norway, with additional support from the Norwegian Institute of Nature Research and the Nordic Centre of Excellence – How to preserve the tundra under a warming climate (NCoE-Tundra) funded by the Norden Top-Level Initiative "Effect studies and adaptation to climate change". The funders had no role in study design, data collection and analysis, decision to publish, or preparation of the manuscript.

Competing Interests: The authors have declared that no competing interests exist.

* E-mail: ole.p.vindstad@uit.no

Introduction

Insect outbreaks are major disturbance factors in many forest ecosystems, periodically causing defoliation and mortality of trees and other vegetation across vast areas [1]. The potential causes of outbreaks have been intensively studied for decades. More recently, however, there has also been increasing interest in their ecological consequences [2]. One important consequence of outbreaks is that they alter resource availability for many consumer species. For instance, outbreaks can increase food availability for insectivorous birds [3–5], small mammals [6,7] and invertebrates [8,9]. Similarly, understory vegetation may benefit from increased nutrient inputs in the form of frass and insect cadavers [10], and increased availability of light due to removal of foliage in the canopy [11,12]. The resulting increase in understory plant biomass may, in turn, provide resources for species that depend on understory vegetation for food or cover [13,14]. In all of these cases, outbreaks provide consumers with resource pulses, i.e. rare and ephemeral events of resource superabundance [15,16]. For most systems, we still have very limited understanding of how such outbreak-induced resource pulses affect the dynamics of consumer communities.

Saproxylic insects are one group of organisms that may experience particularly dramatic increases in resource availability following outbreaks. Saproxylic insects are, per definition, "dependent, during some part of their life cycle, upon the dead or dying wood of moribund or dead trees, or upon wood-inhabiting fungi or the presence of other saproxylics" [17]. Thus, since outbreaks kill or damage a large number of trees within a short time, they may provide saproxylic insects with resource pulses in the form of massive amounts of dead or dying woody material. Accordingly, saproxylic insects could be expected to increase rapidly in abundance after outbreaks. Indeed, it is well documented that saproxylic insects respond numerically to large-scale forest mortality due to other disturbances, like fire [18–20] and wind [21–23], and recently this has also been reported after outbreaks by tree-killing bark beetle in central Europe [24]. However, the magnitude of such responses is not necessarily a simple function of resource input. One important limiting factor may be the intimate relationship between saproxylic insects and the stage of decay of dead wood. A given saproxylic species can usually exploit dead wood only in a specific stage of decay. Thus, at any point in time after an outbreak, only a subset of the saproxylic community will be able to utilize the dead wood resource pulse. Moreover, the decay stage a saproxylic species can utilize is often relatively short in duration, lasting no more than a few years [25–27]. This is likely to lead to satiation effects, where a species is able to colonize and exploit only a small fraction of the dead wood generated by an outbreak within the time period when the wood represents a usable resource. Consequently, numerical responses to outbreaks both in individual species and the saproxylic community as a whole may in many cases be relatively limited in magnitude, compared to the magnitude of dead wood input. It should also be kept in mind that insect populations may be naturally limited or regulated by numerous other abiotic and biotic factors beyond the abundance of their resources, like climate and natural enemies [28].

Saproxylics are strongly affected by the successional decay of wood, but they also contribute to determining the speed of this decay process. When saproxylic insects penetrate the surface of dead wood, they contribute to its physical degradation, improve aeration of the wood and facilitate the entry of decomposing fungi and bacteria [27,29]. This can increase the rate of decay of the wood [30,31]. Thus, the ability of saproxylic insects to respond numerically to the dead wood left by an insect outbreak may affect the speed with which the dead trees are decomposed, and, ultimately, the resilience of the whole forest ecosystem. Understanding the numerical responses of saproxylic insects is, therefore, an important component of understanding the post-outbreak dynamics of forest ecosystems. However, most studies of how saproxylics respond to large-scale forest disturbances have dealt with fire and wind, whose effects are not necessarily directly comparable to those of insect outbreaks. Wind mainly produces lying dead trees, which tend to have high moisture content, while outbreaks leave mostly standing dead trees, which are prone to desiccation. This difference is expected to cause contrasting successional trajectories in the saproxylic community, particularly in the early stages of succession [27]. Meanwhile, trees killed by fire favour a specialized guild of fire-adapted saproxylics [19]. As far as we are aware, the bark beetle outbreaks in the Bavarian Forest National Park in Germany represent the only case where the responses of saproxylic insects to insect outbreaks have been studied so far [24,32,33]. Thus, the knowledge of this subject is still very limited.

In the present study, we examine the responses of saproxylic beetles to a massive outbreak of the two geometrid moths (Lepidoptera: Geometridae) *Epirrita autumnata* (Bkh.) (autumnal moth) and *Operophtera brumata* (L.) (winter moth) that occurred in the mountain birch (*Betula pubescens* ssp. *czerepanovii* Orlova) forest of the Varanger region in northern Norway (Fig. 1A) during 2002–2009. Mortality of birch trees was high throughout this region and reached more than 95% locally [14,34]. Thereby, the availability of dead wood for saproxylic insects was elevated far above normal levels across a very large area. In order to assess the short-term responses of the saproxylic beetle community in Varanger to this massive dead wood resource pulse, we used window traps to sample beetles in 2011 and 2012, both in severely damaged forest and undamaged forest at the outskirts of the outbreak area. Non-saproxylic beetles were also included in the study, in order to provide a contrast group that is not expected to benefit from increased availability of dead wood. By comparing beetle counts between damaged and undamaged forest, we address the following specific questions:

1. What is the magnitude of the short-term numerical responses of saproxylic beetles (both as individual species and as a community) to the superabundance of dead wood resources generated by the moth outbreak?
2. Are the numerical responses of saproxylic beetles weak compared to the magnitude of dead wood resource input, suggesting that important limiting factors act to constrain beetle responses?
3. Have numerical responses occurred only in saproxylic beetle species that are associated with dead wood in a particular stage of decay, suggesting decay-stage-preferences as a limiting factor for beetle responses?
4. Has the outbreak caused changes in saproxylic beetle community structure (i.e. the relative abundances of different beetle species), as would be expected if only a subset of the species in the community have responded numerically to the outbreak (cf. question 3)? If so, what is the magnitude of this change in community structure?

Materials and Methods

Ethics statement

No specific permissions were required to access any of the sampling locations or deploy any of the sampling methodologies described for this study. Three beetle species that are included on the Norwegian red list [35] were incidentally caught (see below for description of the trapping method) in small numbers during the study, namely *Atheta taxiceroides* (IUCN category: Near threatened), *Denticollis borealis* (IUCN category: Vulnerable) and *Enicmus lundbladi* (IUCN category: Vulnerable). This has been reported to the Norwegain Biodiversity Information Centre (http://www.biodiversity.no/). Table S1 in Appendix S1 gives the number of individuals caught.

Study region

The Varanger region is located at approximately 70°N, 29°E, in the north-eastern part of Finnmark county, northern Norway (Fig. 1A). Mean temperatures for January and July are −12.2 and 12.3°C, respectively, and annual precipitation is 400–500 mm [Rustefjelbma meteorological station (70°23′55″N, 28°11′36″E), 1961–2000 normal period]. The forest in the region is strongly dominated by mountain birch, although aspen (*Populus tremula*) and Scots pine (*Pinus sylvestris*) also occur patchily. Outbreaks by *E. autumnata* and *O. brumata* are the main natural disturbance factors in the north-Fennoscandian mountain birch ecosystem and occur approximately every 10 years [36–38]. Outbreaks vary in intensity, but sometimes cause severe defoliation and high mortality rates of mountain birch across thousands of square km. The duration and severity of the last outbreak cycle was historically unprecedented, and it has been estimated that approximately 10 000 square km of birch forest in northern Fennoscandia was affected by severe defoliation at least once during the period 2002–2008 [34]. The Varanger region was particularly strongly affected due to consecutive severe outbreaks by *E. autumnata* (2002–2004) and *O. brumata* (2005–2009).

Study design and sampling

In spring 2011, two transects were established in areas with slightly different defoliation histories, namely Tana (70°03′ N, 27°45′ E. Defoliated mainly during 2006–2007) and Kirkenes (69°46′ N, 29°20′ E. Defoliated mainly during 2007–2009) (Fig. 1B). Both transects were roughly linear and consisted of 10 sampling stations, interspaced by approximately two km. The first

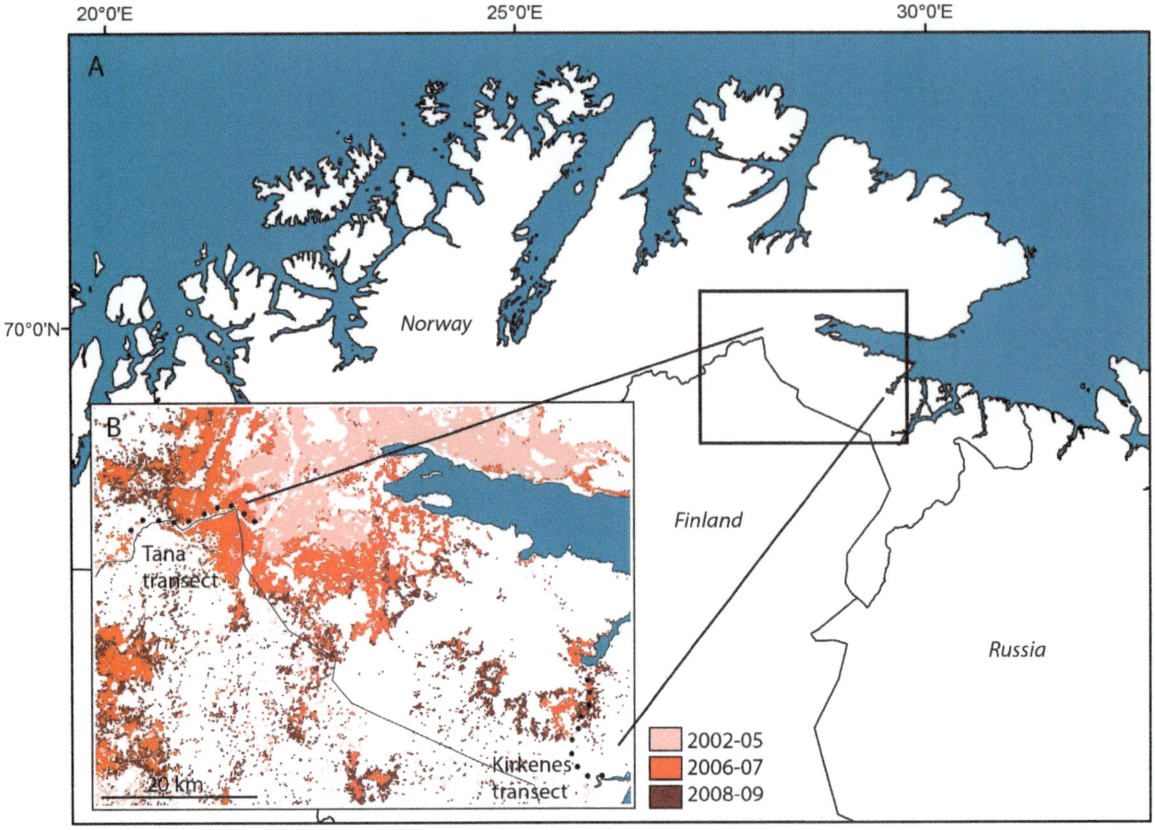

Figure 1. Map of Finnmark County (A) with the study region in Varanger inset (B). Red areas represent birch forest which experienced severe defoliation during the most recent moth outbreak. Different shades indicate the main years of defoliation for different areas (see key). The sampling stations of the two transects are indicated by black dots.

stations in each transect were located deep within forest which had suffered severe mortality due to the outbreak. Both transects then ran towards forest which was located at the outskirts of the outbreak area and had only experienced negligible defoliation (Fig. 1B). Fig S1 in Appendix S1 shows photographs from the individual sampling stations.

To assess the level of forest damage inflicted by the outbreak at the individual sampling stations, we established two 50 m survey lines at each station. The lines started at the centre-point of the station and ran towards the north and south, respectively. Every 7 m along the lines, we selected the closest birch tree with at least one stem (standing or fallen) of more than 1.3 m in height. The damage of the three thickest stems (above a minimum height of 1.3 m) within each selected tree was visually scored according to the criteria in table 1. These registrations were only conducted in 2011, since we had no reason to believe that the level of forest damage would change importantly between the two study-years (i.e. since the outbreak had ended in 2009).

Inspection of damage scores for the individual sampling stations showed that the damage-level of the forest shifted abruptly along both transects (Fig. 2). Station 1–4 in Kirkenes and 1–5 in Tana had almost exclusively dead stems, and were therefore designated as "dead" stations. In contrast, at station 5–10 in Kirkenes and 7–10 in Tana, the great majority of stems were alive and either undamaged or with light loss of foliage. This is representative for the situation in a normal, healthy mountain birch forest. These stations were therefore designated as "living". Station 6 in Tana had a preponderance of dead stems, but also many living ones. The station was designated as dead, since its mean damage score fell closest to the dead stations. The simple dichotomy of dead and living stations was used to represent forest damage in all subsequent analyses. Levels of forest damage at the individual stations were representative for the surrounding landscape (Fig. 1B. [34]). Thus, within a transect, dead and living stations were located within continuous belts of severely damaged and healthy forest, respectively.

Sampling of beetles was conducted by means of large window traps, with two crossed 40×60 cm window panes [39]. All traps were mounted between two adjacent birch stems, with the upper edge of the window panes approximately 1.5–2 m above the ground (Fig. S1 in Appendix S1). We used three traps per sampling station (for a total of 60 traps across the two transects). One trap was located at the centre point of the station, while the other two were located approximately 50 m to the north and south, respectively. In both 2011 and 2012, the traps were mounted in early June and remained operative for approximately two months until early August (thereby covering the peak of insect activity in our high-latitude study system).

All beetle individuals were identified to species level by an expert. Beetle species were thereafter classified according to their degree of association with dead wood, i.e. obligate saproxylic, facultative saproxylic or non-saproxylic, hereafter collectively referred to as dead wood association (DWA) groups. All species were also classified according to larval trophic guild, i.e. predators, fungivores, herbivores, wood-feeding, several (species belonging to more than one guild) or others (guilds that were rare in this study,

Table 1. Classification system used to score the damage of birch stems.

Damage category	Definition	Damage score
Living. Undamaged	Stem with no apparent loss of foliage	1
Living. Lightly damaged	Stem with >50% of foliage reaming compared to healthy stem	2
Living. Severely damaged	Stem with <50% of foliage reaming compared to healthy stem	3
Dead	Dead stem without foliage	4

i.e. scavengers, coprophages and detritus-feeders). Finally, saproxylic species were classified according to host-tree association i.e. associated with birch, not associated with birch and unknown host-tree association. The classification was based mainly on Dahlberg and Stokland [40] and Böhme [41], but additional information was obtained from a number of other sources. Table S1 in Appendix S1 gives a complete list of the sources used.

Statistical analyses

The goal of our analyses was to address the outlined questions about the numerical responses of saproxylic beetles in the study region to the changes in dead wood availability caused by the moth outbreak. The availability of dead wood was represented by our classification of forest damage, i.e. living forest (normal background levels of dead wood) and dead forest (greatly elevated availability of dead wood owing the moth outbreak). Since there was an abrupt shift from living to dead forest along the transects, we also expected that beetle communities would show rather sharp (i.e. non-gradual) spatial changes when moving from one forest type to the other. Therefore, we assumed that spatial changes in the beetle community between living and dead forest were adequately modelled by a categorical living-dead contrast. The possibility of additional spatial patterns in the data was assessed by plotting station-specific beetle counts as well as topographical parameters along the transects.

Forest damage (dead or living) was the focal predictor variable in all analyses. In addition, all models included the categorical predictors year (2011 or 2012) and location (Kirkenes or Tana), to account for variation in beetle communities between study years and transects. Within each year, beetle counts were pooled across the entire trapping season and the three individual traps within each station, resulting in a single datapoint per station, per year. All analyses were conducted separately for the three DWA groups, owing to their different degrees of association with dead wood. For the two saproxylic groups, the analyses only included species which are known to be associated with birch and which could therefore be expected to respond numerically to the dead birch wood left by the outbreak.

We performed quantitatively oriented univariate analyses to assess the magnitude of the numerical responses of saproxylic beetles to the dead wood generated by the moth outbreak. First, we assessed whether the total abundances of the different DWA groups (pooled abundance across all species within each group) differed between dead and living forest. This was done by fitting log-linear models with the station-specific individual-counts of each DWA group as response variables, and forest damage, year, location and their two-way interactions as predictors. All models employed quasi-likelihood correction for overdispersion. We used the Akaike Information Criterion, corrected for small sample size and overdispersion ($QAIC_c$), to select the best model for each DWA group. The minimal models included only the main effect of the focal predictor forest damage.

We also investigated if the proportion that each DWA group constituted, out of the total number of individuals across all groups, differed between dead and living forest. This was done by fitting logistic regression models with the station-specific propor-

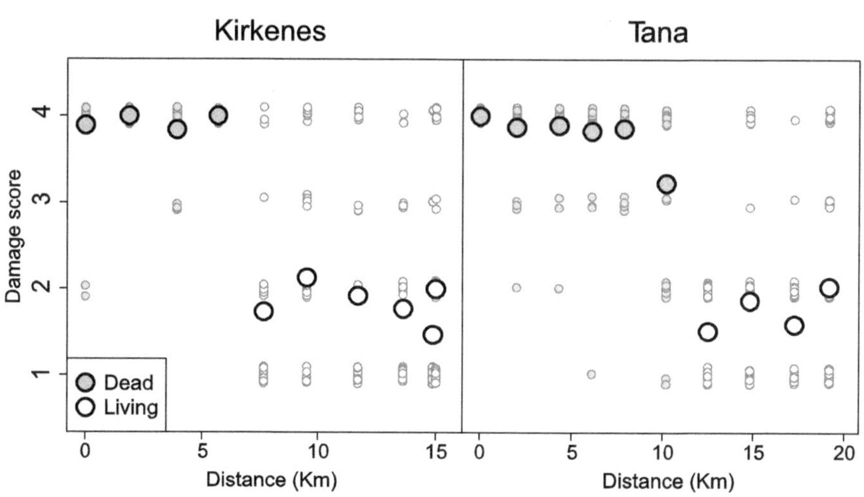

Figure 2. Station-specific damage scores of birch stems along the two sampling transects. The X axes represent the airline distances from the first sampling station in each transect. Large circles represent mean damage scores for each station, while small circles represent scores for individual stems. A small amount of noise has been added to the stem-specific scores to help distinguish individual datapoints. See table 1 for explanation of damage scores.

tions of each DWA group as response variables. We employed the same predictor variables and model selection strategy for this analysis as for the analysis of the total counts of DWA groups (above).

To assess numerical responses to dead wood availability in the most important saproxylic beetle species, we fitted log-linear models to the station-specific counts of all species which were judged to be sufficiently abundant for statistical modelling (total count >80 individuals). The predictor variables and model selection strategy for these species-specific models were the same as for the models fitted to the total counts of the DWA groups (above).

Finally, we used multivariate methods to investigate how beetle community structure varied between living and dead forest. In order to assess overall community-responses to the outbreak, the multivariate models were run not only for the three DWA groups, but also collectively for all species in the data. All beetle counts were Hellinger transformed before multivariate analysis [42].

We performed redundancy analysis (RDA) (`rda` function in the R package `vegan`) to allow visualization and significance-testing of the effects of forest damage, year and location on beetle community structure. We assumed additive effects of the three predictors, since inclusion of interactions did not change the conclusions from the analysis. The significance of the marginal effects of the predictors (i.e. effects after accounting for all other predictors in the models) was assessed with permutational ANOVA (`anova.cca` function in the R package `vegan`) with a maximum of 9999 permutations. Site scores from the RDAs were displayed on biplots in order to investigate how the beetle communities of the individual sampling stations were organized with respect to the predictor variables.

Variance partitioning (`varpart` function in the R package `vegan`) was used to estimate how much of the variance in beetle community structure was explained by the focal predictor forest damage, as well as year and location. The results of the variance partitioning were visualised with venn diagrams, fitted with the `venneuler` function in the R package with the same name. Negative variance fractions were treated as zeros.

Results

Overall beetle community-structure

A total of 5880 beetle individuals, belonging to 207 different species, were caught during the study (Table 2. Table S1 in Appendix S1 provides a complete species list). Obligate-, facultative- and non-saproxylic species accounted for 28.2%, 12.5% and 58.1% of the individuals and 25.1%, 22.7% and 43.5% of the species, respectively. The degree of association with dead wood is unknown for 1.2% of the individuals and 8.7% of the species. Predators and fungivores were the most abundant and species-rich trophic guilds among obligate saproxylic species, while predators and herbivores were most abundant and species-rich among non-saproxylic species (Table 2). Predators also constituted the highest number of individuals and species among facultative saproxylics. Facultative saproxylics also included many individuals and species that were fungivores or belonged to several trophic guilds, as well as many individuals with unknown trophic guild (most of which belonged to the species *Liotrichus affinis*). Most of the obligate saproxylic individuals (85.5%) and species (78.8%) caught were associated with birch. For facultative saproxylics, most of the individuals (70.4%) and roughly half of the species (48.9%) were birch associated.

Effects of dead wood availability on the total abundances and proportions of DWA groups

Outputs from the log-linear models (see table S2 and S3 in Appendix S1 for model selection and parameter estimates) for all three DWA groups were heavily influenced by large-scale spatial gradients in beetle counts along the transects (Fig. 3A–C). These gradients did not appear to have a consistent relationship with forest damage (i.e. their direction with respect to forest damage shifted between the two transects). Nevertheless, the results of the modelling were consistent with our expectation of a positive numerical response to increased dead wood availability for obligate saproxylic species, but not for facultative saproxylic species. In particular, the predicted abundances of all three DWA groups were higher in dead than in living forest in Kirkenes. However, this effect was most pronounced for obligate saproxylics (Fig. 3A), which had roughly 3.5 times higher predicted abundance in dead forest in this transect, while the predicted abundances of facultative- and non-saproxylics were only about 1.5 and 2 times higher, respectively, in dead forest (Fig. 3B, C). In Tana, the patterns for facultative- and non-saproxylics showed a reversal with respect to forest damage, with predicted abundances that were roughly one third and half as high in dead as in living forest (Fig. 3B, C). Meanwhile, the predicted abundance of obligate saproxylics was very similar for the two forest types (Fig. 3A). These patterns were consistent across both study-years (i.e. model selection did not support damage × year interactions for any DWA group), although the overall abundances of all three groups were considerably higher in 2012. Thus, when the three DWA groups were contrasted, it would appear that obligate saproxylics were favoured by dead forest. This impression was supported by logistic modelling of the relative proportions of the DWA groups (see table S3 and S4 in Appendix S1 for model selection and parameter estimates). The predicted proportion of obligate saproxylic individuals was about 10% higher in dead than in living forest (odds ratio: 1.87. 95% CI: 1.22–2.87. Fig. 3D). Meanwhile, there was a roughly 4% decrease in the predicted proportion of non-saproxylic individuals in this forest type (odds ratio: 0.83. 95% CI: 0.59–1.16. Fig. 3E). These patterns were consistent both across locations and years (i.e. model selection did not support damage × year or damage × location interactions for these two groups). The predicted proportion of facultative saproxylic individuals was very similar in living and dead forest in 2011(odds ratio: 1.03. 95% CI: 0.51–2.07. Fig. 3F). However, the model selection supported the damage × year interaction for this DWA group, causing the predicted proportion of facultative saproxylics to be almost 10% lower in dead forest in 2012 (odds ratio: 0.39. 95% CI: 0.25–0.62. Fig. 3F).

Effects of dead wood availability on individual beetle species

Three birch associated obligate saproxylic, one birch associated facultative saproxylic and four non-saproxylic species were judged to be abundant enough for individual log-linear modelling (Fig. 4. See table S5 and S6 in Appendix S1 for model selection and parameter estimates). These species collectively constituted 69.1% of all individuals in the data, 63.0% of all birch associated obligate saproxylic individuals, 81.2% of all birch associated facultative saproxylic individuals and 80.5% of all non-saproxylic individuals. The obligate saproxylic *Malthodes guttifer* was also abundant enough for modelling, but not associated with birch. Nevertheless, we analysed this species for the sake of comparison with the other saproxylics.

Table 2. The total number of individuals and species (in brackets) caught during the study for each DWA group and larval trophic guild.

	Dead wood association (DWA) group				
Trophic guild	**Obligate sx.**	**Facultative sx.**	**Non-sx.**	**Unknown**	**Sum**
Predator	597 [15]	215 [25]	1859 [44]	29 [9]	2700 [93]
Fungivore	820 [20]	63 [14]	65 [16]	36 [6]	984 [56]
Herbivore	-	-	1251 [13]	-	1251 [13]
Wood-feeding	71 [5]	-	-	-	71 [5]
Several	167 [11]	33 [5]	1 [1]	-	201 [17]
Other	-	-	86 [9]	-	86 [9]
Unknown	4 [1]	422 [3]	155 [7]	6 [3]	587 [14]
Sum	1659 [52]	733 [47]	3417 [90]	71 [18]	5880 [207]

Two of the birch associated obligate saproxylic species, namely *Elateroides dermestoides* (Fig. 4A) and *Rabocerus foveolatus* (Fig. 4B), showed evidence for a positive numerical response to dead wood availability, with higher predicted abundances in dead than living forest in both Kirkenes and Tana (*E. dermestoides*: roughly 25 and 3 times more abundant in dead forest in Kirkenes and Tana, respectively. *R. foveolatus*: roughly 2.5 times more abundant in dead forest in both transects). Curiously, the non-saproxylic species *Oxytelus laqueatus* (Fig. 4F) showed a similar pattern (roughly 7.5 times more abundant in dead forest in both transects). The obligate saproxylic *Podistra schoenherri* (Fig. 4C) and the facultative saproxylic *L. affinis* (Fig. 4E) (both birch associated) did not appear to respond consistently to dead wood availability, with predicted abundances being higher in dead forest in Kirkenes and lower in

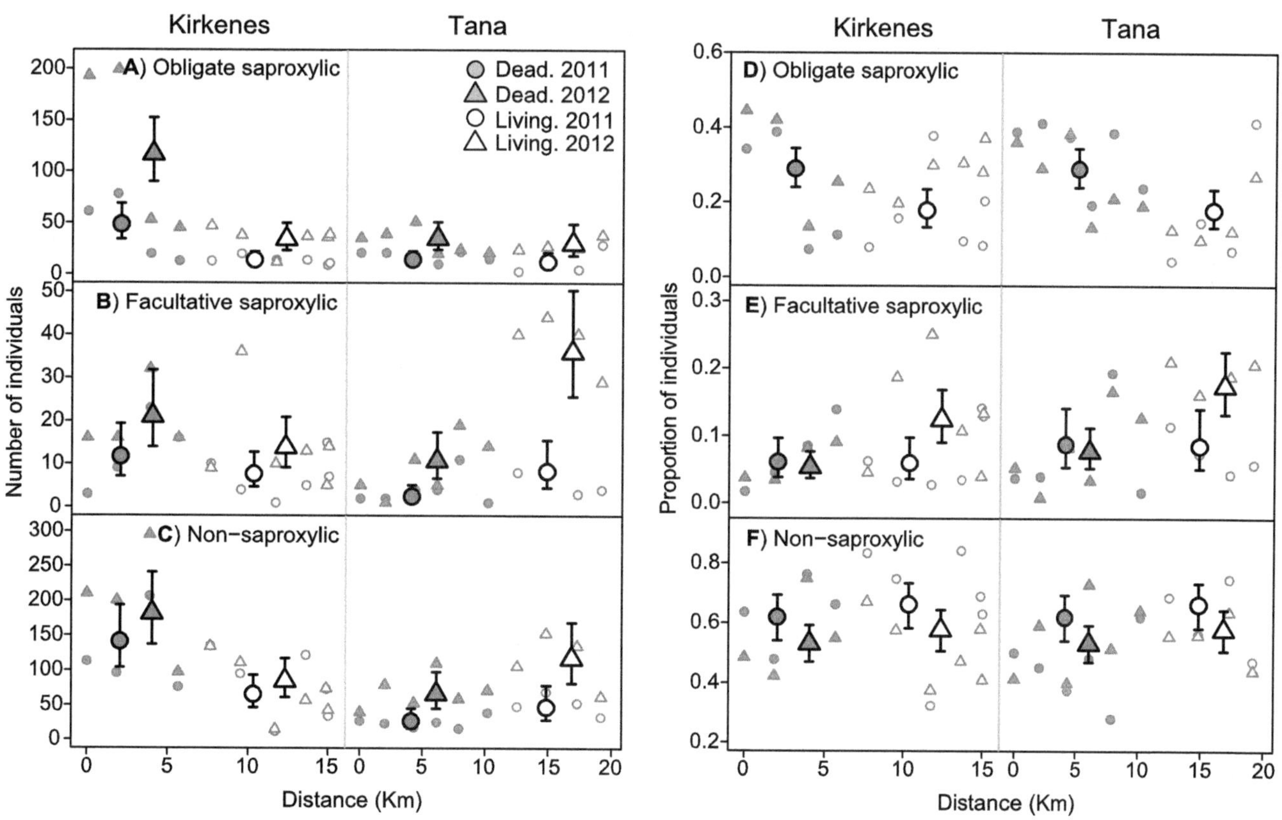

Figure 3. Variation in abundances and proportions of DWA groups along the two sampling transects. Large symbols: predicted abundances from log-linear models (A–C) and proportions from logistic models (D–F) in dead and living forest in Kirkenes and Tana in 2011 and 2012. Error bars represent 95% confidence intervals. In cases where year effects were not supported by the model selection, only a single predicted value for each damage × location combination is presented, representing the overall estimate across 2011 and 2012. Small symbols: original datapoints, i.e. the observed station-specific counts (A–C) and proportions (D–F) of obligate- facultative- and non-saproxylic beetle species in Kirkenes and Tana in 2011 and 2012, plotted against the airline distance from the first station in each transect.

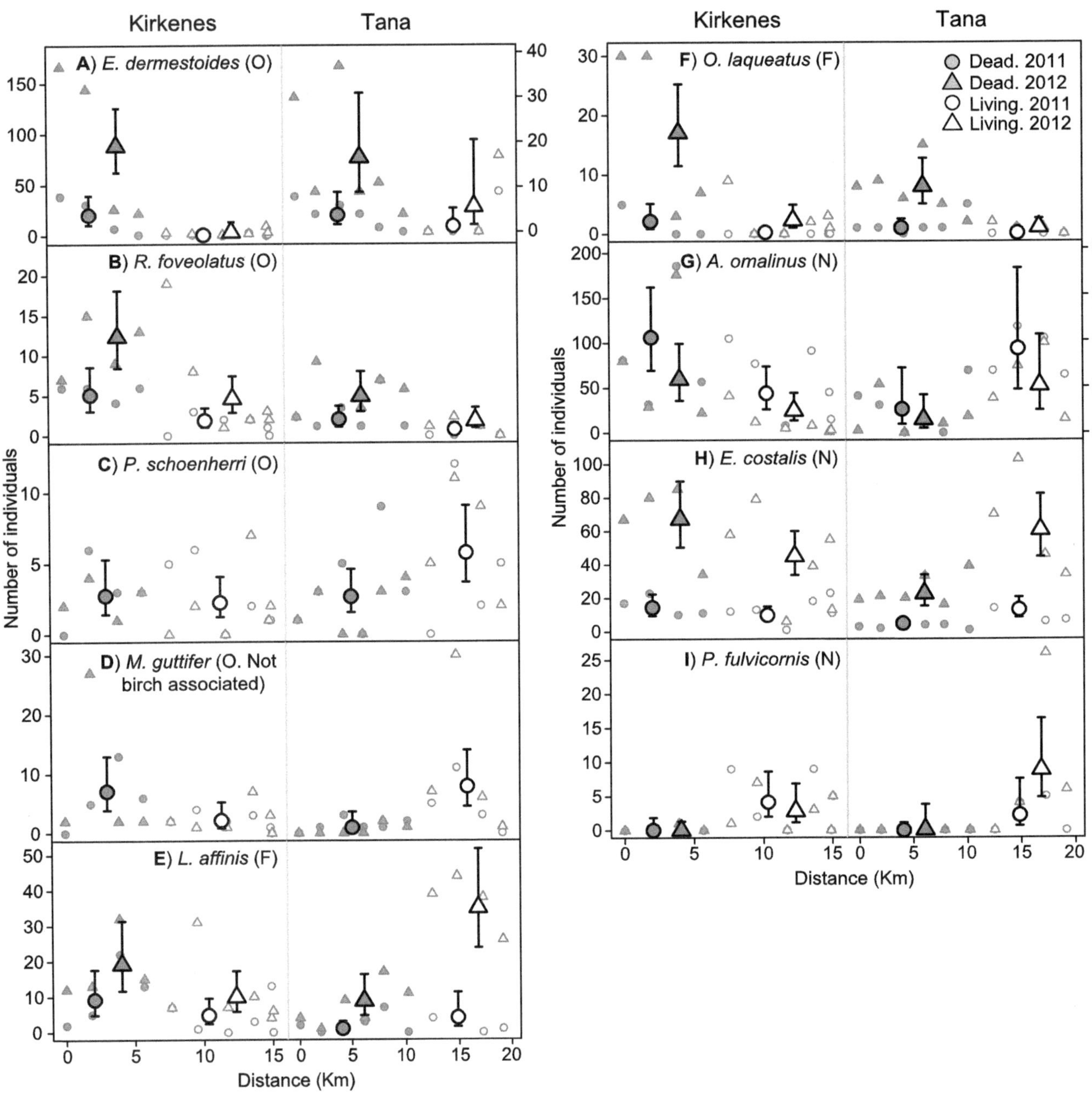

Figure 4. Variation in abundances of individual beetles species along the two sampling transects. Large symbols: predicted abundances from log-linear models in dead and living forest in Kirkenes and Tana in 2011 and 2012. Error bars represent 95% confidence intervals. In cases where year effects were not supported by the model selection, only a single predicted value for each damage × location combination is presented, representing the overall estimate across 2011 and 2012. Small symbols: original datapoints, i.e. the observed station-specific counts of the beetle species in Kirkenes and Tana in 2011 and 2012, plotted against the airline distance from the first station in each transect. The degree of dead wood association for each species is given in parenthesis behind its name. O= Obligate saproxylic. F= Facultative saproxylic. N= Non-saproxylic. All saproxylic species except *Malthodes guttifer* are known to be associated with birch. Note different scales on the Y axis in the plots for Kirkenes and Tana for *Elateroides dermestoides* (A) and *Anthophagus omalinus* (G).

this forest type in Tana. A similar reversal of predicted abundance patterns between transects were evident for the not birch associated obligate saproxylic species *M. guttifer* (Fig. 4D) and the non-saproxylic species *Anthophagus omalinus* (Fig. 4G) and *Eanus costalis* (Fig. 4H). Finally, the non-saproxylic species *Polydrusus fulvicornis* (Fig. 4I) was almost absent in dead forest but relatively abundant in living forest in both transects.

Effects of dead wood availability on beetle community structure

Forest damage explained none of the variance in community structure for facultative saproxylics (see below) and we therefore had no reason to expect that damage would yield significant effects in a RDA for this DWA group. Accordingly, we did not conduct RDA for facultative saproxylics. RDA site scores for living and dead forest tended to be separated in biplots for both obligate- and non-saproxylic species, suggesting that these DWA groups showed consistent differences in community structure between the two forest types (Fig. 5). This separation was quite distinct for non-saproxylic species (Fig. 5B), but less clear for obligate saproxylics (Fig. 5A), with considerable overlap between living and dead stations for the latter group. Nevertheless, the effect of forest damage was highly significant in permutation-based ANOVAs for both DWA groups ($P = 0.005$ in both cases). When all beetle species were considered collectively, the separation between living and dead stations became even clearer, with hardly any overlap between the two forest types (Fig. 5C). Also in this case, permutation-based ANOVA yielded a highly significant effect of forest damage ($P = 0.005$). The degree of separation between site scores for Kirkenes and Tana and 2011 and 2012 was variable among the species groups, but all permutation-based ANOVA tests for location and year were significant at $P = 0.030$ or lower.

Forest damage, location and year collectively explained 16.4% of the variance in beetle community structure for obligate saproxylic species, 27.7% for non-saproxylic species and 20.1% for all species (Table 3, Fig. 6). For facultative saproxylics, damage failed to account for any variance, while location and year collectively explained 9.8%. In all cases, there was little overlap between the variance explained by the individual predictors, so that their total and unique contributions were largely similar. We therefore report only total contributions below. Damage was the most important predictor for obligate saproxylics, accounting for 11.1% of the variance. Year was most important for facultative saproxylics, accounting for 7.4%. Year was also most important for non-saproxylics, accounting for 17.7%, while damage made a comparatively modest contribution of 7.6%. As expected from the results for the individual DWA groups, the relative contributions of year and damage became more even when all beetle species were considered collectively, with year explaining 10.2% and damage explaining 7.3%. Location made only a small contribution for all groups, explaining less than 4.0% of the variance in all cases.

Discussion

Insect outbreaks can rapidly generate more dead woody material than any other ecosystem process (with the possible exception of wind and fire), but little is known of how saproxylic organisms respond to these disturbance events, despite the fact that saproxylics are intimately dependent on dead wood and play an important role in the ecosystem service of decomposition [43,44]. In this study we have contributed to filling this knowledge-gap by characterizing short-term changes in the abundance and community structure of saproxylic beetles following a severe moth outbreak in sub-arctic birch forest.

We expected that the outbreak would generate a dead wood resource pulse and thereby trigger positive numerical responses in the saproxylic beetle community. Importantly, we also expected that such responses would be limited to a subset of the community, owing to the ability of given saproxylic species to exploit only certain dead wood decay-stages, and that increases in responding species would be constrained by limitations to the amount of dead wood that can be colonized within a given amount of time (i.e. satiation effects). Our results appear to match these expectations. By contrasting the overall abundances of obligate-, facultative- and non-saproxylic species, we did indeed find evidence for increased abundance of obligate saproxylics in outbreak-affected forest. However, this increase corresponded only to a roughly 10% higher proportion of obligate saproxylic individuals in dead forest, while the increase in the dead wood resource may have been as much as hundredfold.

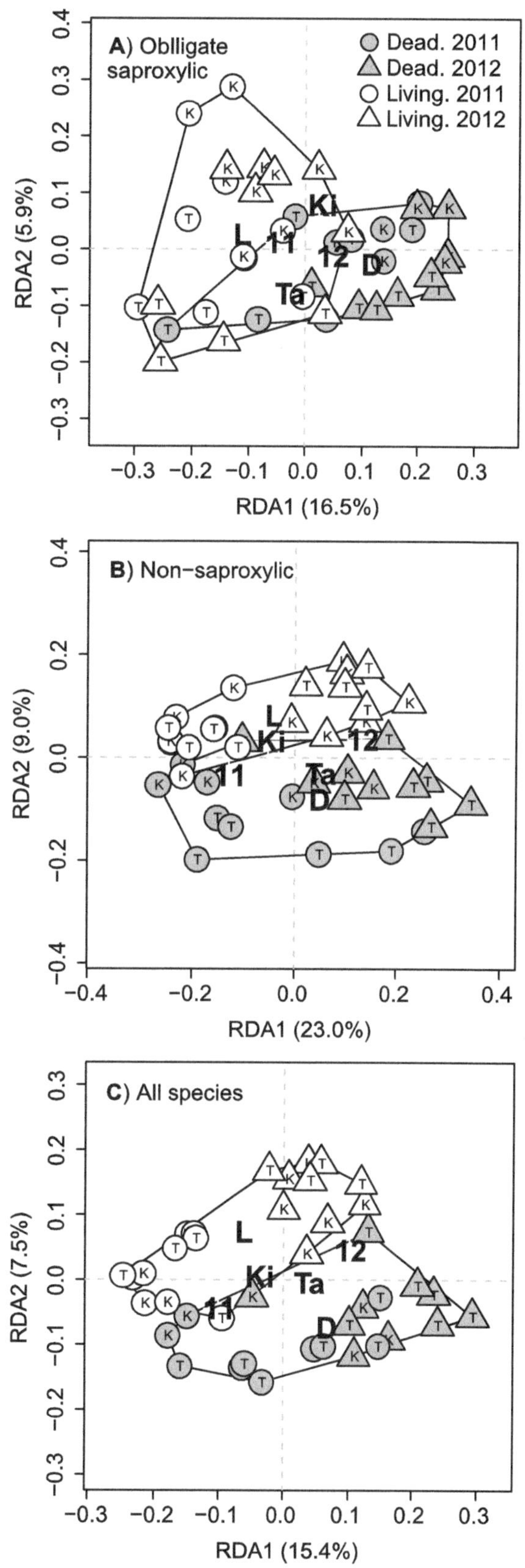

Figure 5. Biplots relating the predictors forest damage, location and year to variation in beetle community structure. A: Obligate saproxylic species. B: Non-saproxylic species. C: All beetle species collectively. Circles and triangles represent sites scores

(corresponding to individuals sampling stations) from the RDAs. The letters on the symbols indicate the transect that each station belongs to. K = Kirkenes. T = Tana. Centroids for the levels of the three predictor variables are indicated by large letters on the plots. Forest damage: D = Dead. L = living. Location: Ki = Kirkenes. Ta = Tana. Year: 11 = 2011. 12 = 2012.The two polygons in each plot surround all stations belonging to dead and living forest. Note that the site scores are shown as principal coordinates [65], implying that the distances between points are two-dimensional approximations of the Hellinger distances between them. The percentage of the total variation that is accounted for by the RDA axes is given in the axis labels.

These findings may suggest that the saproxylic beetle community in the study region has a limited ability to respond to increased availability of dead wood. Our results for individual beetle species suggest species-specific decay-stage-preferences as one potential reason for this limitation. Most of the numerical response of obligate saproxylics was driven by the two species *E. dermestoides* and *R. foveolatus*. Both species exploit recently dead or dying trees and have therefore probably enjoyed increased resource-availability during the 3–5 years separating the conclusion of the outbreak and our study. Meanwhile, the third birch associated obligate saproxylic species that was analysed (*P. schoenherri*) showed no response to the outbreak. This species utilizes dead wood in more advanced stages of decay and is thereby unlikely to have benefitted from the outbreak so far. Thus, our findings match the prediction that species-specific decay-stage-preferences will allow only a subset of the saproxylic beetle community to respond to an outbreak at any given time. Unfortunately, the knowledge of decay-stage-preferences for most species in our data is too limited to allow more rigorous evaluation of this hypothesis.

Our study gives no direct indication of whether the numerical response of the saproxylic beetle community is constrained by satiation effects. However, this seems likely to be the case. Within a given area, several decades can pass between moth outbreaks that cause substantial forest mortality. It is therefore unlikely that saproxylic beetles in the study region are specifically adapted to efficiently exploit the extreme amounts of dead wood that outbreaks generate. Satiation effects could result e.g. from limitations to the amount of dead wood that adult beetles manage to colonize and reproduce in within a given time, including the number of eggs that they have available. Reproductive rates could be further constrained by the cold climate of the study region, which may prolong larval development of the decomposing insect community [45]. Such limitations on reproductive output may be decisive for the magnitude of numerical responses to the outbreak, since these responses are probably driven mainly by reproduction. Behavioural aggregation of beetles to the dead forest seems unlikely to be important, since the outbreak area is very large (Fig. 1B) and a massive number of individuals would have to be drawn in from adjacent forest areas in order to raise densities appreciably. Moreover, the spatial extent of the outbreak area is probably far larger than the dispersal capacity of most beetle

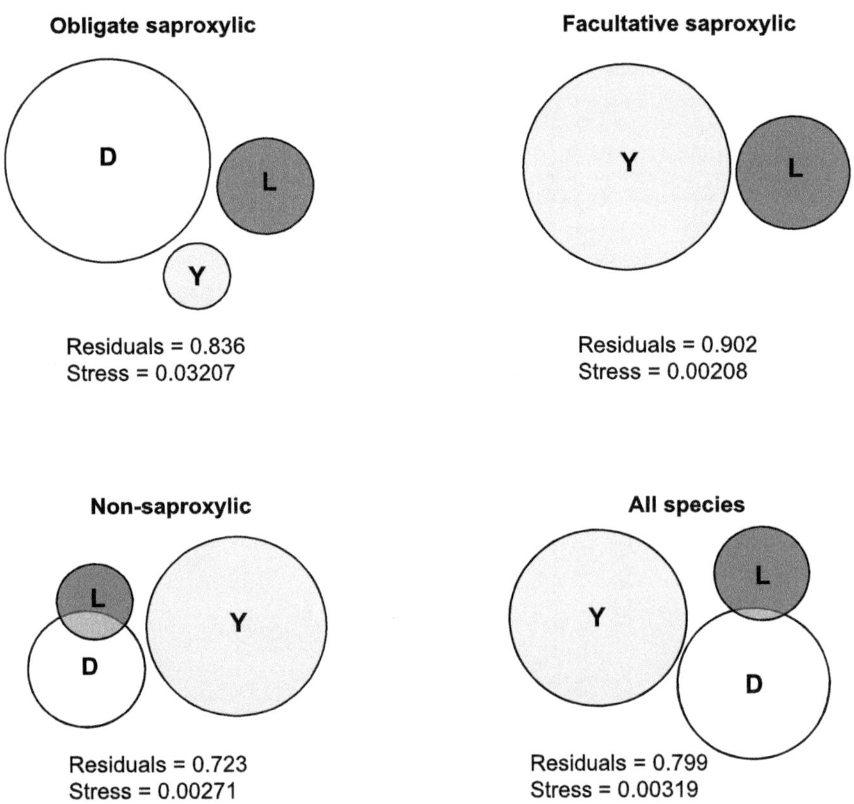

Figure 6. Contributions of forest damage, location and year to explaining variance in beetle community structure. D = damage, L = location and Y = year. The amount of variance explained by each predictor is represented by the relative size of the circles in the venn diagrams. Overlapping areas indicate variance that is jointly explained by two of the predictors, while non-overlapping areas indicate variance that is uniquely explained by a single predictor. The residuals represent the proportion of the total variance in the beetle data that cannot be explained by the predictors. The values of the `venneuler` stress statistic [66] are below the 0.01 threshold ($s_{.01} = 0.056$) in all cases, indicating that there is a good match (i.e. fit) between the venn diagrams and the actual variance components (Table 3).

Table 3. Variance partitioning table showing how the predictor variables forest damage, location and year contribute to explaining the variance in beetle community structure for obligate-, facultative- and non-saproxylic (sx.) beetles, and for all beetle species collectively.

	Df	Adjusted R^2			
Predictors		**Obligate sx.**	**Facultative sx.**	**Non-sx.**	**All species**
Damage	1	0.111	0	0.076	0.073
Location	1	0.038	0.024	0.035	0.030
Year	1	0.016	0.074	0.177	0.102
Damage + Location	2	0.149	0.024	0.100	0.099
Damage + Year	2	0.126	0.074	0.253	0.175
Location + Year	2	0.054	0.098	0.212	0.132
All	3	0.164	0.098	0.277	0.201
Individual fractions					
Damage| Location + Year	1	0.111	0	0.063	0.069
Location | Damage + Year	1	0.038	0.024	0.022	0.026
Year | Damage + Location	1	0.016	0.074	0.177	0.102
Residuals		0.836	0.902	0.723	0.799

species in the data. Notably though, since the level of forest damage inflicted by the outbreak is spatially heterogeneous, aggregation on smaller spatial scale may affect local density patterns within the outbreak area.

Climatic constraints on beetle diversity in our sub-arctic study region may also help explain the seemingly weak overall response of the saproxylic beetle community. In total, we encountered 52 obligate- and 47 facultative saproxylic species. This is considerably less than what is typically reported from studies with comparable sampling efforts in more southerly deciduous forests [46–48] Thus, many species which may have responded to an outbreak further south may simply be absent from our system. Some support for this suggestion is provided by the results of Wermelinger *et al.* [21], who studied short-term responses of saproxylic beetles to a large-scale windthrow event in Switzerland. Within 6 years of the windthrow, they observed much stronger increases in the abundance of saproxylic beetles than we did in the present study. This was largely due to species in the families Cerambycidae and Buprestidae and the subfamily Scolytinae (all of which contain birch associated species). These three groups were either absent (Buprestidae) or very rare (Cerambycidae: 1 species, 11 individuals. Scolytinae: 4 species, 16 individuals) in our data, presumably because our study region lies beyond the northern distributional limit of most of their species [25]. Our system may thereby lack many species that could potentially have responded numerically to a dead wood resource pulse. Another factor to consider is that our study region represents the northern distributional limit for most of the beetle species we encountered. Many species may thus be close to their tolerance limits for abiotic factors, like temperature, and could therefore be highly sensitive to stochastic year-to-year environmental variation [49]. This might also impair their ability to respond effectively to increases in resource availability.

It should be noted that our interpretation of window trap catches in terms of beetle numerical responses has some potential pitfalls. First, local sun exposure can potentially affect beetle activity levels, so that beetle counts are inflated in exposed (i.e. warmer) conditions. Sun exposure may also affect actual beetle abundance, since many saproxylic species have preferences for exposed or shaded conditions [50,51]. In our case, stations in dead forest (without foliage in the canopy) may have been more sun exposed than stations in living forest (with an intact canopy). However, total beetle counts were not consistently higher in dead forest (Fig. 3), suggesting that sun exposure did not inflate beetle counts in our case. Similarly, Kaila *et al.* [39] and Sverdrup-Thygeson and Ims [52] had similar total beetle counts in exposed and shaded conditions. With respect to beetle habitat preferences, we are not aware that any of the most abundant species in our data (saproxylic or non-saproxylic) are considered to have strong preferences relating to sun exposure. Thus, we have no reason to expect that sun exposure was an important driver of the main abundance patterns in our study. Also, we should emphasize that even healthy mountain birch forest is quite open, with mostly small and well-spaced trees (This is well illustrated by Fig. S1 in Appendix S1). Thus, the degree of sun exposure may not have differed radically between stations in living and dead forest. Second, saproxylic beetles may respond to variation in dead wood availability only on certain spatial scales, and window traps may fail to detect these responses if the volume of dead wood around the traps has not been quantified at the relevant scale [53–55]. However, in our case, the volume of dead wood was extremely high not only in the immediate vicinity of the traps, but also throughout the outbreak area as a whole, i.e. on a scale of thousands of hectares around the traps. Thus, in effect, our design captured elevated availabilities of dead wood on a wide range of spatial scales, ranging from locally around the traps and up to the landscape level. Thereby, we should have been able to detect beetle responses to the outbreak, regardless of whether beetles responded to dead wood availability on small or large spatial scales. Third, dispersal-driven "spillover" of saproxylic beetles from dead to living forest may have elevated beetles densities in living forest. This would invalidate our conclusion about weak numerical responses to the outbreak based on comparing beetle catches between the two forest types. However, such spillover would probably have been manifested as a gradual decline in saproxylic beetle counts when moving progressively further away from dead forest. This did not appear to be the case in our data. For beetle counts to be uniformly elevated throughout living forest there would have to be high rates of dispersal over distances of at least 10 km (the approximate distance between the last living and dead stations in both transects). This seems improbable, although the dispersal capacity of the studied beetles is too poorly known to rule out the possibility either. Thus, in summary, although our results must be cautiously interpreted, we still consider it to be a reasonable conclusion that saproxylic beetles have so far responded weakly to the dead wood left by the outbreak.

If the ability of the saproxylic beetle community to respond numerically to dead wood resource pulses is indeed limited, this could have important implications for the decomposition of trees killed by the outbreak. Saproxylic beetles contribute to wood decay by direct consumption and physical degradation of woody material, and by facilitating the entry and activity of decomposing fungi and bacteria [27,29–31,43,44]. However, if saproxylic beetles respond weakly to dead wood input, such decay processes may so far have played a minor role in decomposing the trees killed by the outbreak. Thereby, the decay of these trees may be retarded compared to dead wood in healthy mountain birch forest. This possibility must be explored in future studies, since we did not quantify the decay of dead stems along the transects. Nearly all stems were still standing in 2012 (pers. obs. by the authors), but this tells us little about decay rates, since birch stems have more than 80% probability of remaining standing throughout the first decade after their death (Mäkinen *et al.* [56] studying *B. pubescens* and *B. pendula* in southern and central Finland). Studies that contrast decay and insect-utilization of dead trunks between outbreak areas and healthy forest would be instructive, as would studies of fungal and other microbial decomposer communities. Penetration of bark by saproxylic invertebrates can be important for the colonization of dead trunks by bacteria and fungi [29], and weak numerical responses of saproxylic beetles to outbreaks may thereby retard colonization or alter community-composition in these microbial decomposers. Alternatively, the relative importance of microbial decomposition may increase in the absence of strong numerical responses from saproxylic beetles. Studies addressing these questions would be of great interest for understanding post-outbreak decomposition dynamics.

The indication of limitations in the short-term numerical response from the saproxylic beetle community to increased dead wood amounts is also relevant in a management perspective. Several studies, especially in Northern Europe, have addressed the negative effects the decrease in background levels of dead wood in managed boreal forest have on the saproxylic beetle communities [52,57,58]. In order to alleviate these effects, new management guidelines and restoration efforts are suggested to increase the dead wood amounts [53,59–61]. In this process, a firm understanding of the responses of saproxylic beetles across a wide span of spatial and temporal scales, and in different forest ecosystems, is vital.

The separation of sites in living and dead forest in the RDA for obligate saproxylics indicates that the outbreak has induced changes in community structure (i.e. changes in the relative abundances of different species) among saproxylic beetles. This fits well with our expectation that only some saproxylic species should respond numerically to the outbreak, and the empirical observation of species-specific response patterns. A more unexpected finding is the clearer separation (i.e. larger community changes) of dead and living sites for non-saproxylic species. This may suggest that there has been greater variation in species responses to the outbreak among non-saproxylics than obligate saproxylics. This seems plausible, since the non-saproxylic group contains more species and may therefore encompass a higher diversity of biological traits (e.g. microhabitat requirements and feeding specialisations). It is noteworthy in this respect that there were clear examples of both positive (i.e. *O. laqueatus*) and negative (i.e. *P. fulvicornis*) responses to the outbreak among the individually analysed non-saproxylic species, while this was not the case among saproxylic species. The mechanisms by which non-saproxylic species may have benefitted from the outbreak are unclear. With respect to *O. laqueatus*, this species has been recorded in many types of decaying organic matter and also from sap runs [62]. Thus, we cannot rule out that this species has somehow benefitted from increased availability of dead wood, even though its substrate preferences seem to be too wide to justify classifying it as saproxylic. The finding of clear community changes in non-saproxylics means that our inferences about numerical responses to the outbreak, which were partly based on contrasting the three DWA groups, must be treated with some caution. Clearly, non-saproxylics cannot be viewed as an unaffected control group with respect to the outbreak. Indeed, the pervasive impact of the outbreak on the whole ecosystem, including understory vegetation, herbivores [14,63] and abiotic conditions [64], probably leaves few biological taxa unaffected. However, this does not change our main conclusion about weak numerical responses in the saproxylic beetle community.

Although the RDA indicates that beetle community structure was affected by the outbreak, it should be emphasised that forest damage explained a small fraction of the overall variance in the beetle data (at most 11.1%, in the case of obligate saproxylics). Thus, although the outbreak represents a dramatic disturbance, which has changed both the biotic and abiotic environment, its effect on many beetle species may be overruled by other environmental variables. This conclusion is supported by the fact that counts of the DWA groups and many beetle species showed large-scale spatial gradients that appeared to be independent of forest damage. Notably though, these gradients may to some extent also represent variation in trapping efficiency or beetle activity, due to variation in e.g. wind speed or sun-exposure along the transects. Both transects have considerable heterogeneity in topographical variables, such as altitude and aspect (Fig. S2 in Appendix S1), and this may have affected both trapping efficiencies and beetle activity patterns.

Conclusions

Three to five years after an extensive moth outbreak, the saproxylic beetle community in the sub-arctic mountain birch forest appears to have mounted a weak but detectable numerical response to the increased availability of dead wood in the system. This response has mainly been driven by a few very abundant species, which utilize dead wood in early stages of decay. Although further monitoring is required to assess the full extent of beetle responses to the outbreak, our present results would suggest the possibility that saproxylic beetles increase in abundance to slowly to play a major role in decomposing the enormous quantities of dead wood left by moth outbreaks in this system. Accordingly, the role of microbial decomposer communities in post-outbreak decomposition dynamics represents a prime topic for future research.

Supporting Information

Appendix S1 Supporting tables and figures. Table S1 List of all beetle species encountered in the study, with information on association with dead wood birch. Table S2 Results of model selection for log-linear models for DWA groups. Table S3 Coefficients from selected log-linear and logistic models for DWA groups. Table S4 Results of model selection for logistic models for DWA groups. Table S5 Results of model selection for log-linear models for individual beetle species. Table S6 Coefficients from selected log-linear models for individual beetle species. Figure S1 Photographs from the individual sampling stations in Kirkenes and Tana in early June 2011. Figure S2 Station-specific scores for topographic variables.

Acknowledgments

We are grateful to Sindre Ligaard for identification of beetle species. We would also like to thank two anonymous reviewers for valuable comments that helped to improve the manuscript.

Author Contributions

Conceived and designed the experiments: JUJ RAI. Performed the experiments: OPLV SS LK. Analyzed the data: OPLV MB. Contributed reagents/materials/analysis tools: AST. Wrote the paper: OPLV.

References

1. Barbosa P, Letourneau D, Agrawal A (2012) Insect Outbreaks Revisited: Wiley.
2. Yang LH (2012) The Ecological Consequences of Insect Outbreaks. Insect Outbreaks Revisited: John Wiley & Sons, Ltd. pp. 197–218.
3. Haney JC (1999) Numerical response of birds to an irruption of elm spanworm (Ennomos subsignarius [Hbn.]; Geometridae: Lepidoptera) in old-growth forest of the Appalachian Plateau, USA. Forest Ecology and Management 120: 203–217.
4. Drever MC, Goheen JR, Martin K (2009) Species-Energy Theory, Pulsed Resources, and Regulation of Avian Richness during a Mountain Pine Beetle Outbreak. Ecology 90: 1095–1105.
5. Hogstad O (2005) Numerical and functional responses of breeding passerine species to mass occurrence of geometrid caterpillars in a subalpine birch forest: a 30-year study. Ibis 147: 77–91.
6. Marcello GJ, Wilder SM, Meikle DB (2008) Population dynamics of a generalist rodent in relation to variability in pulsed food resources in a fragmented landscape. Journal of Animal Ecology 77: 41–46.
7. Vandegrift KJ, Hudson PJ (2009) Response to enrichment, type and timing: small mammals vary in their response to a springtime cicada but not a carbohydrate pulse. Journal of Animal Ecology 78: 202–209.
8. Reeve JD (1997) Predation and bark beetle dynamics. Oecologia 112: 48–54.
9. Eveleigh ES, McCann KS, McCarthy PC, Pollock SJ, Lucarotti CJ, et al. (2007) Fluctuations in density of an outbreak species drive diversity cascades in food webs. Proceedings of the National Academy of Sciences 104: 16976–16981.
10. Yang LH (2004) Periodical Cicadas as Resource Pulses in North American Forests. Science 306: 1565–1567.
11. Eschtruth AK, Cleavitt NL, Battles JJ, Evans RA, Fahey TJ (2006) Vegetation dynamics in declining eastern hemlock stands: 9 years of forest response to hemlock woolly adelgid infestation. Canadian Journal of Forest Research 36: 1435–1450.
12. Gandhi KK, Herms D (2010) Direct and indirect effects of alien insect herbivores on ecological processes and interactions in forests of eastern North America. Biological Invasions 12: 389–405.
13. Bell JL, Whitmore RC (1997) Bird Populations and Habitat in Bacillus thuringiensis and Dimilin-treated and Untreated Areas of Hardwood Forest. American Midland Naturalist 137: 239–250.
14. Jepsen J, Biuw M, Ims R, Kapari L, Schott T, et al. (2013) Ecosystem Impacts of a Range Expanding Forest Defoliator at the Forest-Tundra Ecotone. Ecosystems 16: 561–575.
15. Yang LH, Bastow JL, Spence KO, Wright AN (2008) What can we learn from resource pulses? Ecology 89: 621–634.
16. Yang LH, Edwards KF, Byrnes JE, Bastow JL, Wright AN, et al. (2010) A meta-analysis of resource pulse-consumer interactions. Ecological Monographs 80: 125–151.
17. Speight MCD (1989) Saproxylic invertebrates and their conservation: Council of Europe.
18. Muona J, Rutanen I (1994) The short-term impact of fire on the beetle fauna in bored coniferous forest. Annales Zoologici Fennici 31: 109–121.
19. Saint-Germain M, Drapeau P, Hébert C (2004) Landscape-Scale Habitat Selection Patterns of Monochamus scutellatus (Coleoptera: Cerambycidae) in a Recently Burned Black Spruce Forest. Environmental Entomology 33: 1703–1710.
20. Boulanger Y, Sirois L (2007) Postfire Succession of Saproxylic Arthropods, with Emphasis on Coleoptera, in the North Boreal Forest of Quebec. Environmental Entomology 36: 128–141.
21. Wermelinger B, Duelli P, Obrist MK (2002) Dynamics of saproxylic beetles (Coleoptera) in windthrow areas in alpine spruce forests. Forest snow and landscape research 77: 133–148.
22. Bouget C, Duelli P (2004) The effects of windthrow on forest insect communities: a literature review. Biological Conservation 118: 281–299.
23. Grimbacher P, Stork N (2009) How do beetle assemblages respond to cyclonic disturbance of a fragmented tropical rainforest landscape? Oecologia 161: 591–599.
24. Müller J, Noss RF, Bussler H, Brandl R (2010) Learning from a "benign neglect strategy" in a national park: Response of saproxylic beetles to dead wood accumulation. Biological Conservation 143: 2559–2569.
25. Ehnström B, Axelsson R (2002) Insektsgnag i bark och ved: ArtDatabanken SLU.
26. Langor DW, Spence JR, Hammond HJ (2004) Saproxylic beetles (Coleoptera) using Populus in boreal aspen stands of western Canada: spatiotemporal variation and conservation of assemblages. Canadian Journal of Forest Research 34: 1–19.
27. Jogeir N. Stokland JS, Jonsson BG (2012) Biodiversity in Dead Wood: Cambridge University Press.
28. Price PW, Denno RF, Eubanks MD, Finke DL, Kaplan I (2011) Insect Ecology: Behavior, Populations and Communities: Cambridge University Press.
29. Schowalter TD, Caldwell BC, Carpenter SE, Griffiths RP, Harmon ME, et al. (1992) Decomposition of fallen trees: effects of initial conditions and heterotroph colonization rates. In: Singh KP, Singh JS, editors. Tropical Ecosystems: Ecology and Management. New Delhi: Wiley Eastern Limited. pp. 373–383.
30. Zhong H, Schowalter TD (1989) Conifer bole utilization by wood-boring beetles in western Oregon. Canadian Journal of Forest Research 19: 943–947.
31. Muller MM, Varama M, Heinonen J, Hallaksela AM (2002) Influence of insects on the diversity of fungi in decaying spruce wood in managed and natural forests. Forest Ecology and Management 166: 165–181.
32. Müller J, Buβler H, Goβner M, Rettelbach T, Duelli P (2008) The European spruce bark beetle Ips typographus in a national park: from pest to keystone species. Biodiversity and Conservation 17: 2979–3001.
33. Lehnert LW, Bässler C, Brandl R, Burton PJ, Müller J (2013) Conservation value of forests attacked by bark beetles: Highest number of indicator species is found in early successional stages. Journal for Nature Conservation 21: 97–104.
34. Jepsen JU, Hagen SB, Karlsen SR, Ims RA (2009) Phase-dependent outbreak dynamics of geometrid moth linked to host plant phenology. Proceedings of the Royal Society B-Biological Sciences 276: 4119–4128.
35. Kålås JA, Viken Å, Henriksen S, Skjelseth S (2010) The 2010 Norwegian Red List for Species. Norway: Norwegian Biodiversity Information Centre.
36. Tenow O (1972) The outbreaks of Oporinia autumnata Bkh. and Operophtera spp. (Lep., Geometridae) in the Scandinavian mountain chain and northern Finland 1862–1968. Zoologiska bidrag från Uppsala, Supplement, 2, 1–107.
37. Bylund H (1999) Climate and the population dynamics of two insect outbreak species in the north. Ecological Bulletins 47: 54–62.
38. Ruohomäki K, Tanhuanpää M, Ayres MP, Kaitaniemi P, Tammaru T, et al. (2000) Causes of cyclicity of Epirrita autumnata (Lepidoptera, Geometridae): grandiose theory and tedious practice. Population Ecology 42: 211–223.
39. Kaila L, Martikainen P, Punttila P (1997) Dead trees left in clear-cuts benefit saproxylic Coleoptera adapted to natural disturbances in boreal forest. Biodiversity & Conservation 6: 1–18.
40. Dahlberg A, Stokland JN (2004) Vedlevande arters krav på substrat: sammanställning och analys av 3 600 arter: Skogsstyrelsen.
41. Böhme J (2005) Die Käfer Mitteleuropas: Katalog. München: Elsevier Spektrum Akademischer Verlag.
42. Legendre P, Gallagher ED (2001) Ecologically Meaningful Transformations for Ordination of Species Data. Oecologia 129: 271–280.
43. Ulyshen MD (2013) Strengthening the case for saproxylic arthropod conservation: a call for ecosystem services research. Insect Conservation and Diversity 6: 393–395.
44. Ulyshen MD, Wagner TL (2013) Quantifying arthropod contributions to wood decay. Methods in Ecology and Evolution 4: 345–352.
45. Gullan PJ, Cranston PS (2010) The Insects: An Outline of Entomology: Wiley.
46. Martikainen P, Kaila L (2004) Sampling saproxylic beetles: lessons from a 10-year monitoring study. Biological Conservation 120: 171–181.
47. Menke N (2006) Untersuchungen zur Struktur und Sukzession der saproxylen Käferfauna (Coleoptera) an Eichen- und Buchentotholz: Georg-August-Universität Göttingen.
48. Sverdrup-Thygeson A, Birkemoe T (2009) What window traps can tell us: effect of placement, forest openness and beetle reproduction in retention trees. Journal of Insect Conservation 13: 183–191.
49. Brown JH, Mehlman DW, Stevens GC (1995) Spatial Variation in Abundance. Ecology 76: 2028–2043.
50. Jonsell M, Weslien J, Ehnström B (1998) Substrate requirements of red-listed saproxylic invertebrates in Sweden. Biodiversity & Conservation 7: 749–764.
51. Lindhe A, Lindelöw Å, Åsenblad N (2005) Saproxylic Beetles in Standing Dead Wood Density in Relation to Substrate Sun-exposure and Diameter. Biodiversity & Conservation 14: 3033–3053.

52. Sverdrup-Thygeson A, Ims RA (2002) The effect of forest clearcutting in Norway on the community of saproxylic beetles on aspen. Biological Conservation 106: 347–357.
53. Gibb H, Hjältén J, P. Ball J, Atlegrim O, Pettersson RB, et al. (2006) Effects of landscape composition and substrate availability on saproxylic beetles in boreal forests: a study using experimental logs for monitoring assemblages. Ecography 29: 191–204.
54. Bergman K-O, Jansson N, Claesson K, Palmer MW, Milberg P (2012) How much and at what scale? Multiscale analyses as decision support for conservation of saproxylic oak beetles. Forest Ecology and Management 265: 133–141.
55. Sverdrup-Thygeson A, Gustafsson L, Kouki J (2014) Spatial and temporal scales relevant for conservation of dead-wood associated species: current status and perspectives. Biodiversity and Conservation 23: 513–535.
56. Mäkinen H, Hynynen J, Siitonen J, Sievänen R (2006) Predicting the Decomposition of Scots Pine, Norway Spruce, and Birch Stems in Finland. Ecological Applications 16: 1865–1879.
57. Martikainen P, Siitonen J, Punttila P, Kaila L, Rauh J (2000) Species richness of Coleoptera in mature managed and old-growth boreal forests in southern Finland. Biological Conservation 94: 199–209.
58. Jonsson B, Kruys N, Ranius T (2005) Ecology of species living on dead wood-Lessons for dead wood management. Silva Fennica 39: 289–309.
59. Hyvärinen E (2006) Green-tree retention and controlled burning in restoration and conservation of beetle diversity in boreal forests [Ph. D]: University of Joensuu.
60. Gustafsson L, Kouki J, Sverdrup-Thygeson A (2010) Tree retention as a conservation measure in clear-cut forests of northern Europe: a review of ecological consequences. Scandinavian Journal of Forest Research 25: 295–308.
61. Lindenmayer DB, Franklin JF, Lõhmus A, Baker SC, Bauhus J, et al. (2012) A major shift to the retention approach for forestry can help resolve some global forest sustainability issues. Conservation Letters 5: 421–431.
62. Buckland PI, Buckland PC (2006) BugsCEP Coleopteran Ecology Package. IGBP PAGES/World Data Center for Paleoclimatology Data Contribution Series # 2006-116. NOAA/NCDC Paleoclimatology Program, Boulder CO, USA. URL: http://www.ncdc.noaa.gov/paleo/insect.html.
63. Karlsen S, Jepsen J, Odland A, Ims R, Elvebakk A (2013) Outbreaks by canopy-feeding geometrid moth cause state-dependent shifts in understorey plant communities. Oecologia: 1–12.
64. Kaukonen M, Ruotsalainen AL, Wäli PR, Männistö MK, Setälä H, et al. (2012) Moth herbivory enhances resource turnover in subarctic mountain birch forests? Ecology 94: 267–272.
65. Greenacre MJ (2010) Biplots in Practice: Fundación BBVA.
66. Wilkinson L (2012) Exact and Approximate Area-Proportional Circular Venn and Euler Diagrams. IEEE Transactions on Visualization and Computer Graphics 18: 321–331.

15

Estimating Species Richness and Modelling Habitat Preferences of Tropical Forest Mammals from Camera Trap Data

Francesco Rovero[1,2]*, **Emanuel Martin**[2,3], **Melissa Rosa**[4], **Jorge A. Ahumada**[4], **Daniel Spitale**[1]

1 Tropical Biodiversity Section, MUSE - Museo delle Scienze, Trento, Italy, **2** Udzungwa Ecological Monitoring Centre, Udzungwa Mountains National Park, Mang'ula, Tanzania, **3** Sokoine University of Agriculture, Morogoro, Tanzania, **4** Tropical Ecology Assessment & Monitoring (TEAM) Network, Betty & Gordon Moore Center for Science and Oceans, Conservation International, Washington, D.C., United States of America

Abstract

Medium-to-large mammals within tropical forests represent a rich and functionally diversified component of this biome; however, they continue to be threatened by hunting and habitat loss. Assessing these communities implies studying species' richness and composition, and determining a state variable of species abundance in order to infer changes in species distribution and habitat associations. The Tropical Ecology, Assessment and Monitoring (TEAM) network fills a chronic gap in standardized data collection by implementing a systematic monitoring framework of biodiversity, including mammal communities, across several sites. In this study, we used TEAM camera trap data collected in the Udzungwa Mountains of Tanzania, an area of exceptional importance for mammal diversity, to propose an example of a baseline assessment of species' occupancy. We used 60 camera trap locations and cumulated 1,818 camera days in 2009. Sampling yielded 10,647 images of 26 species of mammals. We estimated that a minimum of 32 species are in fact present, matching available knowledge from other sources. Estimated species richness at camera sites did not vary with a suite of habitat covariates derived from remote sensing, however the detection probability varied with functional guilds, with herbivores being more detectable than other guilds. Species-specific occupancy modelling revealed novel ecological knowledge for the 11 most detected species, highlighting patterns such as 'montane forest dwellers', e.g. the endemic Sanje mangabey (*Cercocebus sanjei*), and 'lowland forest dwellers', e.g. suni antelope (*Neotragus moschatus*). Our results show that the analysis of camera trap data with account for imperfect detection can provide a solid ecological assessment of mammal communities that can be systematically replicated across sites.

Editor: Danilo Russo, Università degli Studi di Napoli Federico II, Italy

Funding: This work was funded with the generous contribution of the Betty and Gordon Moore Foundation (http://www.moore.org/). Additional funding was provided by MUSE - Museo delle Scienze (http://www.muse.it/). The funders had no role in study design, data collection and analysis, decision to publish, or preparation of the manuscript.

Competing Interests: The authors have declared that no competing interests exist.

* Email: francesco.rovero@muse.it

Introduction

Profiling large-bodied animal communities, such as mammals, fundamentally implies assessing species richness and composition. Determining a state variable of species' abundance is also required to make inferences on species distribution, habitat associations, and trends over time [1–4]. In this context, medium-to-large mammals in tropical forests are of priority because they represent a rich and functionally diversified component of this biome, and yet they are universally threatened by hunting, and habitat loss and fragmentation [5–8]. The removal, or decrease in abundance, of tropical mammals will likely impact forest dynamics [11,12] due to their direct involvement in seed predation, seed dispersal, herbivore control, nutrient cycling and other ecosystem functions [9,10]. Systematic assessments that allow inference of tropical forest mammal communities in space and time remain limited and a chronic gap persists in standardized data collection.

The Tropical Ecology, Assessment and Monitoring (TEAM) network was set-up to fill this gap by establishing a network of field stations, scientists and partners across the tropics for long-term monitoring of mammal communities using a standardized and annually repeated sampling protocol [13]. The excellent potential of TEAM network data for answering questions on the status and trends of mammals has already been shown both through the first pan-tropical analysis from seven sites, which compared communities' richness and composition against forest area and fragmentation [8], as well as the first assessment of temporal changes at one particular site in Costa Rica using dynamic occupancy analysis [14]. In the present study, we used data from the first TEAM site established in Africa in 2009, the Udzungwa Mountains of south-central Tanzania, to propose a standardized approach for assessing the community of medium-to-large mammals detected through camera-trapping during the first, baseline year of the long-term programme.

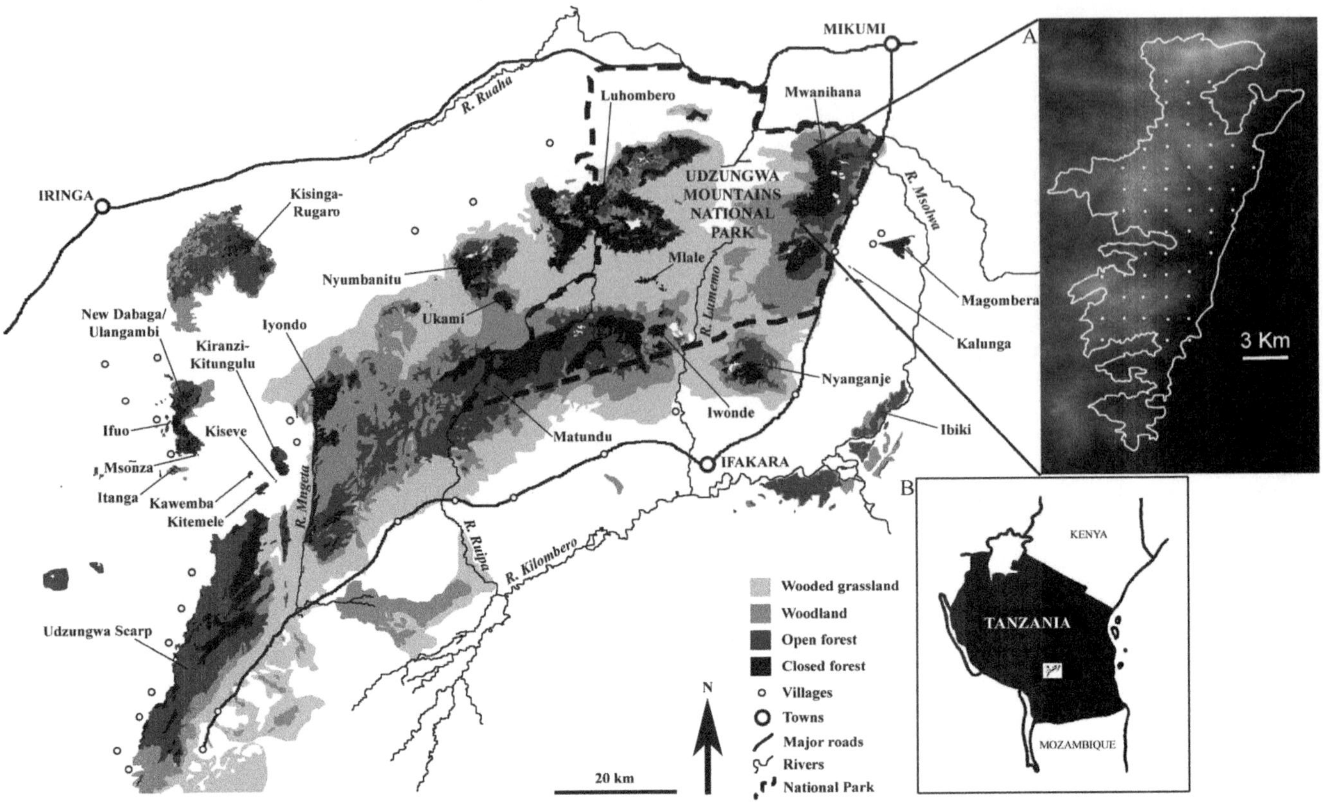

Figure 1. Map of the study area, the Udzungwa Mountains of south-central Tanzania. The map shows the main habitat types and blocks with closed-canopy forest (adapted from [53]). The study forest was Mwanihana in the northeastern portion of the range, which is zoomed in inset (**A**) where the 60 camera trap sites are shown as white dots and the background is a Digital Elevation Model (dark is lower elevation); (**B**) shows the position of Udzungwa in Tanzania.

The use of camera trapping for wildlife studies has increased exponentially in the last decade as it is an efficient, cost-effective and easily replicable tool to study and monitor ground-dwelling terrestrial mammals and birds [15,16]. Camera trapping is particularly suited to collect standardized data because sampling effort can be easily controlled for and sampling design can be replicated across time and space [15]. In addition, sampling can be considered as multiple occasions during a discrete season, hence data are suited for analyses that account for imperfect detection, such as occupancy [17,18]. Occupancy (ψ) is defined as the proportion of area, patches or sites occupied by a species [19] and can be used as a surrogate for abundance [20]. Detection probability (p) is defined as the likelihood of detecting an individual, or species, during a sampling occasion [18]. With the inclusion of covariates, occupancy models provide a robust statistical framework for testing scientific hypotheses. For example, one can test for differences in occupancy rates between study sites that contrast by habitat type, hunting level, distance to key resources, climate conditions and vegetation features [21,22]. In addition, the same approach used for occupancy analysis can also be used for estimating species richness and accumulation [23].

The Udzungwa Mountains are an area of outstanding importance for biodiversity endemism and conservation in Africa [24], and are particularly rich in forest dwelling mammals [25]. Through our assessment we aimed to (1) evaluate sampling effort and estimate species' richness, (2) determine drivers of variation in species richness and detection probability (p) using an occupancy framework [17], (3) estimate species' occupancy (ψ), and (4) determine the best habitat and human disturbance predictors of both ψ and p to identify major patterns of species' responses to these predictors.

Materials and Methods

Ethics Statement

Data collection used non-invasive, remotely set camera traps and hence did not involve direct contact or interaction with the animals. Fieldwork was done under research permit number 2009-139-NA-2009-49 to FR, issued by the Tanzania Commission for Science and Technology (COSTECH).

Study area

The Udzungwa Mountains of south-central Tanzania (over 10,000 km^2; 7°40′–8°40′S, 35°10′–36°50′E) are a mosaic of moist forest blocks interspersed with drier habitats. The study was conducted in Mwanihana forest, which at 180 km^2 is one of the largest forests in the area and with the widest, continuous forest elevation range (290–2250 m a.s.l.; Fig. 1). The forest is inside the Udzungwa Mountains National Park (1990 km^2). The eastern border of the forest coincides with the eastern boundary of the park. The forest habitat broadly ranges east-west from lowland, deciduous forest to montane, evergreen forest [26]. The lower elevation habitats, which include deciduous, semi-deciduous and riverine evergreen forest, have been degraded historically and have large portions of secondary, regenerating vegetation. The interior forest is mainly undisturbed with large chunks of pristine, closed-canopy moist forest. Anthropogenic disturbance in the form of firewood collection occurred at the lower elevations, a practice likely coupled with illegal bush meat hunting done using snares. The upper elevation zone has lower canopy and bamboo forest with rocky and very steep areas, especially in the northern part. Total rainfall in Mwanihana forest is around 1500 mm per year (data from Udzungwa Mountains National Park); rainfall measured at 1200 m a.s.l. by an automatic rainfall gauge was 1387 and 1451 mm in 2011 and 2012, respectively (FR/TEAM Network, unpublished data). The dry season spans from June to November, while two rainy seasons occur during November-June. In 2012, mean monthly air temperature at 1200 m a.s.l. ranged 17.2–22.6°C. (FR/TEAM Network, unpublished data).

Data collection

Camera trapping was conducted from July to November 2009, as the baseline year of the TEAM programme [13]. We used digital cameras (Reconyx RM45, Reconyx Inc., Holmen, Wisconsin, USA) set to take photos without delay between consecutive triggers. Using ESRI's ArcGIS 10 software, we designed a regular grid of 60 camera trap locations at a density of one camera per 2 km^2, and placed at random across the forest. We then conducted a ground survey to select the final camera positions, and locations that fell in excessively steep, open canopy or rocky areas were repositioned no more than 100 m from the original location (Fig. 1).

Camera traps were positioned so the field of view included an active wildlife trail and then secured to a tree about 2–3 m away from the trail at an average height of 50 cm and left running for 30 days. Since cameras can run automatically over such period, we did not check them to avoid unnecessary disturbance. Due to limits in the number of cameras available and the time needed for the field team to set cameras, we sampled the 60 points by deploying three consecutive arrays of 20 camera traps (south, central and northern Mwanihana, respectively). Therefore, the data collection lasted 133 days from July 24 to December 4, 2009.

At sampling completion, memory cards were recovered and images were identified using specialized software (DeskTEAM, [27]; see also www.teamnetwork.org/en/help-deskteam). A single taxonomic authority [28] was used across all TEAM sites for species identification, except for species that were not included in this reference or species that were re-assessed. The validated and publicly available data were downloaded from the TEAM portal (data package id: TV-20111116005138_3515).

Data analysis

We derived standard descriptors of mammal community by filtering the image records for each species of mammal to derive the number of events per hour, hence avoiding that multiple images of the same individual pausing in front of the camera trap were scored as multiple events [29]. We then computed a relative abundance index (RAI) as the number of events divided by sampling effort and multiplied by 100 (i.e. events per 100 days of camera trapping). We also computed the naïve occupancy as the number of camera trap sites occupied on sites sampled.

We derived a number of spatial environmental covariates deemed relevant to explain both the spatial variations of species' richness and occupancy of selected species using geoprocessing tools available in ArcGIS. We calculated the following variables: (1) distance from eastern park border ('border'), (2) distance from forest edge ('edge'), (3) forest habitat type, i.e. montane forest and lowland forest ('habitat'), (4) slope and (5) distance to rivers. The distance from each camera trap point to the nearest 'border', 'edge', or river segment was calculated in ArcGIS. Forest habitat type was mapped using a supervised classification approach on Landsat TM and ETM+ satellite imagery (30 m resolution). Habitats were categorized into three forest types: 1) Montane, 2) Deciduous, and 3) Regenerating. Forest habitat type was then extracted for each camera trap point in ArcGIS. 'Border' correlated highly with elevation at camera trap sites given the forest morphology of an east-west escarpment (Pearson's $r = 0.802$, $P<0.001$) and it is considered a proxy of decreasing anthropogenic disturbance, which may be mainly associated to firewood collection and pole/timber cutting [30]. After checking that no collinearity existed among the covariates used, these were standardized to have mean 0 and unit variance before estimating the model coefficients.

As a fundamental measure of the community structure, we analysed species richness under three different perspectives and with different aims. (1) Species accumulation curve with cumulative camera trap days was used to check if data collection lasted a sufficient number of days to virtually capture the total number of species. The order in which samples (they consisted of number of events per day) were included in the curve was randomized 1000 times and results were used to derive 95% confidence intervals around the mean [31]. Even though this approach ignores imperfect detection of individual species, it is useful for comparison with other studies [32,33]. (2) Analysis of species richness that accounts for imperfect detection was studied using the model by Dorazio and colleagues [23], which requires repeated temporal replications to resolve the ambiguity between species absence and non-detection when species are unobserved at sample locations. This Bayesian approach combines community-level and species-level attributes in the same framework, allowing either community-level or species-level parameters to be evaluated. Such flexibility is not matched by other methods for estimating species richness [23]. The frequentist approach to the same problem is possible, but computationally intensive to implement [23]. The model was specified in BUGS language and fitted to data using WinBUGS and the package 'R2WinBUGS' in R software [34,35]. Simulations were executed with five Markov chains; 55,000 iterations for each chain, discarding 5,000 iterations at the beginning (burn-in) and setting the thinning rate to 50. This returned 5,000 samples from the posterior distributions. (3) We used the occupancy analysis framework to investigate possible relationships between species richness and environmental covariates [19,36]. In particular, we compared two sets of models: (1) testing the effect of environmental covariates on the occupancy of all the species (species richness), and (2) testing if trophic guild (carnivores, herbivores, omnivores, insectivores) and body mass (data from [37]) were related to detection probability. Akaike Information Criterion (AIC) was used to rank all the candidate models and calculate their Akaike weights [38]. To achieve intra-guild homogeneity, we discarded elephant (*Loxodonta africana*) and buffalo (*Syncerus caffer*) among the herbivores for their large body mass and movement habits (i.e. they periodically move into the forests from drier habitats in the park). Among the carnivores, the bushy-tailed mongoose (*Bdeogale crassicauda*) was discarded because it is a common, non-elusive, and partially omnivorous species; hence it effectively represents an outlier in the carnivore guild.

We also used occupancy [17] as the species-specific state variable of abundance to assess differences across species under an unbiased framework and determine covariates of both occupancy and detection probability for a set of species. We used scripts already developed [8] and implemented in R to arrange the TEAM data (http://www.teamnetwork.org/) into a list of species' occupancy matrices. Data for each species were arranged as

Table 1. List of mammals camera trapped in Mwanihana forest, Udzungwa Mountains, Tanzania, reporting species traits (body mass from Smith et al. (2003) and functional guild from IUCN (2013)) and three raw indices of abundance: hourly events, relative abundance index (RAI) and naïve occupancy (number of sites that are positive to species' presence divided by the total number of sites sampled).

#	Taxonomic group	Common name	Latin name	Mass (kg)	Functional guild	Events per hour	RAI (events/100 days)	Naïve occupancy
1	Afrotheria	Tree hyrax	*Dendrohyrax arboreus*	2.95	Omniv	23	1.27	0.241
2		African elephant	*Loxodonta africana*	3900	Herbiv	11	0.61	0.121
3		Four-toad sengi	*Petrodromus tetradactylus*	0.19	Insectiv	3	0.17	0.017
4		Chequered sengi	*Rhynchocyon cirnei*	0.49	Insectiv	4	0.22	0.052
5		Grey-faced sengi	*Rhynchocyon udzungwensis*	0.80	Insectiv	45	2.48	0.259
6	Carnivores	Marsh mongoose	*Atilax paludinosus*	3.30	Carniv	3	0.17	0.052
7		Bushy-tailed mongoose	*Bdeogale crassicauda*	1.55	Carniv	130	7.15	0.741
8		African civet	*Civettictis civetta*	12	Omniv	1	0.06	0.017
9		Lowe's servaline genet	*Genetta servalina lowei*	1.06	Omniv	18	0.99	0.259
10		Honey badger	*Mellivora capensis*	8.50	Carniv	7	0.39	0.103
11		Banded mongoose	*Mungos mungo*	1.93	Insectiv	2	0.11	0.034
12		African palm civet	*Nandinia binotata*	1.90	Carniv	2	0.11	0.034
13		Leopard	*Panthera pardus*	52	Carniv	8	0.44	0.052
14	Primates	Sanje mangabey	*Cercocebus sanjei*	8	Omniv	73	4.02	0.517
15		Sykes' monkey	*Cercopithecus mitis*	5	Omniv	22	1.21	0.241
16		Yellow baboon	*Papio cynocephalus*	18.4	Omniv	3	0.17	0.052
17		Udzungwa red colobus	*Procolobus gordonorum*	10	Omniv	5	0.28	0.069
18		Angolan colobus	*Colobus angolensis*	8.6	Omniv	1	0.06	0.017
19	Rodents	Giant pouched-rat	*Cricetomys gambianus*	1.24	Omniv	276	15.18	0.534
20		Cape porcupine	*Hystrix africaeaustralis*	14.94	Omniv	11	0.61	0.086
21		Tanganyika mountain squirrel	*Paraxerus vexillarius*	0.68	Herbiv	46	2.53	0.328
22	Ungulates	Harvey's duiker	*Cephalophus harveyi*	12	Herbiv	367	20.19	0.862
23		Abbott's duiker	*Cephalophus spadix*	56	Herbiv	60	3.30	0.466
24		Suni	*Neotragus moschatus*	6.5	Herbiv	114	6.27	0.448
25		Bush pig	*Potamochoerus larvatus*	48.78	Omniv	18	0.99	0.190
26		African buffalo	*Syncerus caffer*	580	Herbiv	4	0.22	0.052

Trophic guilds: Herbiv = herbivores, Omniv = omnivores, Insectiv = insectivores, Carniv = carnivores.

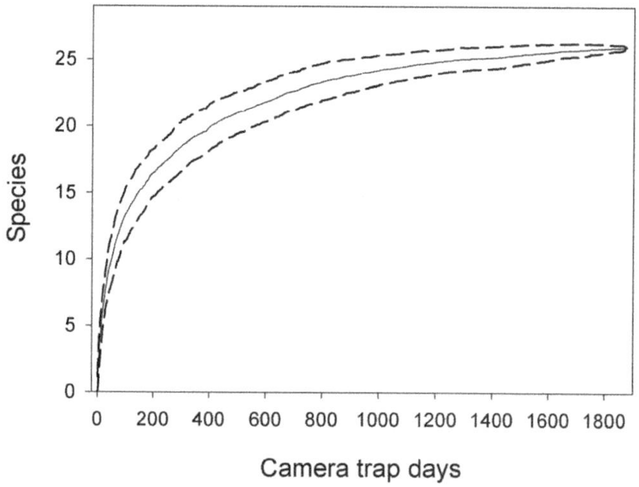

Figure 2. Species accumulation curve for the community of medium-to-large mammals detected by camera trapping. Detection of species is randomized 1000 times and results used to derive the 95% confidence intervals of the mean.

matrices of sites by surveys (i.e. sampling occasion). Each entry indicated if the species was observed at site i on survey j or not. If the species was observed at site i on survey j, then the entry was given a score of 1. If the species was not observed, then the entry was given a score of 0. NA indicated site i was not sampled on survey j. The species-specific occupancy matrix had a resolution of five days.

We used these matrices as the input for the single-season occupancy model [19]. We modelled both estimated occupancy (Ψ) and detection probability (p) with and without covariates. A common set of models was used for all the species. In addition to the null model, that assumes constant Ψ and p (i.e. $\Psi(.)$, $p(.)$), for other models p was allowed to vary by distance to border and distance to edge. In both cases, our hypothesis was that animals would be more elusive near the border and/or the edge because of greater disturbance [30]. Four covariates for Ψ were the following: (1) 'border', (2) 'edge', (3) 'river' and (4) 'habitat'. Numerical covariate were standardized into z-scores and included both individually and in combination. We used the Akaike Information Criterion (AIC) to rank candidate models and calculate their Akaike weights [38]. In the case of top-ranked models with similar AIC (and weight >0.01), we applied a model-averaging technique to estimate occupancy from these multiple models [38]. Occu-

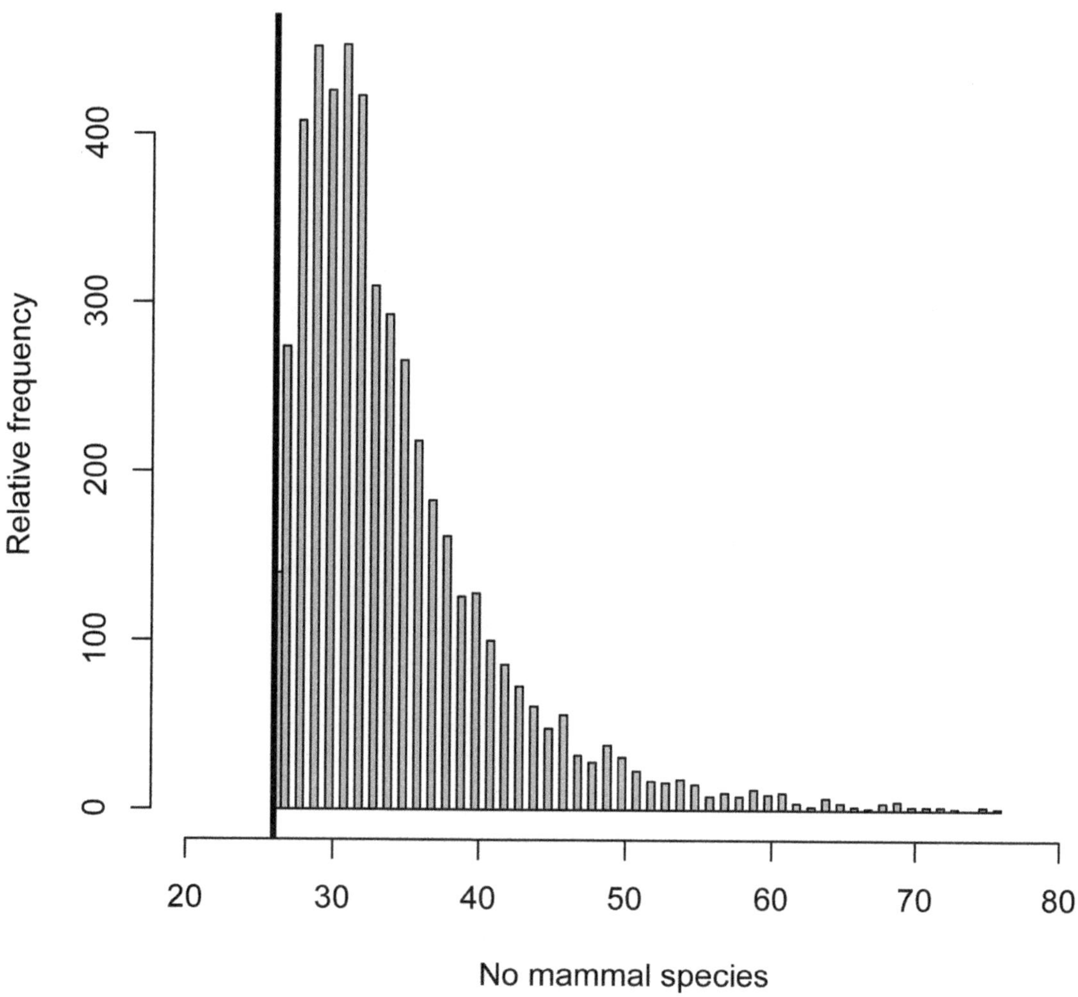

Figure 3. Posterior distribution of species richness. The analysis follows Dorazio et al. (2006). The posterior probability that the community comprises only 26 species (vertical line is the observed species richness) is essentially zero, and the estimated median and mean values of species richness are 32.0 (±7.04 SD) and 34.3, respectively (26–54 CRI 95%; CRI = credible intervals).

Table 2. Summary of model selection outcome for predictions of mammal species' richness in the Udzungwa Mountains of Tanzania.

Model	Number of parameters	AIC	Delta	AIC weight	Cumulative weight
ψ(.) *p*(mass, guild)	6	1151.59	0.00	0.410	0.410
ψ(.) *p*(guild)	5	1152.57	0.98	0.250	0.660
ψ(habitat, species) *p*(mass, guild)	7	1153.59	2.00	0.150	0.810
ψ(border) *p*(guild)	6	1154.56	2.97	0.093	0.910
ψ(habitat) *p*(guild)	6	1154.56	2.98	0.093	1.000
ψ(.) *p*(.)	2	1323.63	172.04	0.000	1.000

The top-ranked models are shown (delta AIC <3) followed by the null model.

pancy analysis was performed using the package 'unmarked' in R [39]. The relative importance of the model parameters were calculated with the R package AICmodavg [40]. Once we identified the best occupancy model (or the average of the best models), we mapped occupancy probability across Mwanihana forest by deriving occupancy estimates from covariates computed on a spatial grid with a cell size of 100 m.

Results

Of the 60 camera traps set, two malfunctioned and the remaining 58 accumulated 1,818 camera days (mean 31.34) yielding 10,647 images of mammals. The list of 26 species recorded and standard descriptors are reported in Table 1. The range of species captured per camera was 3–10 (median 6). Four species were recorded with >100 events in this order: (1) Harvey's duiker (*Cephalophus harveyi*), (2) giant-pouched rat (*Cricetomys gambianus*), (3) bushy-tailed mongoose and (4) suni. Six species scored >20 and ≤100 events: (1) Abbott's duiker (*Cephalophus spadix*), (2) Tanganyika mountain squirrel (*Paraxerus vexillarius*), (3) grey-faced sengi (*Rhynchocyon udzungwensis*), (4) Sykes' monkey (*Cercopithecus mitis*) and (5) tree hyrax (*Dendrohyrax validus*). The remaining 16 species scored ≤20 events, of which 10 species scored ≤5 events. The accumulation of species detected with sampling effort was initially steep, but by 1,000 camera days the majority of species were detected (24 species, or 92%; Fig. 2). The estimated size of the community according to [23] exceeds the number of species observed in the sample by a substantial margin, with median and mean values being 32 and 34.3, respectively (Fig. 3).

The modelling of species richness using a sub-set of 23 species revealed no support for the null model, with several models having lower AIC, and five that were top-ranked with delta AIC<3 (Table 2). Model averaging using these first five models shows that no environmental covariates affected relative species richness. However, there is a significant influence of the functional guild on the detection probability (Table 3). Herbivores had the highest detection probability (0.52±0.03 SE), followed by omnivores (0.20±0.02 SE), insectivores (0.09±0.02 SE) and carnivores (0.06±0.02 SE; Fig. 4).

We could fit occupancy models for the 11 most recorded species. We initially considered 14 species with ≥10 events or naïve occupancy ≥0.1; however, for three of these (*Genetta servalina*, *Loxodonta africana*, *Hystrix africaeaustralis*) the models did not converge. For these 11 species, ψ ranged from 0.25–0.86 and p ranged from 0.10–0.51. The null model was not supported for any of these species, and at least one of the covariates considered affected significantly or marginally significantly ψ and p (Table 4). Details of model selection for each species are shown in Table S1.

The main patterns of predicted ψ and the functional relationships of ψ with the dominant covariate represented by the four species shown in Fig. 5 are as follows:

(1) As a 'montane forest dweller', Sanje mangabey's ψ is positively associated with montane forest habitat and not affected by any of the other variables. Hence, predicted occupancy falls in two values of 0.34±0.12 SE in lowland, deciduous forest and 0.76±0.10 SE in montane, evergreen forest.

(2) In contrast with the above, suni is a 'lowland forest dweller', with ψ being negatively related to montane forest habitat; hence,

Table 3. Summary of model averaging for the effect of environmental covariates on species richness (ψ) and detection probability (*p*) of the mammal community in the Udzungwa Mountains of Tanzania.

Model	Estimate	SE	*Z*	*P*(>|z|)
p(mass)	−0.136	0.08	1.704	0.089
p(herbivores)	2.860	0.31	9.345	<0.001
p(insectivores)	0.380	0.37	1.040	0.298
p(omnivores)	1.327	0.30	4.497	<0.001
ψ(habitat - montane)	2.814	292.60	0.010	0.992
ψ(border)	1.273	628.75	0.002	0.998

See Table 2 for the covariates modelled with both ψ and *p*.

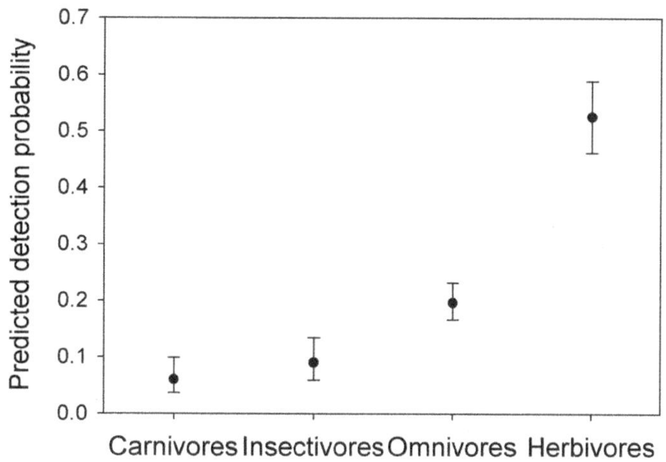

Figure 4. Detection probability by functional guild. Values are from the model averaging of relative species richness of the mammal community in the Udzungwa Mountains of Tanzania. Bars are 95% confidence intervals.

indicating preference for lowland forest. In addition, the species' ψ is marginally affected by distance to park border, with predicted occupancy decreasing in the proximity of park border relative to more interior zones of lowland forest. Its detection probability also significantly increased with distance to edge.

(3) The Harvey's duiker is a typical 'edge lover' species, as ψ is negatively affected by distance to edge, which is clearly seen in the spatially-explicit model. Therefore, the species avoids interior forest, with predicted ψ declining sharply and non-linearly after 1.5–2 km from the forest edge.

(4) An opposite pattern is shown by the grey-faced sengi, which seems to be an 'edge avoider' with ψ being positively affected by distance from edge and preference for montane habitat and both associations being marginally significant ($0.05<P<0.1$).

For approximately half of the species, detection probability varied with distance to the park border and/or distance to the forest edge (Online Resource 1). We portray two limiting cases (Fig. 6): (1) the bushy-tailed mongoose, where p decreased linearly with the distance to park border, and (2) the Abbott's duiker, where p exponentially increased with distance to border.

Discussion

Our study shows how camera trap data collected using a robust, standardized field methodology, and analysed with statistical approaches that account for imperfect detection and incorporate ecological factors, can provide a robust baseline assessment of mammal communities in tropical forests. In particular, we estimated species richness using a sampling protocol requiring repeated observations at sample locations. This approach provided information needed to resolve the ambiguity between species absence and non-detection. The Bayesian analysis used [23] is a flexible alternative to the classic frequentist approach, which is computationally complex, and combines community-level and species-level attributes in the same modelling framework. We also determined how simple ecological covariates such as gross habitat, distance to forest edge and distance to park border explain the occupancy of most species in the community. Finally, we provided a framework for deriving spatially-explicit, fine resolution models of estimated species occupancy in relation to covariates, which represent a valuable tool for conservation management of threatened and/or poorly known species.

The efficiency of camera trapping for inventorying species has already been indicated by other studies of tropical mammal communities [33,41]. In the Udzungwas, additional camera trapping effort and scattered sighting reports indicate that at least four species have been 'missed' by the present survey (bushbuck *Tragelaphus scriptus*, spotted hyena *Crocuta crocuta*, cane rat *Thryonomys swinderianus* and serval cat *Leptailurus serval*; FR unpublished data). Other small, elusive carnivores may also be present in the target forest [42]. This observation is supported by our models, which estimate that >30 species occur. It is worth noting that the classic species richness estimators, parametric and non-parametric, asymptotic and non-asymptotic, rely on extrapolations of the species accumulation curve and do not account for imperfect detection [31,43]. The explicit incorporation of detection probability in the models we used is particularly important in estimating species richness of communities that contain a preponderance of rare, or difficult to detect, species [44]. In these cases, using traditional approaches may yield incorrect inferences if heterogeneity in detectability exists among species or if the effects of environmental covariates on occurrence differ among species.

In terms of species composition, the pool of ten most-detected species (>20 events) reveals the relative high occurrence of a number of species that are poorly known, and poorly detected using alternative methods. For example, the Abbott's duiker is a IUCN-Endangered 'giant' duiker endemic to and found only in a handful of montane forests in Tanzania, including the Udzungwa population considered to be the stronghold, and was the third most common species in terms of occupancy, $\psi = 0.72$ [28,45]. Similarly, the fourth most common Sanje mangabey ($\psi = 0.62$) is a predominantly terrestrial forest monkey endemic to only two forests in the Udzungwa mountains and classified as Endangered [28]. Being terrestrial and elusive, this monkey is poorly sighted from line-transects despite living in large groups of up to 50 individuals [46], and hence it is so far regarded as rarer than our data reveal. Among other commonly detected species, the relatively high ranking of tree hyrax ($\psi = 0.48$) is also surprising given this is known as an arboreal dweller [47]. Our data show that tree hyraxes spend more time on the ground than previously thought. It is also worth mentioning the 11th position in the occupancy ranking of the Udzungwa-endemic and IUCN Vulnerable grey-faced sengi (or elephant-shrew), a species described in 2008 which is very rarely seen despite being diurnal,

Table 4. Summary of species-specific occupancy estimates for 11 mammals that had adequate detection for the analysis, ordered by decreasing estimated occupancy (ψ).

Species	Naïve ψ	ψ	SE(ψ)	p	SE(p)	p(border)	p(edge)	ψ(border)	ψ(edge)	ψ(river)	ψ(habitat [type])
Cephalophus harveyi	0.862	0.876	0.080	0.432	0.046	(+)			–	(–)	
Bdeogale crassicauda	0.741	0.815	0.069	0.298	0.039	–					
Cephalophus spadix	0.466	0.716	0.094	0.171	0.040	+	+				
Cercocebus sanjei	0.517	0.615	0.107	0.234	0.057						+[montane]
Cricetomys gambianus	0.534	0.539	0.118	0.505	0.033					+	
Cercopithecus mitis	0.241	0.503	0.198	0.096	0.039		–				
Dendrohyrax validus	0.241	0.480	0.116	0.106	0.039		(+)			+	
Neotragus moschatus	0.448	0.472	0.105	0.374	0.057		+	(+)			–[montane]
Paraxerus vexillarius	0.328	0.400	0.100	0.203	0.040				+		
Potamochoerus larvatus	0.190	0.387	0.150	0.098	0.055	(–)					
Rhynchocyon udzungwensis	0.259	0.278	0.114	0.298	0.048				(+)		(+) [montane]

This is the average value of predicted occupancy at the 58 camera trap localities from the final models. Significant outcomes of the effects of covariates on ψ and detection probability (p) are also indicated with their directionality (positive/negative effect), which is in parenthesis when the significance is marginal ($0.05<P<0.1$). Naïve occupancy values are also shown to appreciate the differences with ψ, which are particularly remarkable for species with $p<0.2$.

and for which Mwanihana holds approximately half of the global population [48]. The pool of least-detected species contains a number of truly arboreal mammals, typically the two colobine monkeys that are common in high densities across the forest [46], which are completely explained by their habit. Besides these 'exceptions', the other least-detected species are a diverse suite of less common (e.g. bush pig, Lowe's servaline genet), or rare animals for the target forest (e.g. leopard, marsh mongoose), in addition to species that are mainly found in savannah and/or in the deciduous woodland occurring at the lower edge of Mwanihana forest (e.g. yellow baboon, African civet, banded mongoose, honey badger).

It is not surprising that we did not find any significant pattern of variation of estimated species richness across camera trap sites because Mwanihana forest has continuous forest cover without drastic habitat changes, except for the gradual variation in habitat type that broadly follows altitudinal and edge versus interior gradients. Whilst the species-specific occupancy models do highlight clear patterns of ecological preference by a suite of species, these preferences do not hold across the whole community. Interestingly, we found that the trophic guilds have significantly different detection probabilities. The low detectability of carnivores and insectivores matches their generally greater elusiveness relative to omnivores and herbivores. In contrast to our expectations, detection probability of species decreases with body mass, although the relationship is marginally significant (Table 3). Previous studies examining the effect of body mass on the animal detection process by camera traps suggest that small species are more likely to be missed due to the sensitivity and dimensions of the detection zone of the camera sensor [33,49]: However, this aspect did not appear to have a statistical effect within the range of body mass in our study, perhaps because of the high sensitivity of the camera model we used. The relationship we found may rather reflect inter-specific behavioural differences, with larger species being less detected because of their greater elusiveness.

The species-specific occupancy analysis generally revealed novel ecological knowledge for roughly half of the species included in the analysis, excluding the strictly arboreal ones and those that are not typical forest-dwellers (see considerations above). The need to include corrections for imperfect detection in the modelling process is clearly shown by the remarkable variation of p among species (range 0.096–0.505; see Table 4). Because of this variation, the difference between naïve and estimated occupancy is also varying, and for the least-detectable species ($p<0.2$), the increment between naïve and estimated occupancy is 54–109% the naïve occupancy (see Table 4). The importance of allowing ψ and p to vary with covariates is shown by the fact that the null model was the least supported for any species. This is shown by a number of previous studies that investigated habitat associations from camera trapping data in an occupancy framework [21,22,36,50]. To achieve similar inference for the remaining half of the species ($p<0.1$), a large number of sites should be surveyed [44]. Alternatively, one could pool data for more than one season under the assumption of a closed community (e.g. occupancy status does not change among survey seasons [22]).

The four limiting cases we highlighted show the particularly relevant ecological and conservation implications of our approach. For example, ecological knowledge on the Sanje mangabey was limited to results from a single, long-term focal group study located in the lower part of the forest [46,51] before our analyses. There also was a lack of general understanding of their occurrence across the entire forest, which includes about half of the global population. Our results indicate that the species' occupancy in montane forest is more than double than in lowland forest, which

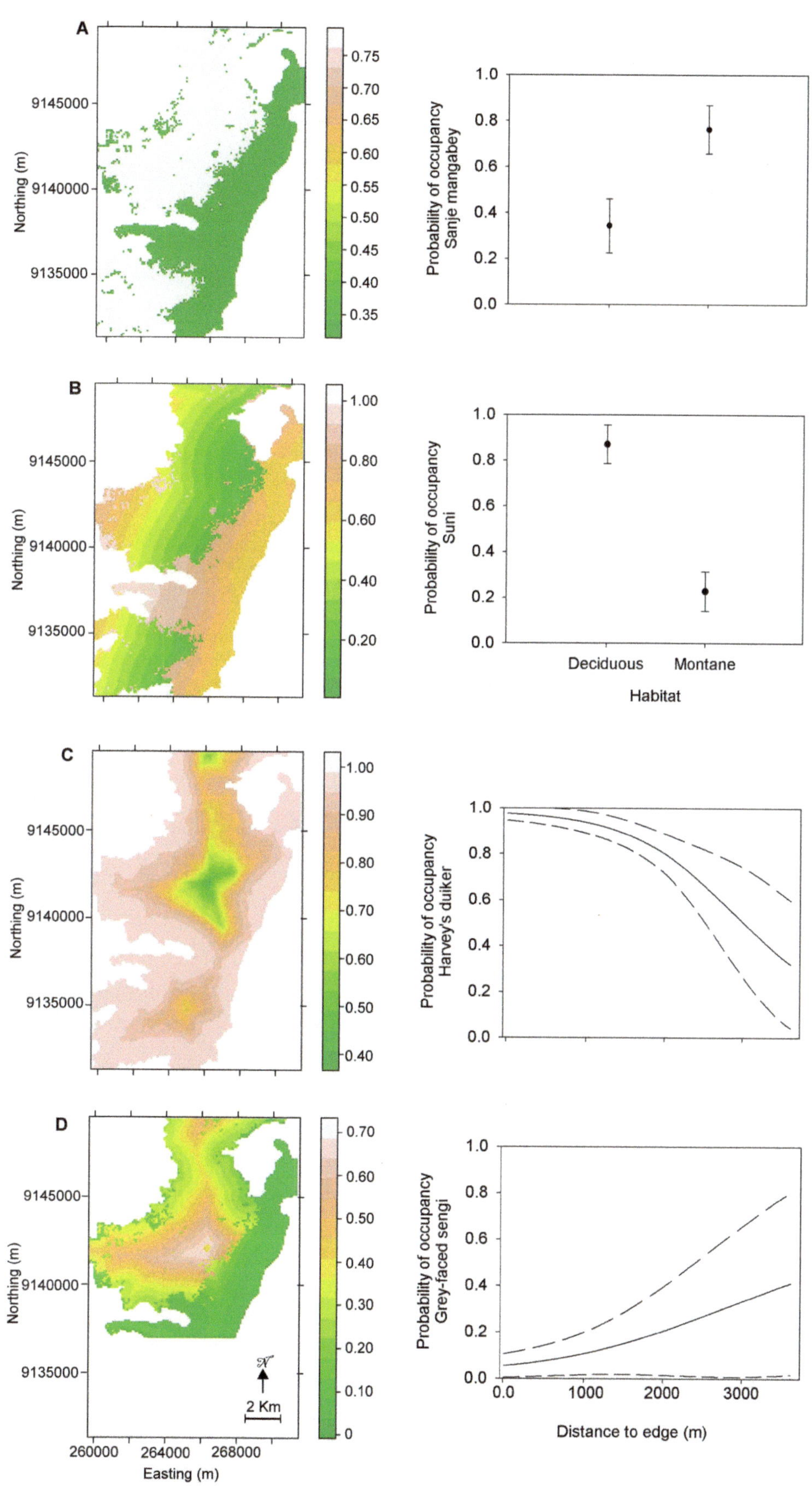
A
B
C
D
Northing (m)
9145000
9140000
9135000
260000
264000
268000
Easting (m)
2 Km
0.75
0.70
0.65
0.60
0.55
0.50
0.45
0.40
0.35
1.00
0.80
0.60
0.40
0.20
0.90
0.30
0.10
0
Probability of occupancy Sanje mangabey
Probability of occupancy Suni
Probability of occupancy Harvey's duiker
Probability of occupancy Grey-faced sengi
1.0
0.8
0.6
0.4
0.2
0.0
Deciduous
Montane
Habitat
1000
2000
3000
Distance to edge (m)

Figure 5. Spatially-explicit occupancy models. Maps of predicted occupancy (left) and functional relationship between the most relevant covariate and ψ (right, with confidence intervals indicated by dashed lines) for four mammals in the Udzungwa Mountains of Tanzania, representing limiting cases in occupancy pattern: (**A**) Sanje mangabey, a montane evergreen forest species; (**B**) suni, a lowland deciduous forest species; (**C**) Harvey's duiker, an edge-lover and disturbance-tolerant species; (**D**) grey-faced sengi, an edge-avoider and disturbance-sensitive species.

in turn suggests the vulnerability of this species to both human-induced (e.g. logging and forest degradation), stochastic (e.g. fires) and climate change impacts. The limit of using a categorical and broad classification of habitat type for this and other habitat-sensitive species may be overcome in future studies by collecting fine-scale vegetation and human disturbance data at camera trap sites for consideration in the modelling [22,52]. Similar considerations apply to the results for the grey-faced sengi, whose preference for forest interior and edge avoidance matches the results from a recent focal study on habitat associations [22].

While the forest antelope community has been previously studied using camera trapping [29,52] the fine grain occupancy models we derived shed new light into the occurrence of these species. Suni and Harvey's duiker occur predominantly in the lower forest with the latter occurring across the forest edges. This is relevant to the need to protect the full array of forest cover, including the lower elevation areas, which border densely populated settlements. The preference of Harvey's duiker for edges also indicates its suitability as an indicator of connectivity between forest blocks across marginal, often riverine habitat, which is important in highly heterogeneous areas such as the Udzungwas.

Despite a minority of species whose detection probability did not vary significantly with the covariates used (e.g. distance from edge and from border), the general finding is assuming that constant detection is broadly incorrect. Care needs to be taken when choosing covariates for p to ensure they are meaningful, which may be related to assumptions on the differences in the density of vegetation across camera trap sites. This assumption, in turn, may affect the efficiency of camera traps to capture an image of passing animals. In addition, variation in detectability may be due to differential animals' shyness in relation to human disturbance and/or density of vegetation on the forest floor compressing the field of view of camera traps. These results may indicate a pattern of lower detectability in areas that are closer to human disturbance (e.g. border) and/or habitat 'disturbance' (e.g. border and edge), where forest floor vegetation is generally denser due to higher canopy degradation than in forest interior, and include animals that are more shy. The few cases of a negative relationship between p and one of the two covariates may also be explained by species-specific habits. For instance, the Sykes' monkey's p is negatively related to edge, which fits with the habit of this opportunistic primate to move easily among dense, degraded and regenerating vegetation in forest edge [30]. Similar considerations may also be valid for the bushy-tailed mongoose, a small, nocturnal and opportunistic carnivore that is often sighted by the park border and forest edge (FR unpublished data).

Conclusions

Our study applied a robust analytical framework to profiling tropical mammal communities detected by the standard camera trapping protocol adopted by the TEAM Network. With the network currently made of 16 sites across three continents and progressively expanding (http://www.teamnetwork.org), and a number of studies adopting similar designs outside the network [36], there emerges a growing need for standardized analytical procedures to facilitate and enhance the sound use of the large data-sets being accumulated. In turn, detailed and site-specific baseline analysis will help interpreting patterns of community composition and changes from multi-site comparisons [8]. Similarly, with data collected from a number of sites for >5 years, baselines such as ours are relevant to the interpretation of temporal trends in species and community occupancy, for which robust and standardized analytical procedures have recently been proposed, including the Wildlife Picture Index [2,14].

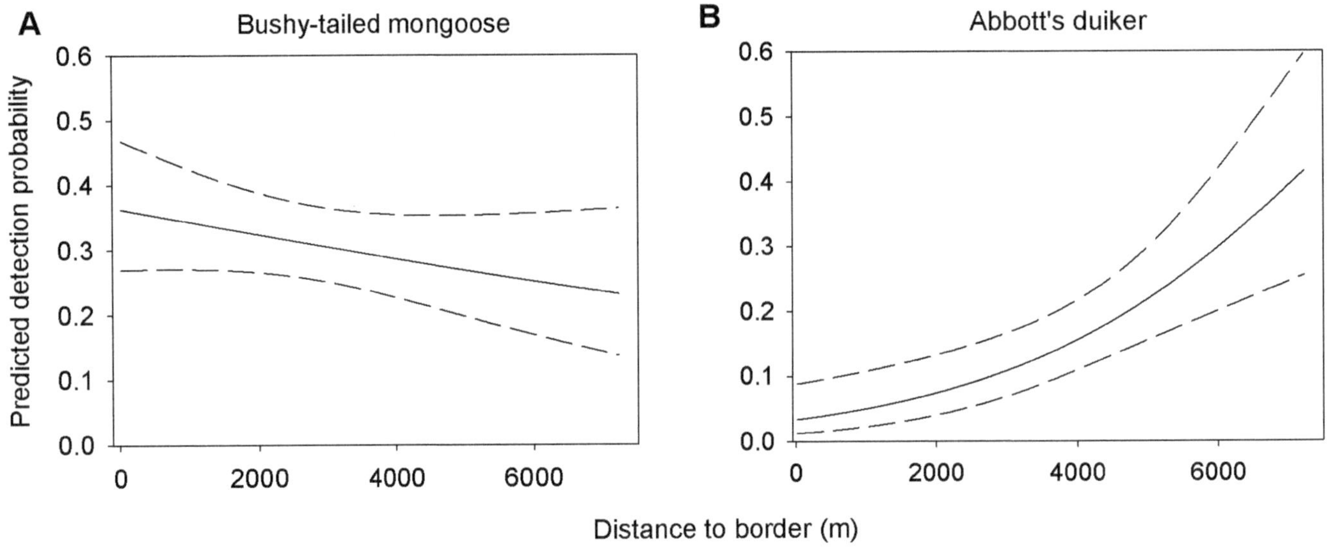

Figure 6. Graphs of predicted detection probability. Values are modelled with distance from the park border for (**A**) the bushy-tailed mongoose, and (**B**) the Abbott's duiker, in the Udzungwa Mountains of Tanzania. Confidence intervals are indicated by dashed lines.

The ultimate relevance of standardizing tropical mammal community assessments rests in the need to develop indicators for distribution and abundance of pan-tropical species, as outlined by the Convention on Biological Diversity [2,14]. In this context, our study offers an example of how analysis of species' richness in occupancy framework, focal species' occupancy and their spatial variation relative to a suite of covariates, represents a useful approach for comparing data from several sites, and hence for deriving indicators for these global targets.

Acknowledgments

We are grateful to two anonymous reviewers for very valuable comments, and to Simone Tenan and Claudia Barelli for comments and editing on an earlier version of the manuscript. Tanzania Wildlife Research Institute, Tanzania Commission for Science and Technology, Forestry and Beekeeping Division and Tanzania National Parks provided the research permits. Staff at the Udzungwa Mountains National Park and the Udzungwa Ecological Monitoring Centre provided critical logistic support. Several field assistants, especially R. Mwakisoma, S. Shinyambala and A. Wissel, helped data collection invaluably.

Author Contributions

Conceived and designed the experiments: FR JAA. Performed the experiments: FR EM MR. Analyzed the data: FR MR JAA DS. Contributed to the writing of the manuscript: FR EM MR JAA DS.

References

1. Ceballos G, Ehrlich PR (2006) Global mammal distributions, biodiversity hotspots, and conservation. Proc Natl Acad Sci USA 103: 19374–19379.
2. O'Brien TG, Baillie JEM, Krueger L, Cuke M (2010) The Wildlife Picture Index: monitoring top trophic levels. Anim Conserv 13: 335–343.
3. Hoffmann M, Hilton-Taylor C, Angulo A, Böhm M, Brooks TM, et al. (2010) The impact of conservation on the status of the world's vertebrates. Science 330: 1503–1509.
4. Rondinini C, Rodrigues ASL, Boitani L (2011) The key elements of a comprehensive global mammal conservation strategy. Phil Trans R Soc B 366: 2591–2597.
5. Ceballos G, Ehrlich P, Soberon J, Salazar I, Fay J (2005) Global mammal conservation: what must we manage? Science 309: 603–607.
6. Schipper J, Chanson JS, Chiozza F, Cox NA (2008) The status of the world's land and marine mammals: diversity, threat, and knowledge. Science 322: 225–30.
7. Visconti P, Pressey RL, Giorgini D, Maiorano L, Bakkenes M, et al. (2011) Future hotspots of terrestrial mammal loss. Phil Trans R Soc B 366: 2693–2702.
8. Ahumada JA, Silva CEF, Gajapersad K, Hallam C, Hurtado J, et al. (2011) Community structure and diversity of tropical forest mammals: data from a global camera trap network. Philos T R Soc B 366: 2703–2711.
9. Struhsaker TT (1997) Ecology of an African rain forest: logging in Kibale and the conflict between conservation and exploitation. Gainesville: University Press of Florida. 432 p.
10. Weber W (2001) African rain forest ecology and conservation: an interdisciplinary perspective. New Haven: Yale University Press. 608 p.
11. Hooper DU, Chapin III FS, Ewel JJ, Hector A, Inchausti P, et al. (2005) Effects of biodiversity on ecosystem functioning: a consensus of current knowledge. Ecol Monogr 75: 3–35.
12. Asquith NM, Mejía-Chang M (2005) Mammals, edge effects, and the loss of tropical forest diversity. Ecology 86: 379–390.
13. TEAM Network (2011) Terrestrial vertebrate monitoring protocol. v 3.1. TEAM Standardized Monitoring Protocols. http://www.teamnetwork.org/protocols/bio/terrestrial-vertebrate.
14. Ahumada J, Hurtado J, Lizcano D (2013) Monitoring the status and trends of tropical forest, terrestrial vertebrate communities from camera trap data: a tool for conservation. PLOS ONE 8: e73707.
15. O'Connell AF, Bailey LL (2011) Inference for occupancy and occupancy dynamics. In: O'Connell AF, Nichols JD, Karanth KU, editors. Camera traps in animal ecology: methods and analysis. New York: Springer. pp. 191–205.
16. Rovero F, Zimmermann F, Berzi D, Meek P (2013) "Which camera trap type and how many do I need?" A review of camera features and study designs for a range of wildlife research applications. Hystrix 24: 1–9.
17. MacKenzie DI, Nichols JD, Lachman G, Droege S, Royle JA, et al. (2002) Estimating site occupancy rates when detection probabilities are less than one. Ecology 83: 2248–2255.
18. O'Connell AF, Nichols JD, Karanth KU (2011) Camera traps in animal ecology: methods and analysis. New York: Springer. 280 p.
19. MacKenzie DI, Nichols JD, Royle JA, Pollock KH, Bailey LL, et al. (2006) Occupancy estimation and modeling: Inferring patterns and dynamics of species occurrence. San Diego: Elsevier. 344 p.
20. Mackenzie DI, Nichols JD (2004) Occupancy as a surrogate for abundance estimation. Anim Biodivers Conservn 27: 461–467.
21. Linkie M, Dinata Y, Nugroho A, Haidir IA (2007) Estimating occupancy of a data-deficient mammalian species living in tropical rainforests: sun bears in the Kerinci Seblat region, Sumatra. Biol Conserv 137: 20–27.
22. Rovero F, Collett L, Ricci S, Martin E, Spitale D (2013) Distribution, occupancy and habitat associations of the grey-faced sengi (*Rhynchocyon udzungwensis*) as revealed by camera traps. J Mammal 94: 792–800.
23. Dorazio RM, Royle JA, Sodestrom B, Glimskar A (2006) Estimating species richness and accumulation by modeling species occurrence and detectability. Ecology 87: 842–854.
24. Rovero F, Menegon M, Fjeldså J, Collett L, Doggart N, et al. (in press). Targeted vertebrate surveys enhance the faunal importance and improve explanatory models within the Eastern Arc Mountains of Kenya and Tanzania. Diversity and Distributions.
25. Rovero F, De Luca DW (2007) Checklist of mammals of the Udzungwa Mountains of Tanzania. Mammalia 71: 47–55.
26. Lovett JC., Marshall AR, Carr J (2006) Changes in tropical forest vegetation along an altitudinal gradient in the Udzungwa Mountains National Park, Tanzania. Afr J Ecol 44: 478–490.
27. Fegraus EH, Lin K, Ahumada J, Baru C, Chandra S, et al. (2011) Data acquisition and management software for camera trap data: A case study from the TEAM Network. Ecological Informatics 6: 345–353.
28. IUCN (2013) The IUCN red list of threatened species. http://iucnredlist.org.
29. Rovero F, Marshall AR (2009) Camera trapping photographic rate as an index of density in forest ungulates. J Appl Ecol 46: 1011–1017.
30. Rovero F, Mtui A, Kitegile A, Nielsen M (2012) Hunting or habitat degradation? Decline of primate populations in Udzungwa Mountains, Tanzania: An analysis of threats. Biol Conserv 146: 89–96.
31. Gotelli NJ, Colwell RK (2001) Quantifying biodiversity: procedures and pitfalls in the measurement and comparison of species richness. Ecol Lett 4: 379–391.
32. Silveira L, Jácomo ATA, Diniz-Filho JAF (2003) Camera trap, line transect census and track surveys: a comparative evaluation. Biol Conserv 114: 351–355.
33. Tobler MW, Carrillo-Percastegui SE, Leite Pitman R, Mares R, Powell G (2008) An evaluation of camera traps for inventorying large-and medium-sized terrestrial rainforest mammals. Anim Conserv 11: 169–178.
34. Sturtz S, Ligges U, Gelman A (2005) R2WinBUGS: a package for running WinBUGS from R. J Stat Softw 12: 1–16.
35. R Development Core Team (2013) R: a language and environment for statistical computing. R Foundation for Statistical Computing, Vienna, Austria. http://www.R-project.org.
36. Kinnaird MF, O'brien TG (2012) Effects of private-land use, livestock management, and human tolerance on diversity, distribution, and abundance of large african mammals. Conserv Biol 26: 1026–1039.
37. Smith F, Lyons S, Ernest S, Jones K, Kauffman D, et al. (2003) Body mass of late quaternary mammals. Ecology 84: 3403.
38. Burnham KP, Anderson DR (2002) Model selection and multi-model inference. A practical information–theoretic approach. New York: Springer. 488 p.
39. Fiske I, Chandler R (2011) Unmarked: An R package for fitting hierarchical models of wildlife occurrence and abundance. J Stat Softw 43: 1–23.
40. Mazerolle MJ (2012) AICcmodavg: model selection and multimodel inference based on (Q)AIC(c). R package version 1.26. http://CRAN.R-project.org/package¼AICcmodavg.
41. Rovero F, Tobler M, Sanderson J (2010) Camera-trapping for inventorying terrestrial vertebrates. In: Eymann J, Degreef J, Häuser C, Monje JC, Samyn Y, VandenSpiegel D, editors. Manual on field recording techniques and protocols for all taxa biodiversity inventories and monitoring. Abc Taxa. pp. 100–128.
42. De Luca DW, Mpunga NE (2005) Small carnivores of the Udzungwa Mountains: presence, distributions and threats. Small Carnivore Conservation 32: 1–7.
43. Dorazio RM, Gotelli NJ, Ellison AM (2011) Modern methods of estimating biodiversity from presence-absence surveys. In: Grillo O, Venora G. Editors. Biodiversity Loss in a Changing Planet. InTech. 318 p.
44. Dorazio RM, Royle JA (2005) Estimating size and composition of biological communities by modeling the occurrence of species. J Am Stat Assoc 100: 389–398.

45. Rovero F, Davenport TRB, Jones T (2013) Abbott's duiker. In: Kingdon JS, Hoffmann M, editor. The Mammals of Africa. Vol. 6. Pigs, Deer, Giraffe, Bovids, and Hippos. Bloomsbury: Academic Press. pp. 285–288.
46. Rovero F, Marshall AR, Jones T, Perkin A (2009) The primates of the Udzungwa Mountains: diversity, ecology and conservation. J Anthropol Sci 87: 93–126.
47. Roberts D, Topp-Jørgensen E, Moyer D (2013) Eastern tree hyrax. In: Kingdon JS, Hoffmann M, editors. The Mammals of Africa. Vol. 1. Introductory chapters and Afrotheria. Bloomsbury: Academic Press. pp. 158–161.
48. Rovero F, Rathbun GB, Perkin A, Jones T, Ribble DO, et al. (2008) A new species of giant sengi or elephant-shrew (genus *Rhynchocyon*) highlights the exceptional biodiversity of the Udzungwa Mountains of Tanzania. J Zool 274: 126–133.
49. Rowcliffe MJ, Carbone C, Jansen PA, Kays R, Kranstauber B (2011) Quantifying the sensitivity of camera traps: an adapted distance sampling approach. Method Ecol Evol 2: 464–476.
50. Tobler MW, Carrillo-Percastegui SE, Powell G (2009) Habitat use, activity patterns and use of mineral licks by five species of ungulate in south-eastern Peru. J Trop Ecol 25: 261–270.
51. Ehardt CL, Jones TP, Butynski TM (2005) Protective status, ecology and strategies for improving conservation of *Cercocebus sanjei* in the Udzungwa Mountains, Tanzania. Int J Primatol 26: 557–583.
52. Bowkett AE, Rovero F, Marshall AR (2008) The use of camera-trap data to model habitat use by antelope species in the Udzungwa Mountain forests, Tanzania. Afr J Ecol 46: 479–487.
53. Marshall AR, Jørgensbye HIO, Rovero F, Platts PJ, White PCL, et al. (2010) The species-area relationship and confounding variables in a threatened monkey community. Am J Primatol 72: 325–336.

16

Species Divergence and Phylogenetic Variation of Ecophysiological Traits in Lianas and Trees

Rodrigo S. Rios[1], Cristian Salgado-Luarte[1], Ernesto Gianoli[1,2]*

Departamento de Biología, Universidad de La Serena, La Serena, Chile, **2** Departamento de Botánica, Universidad de Concepción, Concepción, Chile

Abstract

The climbing habit is an evolutionary key innovation in plants because it is associated with enhanced clade diversification. We tested whether patterns of species divergence and variation of three ecophysiological traits that are fundamental for plant adaptation to light environments (maximum photosynthetic rate [A_{max}], dark respiration rate [R_d], and specific leaf area [SLA]) are consistent with this key innovation. Using data reported from four tropical forests and three temperate forests, we compared phylogenetic distance among species as well as the evolutionary rate, phylogenetic distance and phylogenetic signal of those traits in lianas and trees. Estimates of evolutionary rates showed that R_d evolved faster in lianas, while SLA evolved faster in trees. The mean phylogenetic distance was 1.2 times greater among liana species than among tree species. Likewise, estimates of phylogenetic distance indicated that lianas were less related than by chance alone (phylogenetic evenness across 63 species), and trees were more related than expected by chance (phylogenetic clustering across 71 species). Lianas showed evenness for R_d, while trees showed phylogenetic clustering for this trait. In contrast, for SLA, lianas exhibited phylogenetic clustering and trees showed phylogenetic evenness. Lianas and trees showed patterns of ecophysiological trait variation among species that were independent of phylogenetic relatedness. We found support for the expected pattern of greater species divergence in lianas, but did not find consistent patterns regarding ecophysiological trait evolution and divergence. R_d followed the species-level pattern, i.e., greater divergence/evolution in lianas compared to trees, while the opposite occurred for SLA and no pattern was detected for A_{max}. R_d may have driven lianas' divergence across forest environments, and might contribute to diversification in climber clades.

Editor: Sylvain Delzon, INRA - University of Bordeaux, France

Funding: This study was funded by the FONDECYT grant 1100585 (http://www.conicyt.cl/fondecyt/). FONDECYT is the Chilean Agency of Research Funding. The funders had no role in study design, data collection and analysis, decision to publish, or preparation of the manuscript.

Competing Interests: The authors have declared that no competing interests exist.

* E-mail: egianoli@userena.cl

Introduction

Climbing plants, in particular woody vines (lianas), are a distinctive component of mature forests in both tropical and temperate regions [1–3]. Data from long-term plots indicate that the dominance of lianas relative to trees is increasing in tropical forests [4,5]. Moreover, liana abundance is negatively associated with tree carbon storage in tropical forests [6,7]. The climbing habit has independently arisen numerous times throughout plant evolution [1,8], and it seems to be a key innovation in angiosperms: climbing plant lineages have greater species richness than their non-climbing sister groups [9]. Thus, evidence from both ecological and macroevolutionary patterns suggests a performance advantage of lianas over trees.

Explanatory factors for the increased abundance and biomass of lianas in tropical forests include increasing forest disturbance, which increases local resource availability, and rising levels of atmospheric CO_2 [4]. Moreover, increased abundance of lianas in seasonal forests during the dry season, as compared to trees, has been related to their increased efficiency in water uptake and transport, and higher photosynthetic rates ([10–12]; but see [13]). Thus, data suggest that lianas are better than trees at exploiting resource pulses. When providing functional arguments for the key innovation of the climbing habit (*sensu* [14]), Gianoli [9] suggested that ecological specialization may arise as a consequence of an hypothetically expanded light niche of lianas in the forest, which would result from the co-occurrence of unsupported (creeping) and supported (climbing) individuals that go up and down the forest canopy. This would maximize interactions with a wide array of antagonistic and mutualistic species [15,16] that, in turn, might promote diversification [17]. It is increasingly recognized that purported evolutionary key innovations may be tested at an ecological time scale [14,18–20].

Ecophysiological traits are fundamental components of plant adaptation to the environment [21,22]. Specifically, A_{max} (maximum photosynthetic rate), R_d (dark respiration rate) and SLA (specific leaf area) play a key role in the phenotypic adjustment to heterogeneous light environments in both lianas and trees [23–26]. Thus, they reflect the balance between carbon gain (A_{max}) and carbon use (R_d), and the allocation of leaf biomass to light interception (SLA), which together determine plant growth and performance across light environments [27,28]. Importantly, variation in plant functional traits observed at the population level is likely to be paralleled by evolutionary divergences under contrasting environments [29]. Moreover, the analysis of the phylogenetic structure of communities can provide insights to our understanding of trait evolution [30]. Recent studies have addressed phylogenetic variation in ecophysiological traits in climbing plants and trees [31–34], but their approach has been either exploratory (aiming to report global patterns) or methodological (testing new analytical tools); to our knowledge, a hypothesis-driven analysis is wanting.

Using data reported for several liana and tree species coexisting in tropical and temperate forests, and focusing on three key ecophysiological traits involved in plant adaptation across forest light gradients: A_{max}, R_d and SLA, we herein compare lianas and trees in terms of trait evolutionary rates, phylogenetic diversity, phylogenetic trait diversity, and the phylogenetic signal. Thus, we compared the rate at which variance in the traits is accumulated among species per unit time at the tips of the phylogenetic tree [35,36]. We also evaluated how similar is the average pair of species of lianas and trees both in terms of mean phylogenetic distance and trait variation [37]. We finally evaluated in lianas and trees the tendency for phylogenetically related species to resemble each other, i.e., the phylogenetic signal [38]. We tested the hypotheses that if the climbing habitat enhances clade diversification [9], and ecological divergence is the process underlying this pattern, then lianas should show higher trait evolutionary rates and greater species and trait divergence than trees under common environmental scenarios.

Materials and Methods

Data collection

We searched the literature for field studies in forest ecosystems where lianas and trees were analyzed for at least one of three ecophysiological traits: A_{max} on an area basis, R_d, and SLA. We only chose those studies carried out in mature forests, where light heterogeneity across microsites is the greatest [3]. We only included native species because they have a long history of adaptation to the environment. We focused on angiosperms because of the availability of tools to reconstruct their phylogenetic history and estimate trait evolution (see below). The final data set included a pool of 63 liana species and 71 tree species belonging to four tropical forests (Gamboa, Panama; San Lorenzo, Panama; Riberalta, Bolivia; Xishuangbanna, China) and three temperate forests (Yakushidake, Japan; Beltsville, USA; Puyehue, Chile). We pooled species from all sites into growth forms, thus we had one liana "super-community" and one tree "super-community". Phylogenetic analyses were conducted on these super-communities (see below). Detailed information, including study species, traits, sites, and data sources, is available in Supporting Information S1.

Phylogeny reconstruction

We produced a phylogeny of all species using a backbone tree based on the angiosperm megatree provided by the Phylodiversity Network in cooperation with the Angiosperm Phylogeny Group (APG; http://www.mobot.org/MOBOT/research/APweb/). Our tree was generated using Phylomatic (http://www.phylodiversity.net/phylomatic/phylomatic.html), a program that returns a working phylogenetic tree after matching the genus and family names of study species to those contained in the angiosperm phylogeny [39]. Comparative inferences require branch lengths for the tree, which were calculated based on the branch length adjustment algorithm (BLADJ) implemented in Phylocom v. 4.2 (www.phylodiversity.net/phylocom) [40]. This algorithm fixes a subset of nodes in the tree to specified ages and evenly distributes the ages to the remaining nodes. Age estimates for major nodes in our tree were taken from [41]. To avoid inaccuracies in tree calibration and to have an updated version of our tree, we corrected the *ages* file with age estimates in [41] included in Phylocom. Corrections followed procedures suggested recently [42]. We also checked and updated age estimates of internal order-level clades according to a net diversification rate estimate of angiosperms [43]. The few polytomies in the working tree were resolved randomly using the *multi2di* function in R. Values of functional traits of closely related species resulting from such random resolutions were very similar, so results of the final comparative tests were highly robust to topological uncertainty. These and all subsequent analyses were conducted using the R statistical environment version 3.0.2 [44]. Reconstructed phylogenetic trees with associated trait variation are shown in Figures 1–3.

Rate of trait evolution

To assess differences in ecophysiological trait evolution between liana and tree species, we compared estimates of evolutionary rate for A_{max}, R_d, and SLA. To this end, we first used stochastic character mapping, a Bayesian method that uses Monte Carlo simulations to sample the posterior probability distribution of ancestral states and timings of transitions on phylogenetic branches under a Markov process of evolution [45,46]. We built stochastic character-mapped reconstructions for each trait/growth form combination using the *make.simmap* function in the phytools package of R [36]. We thus simulated character history evolution of all three traits in relation to growth form as an initial step, as suggested recently [35,47]. To test the hypothesis that a discrete character state had influenced the rate of a continuous character, one should first stochastically map the discrete trait (e.g., climbing habit), and then test if one state of the discrete character has a different evolutionary rate for the continuous trait of interest (e.g., ecophysiological trait) than the other discrete state [35,47].

The resulting reconstructions of trait states and phylogeny represented a set of phylogenetic topologies, branch lengths and growth forms sampled in proportion to their posterior probabilities. Reconstructions were then used in subsequent analyses as a way of integrating over uncertainty in phylogeny and ancestral states. Finally, we fitted the evolutionary models of character history on the trees to trait data using a likelihood method [35]. This is a maximum likelihood approach that estimates rates of evolution (σ^2). The parameter σ^2 was calculated using the function *brownie.lite* in the *phytools* package [36]. σ^2 is interpreted as the Brownian motion process most likely to have produced the data at the tips of the tree, i.e., the rate at which variance is accumulated among species per unit time. 95% confidence intervals were calculated for each σ^2 to infer differences between lianas and trees in ecophysiological trait evolution.

Trait and phylogenetic diversity

To compare the phylogenetic relatedness among liana species against the phylogenetic relatedness among tree species we used measures of phylogenetic structure. Specifically, we calculated, based on a phylogenetic distance matrix, the mean phylogenetic distance (MPD) and the standardized effect size of the mean phylogenetic distance (SES_{MPD} [48]) between pairs of species for each group. Interspecific phylogenetic distance matrices were obtained from the reconstructed tree of phylogenetic relationships among taxa using the *cophenetic* function in R and unweighted pair-group average (UPGMA) as the clustering method.

Standardized effect sizes describe the difference between average phylogenetic distances in the observed super communities or groups (lianas and trees) compared to null distributions generated for each group with randomization procedures, standardized by the standard deviation of phylogenetic distances in the null data [48]. We compared observed mean distances (branch length) against a null model generated by calculating 999 times the mean phylogenetic distance between 8911 random pairs of species (without replacement) drawn from the matrix of

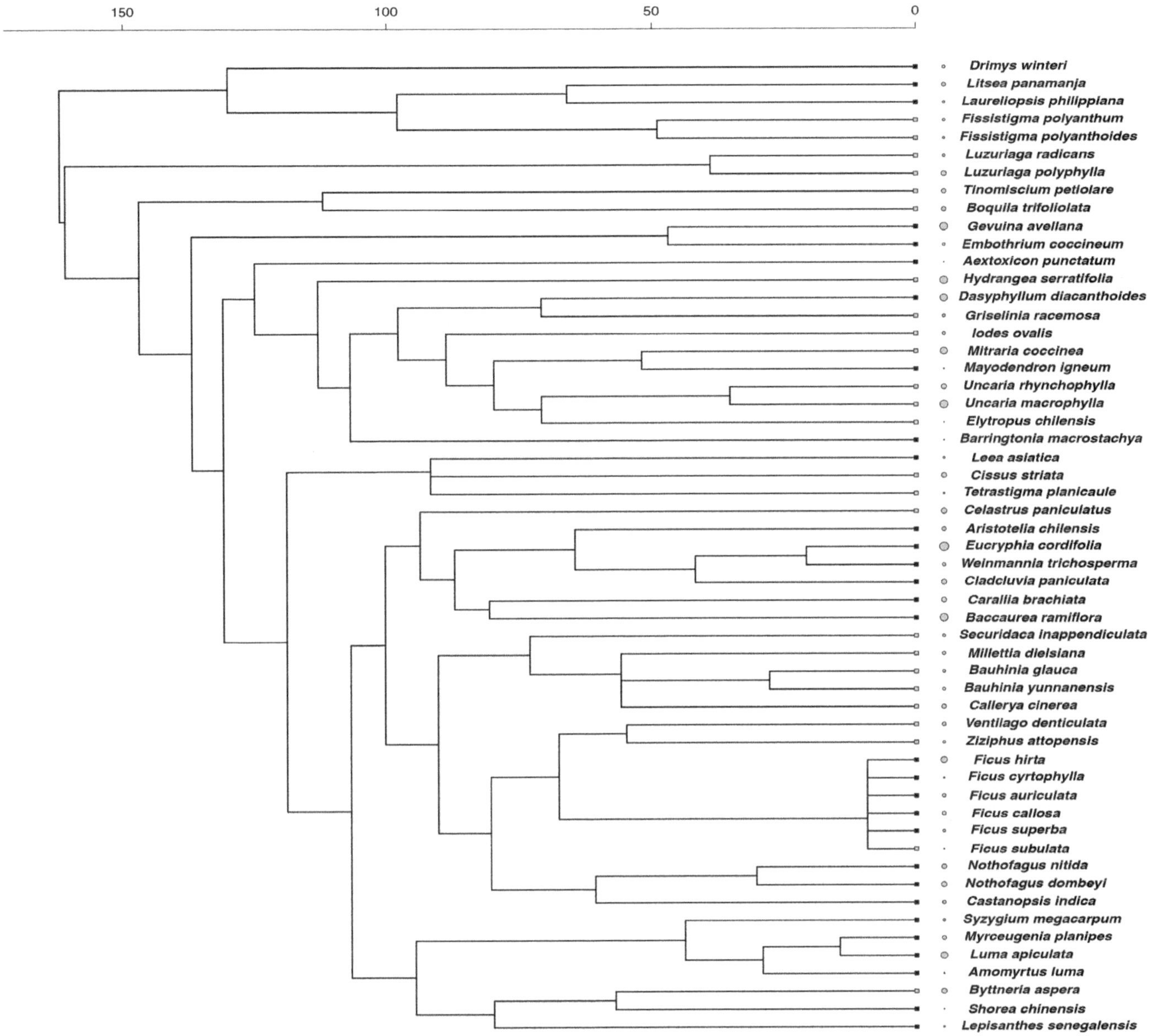

Figure 1. Phylogenetic relationships among tree and liana species and species values of dark respiration rate (R_d). Grey circle size represents the proportional magnitude of the trait across species. Square tip symbols represent climbing habit (grey squares = liana, black squares = tree). Timescale is in millions of years before present.

phylogenetic distances between all liana and tree species. In all cases 999 iterations were found to be suitable for our randomization procedures as they were sufficient to attain convergence. The null model was constructed by reshuffling the distance of species labels across the phylogenetic tree using the *ses.mpd* function and the *taxa.labels* algorithm of the *picante* package of R. Positive values of SES_{MPD} (*mpd.obs.z*) and high quantiles (*p*-values >0.95) indicate significant phylogenetic evenness, while negative values of SES_{MPD} and low quantiles (*p*-values <0.05) indicate significant phylogenetic clustering [48]; these outcomes correspond to scenarios where species are more distantly or more closely related than expected by chance, respectively [48,49]. Authors often refer to (weak) evenness or clustering when *p*-values are slightly lower than 0.95 or slightly higher than 0.05, respectively (e.g., [50,51]). Finally, to assess how similar are the average pair of species within each group in terms of ecophysiological traits; we used the SES_{MPD} as a trait diversity measure. This was done by replacing the phylogenetic distance matrix in the analysis with a trait distance matrix, and proceeding accordingly to calculate standardized values of mean phylogenetic trait distance (SES_{MTD}). These results are interpreted in the same way as those of SES_{MPD} with regard to phylogenetic evenness or clustering [48].

Phylogenetic signal

To quantify the degree to which phylogenetic relatedness predicts the similarity of species in functional traits for both trees and lianas, we calculated separately phylogenetic signal for A_{max}, R_d, and SLA. Phylogenetic signal indicates to what extent phenotypic expression is explained by the lineage to which a

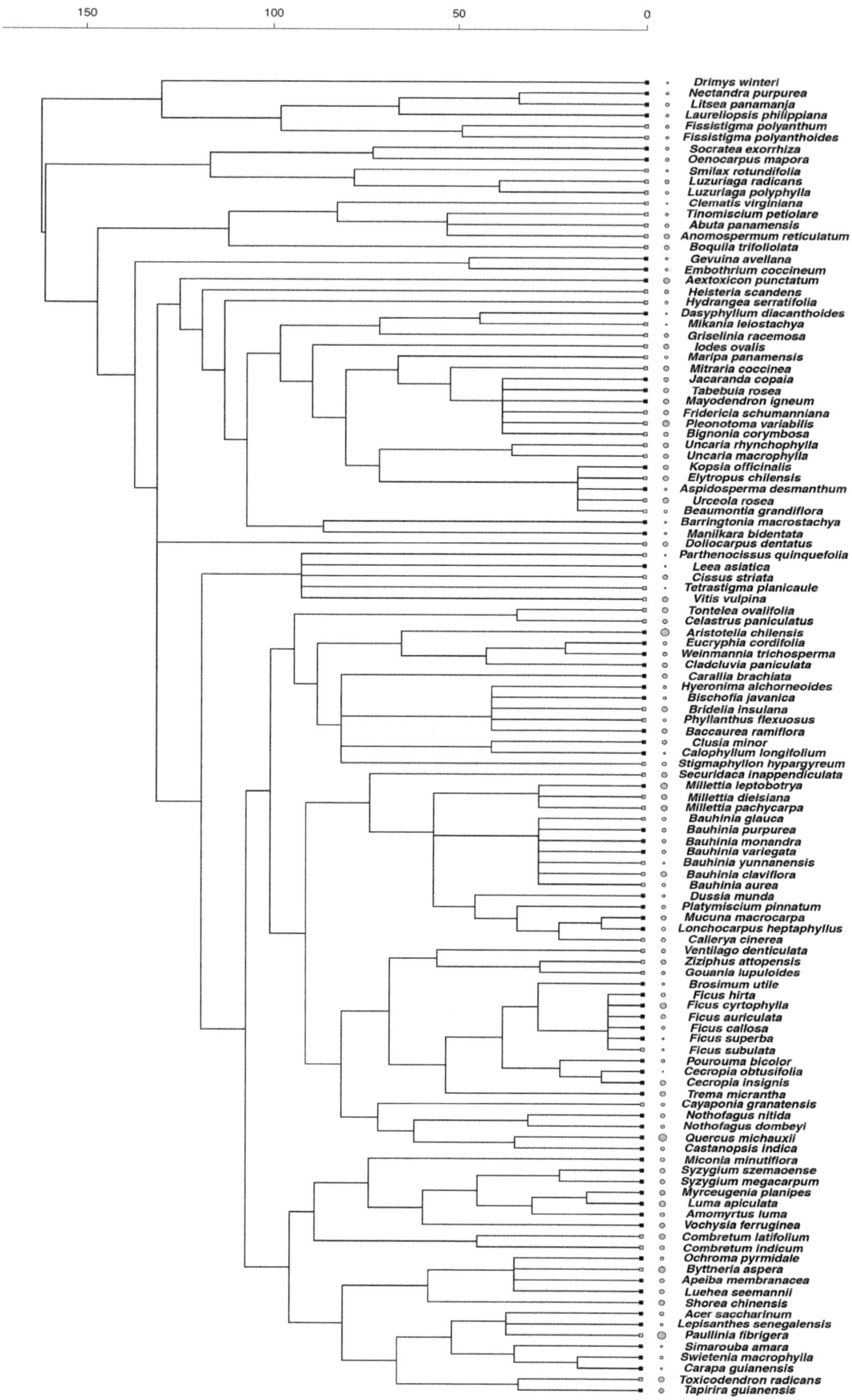
150
100
50
0
Drimys winteri
Nectandra purpurea
Litsea panamanja
Laureliopsis philippiana
Fissistigma polyanthum
Fissistigma polyanthoides
Socratea exorrhiza
Oenocarpus mapora
Smilax rotundifolia
Luzuriaga radicans
Luzuriaga polyphylla
Clematis virginiana
Tinomiscium petiolare
Abuta panamensis
Anomospermum reticulatum
Boquila trifoliolata
Gevuina avellana
Embothrium coccineum
Aextoxicon punctatum
Heisteria scandens
Hydrangea serratifolia
Dasyphyllum diacanthoides
Mikania leiostachya
Griselinia racemosa
Iodes ovalis
Maripa panamensis
Mitraria coccinea
Jacaranda copaia
Tabebuia rosea
Mayodendron igneum
Fridericia schumanniana
Pleonotoma variabilis
Bignonia corymbosa
Uncaria rhynchophylla
Uncaria macrophylla
Kopsia officinalis
Elytropus chilensis
Aspidosperma desmanthum
Urceola rosea
Beaumontia grandiflora
Barringtonia macrostachya
Manilkara bidentata
Doliocarpus dentatus
Parthenocissus quinquefolia
Leea asiatica
Cissus striata
Tetrastigma planicaule
Vitis vulpina
Tontelea ovalifolia
Celastrus paniculatus
Aristotelia chilensis
Eucryphia cordifolia
Weinmannia trichosperma
Cladcluvia paniculata
Carallia brachiata
Hyeronima alchorneoides
Bischofia javanica
Bridelia insulana
Phyllanthus flexuosus
Baccaurea ramiflora
Clusia minor
Calophyllum longifolium
Stigmaphyllon hypargyreum
Securidaca inappendiculata
Millettia leptobotrya
Millettia dielsiana
Millettia pachycarpa
Bauhinia glauca
Bauhinia purpurea
Bauhinia monandra
Bauhinia variegata
Bauhinia yunnanensis
Bauhinia claviflora
Bauhinia aurea
Dussia munda
Platymiscium pinnatum
Mucuna macrocarpa
Lonchocarpus heptaphyllus
Callerya cinerea
Ventilago denticulata
Ziziphus attopensis
Gouania lupuloides
Brosimum utile
Ficus hirta
Ficus cyrtophylla
Ficus auriculata
Ficus callosa
Ficus superba
Ficus subulata
Pourouma bicolor
Cecropia obtusifolia
Cecropia insignis
Trema micrantha
Cayaponia granatensis
Nothofagus nitida
Nothofagus dombeyi
Quercus michauxii
Castanopsis indica
Miconia minutiflora
Syzygium szemaoense
Syzygium megacarpum
Myrceugenia planipes
Luma apiculata
Amomyrtus luma
Vochysia ferruginea
Combretum latifolium
Combretum indicum
Ochroma pyrmidale
Byttneria aspera
Apeiba membranacea
Luehea seemannii
Shorea chinensis
Acer saccharinum
Lepisanthes senegalensis
Paullinia fibrigera
Simarouba amara
Swietenia macrophylla
Carapa guianensis
Toxicodendron radicans
Tapirira guianensis

Figure 2. Phylogenetic relationships among tree and liana species and species values of maximum photosynthetic rate (A_{max}). Grey circle size represents the proportional magnitude of the trait across species. Square tip symbols represent climbing habit (grey squares = liana, black squares = tree). Timescale is in millions of years before present.

species belongs, and it can be compared among clades and among traits [52]. We quantified phylogenetic signal using both Blomberg's *K* [38] and Pagel's λ [53] statistics for quantitative traits. To calculate these parameters, we first pruned two separate phylogenies, one for the group of lianas and one for the group of trees, using the original tree as a base phylogeny. Then we pruned a tree for each group-trait combination independently, removing taxa for which trait information was not available. The number of species included in each trait/plant growth habit analysis ranged from 26 (R_d/lianas) to 67 (A_{max}/trees), thus meeting the N>20 threshold to achieve good statistical power [38].

Values of $K = 1$ imply that a trait shows exactly the amount of phylogenetic signal expected under a null, stochastic model of character evolution (Brownian motion evolution) [38]. *K*-values > 1 and <1 imply that close relatives are more similar and less similar, respectively, than expected under a Brownian motion model of trait evolution [38]. If *K* does not differ from zero it is concluded that the trait has no phylogenetic signal. Statistical significance of *K* [38] was assessed via permutation tests with 1000 randomizations. The significance of the phylogenetic signal was based on the variance of phylogenetically independent contrasts relative to tip shuffling randomization implemented by the *phylosignal* function of the *picante* package in R [48]. *P*-values were determined by comparing the variance of standardized independent contrasts for the tip values against variances for randomized data.

The parameter λ scales tree structure in terms of expected variances and covariances in trait change [54]. Thus, λ is a phylogenetic transformation that maximizes the likelihood of the data given a Brownian motion model [54]. When $\lambda = 1$, the trait is consistent with a Brownian motion evolution based on branch lengths represented by the variance-covariance in trait change. Values between 0 and 1 indicate less phylogenetic signal than expected under a Brownian motion model, while values >1 indicate more signal than expected, although λ is not always defined for values greater than one [54]. Values of λ were estimated using the *fitContinuous* function of the *geiger* package. To determine the significance of λ as an indicator of phylogenetic signal, we compared the maximum likelihood estimate of λ against the maximum likelihood of models when $\lambda = 1$ using likelihood ratio tests (LRT).

Results and Discussion

Rate of trait evolution

In general, lianas and trees presented homogenous evolution of ecophysiological traits. In all cases evolutionary rates, as estimated by σ^2, were not significantly different from a single-rate Brownian motion process of evolution (Table 1). Parameter estimate values of σ^2, however, did differ between lianas and trees in two of the three ecophysiological traits considered (Table 1). The evolutionary rate for dark respiration rate (R_d) in lianas was 1.8 times greater than in trees. In the case of the biomass allocation trait (specific leaf area, SLA), the evolutionary rate was 1.2 times greater in trees than in lianas. Evolutionary rates for maximum photosynthetic rate (A_{max}) did not differ between lianas and trees; overall, this trait showed the lowest evolutionary rate among the traits considered (Table 1).

The patterns observed suggest that for all ecophysiological traits a change along any given branch in the phylogeny is independent of both previous changes and changes in other branches of the reconstructed tree. Evolutionary rates in both lianas and trees did not differ from a single-rate Brownian motion model of evolution, which assumes that variance among species in the phylogenetic tree accumulates as function of their time of independent evolution [55]. Thus, it cannot be ruled out that ecophysiological traits evolve at a constant rate over time. A Brownian motion process, however, is not equal to a neutral model of evolution. Brownian motion simply describes the distribution of observed trait changes and may be consistent with adaptive models of evolution [35,56]. Therefore, natural selection could be a plausible force behind the alteration in rate change of traits in relation to growth form (climbers vs. non-climbers).

Evolutionary rates (σ^2) differed between lianas and trees in two of the three ecophysiological traits considered, but in opposite trends. Thus, R_d evolved at a higher rate in lianas, while SLA evolution occurred at a higher rate in trees. This suggests that the outcome of modifications in the selective regime related to the climbing habit depends on the particular plant traits that are under selection (gas-exchange traits vs. biomass allocation traits). Gas-exchange traits have been shown to be of selective value for the exploitation of light availability in mature forests for trees [57], vines [58] and ferns [59]. Our findings suggest that climbers are more evolutionary responsive with regard to R_d than trees. Assuming that (adaptive) ecological speciation is the process behind species divergence in this trait [20], the next step would be to address whether this results from a greater magnitude of selection on R_d or from greater trait heritability [21,60]. Conversely, SLA showed a higher greater evolutionary rate across tree species. This somewhat supports the view of SLA as an essential attribute for tree performance and carbon gain [57,61].

Trait and phylogenetic diversity

We found that mean phylogenetic distance (MPD, non-standardized values) was greater among liana species (259.9 Myr) than among tree species (229.6 Myr). Moreover, there was a clear-cut difference between lianas and trees in the standardized mean phylogenetic distance among species (SES_{MPD}). Whereas lianas showed greater distances between species relative to the null model ($SES_{MPD} = 2.271$; *p*-value = 0.99), i.e., phylogenetic evenness, trees showed a pattern of phylogenetic clustering ($SES_{MPD} = -3.622$; *p*-value = 0.006).

Lianas and trees differed in their patterns of trait diversity. For one of the gas-exchange traits (R_d), lianas showed phylogenetic evenness ($SES_{MTD} = 1.266$, *p*-value = 0.893), which means that trait dissimilarity among liana species was higher than expected by chance, while trees showed phylogenetic clustering ($SES_{MTD} = -1.863$, *p*-value = 0.039), indicating that tree species were more phenotypically similar than expected by chance (Figure 1). In contrast, for the biomass allocation trait (SLA), lianas exhibited phylogenetic clustering ($SES_{MTD} = -1.194$, *p*-value = 0.122) and trees showed phylogenetic evenness ($SES_{MTD} = 1.193$, *p*-value = 0.877) (Figure 2). Finally, the other gas-exchange trait, A_{max}, did not show phylogenetic structure in both lianas ($SES_{MTD} = 0.096$, *p*-value = 0.536) and trees ($SES_{MTD} = -0.598$, *p*-value = 0.277) (Figure 3).

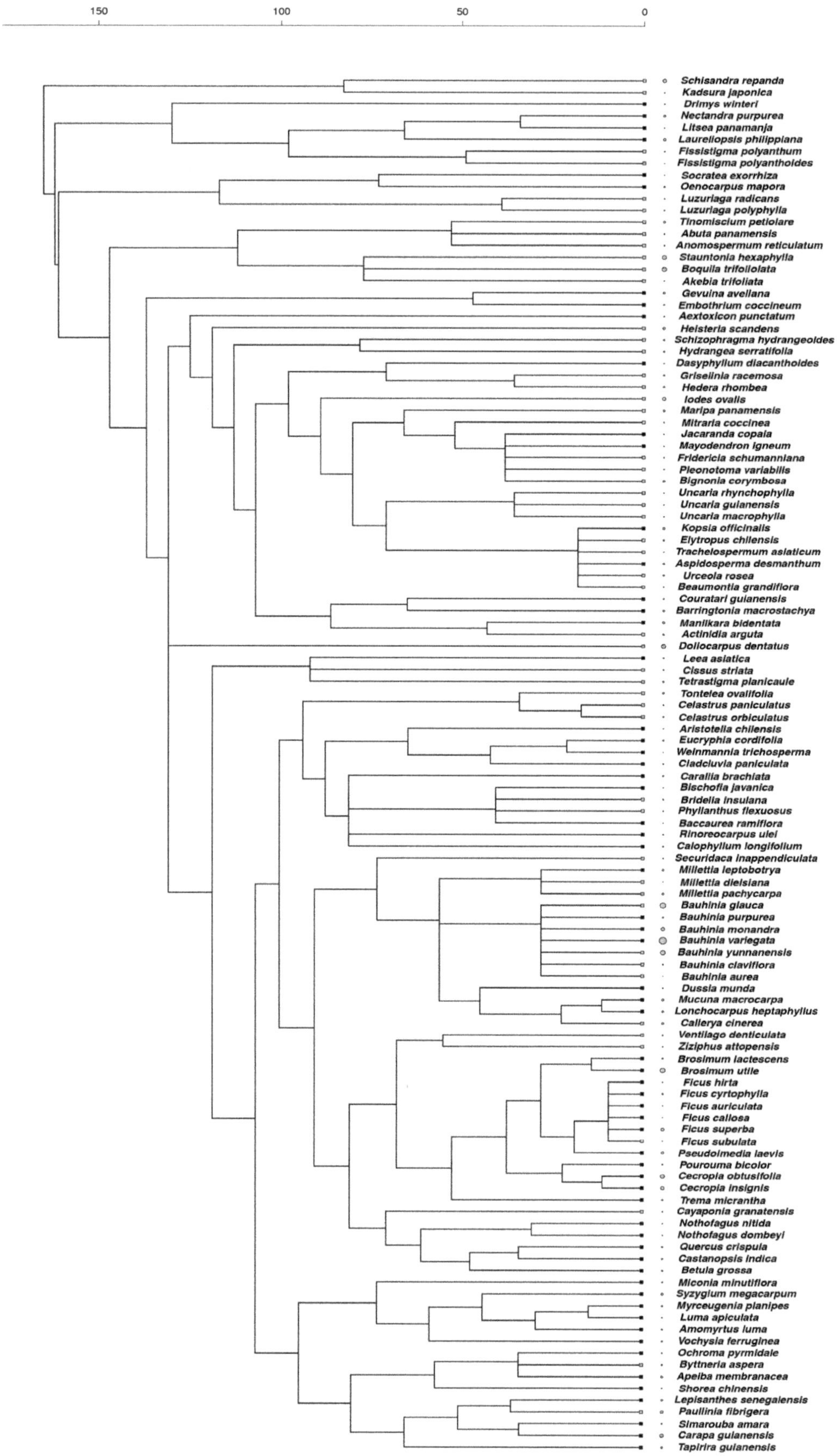
150
100
50
0
Schisandra repanda
Kadsura japonica
Drimys winteri
Nectandra purpurea
Litsea panamanja
Laureliopsis philippiana
Fissistigma polyanthum
Fissistigma polyanthoides
Socratea exorrhiza
Oenocarpus mapora
Luzuriaga radicans
Luzuriaga polyphylla
Tinomiscium petiolare
Abuta panamensis
Anomospermum reticulatum
Stauntonia hexaphylla
Boquila trifoliolata
Akebia trifoliata
Gevuina avellana
Embothrium coccineum
Aextoxicon punctatum
Heisteria scandens
Schizophragma hydrangeoides
Hydrangea serratifolia
Dasyphyllum diacanthoides
Griselinia racemosa
Hedera rhombea
Iodes ovalis
Maripa panamensis
Mitraria coccinea
Jacaranda copaia
Mayodendron igneum
Fridericia schumanniana
Pleonotoma variabilis
Bignonia corymbosa
Uncaria rhynchophylla
Uncaria guianensis
Uncaria macrophylla
Kopsia officinalis
Elytropus chilensis
Trachelospermum asiaticum
Aspidosperma desmanthum
Urceola rosea
Beaumontia grandiflora
Couratari guianensis
Barringtonia macrostachya
Manilkara bidentata
Actinidia arguta
Doliocarpus dentatus
Leea asiatica
Cissus striata
Tetrastigma planicaule
Tontelea ovalifolia
Celastrus paniculatus
Celastrus orbiculatus
Aristotelia chilensis
Eucryphia cordifolia
Weinmannia trichosperma
Cladcluvia paniculata
Carallia brachiata
Bischofia javanica
Bridelia insulana
Phyllanthus flexuosus
Baccaurea ramiflora
Rinoreocarpus ulei
Calophyllum longifolium
Securidaca inappendiculata
Millettia leptobotrya
Millettia dielsiana
Millettia pachycarpa
Bauhinia glauca
Bauhinia purpurea
Bauhinia monandra
Bauhinia variegata
Bauhinia yunnanensis
Bauhinia claviflora
Bauhinia aurea
Dussia munda
Mucuna macrocarpa
Lonchocarpus heptaphyllus
Callerya cinerea
Ventilago denticulata
Ziziphus attopensis
Brosimum lactescens
Brosimum utile
Ficus hirta
Ficus cyrtophylla
Ficus auriculata
Ficus callosa
Ficus superba
Ficus subulata
Pseudolmedia laevis
Pourouma bicolor
Cecropia obtusifolia
Cecropia insignis
Trema micrantha
Cayaponia granatensis
Nothofagus nitida
Nothofagus dombeyi
Quercus crispula
Castanopsis indica
Betula grossa
Miconia minutiflora
Syzygium megacarpum
Myrceugenia planipes
Luma apiculata
Amomyrtus luma
Vochysia ferruginea
Ochroma pyrmidale
Byttneria aspera
Apeiba membranacea
Shorea chinensis
Lepisanthes senegalensis
Paullinia fibrigera
Simarouba amara
Carapa guianensis
Tapirira guianensis

Figure 3. Phylogenetic relationships among tree and liana species and species values of specific leaf area (SLA). Grey circle size represents the proportional magnitude of the trait across species. Square tip symbols represent climbing habit (grey squares = liana, black squares = tree). Timescale is in millions of years before present.

First, in agreement with the study hypotheses, we found greater phylogenetic distance among species within the group of lianas (63 species) than within the group of trees (71 species). This agrees with a recent study in Australian rainforests, where standardized values of mean phylogenetic distance indicated that climbers show weak evenness or no phylogenetic structure, while trees/shrubs show weak to significant phylogenetic clustering [50]. Second, the average phenotypic distance among species for one gas-exchange trait (but not for the other two ecophysiological traits) was greater in the phylogenetic tree of lianas than in that of trees. The environmental gradient experienced by the study species was roughly the same for lianas and trees because data were obtained from sites where trees and lianas coexisted. Therefore, these patterns of (partial) increased phenotypic distance and greater phylogenetic divergence in lianas are consistent with the notion that lianas have a greater differentiation potential than trees [9]. Several plant attributes have been associated with evolutionary rates in angiosperms. For instance, it has been shown that trees and shrubs have lower rates of molecular evolution than herbaceous plants [62,63], and that taller plants have lower rates of molecular evolution [64]; in both cases the outcome is thought to be linked to differences in generation time, which in turn is related to mutation rate. In our study all climbers were woody species (lianas) so herbaceousness should not be a confounded factor. However, there is no available information to reject the possibility that there were longer generation times in the group of trees (see General Conclusions). As to the plant height factor, it is a rather problematic issue, because trees are usually taller than lianas in terms of freestanding height, but if total length is considered, then canopy lianas may be taller. Both issues deserve further scrutiny.

In the realm of community phylogenetics, patterns of phylogenetic evenness in resource-use traits are often interpreted to reflect niche differentiation processes [49]. If trait-based niche differentiation facilitates evolutionary responses to divergent selection, which in turn may lead to incipient speciation [65], then our results of phylogenetic evenness in a liana ecophysiological trait might be linked to the ecological/evolutionary processes that underlie the key innovation of the climbing habit in plants [9]. Whereas results of trait phylogenetic distance suggest that R_d may have played a role as driver of lianas' adaptive divergence, SLA showed greater phenotypic divergence among tree species, as was shown for rates of trait evolution (see above). However, this trait distribution pattern across the phylogenetic tree was not accompanied by an overall greater phylogenetic distance among tree species. This might be interpreted as SLA contributing to tree adaptation to environmental challenges at local scales but do not driving taxonomic divergence across clades.

Phylogenetic signal

Overall, lianas and trees presented mixed but comparable patterns of phylogenetic signal (or lack thereof) in ecophysiological traits (Table 2). In lianas, A_{max} showed no significant phylogenetic signal, but values were lower than expected under a Brownian model of evolution (with $K<1$ and $\lambda<1$). R_d showed mixed results, with significant phylogenetic signal indicated by K, and lower than expected under a Brownian model of evolution, but no significant signal as indicated by λ (= 1). There was no phylogenetic signal detected for specific leaf area (SLA) using K but strong signal using λ, and lower than expected under a Brownian model of evolution (Table 2). In trees, whereas no phylogenetic signal was found for both A_{max} and R_d as indicated by K and λ, a significant phylogenetic signal was found for SLA when K is considered (with $K<1$) but no signal was detected by λ, with values lower than expected under a Brownian model of trait evolution ($\lambda \approx 0$) (Table 2). Summarizing, in all cases both lianas and trees tended to show patterns of ecophysiological trait variation among species that were independent of phylogenetic relatedness.

Our results are consistent with the general pattern that physiological traits tend to show low values of phylogenetic signal [38]. A global analysis of trait variation in climbing plants reported that SLA showed no phylogenetic signal [34], as found in the present study. Likewise, in agreement with our results, a global-scale study in Angiosperms reported that A_{max} (on an area basis) showed no consistent phylogenetic signal [32]. Conversely, a genus-level study in trees [31] found that A_{max} exhibited significant phylogenetic signal, which seemingly opposes our findings. However, this study used a metric other than Blomberg's K and Pagel's λ, and given that different indices of phylogenetic signal often lead to contrasting outcomes ([68]; and Table 2), these results are not necessarily contrary to those reported here. Another group of ecophysiological traits that could have been studied to seek phylogenetic and evolutionary differences between trees and lianas is that of hydraulic characters. Regarding hydraulic traits, lianas have wider and longer vessels compared to trees, features that enable them to supply a large leaf area with a relatively small allocation to xylem tissue [33,66,67]. However, xylem vessel length did not show significant phylogenetic signal in a recent global analysis including lianas, shrubs and trees [33].

Results indicate that, in both lianas and trees, ecophysiological traits related to light use and carbon economy have undergone evolutionary trajectories different to those expected after phylogenetic relationships, assuming a Brownian motion model of trait evolution [68]. These phylogenetic signal results do not match the patterns of trait divergence and trait evolutionary rates found here. Although under some circumstances (e.g., fluctuating selection in related lineages) a negative association between K and evolutionary rate may be found [52,55], it is generally considered that changes in trait evolutionary rates –and ensuing phenotypic divergence– does not influence phylogenetic signal for continuous characters [52,55].

General conclusions

Lianas and trees differ in a number of anatomical, physiological, morphological and life history traits [69–71]. Among the main differences, trees show a greater allocation of biomass (and carbon) to stems and lianas have lower costs of height gain and larger total leaf area potential. Moreover, compared to shrubs and trees, lianas have lower leaf mass per area (LMA, the inverse of SLA), higher foliar N and higher mass-based photosynthetic rate, which is consistent with the characterization of lianas as fast metabolism/rapid turnover species [71]. This could be related to hypothetical differences in generation time between lianas and trees that could explain their differential evolutionary rates, as shown here. Nonetheless, when it comes to explain species distribution across the light gradient in forests [72], the life history trade-off between juvenile growth and survival is observed alike in trees and lianas [73].

Table 1. Parameter estimates and 95% confidence intervals (CI) of the evolutionary rate (σ^2) for ecophysiological traits in lianas and trees.

	Lianas				Trees			
Trait	σ^2	95% CI	χ^2	*p*-value	σ^2	95% CI	χ^2	*p*-value
A_{max}	0.075	0.072–0.079	−6.8	0.948	0.075	0.072–0.078	−10.6	0.995
R_d	**0.148**	0.132–0.165	−4.1	0.999	**0.083**	0.076–0.091	−13.7	0.991
SLA	**0.093**	0.087–0.097	−11.8	0.999	**0.103**	0.099–0.108	−11.2	0.999

A_{max} = maximum photosynthetic rate; R_d = dark respiration rate; SLA = specific leaf area. Values of σ^2 represent rates at which variance accumulates among species per unit time through a phylogeny with branch lengths in units of millions of years. *P*-values are for likelihood ratio tests against the chi-square distribution between a single-rate (homogenous) and a heterogeneous Brownian motion process. Bold cells indicate σ^2 values for which 95% CI do not overlap between lianas and trees.

Table 2. Phylogenetic signal, quantified as Blomberg's *K* and Pagel's λ, for three ecophysiological traits in lianas and trees.

	Lianas				Trees			
Trait	*K*	*p*-value	λ	*p*-value	*K*	*p*-value	λ	*p*-value
A_{max}	0.518	0.157	0.338	0.281	0.308	0.459	0.223	0.503
R_d	**0.874**	0.015	1.000	0.096	0.284	0.573	0.125	0.713
SLA	0.511	0.166	**<0.001**	<0.001	**0.638**	0.001	0.096	0.282

A_{max} = maximum photosynthetic rate; R_d = dark respiration rate; SLA = specific leaf area.
Significant phylogenetic signals are shown in bold.

Phylogenetic information is increasingly used to test macroevolutionary hypotheses of trait evolution [74–76]. The study hypotheses, arising from the macroevolutionary pattern of increased taxonomic diversification in lianas [9], received mixed support. Overall, mean phylogenetic distance among liana species was larger than that of trees. Lianas showed a higher evolutionary rate for a gas-exchange trait (R_d), but the biomass allocation trait (SLA) evolved at a higher rate in trees. Likewise, average trait divergence across the phylogenetic tree was greater in lianas for R_d but it was greater in trees for SLA. Therefore, although we have found support for the expected pattern of increased species divergence in lianas compared to trees, we did not find consistent patterns regarding ecophysiological trait evolution and divergence. R_d followed the species-level patterns, i.e., greater divergence/evolution in lianas compared to trees, while the opposite was found for SLA. R_d may have driven lianas' divergence across forest environments and, furthermore, might contribute to the pattern of increased diversification in climber clades.

Acknowledgments

We thank M. Rivadeneira for methodological advice, A. Saldaña for data acquisition, and Gavin Thomas and 3 anonymous reviewers for thoughtful suggestions that substantially improved an earlier version of this manuscript.

Author Contributions

Conceived and designed the experiments: EG CS-L RSR. Performed the experiments: EG CS-L RSR. Analyzed the data: EG CS-L RSR. Contributed reagents/materials/analysis tools: EG CS-L RSR. Wrote the paper: EG CS-L RSR.

References

1. Gentry AH (1991) The distribution and evolution of climbing plants. In: Putz FE, Mooney HA, eds. The biology of vines. Cambridge: Cambridge University Press.pp 3–49.
2. Schnitzer SA, Bongers F (2002) The ecology of lianas and their role in forests. Trends Ecol Evol 17: 223–230.
3. Gianoli E, Saldaña A, Jiménez-Castillo M, Valladares F (2010) Distribution and abundance of vines along the light gradient in a southern temperate rainforest. J Veg Sci 21: 66–73.
4. Schnitzer SA, Bongers F (2011) Increasing liana abundance and biomass in tropical forests: emerging patterns and putative mechanisms. Ecol Lett 14: 397–406.
5. Phillips OL, Martínez RV, Arroyo L, Baker TR, Killeen T, et al. (2002) Increasing dominance of large lianas in Amazonian forests. Nature 418: 770–774.
6. Durán SM, Gianoli E (2013) Carbon stocks in tropical forests decrease with liana density. Biol Lett 9: 20130301.
7. van der Heijden GMF, Schnitzer SA, Power JS, Phillips OL (2013) Liana impacts on carbon cycling, storage and sequestration in tropical forests. Biotropica 45: 682–692.
8. Gianoli E (2014) Evolutionary implications of the climbing habit in plants. In: Schnitzer SA, Bongers F, Burnham RJ, Putz FE, eds. Ecology of lianas. New York: Wiley-Blackwell. *in press.*
9. Gianoli E (2004) Evolution of a climbing habit promotes diversification in flowering plants. Proc R Soc Lond B 271: 2011–2015.
10. Schnitzer SA (2005) A mechanistic explanation for global patterns of liana abundance and distribution. Am Nat 166: 262–276.
11. Cai Z-Q, Schnitzer SA, Bongers F (2009) Seasonal differences in leaf-level physiology give lianas a competitive advantage over trees in a tropical seasonal forest. Oecologia 161: 25–33.
12. Zhu S-D, Cao K-F (2009) Hydraulic properties and photosynthetic rates in co-occurring lianas and trees in a seasonal tropical rainforest in southwestern China. Plant Ecol 204: 295–304.
13. van der Sande MT, Poorter L, Schnitzer SA, Markesteijn L (2013) Are lianas more drought tolerant than trees? A test for the role of hydraulic architecture and other stem and leaf traits. Oecologia 172: 961–972.
14. Heard SB, Hauser D (1995) Key evolutionary innovations and their ecological mechanisms. Hist Biol 10: 151–173.
15. Gentry AH (1991) Breeding and dispersal systems of lianas. In: Putz FE, Mooney HA, eds. The biology of vines. Cambridge: Cambridge University Press.pp 393–423.
16. Riveros M, Smith-Ramírez C (1995) Patrones de floración y fructificación en bosques del sur de Chile. In: Armesto JJ, Villagrán C, Arroyo MK, eds. Ecología de los bosques nativos de Chile. Santiago: Editorial Universitaria. pp. 235–250.
17. Futuyma DJ, Agrawal AA (2009) Macroevolution and the biological diversity of plants and herbivores. Proc Natl Acad Sci USA 106: 18054–18061.
18. Hunter JP (1998) Key innovations and the ecology of macroevolution. Trends Ecol Evol 13: 31–36.
19. Schluter D (2000) The ecology of adaptive radiation. Oxford: Oxford University Press. 288 p.
20. Funk DJ, Nosil P, Etges WJ (2006) Ecological divergence exhibits consistently positive associations with reproductive isolation across disparate taxa. Proc Natl Acad Sci USA 103: 3209–3213.
21. Ackerly DD, Dudley SA, Sultan SE, Schmitt J, Coleman JS, et al. (2000) The evolution of plant ecophysiological traits: recent advances and future directions. Bioscience 50: 979–995.
22. Wright IJ, Reich PB, Westoby M, Ackerly DD, Baruch Z, et al. (2004) The worldwide leaf economics spectrum. Nature 428: 821–827.
23. Poorter L, Bongers F (2006) Leaf traits are good predictors of plant performance across 53 rain forest species. Ecology 87: 1733–1743
24. Santiago LS, Wright SJ (2007) Leaf functional traits of tropical forest plants in relation to growth form. Functional Ecology 21: 19–27.
25. Valladares F, Niinemets Ü (2008) Shade tolerance, a key plant feature of complex nature and consequences. Annu Rev Ecol Evol Syst 39: 237–257.
26. Gianoli E, Saldaña A, Jiménez-Castillo M (2012) Ecophysiological traits may explain the abundance of climbing plant species across the light gradient in a temperate rainforest. PLoS ONE 7(6): e38831.
27. Lambers H, Chapin FS, Pons TL (1998) Growth and allocation. In: Lambers H, Chapin FS, Pons TL, eds. Plant physiological ecology. New York: Springer. pp. 299–351.
28. Pearcy RW (2007) Responses of plants to heterogeneous light environments. In: Pugnaire FI, Valladares F, eds. Functional plant ecology. Boca Raton: CRC Press. pp. 213–257.
29. Ackerly DD (1999) Comparative plant ecology and the role of phylogenetic information. In: Press MC, Scholes JD, Barker MG, eds. Physiological plant ecology. Oxford: Blackwell Science.pp 391–413.
30. Vamosi SM, Heard SB, Vamosi JC, Webb CO (2009) Emerging patterns in the comparative analysis of phylogenetic community structure. Mol Ecol 18: 572–592.
31. Zheng L, Ives AR, Garland T, Larget BR, Yu Y, et al. (2009) New multivariate tests for phylogenetic signal and trait correlations applied to ecophysiological phenotypes of nine *Manglietia* species. Funct Ecol 23: 1059–1069.
32. Walls RL (2011) Angiosperm leaf vein patterns are linked to leaf functions in a global-scale data set. Am J Bot 98: 244–253.
33. Jacobsen AL, Pratt RB, Tobin MF, Hacke UG, Ewers FW (2012) A global analysis of xylem vessel length in woody plants. Am J Bot 99: 1583–1591.
34. Gallagher RV, Leishman MR (2012) A global analysis of trait variation and evolution in climbing plants. J Biogeogr 39: 1757–1771.
35. O'Meara BC, Ané C, Sanderson MJ, Wainwright PC (2006) Testing for different rates of continuous trait evolution using likelihood. Evolution 60: 922–933.
36. Revell LJ (2012) phytools: an R package for phylogenetic comparative biology (and other things). Methods Ecol Evol 3: 217–223.
37. Webb CO, Ackerly DD, McPeek MA, Donoghue MJ (2002) Phylogenies and community ecology. Annu Rev Ecol Syst 33: 475–505.
38. Blomberg SP, Garland T, Ives AR (2003) Testing for phylogenetic signal in comparative data: behavioral traits are more labile. Evolution 57: 717–745.
39. Webb CO, Donoghue MJ (2005) Phylomatic: tree assembly for applied phylogenetics. Mol Ecol Notes 5: 181–183.
40. Webb CO, Ackerly DD, Kembel SW (2008) Phylocom: software for the analysis of phylogenetic community structure and trait evolution. Bioinformatics 24: 2098–2100.
41. Wikstrom N, Savolainen V, Chase MW (2001) Evolution of the angiosperms: calibrating the family tree. Proc R Soc Lond B 268: 2211–2220.
42. Gastauer M, Meira-Neto JAA (2013) Avoiding inaccuracies in tree calibration and phylogenetic community analysis using Phylocom 4.2. Ecol Inform 15: 85–90.
43. Magallon M, Castillo S (2009) Angiosperm diversification through time. Am J Bot 96: 349–365.
44. R Core Team (2013) R: A language and environment for statistical computing. R Foundation for Statistical Computing, Vienna, Austria. URL http://www.R-project.org/.
45. Nielsen R (2002) Mapping mutations on phylogenies. Syst Biol 51: 729–739.
46. Huelsenbeck JP, Nielsen R, Bollback JP (2003) Stochastic mapping of morphological characters. Syst Biol 52: 131–158.

47. Revell LJ (2013) A comment on the use of stochastic character maps to estimate evolutionary rate variation in a continuously valued trait. Syst Biol 62: 339–345.
48. Kembel SW, Cowan PD, Helmus MR, Cornwell WK, Morlon H, et al. (2010) Picante: R tools for integrating phylogenies and ecology. Bioinformatics 26: 1463–1464.
49. Kraft NJB, Ackerly DD (2010) Functional trait and phylogenetic tests of community assembly across spatial scales in an Amazonian forest. Ecol Monogr 80: 401–422.
50. Kooyman RM, Rossetto M, Sauquet H, Laffan SW (2013) Landscape patterns in rainforest phylogenetic signal: isolated islands of refugia or structured continental distributions? PLoS ONE 8(12): e80685.
51. Verdú M, Pausas JG (2007) Fire drives phylogenetic clustering in Mediterranean Basin woody plant communities. J Ecol 95: 1316–1323.
52. Ackerly DD (2009) Conservatism and diversification of plant functional traits: evolutionary rates versus phylogenetic signal. Proc Natl Acad Sci USA 106: 19699–19706.
53. Pagel M (1999) Inferring the historical patterns of biological evolution. Nature 401: 877–884.
54. Freckleton RP, Harvey PH, Pagel M (2002) Phylogenetic analysis and comparative data: a test and review of evidence. Am Nat 160: 712–726.
55. Revell LJ, Harmon LJ, Collar DC (2008) Phylogenetic signal, evolutionary process, and rate. Syst Biol 57: 591–601.
56. Hansen TF, Martins EF (1996) Translating between microevolutionary process and macroevolutionary patterns: The correlation structure of interspecific data. Evolution 50: 1404–14017.
57. Salgado-Luarte C, Gianoli E (2012) Herbivores modify selection on plant functional traits in a temperate rainforest understory. Am Nat 180: E42-E53.
58. Gianoli E, Saldaña A (2013) Phenotypic selection on leaf functional traits of two congeneric species in a temperate rainforest is consistent with their shade tolerance. Oecologia 173: 13–21.
59. Saldaña A, Lusk CH, Gonzáles WL, Gianoli E (2007) Natural selection on ecophysiological traits of a fern species in a temperate rainforest. Evol Ecol 21: 651–662.
60. Geber MA, Griffen LR (2003) Inheritance and natural selection on functional traits. Int J Plant Sci 164: S21–S42.
61. Evans JR, Poorter H (2001) Photosynthetic acclimation of plants to growth irradiance: the relative importance of specific leaf area and nitrogen partitioning in maximizing carbon gain. Plant Cell Env 24: 755–767.
62. Dodd ME, Silvertown J, Chase MW (1999) Phylogenetic analysis of trait evolution and species diversity variation among angiosperm families. Evolution 53: 732–744.
63. Smith SA, Donoghue MJ (2008) Rates of molecular evolution are linked to life history in flowering plants. Science 322: 86–89.
64. Lanfear R, Ho SY, Davies TJ, Moles AT, Aarssen L, et al. (2013) Taller plants have lower rates of molecular evolution. Nature Comm 4: 1879.
65. Nosil P, Harmon LJ, Seehausen O (2009) Ecological explanations for (incomplete) speciation. Trends Ecol Evol 24: 145–156.
66. Gartner BL, Bullock SH, Mooney HA, Brown VB, Whitbeck JL (1990) Water transport properties of vine and tree stems in a tropical deciduous forest. Am J Bot 77: 742–749.
67. Ewers FW, Fisher JB (1991) Why vines have narrow stems: histological trends in *Bauhinia*. Oecologia 88: 233–237.
68. Münkemüller T, Lavergne S, Bzeznik B, Dray S, Jombart T, et al. (2012) How to measure and test phylogenetic signal. Methods Ecol Evol 3: 743–756.
69. Cornelissen JHC, Werger MJA, Castro-Diez P, van Rheenen JWA, Rowland AP (1997) Foliar nutrients in relation to growth, allocation and leaf traits in seedlings of a wide range of woody plant species and types. Oecologia 111: 460–469.
70. Cai ZQ, Schnitzer SA, Bongers F (2009) Seasonal differences in leaf-level physiology give lianas a competitive advantage over trees in a tropical seasonal forest. Oecologia 161: 25–33.
71. Wyka TP, Oleksyn J, Karolewski P, Schnitzer SA (2013) Phenotypic correlates of the lianescent growth form: a review. Ann Bot 112: 1667–1681.
72. Wright SJ (2002) Plant diversity in tropical forests: a review of mechanisms of species coexistence. Oecologia 130: 1–14.
73. Gilbert B, Wright SJ, Muller-Landau HC, Kitajima K, Hernández A (2006) Life history trade-offs in tropical trees and lianas. Ecology 87: 1281–1288.
74. Mooers AØ, Vamosi SM, Schluter D (1999) Using phylogenies to test macroevolutionary hypotheses of trait evolution in cranes (Gruinae). Am Nat 154: 249–259.
75. Adams DC, Berns CM, Kozak KH, Wiens JJ (2009) Are rates of species diversification correlated with rates of morphological evolution? Proc R Soc B 276: 2729–2738.
76. Magnuson-Ford K, Otto SP (2012) Linking the investigations of character evolution and species diversification. Am Nat 180: 225–245.

17

Effects of Dwarf Mistletoe on Stand Structure of Lodgepole Pine Forests 21-28 Years Post-Mountain Pine Beetle Epidemic in Central Oregon

Michelle C. Agne[1]*, David C. Shaw[1], Travis J. Woolley[1¤], Mónica E. Queijeiro-Bolaños[2]

1 Department of Forest Engineering, Resources, and Management, Oregon State University, Corvallis, Oregon, United States of America, **2** Departamento de Ecología y Recursos Naturales, Facultad de Ciencias, Universidad Nacional Autónoma de México, Mexico City, DF, Mexico

Abstract

Lodgepole pine (*Pinus contorta*) forests are widely distributed throughout North America and are subject to mountain pine beetle (*Dendroctonus ponderosae*) epidemics, which have caused mortality over millions of hectares of mature trees in recent decades. Mountain pine beetle is known to influence stand structure, and has the ability to impact many forest processes. Dwarf mistletoe (*Arceuthobium americanum*) also influences stand structure and occurs frequently in post-mountain pine beetle epidemic lodgepole pine forests. Few studies have incorporated both disturbances simultaneously although they co-occur frequently on the landscape. The aim of this study is to investigate the stand structure of lodgepole pine forests 21–28 years after a mountain pine beetle epidemic with varying levels of dwarf mistletoe infection in the Deschutes National Forest in central Oregon. We compared stand density, stand basal area, canopy volume, proportion of the stand in dominant/codominant, intermediate, and suppressed cohorts, average height and average diameter of each cohort, across the range of dwarf mistletoe ratings to address differences in stand structure. We found strong evidence of a decrease in canopy volume, suppressed cohort height, and dominant/codominant cohort diameter with increasing stand-level dwarf mistletoe rating. There was strong evidence that as dwarf mistletoe rating increases, proportion of the stand in the dominant/codominant cohort decreases while proportion of the stand in the suppressed cohort increases. Structural differences associated with variable dwarf mistletoe severity create heterogeneity in this forest type and may have a significant influence on stand productivity and the resistance and resilience of these stands to future biotic and abiotic disturbances. Our findings show that it is imperative to incorporate dwarf mistletoe when studying stand productivity and ecosystem recovery processes in lodgepole pine forests because of its potential to influence stand structure.

Editor: Mai-He Li, Swiss Federal Research Institute WSL, Switzerland

Funding: The Richard Strachan Fellowship granted through the College of Forestry at Oregon State University provided financial support for this project. The Western Wildland Environmental Threat Assessment Center provided additional funds for data collection. The authors also benefited from the Joint Fire Science project # 09-1-06-17. The funders had no role in study design, data collection and analysis, decision to publish, or preparation of the manuscript.

Competing Interests: The authors have declared that no competing interests exist.

* Email: Michelle.Agne@oregonstate.edu

¤ Current address: The Nature Conservancy, Flagstaff, Arizona, United States of America

Introduction

Lodgepole pine (*Pinus contorta* Dougl. ex Loud.) forests are widely distributed throughout western North America [1] and are subject to widespread mortality by mountain pine beetle (*Dendroctonus ponderosae* Hopkins) throughout their range. Although this bark beetle is native to lodgepole pine forests, it has caused landscape-scale losses of mature lodgepole pine in British Columbia and the Intermountain West in the last two decades, prompting concern regarding ecosystem function following disturbance events of this magnitude [2]. Recent studies have shown that epidemic disturbance by mountain pine beetle influences many ecosystem functions, including carbon sequestration [3], hydrology and nutrient cycling [4], fire hazard [5], and stand regeneration [6]. Stand structure, which affects many of these processes, is also dramatically influenced by mountain pine beetle outbreaks, undergoing rapid changes through time after an epidemic [7], [8].

The effect of mountain pine beetle epidemics on stand structure is highly dependent upon the time since beetle (TSB), as post-epidemic stands go through several structural phases over time [8]. Mountain pine beetle epidemics typically remove the largest cohort of trees and leave the suppressed and intermediate cohorts, as well as trees with low vigor, thin phloem, and dwarf mistletoe [9], [10]. Stand density is greatly decreased immediately post-epidemic and declined by over 50% as compared with pre-epidemic stands during a recent epidemic in Colorado [11], [12]. However, by 20 to 30 years TSB, stand density has been shown to recover to pre-epidemic conditions in some areas [13], and may surpass the density of stands which have not been recently attacked [14]. Similarly, stand basal area is immediately reduced by up to 70% post-epidemic [11], [12], but reaches about 60% of pre-epidemic basal area by 25–30 years TSB [13] and is predicted to recover fully by 80 years TSB in Colorado lodgepole pine stands [15]. Tree size distribution is immediately skewed toward the small size classes post-epidemic with the removal of the large trees which are the preferred mountain pine beetle host in epidemics [9].

Overall stand height and average diameter at breast height (DBH) are also decreased following a mountain pine beetle epidemic [9], [12].

Forest stand structure of lodgepole pine influences hydrologic function [16], biodiversity [17], stand productivity [18], and ecosystem resistance and resilience to insect outbreaks [11], [19], fire [20], fungal infections [21], as well as complex interactions between biotic and abiotic disturbances [22]. Because stand structure drives many forest processes, interest in the influence of mountain pine beetle epidemics on stand structure has recently increased [6], [13]. However, less attention has been focused on the investigation of other factors which might influence the stand structure of post-mountain pine beetle epidemic environments. To fully understand the processes which occur in a mountain pine beetle-disturbed forest, other factors affecting stand structure must be considered. The presence of compound disturbances (*sensu* [23]) in ecosystems necessitates that multiple disturbance effects are accounted for simultaneously, as summing their individual effects may not represent their combined effects. Although complex interactions surrounding disturbance dynamics have begun to be quantified [24], [25], integration of multiple disturbances remains a key gap in ecosystem modeling [26], [27].

Dwarf mistletoes (*Arceuthobium* spp.) are a group of obligate hemiparasites which obtain the majority of their carbohydrates, nutrients, and water from their hosts [28], [29], leading to host growth loss, lowered vigor, and higher susceptibility to mortality when trees are severely infected [30]. Lodgepole pine dwarf mistletoe (*Arceuthobium americanum* Nutt. ex Engelm.), is a widespread pathogen of lodgepole pine, occurring throughout the range of its host [31]. In addition to growth loss, moderate to severe infection by *A. americanum* often induces host deformities by forming dense branch masses called witches brooms (Figure 1), which concentrate biomass in infected branches and act as nutrient sinks [32]. Although not all dwarf mistletoe species induce witches brooms as a symptom of infection, the individual structure of lodgepole pine crowns infected with *A. americanum* is significantly impacted via this mechanism, leading to shorter live crowns, which are skewed to the lower strata of the canopy [33]. Furthermore, decreases in average lodgepole pine diameter and height, particularly within larger size classes, have been observed with increased dwarf mistletoe infection levels [34], [35], [36], [37]. Lodgepole pine dwarf mistletoe also influences several aspects of overall stand structure in severely infected lodgepole pine forests, in addition to individual crown structure. Stand density has been shown to increase five-fold in severely infected stands [36]. The increase in stand density is attributed to a shift in tree size distribution to smaller size classes, with increased densities of suppressed trees and decreased densities of dominant trees in stands with high levels of dwarf mistletoe as compared with stands without dwarf mistletoe [33], [36], [38].

Although lodgepole pine dwarf mistletoe and mountain pine beetle are both known to individually influence stand structure, previous dwarf mistletoe research in lodgepole pine has not taken time since the previous mountain pine beetle epidemic into account [33], [36], [38], [39]. Similarly, previous studies on mountain pine beetle have not accounted for dwarf mistletoe [6], [11], [13], [15]. However, a recent study demonstrated that the interaction between southwestern dwarf mistletoe (*Arceuthobium vaginatum* subsp. *cryptopodum* Engelm.) and mountain pine beetle influences stand structure in ponderosa pine forests in Colorado [40]. This indicates that the interaction between dwarf mistletoe and bark beetles may influence the ecology of other forest types and should be investigated.

Figure 1. Forest structures of lodgepole pine stands 21–28 years after a mountain pine beetle epidemic. Pictured are stands A) without dwarf mistletoe and B) with severe dwarf mistletoe and high levels of witches' brooming.

Lodgepole pine dwarf mistletoe is frequently found in lodgepole pine stands which have recently experienced a mountain pine beetle epidemic. A random sample of 212 lodgepole pine dominated plots in central Oregon 2–31 years post-mountain pine beetle epidemic indicated that 72% of post-epidemic lodgepole pine stands in the area had some level of dwarf mistletoe infection (unpublished data). Light to moderate dwarf mistletoe infection was found in 53% of the plots while 19% of the plots had severe dwarf mistletoe infection. Although there is evidence that dwarf mistletoe influences stand structure, few studies have demonstrated the effect of dwarf mistletoe on stand structure using randomized sites, so inference is typically limited to sites with severe dwarf mistletoe. The large percentage of post-mountain pine beetle epidemic plots with light to moderate dwarf mistletoe infection indicate that understanding the effects of lower infection levels, in addition to the effects of high severity infections, is important to understanding its influences on stand dynamics. Mountain pine beetle and lodgepole pine dwarf mistletoe co-occur with high frequency, so the response of stand structure to both mountain pine beetle and dwarf mistletoe must be considered simultaneously to understand structural effects occurring on the landscape.

We chose to specifically investigate the interaction between mountain pine beetle and lodgepole pine dwarf mistletoe in stands 21–28 years TSB. We chose this time period because central

Oregon experienced a mountain pine beetle epidemic in the 1980's spatially analogous to that which is currently occurring in British Columbia and the Intermountain West. Therefore, the results from this study could provide valuable insight to future stand structure in other areas post-mountain pine beetle epidemic. Differences exist throughout the range of this forest type, as in central Oregon the lodgepole pine stands often exist as climax, uneven aged stands [41], while in most other regions lodgepole pine is an early successional species that grows primarily in even-aged cohorts [1]. However, mountain pine beetle rarely causes 100% mortality of mature lodgepole pine at the stand level [42] and advance lodgepole pine regeneration is often a large component of the understory, even in seral stands [43]. Uneven stand structure and significant amounts of lodgepole pine regeneration have been noted in British Columbia [44], [45], Colorado [11], [14], Wyoming [6], Idaho, Utah, and Montana [43] lodgepole pine stands after a mountain pine beetle epidemic. The results from this study may inform future conditions in uneven-aged lodgepole pine stands in these areas that have recently experienced widespread mortality.

To address the interaction between dwarf mistletoe and mountain pine beetle we asked: "How does stand structure of lodgepole pine forests 21–28 years post-mountain pine beetle epidemic change with varying levels of dwarf mistletoe infection?" We identified several important metrics for assessment of stand structure: stand density, stand basal area, canopy volume, proportion of lodgepole pine in dominant/codominant, intermediate, and suppressed cohorts, and average height and DBH of trees within cohorts. Each of these characteristics is affected by either dwarf mistletoe, mountain pine beetle, or both of these disturbance agents. We hypothesized that in lodgepole pine stands 21–28 years TSB: 1) stand density increases with increasing dwarf mistletoe severity, 2) stand basal area and canopy volume decrease as dwarf mistletoe severity increases, 3) as dwarf mistletoe severity increases, larger proportions of lodgepole pine are represented by cohorts of suppressed trees and smaller proportions of lodgepole pine are represented by cohorts of dominant and codominant trees, and 4) average height and diameter of all cohorts decreases as dwarf mistletoe severity increases.

Materials and Methods

Study Area

The study area for this research is located in central Oregon within the Deschutes National Forest. The Deschutes National Forest is located on the east side of the Cascade Mountains, covering an area of approximately 728,000 hectares (Figure 2). Stands were chosen within the edaphic and topoedaphic climax lodgepole pine zones according to the plant association guide for the area [41], [46]. In this area, the ecological site characteristics of the climax lodgepole pine type are relatively uniform, characterized by pumice soils and flat to gently rolling topography which often results in cold air drainage [47]. The lodgepole pine zone is located between 1,200 and 1,525 meters elevation with mean annual temperatures ranging from 4.3 to 5.8°C and mean annual precipitation ranging from 38 to 89 cm depending upon the specific plant association [41]. The Wickiup Dam climate station (the most representative climate station for the study area) showed average daily temperatures ranging from −2.2°C in January to 18.3°C in July [48].

Measured stands were restricted to areas 21–28 years post mountain pine beetle epidemic to control for the effect of mountain pine beetle on stand structure. The year of initiation of each mountain pine beetle mortality event was determined using Aerial Detection Survey (ADS) data [49]. Areas with known past management or recent fire were excluded from sampling. Stands were characterized by large amounts of coarse wood, few standing snags, and dense lodgepole pine regeneration. The overstory was typically comprised of lodgepole pine too small to support a mountain pine beetle brood at the time of the previous epidemic that subsequently released after the mortality of the majority of the previous overstory.

A minimum of 70% of live trees at each plot measured were lodgepole pine. Other tree species found at the sites varied with elevation. Ponderosa pine (*Pinus ponderosa* Dougl. ex Laws.) was present at lower elevations, while white fir (*Abies concolor* (Gord. and Glend.) Lindl. ex Hildebr.), grand fir (*Abies grandis* (Dougl. ex D. Don.), mountain hemlock (*Tsuga mertensiana* (Bong.) Carrière), and whitebark pine (*Pinus albicaulis* Engelm.) were located at higher elevations. Engelmann spruce (*Picea engelmanni* Parry ex Engelm.) and western white pine (*Pinus monticola* Dougl. ex D. Don) were occasionally present within the study area as well.

Stand Selection

Stands were selected based on a network of 119 plots established in 2010 and 2011 within post-mountain pine beetle epidemic climax lodgepole pine, 2 to 31 years TSB, in the Deschutes National Forest (unpublished data). The network of plots was designed using a spatially balanced random sampling design [50] with the purpose of broadly characterizing change in stand structure and fuels across lodgepole pine forests after a mountain pine beetle mortality event in central Oregon. However, individual stands were not intensively sampled. Therefore, stand polygons were drawn in ArcGIS 9.3 [51] around the 38 original plots which were 21 to 28 years TSB to more intensively sample the areas in which the original plots were located. Because some stands contained more than one of the original plots, or were not large enough to accommodate our sampling design, a total of 26 stands were available for sampling. Stand boundaries were drawn based on the presence of a climax lodgepole pine plant association, ADS data regarding the most recent mountain pine beetle epidemic, and GIS layers from the Deschutes National Forest regarding past management activities, to ensure that the stand polygons were ecologically consistent with the original plot.

Within each polygon, three GPS points were selected as beginning points for the plots using a spatially balanced random sampling design [50]. A random azimuth used for orientation of the plot was generated for each point. Each point was checked in the field to ensure that the associated plot was dominated by lodgepole pine, had past influence of mountain pine beetle, and had no sign of past management or recent fire. If any of these criteria were not met for a given point, a randomly selected replacement GPS point within the site boundary was used in its place.

Plot Layout and Sampling Protocol

During the summer of 2012, a total of 13 stands were randomly selected from the 26 stands available for sampling. We established three 75 m×10 m (0.075 ha) belt transects randomly located and oriented within each of the 13 stands, for a total of 39 plots (Table S1). This layout was chosen in order to obtain spatially explicit canopy data [52]. Upon establishment of each plot, UTM coordinates were taken at each end of the belt transect using a Trimble unit. From these points, slope and aspect were recorded. Species, vigor rating, crown class (dominant, codominant, intermediate, or suppressed), an X, Y coordinate, and DBH were determined for all live trees (DBH >/ = 5.0 cm) within the plots. In addition, tree height (to highest live crown), height to crown

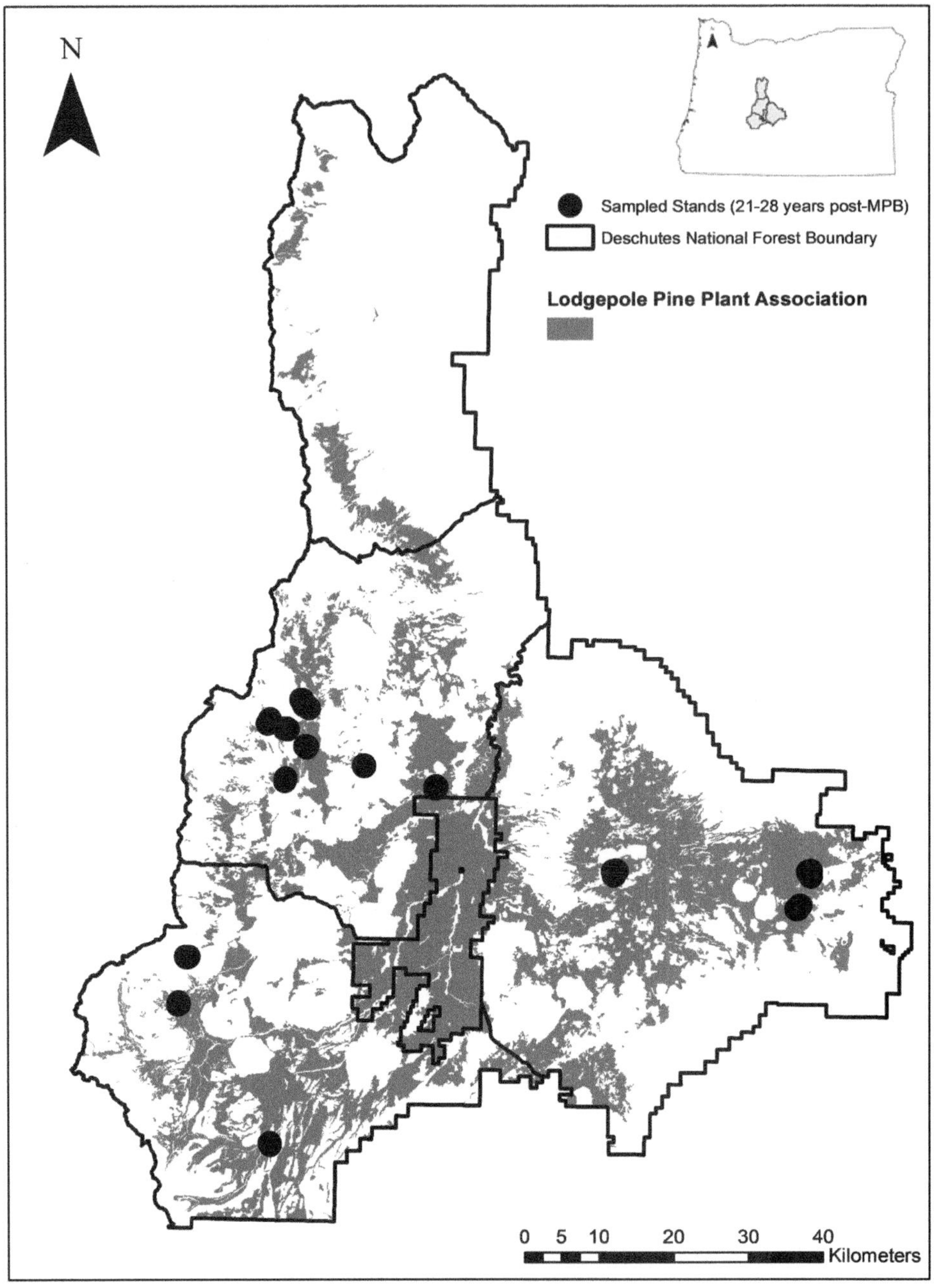

Figure 2. Study area map. Deschutes National Forest, Oregon boundary with sampled stands and lodgepole pine plant associations.

base (defined as the lowest live foliage), and crown width were measured to the nearest 0.1 meters.

Each live tree (DBH≥5 cm) was given a dwarf mistletoe severity rating using the Hawksworth Six-Class Dwarf Mistletoe Rating (DMR) System [53]. This system is based upon a rating of the number of branches visibly infected by dwarf mistletoe within each third of the tree's live crown. Scores range from 0 (no visible infections) to 6 (50% or more of the branches in each third of the tree have visible infections). Brooms influence DMR estimations [54], so we observed crowns with binoculars and based DMR ratings on the presence of dwarf mistletoe plants themselves, rather than associated symptoms such as witches' brooms. Tree crowns are open and clearly visible in this forest type, providing a high level of confidence in the accuracy of our DMR estimations. DMR of all lodgepole pine were averaged over the plot to obtain a plot-level DMR. Hereafter, DMR refers to dwarf mistletoe severity rating at the plot-level, rather than the individual tree-level.

Ethics Statement

No permits were required to complete the field sampling for this study, as all sites were located on public Forest Service land, managers were informed of the research, and no destructive sampling was conducted. No protected species were sampled for this study.

Stand Structure Metric Calculations

Stand density is defined as the total number of stems DBH≥ 5.0 cm/ha, calculated at the plot-level (Table 1). Stand basal area

Table 1. Characteristics of 39 lodgepole pine plots within 13 stands in the Deschutes National Forest, Oregon.

Stand	Plot	DMR	Productivity Class	MPB Mortality Class	Stand Density (stems/ha)	Stand Basal Area (m^2)	Canopy Volume (m^3)
CRL	1	2.78	L	L	1213	11.94	409.9
	2	2.93	L	L	680	6.57	266.4
	3	2.40	L	L	600	15.02	539.1
CRP	1	0	H	H	1120	21.29	1129.3
	2	1.02	H	H	880	16.55	1235.3
	3	0	H	H	1107	23.87	608.0
CRP2	1	1.81	H	L	1107	14.58	337.0
	2	2.45	H	L	893	14.99	313.9
	3	2.94	H	L	1187	15.26	265.9
CUL2	1	2.42	H	L	973	14.67	329.4
	2	2.35	H	M	1040	11.08	252.3
	3	2.38	H	H	827	7.70	316.7
CUL6	1	3.98	M	H	987	12.78	696.9
	2	3.70	M	L	1000	10.67	431.7
	3	2.59	M	L	1760	22.12	931.2
DES	1	0	M	M	600	19.91	1347.1
	2	0	M	M	720	16.18	1216.4
	3	0	M	M	493	15.00	1422.8
EFR	1	0	L	M	600	18.94	764.4
	2	0	L	M	387	19.25	1185.0
	3	0	L	M	893	23.04	640.0
EFR3	1	0	L	H	613	16.00	917.3
	2	0	L	L	373	14.09	905.8
	3	0	L	M	2053	34.09	734.2
LDES	1	0.54	M	M	680	9.64	332.6
	2	0	M	M	1053	21.78	1305.7
	3	0.73	M	M	1013	18.29	813.1
LVLK	1	1.11	H	H	880	7.70	282.4
	2	2.09	H	H	1227	15.08	934.9
	3	0	H	H	1093	7.84	341.3
ODL	1	2.50	L	L	827	16.53	390.5
	2	1.85	L	L	987	12.27	396.4
	3	2.88	L	L	800	11.70	387.7
PAU	1	2.36	H	M	1733	21.87	602.3
	2	2.38	H	H	2493	26.03	609.8

Table 1. Cont.

Stand	Plot	DMR	Productivity Class	MPB Mortality Class	Stand Density (stems/ha)	Stand Basal Area (m^2)	Canopy Volume (m^3)
	3	1.80	H	H	1560	23.48	664.9
SNC	1	1.40	H	H	533	8.02	277.4
	2	1.26	H	H	413	10.51	376.7
	3	0.16	H	H	493	12.99	590.6

Note: Data are calculated from measures of trees DBH >/ = 5.0 cm. DMR = stand-level dwarf mistletoe rating; MPB = mountain pine beetle. Productivity class was determined by previously developed plant associations for the area. MPB Mortality class was determined using Aerial Detection Survey cumulative mortality data where L = 5–15 trees/acre, M = 15–25 trees/acre, H = 25–36 trees/acre killed by mountain pine beetle over the time period of the epidemic.

is defined as the sum of live tree basal area of lodgepole pine DBH>5.0 cm (m^2/ha) in each plot using the formula: 3.142*(DBH/200)^2. Crown volume (m^3) measurements were calculated for live trees (DBH≥5.0 cm) using measurements of crown length, width, and height taken in the field. A crown form factor (CFF) was then applied to each volume to simulate the shape of a lodgepole pine crown [55]. The idealized crown shape "fat cone" (CFF: 0.2945) was used for dense stands (>1000 stems/ha), while the idealized crown shape "paraboloid" (CFF: 0.3927) was used for moderate and open stands (<1000 stems/ha). Individual crown volumes were summed over each plot to obtain total canopy volume (m^3).

Cohorts were defined by tree crown class assigned in the field. Three cohorts were identified: dominant/codominant, intermediate, and suppressed. Very few trees in each plot were classified as "dominant," and these trees were often not much taller than the trees that were classified as "codominant." This lack of distinction suggested that it was not appropriate to refer to dominant and codominant as separate cohorts. Hereafter, they will be grouped as a single cohort. For each cohort, we calculated the proportion of total lodgepole pine (DBH≥5.0 cm) represented by that cohort (proportion in cohort), the average height of the cohort (cohort height), and the average DBH of the cohort (cohort diameter) at the plot-scale (Table S2). We included only lodgepole pine in our cohorts because we were interested in the response of each cohort to dwarf mistletoe rating. Although ponderosa pine, whitebark pine, and Engelmann spruce are known occasional hosts of *A. americanum* [30], we never observed infection of occasional hosts in our plots and consider them non-hosts for the purposes of this study.

Model Selection and Data Analysis

To describe the responses of stand structure parameters to DMR, we used linear mixed models (LMMs) and generalized linear mixed models (GLMMs) in which the response and predictor variables were continuous. Plots were nested within stands to account for potentially high levels of within-stand structural variability. Several covariates were identified as potentially influential in the responses of various aspects of stand structure to DMR. Stand density, site productivity, and mountain pine beetle mortality were all determined to be potentially influential to stand structure, and could have the ability to mask an effect of DMR if not accounted for in the model. Because previous findings have indicated there is a relationship between stand density and DMR [36], stand density could only be used as a covariate if there was no significant relationship between stand density and DMR in our data. We found no evidence of this relationship; therefore, stand density was accounted for as a continuous covariate in our models. Plots were assigned to a site productivity category (low, moderate, or high) using plant association data [46] (Table 1). Plots were also assigned to a mountain pine beetle mortality category (low, moderate, or high) based upon total mountain pine beetle mortality density mapped from ADS from 1979 to 2008 (Table 1). These covariates were assessed for multicollinearity with DMR prior to model fitting.

To ensure that our final fitted models adequately captured ecological relationships of interest while retaining maximum parsimony, we used Bayesian Information Criterion (BIC) to select the most appropriate model(s) from a set of candidate models (Tables S3–S14). Models were fitted using the maximum likelihood method for each response [56]. A candidate model with the lowest BIC value (ΔBIC of 0) was considered to be the most appropriate model, however models with ΔBIC values of less than two were considered to perform equally well. In situations with

Table 2. BIC preferred linear mixed models of stand characteristics.

Response variable	Model with parameter estimates (*SE*)	BIC	Marginal r^2	Conditional r^2	p value	
Stand density	Log(SD) = 6.652 (*0.128*) +0.094 * DMR (*0.066*)	−31.95	0.07	0.39	0.1686	
Stand basal area	SBA = 1.355 (*0.130*) −0.117 * DMR (*0.067*)	53.52	0.1	0.41	0.094	.
Canopy volume	Log(CV) = 6.615 (*0.146*) −0.195 * DMR (*0.074*)	59.11	0.21	0.53	0.0146	*
Proportion dominant	logit(PD) = −0.448 (*0.167*) −0.242 * DMR (*0.089*)	283.08	0.03	0.10	0.0153	*
Proportion intermediate	logit(PI) = −0.899 (*0.146*) −0.018 * SD (*0.011*)	238.02	0.01	0.02	0.1097	
Proportion suppressed	logit(PS) = −1.159 (*0.168*) +0.318 * DMR (*0.087*)	290.06	0.04	0.14	0.0022	**
Cohort height of dominants	CHD = 20.19 (*0.881*) −0.476 * DMR (*0.317*) −0.002 * SD (*0.001*)	175.34	0.31	0.56	0.0501	.
Cohort height of intermediates	CHI = 11.154 (*0.620*) −0.663 * DMR (*0.322*)	179.23	0.13	0.37	0.0501	.
Cohort height of suppressed	CHS = 5.676 (*0.295*) −0.450 * DMR (*0.150*)	113.39	0.26	0.57	0.0061	**
Cohort diameter of dominants	Log(CDD) = 3.37 (*0.055*) −0.0428 * DMR (*0.020*) −0.0002 * SD (*0.00004*)	−39.51	0.42	0.59	0.0074	**
Cohort diameter of intermediates	Log(CDI) = 2.76 (*0.050*) −0.0002 * SD (*0.00004*)	−44.62	0.3	0.56	0.0005	**
Cohort diameter of suppressed	Log(CDS) = 2.06 (*0.048*) −0.015 * DMR (*0.024*)	−33.95	0.02	0.47	0.5235	

Note: Log(SD) = mean of the natural logarithm of stand density; SBA = mean of stand basal area; Log(CV) = mean of the natural logarithm of canopy volume; logit(PD) = log odds that a lodgepole pine is in the dominant/codominant cohort; logit(PI) = log odds that a lodgepole pine is in the intermediate cohort; logit(PS) = log odds that a lodgepole pine is in the suppressed cohort; CHD = mean of the cohort height of dominants; CHI = mean of the cohort height of intermediates; CHS = mean of the cohort height of suppressed; CDD = mean of the natural logarithm of cohort diameter of dominants; CDI = mean of the natural logarithm of cohort diameter of intermediates; CDS = mean of the natural logarithm of cohort diameter of suppressed; DMR = stand level dwarf mistletoe rating; SD = stand density; BIC = Bayesian Information Criterion; ** = $p<0.01$; * = $p<0.05$; . = $p<0.1$.

more than one preferred model ($\Delta BIC<2$), we chose to interpret the preferred model which included DMR as a predictor variable as the primary interest of this investigation is in the impact of dwarf mistletoe on stand structure. BIC weights and evidence ratios were also calculated for each set of candidate models to further assess the weight of evidence for each model in the set [56] (Tables S3–S14).

Assumptions of linearity, homoscedasticity and normality were assessed for all candidate models using standard diagnostics prior to model selection [57]. We logarithmically transformed the response variables canopy volume, cohort diameter of dominant/codominant, cohort diameter of intermediates, and cohort diameter of suppressed to correct for heteroscedasticity. We used LMMs to model the relationship between DMR and stand structure variables when these assumptions were met and refitted the preferred candidate models using the residual maximum likelihood method for final inference. The normality assumption is not met for proportion data, so we used binomial GLMMs to model the response of the proportion in cohort data. We assessed these models for overdispersion prior to model selection. We corrected for overdispersion, when necessary, by adding an individual-level random effect to the model [58]. We performed likelihood ratio tests to determine that the coefficients for the explanatory variables were different from zero.

Models with p-values below an α-level of 0.05 were interpreted to have strong evidence of a relationship. Models with $p<0.10$ were interpreted to have suggestive but inconclusive evidence of a relationship to lower the probability of making a Type II error given our sample size and the inherent variability of this study area. We report 95% confidence intervals for means estimated for LMMs and profile 95% confidence intervals for odds ratios estimated for GLMMs. We calculated marginal and conditional R^2 values to describe goodness of fit for linear mixed models and generalized linear mixed models [59]. The marginal R^2 represents variance explained by fixed factors alone and conditional R^2 represents the variance explained by fixed and random factors. Both measures were calculated to understand the fit of these models. All analyses were performed using the program R, version 2.12.0 [60].

Results

Stand Attributes

For the each of the responses of stand density, stand basal area, and canopy volume, BIC showed that a model including the single continuous predictor variable of DMR was preferred (Tables S3, S4, S5). Although this model was selected for stand density, there was no evidence of a difference in stand density over the range of DMR ($F_{1,25} = 2.01$, $p = 0.1686$) (Table 2, Figure 3). This finding allowed us to consider stand density as a covariate in subsequent analyses. There was suggestive but inconclusive evidence that stand basal area decreased with increasing DMR ($F_{1,25} = 3.04$, $p = 0.094$) (Table 2, Figure 3). There was strong evidence that the natural logarithm of canopy volume decreased with increasing DMR in these stands ($F_{1,25} = 6.890$, $p = 0.0146$) (Table 2). There was an estimated 17.8% (95% CI: 4.1%, 29.4%) decrease in the median canopy volume for each unit increase in DMR (Figure 3).

Cohort Attributes

The models for the proportion of lodgepole pine in the dominant/codominant cohort and the proportion of lodgepole pine in the suppressed cohort used DMR as their only predictor in the BIC preferred models (Tables S6, S7). There was strong evidence that the proportion of lodgepole pine in the dominant/codominant cohort decreased with DMR severity ($X^2_1 = 5.88$, $p = 0.0153$) while the proportion of lodgepole pine in the suppressed cohort increased with DMR severity ($X^2_1 = 9.35$, $p = 0.0022$) (Table 2, Figure 4). Each doubling of plot-level DMR was associated with a decrease in the odds of a given lodgepole pine tree being in the dominant/codominant cohort by

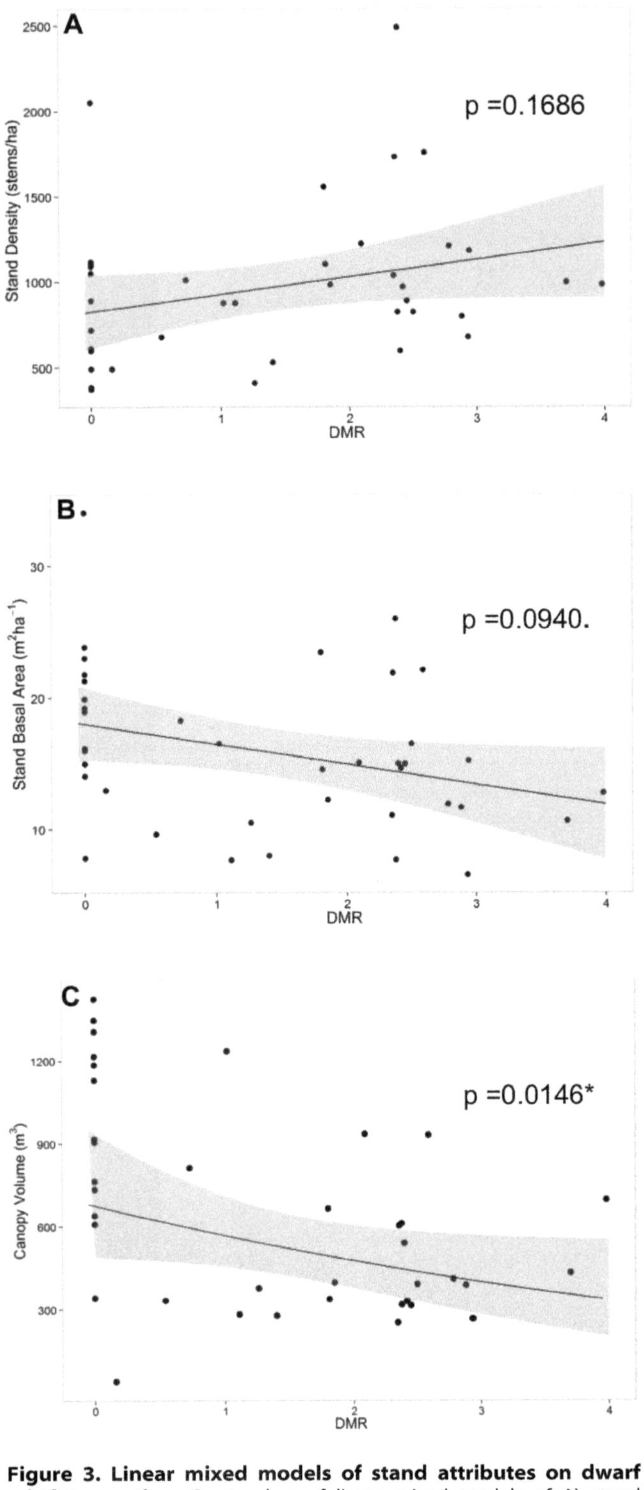

Figure 3. Linear mixed models of stand attributes on dwarf mistletoe rating. Scatterplots of linear mixed models of A) stand density (stems/ha), B) stand basal area (m^2/ha), and C) backtransformed natural logarithm of canopy volume (m^3) on dwarf mistletoe rating (DMR) with 95% confidence intervals.

an estimated 15.4% (95% CI: 5.3, 42.2%). Conversely, each doubling of plot-level DMR was associated with an increase in the odds of a lodgepole pine tree being in the suppressed cohort by an estimated 24.6% (95% CI: 14.3, 64.6%). The proportion of lodgepole pine in the intermediate cohort used stand density as its only predictor variable in the BIC preferred model (Table S8). However, there was no evidence of a significant relationship in the preferred model ($X^2_1 = 2.56$, $p = 0.1097$) (Table 2), nor was there evidence of a significant relationship between the proportion of the stand in the intermediate cohort with DMR ($X^2_1 = 0.46$, $p = 0.4970$) (Figure 4).

Cohort height of intermediates and cohort height of suppressed trees were both best predicted by DMR alone (Tables S9, S10), while the preferred model for cohort height of dominant/codominants included both DMR and stand density as predictors (Table S11). There was suggestive evidence that cohort height of dominant/codominants decreased with DMR after accounting for stand density (Table 2). There was an estimated 0.48 meter decrease (95% CI: 1.13 meter decrease, 0.18 meter increase) in mean dominant/codominant cohort height with each unit increase in DMR, holding stand density at its mean ($F_{1,24} = 4.257$, $p = 0.0501$) (Figure 5). Evidence of a relationship of cohort height of intermediates to DMR was also suggestive ($F_{1,25} = 4.24$, $p = 0.0501$) (Table 2). There was an estimated 0.66 (95% CI: 0, 1.33) meter decrease in the mean cohort height of intermediate lodgepole pine for each unit increase in DMR (Figure 5). However, there was strong evidence of an effect of DMR on cohort height of suppressed trees ($F_{1,25} = 8.975$, $p = 0.0061$) (Table 2). There was an estimated 0.45 (95% CI: 0.14, 0.76) meter decrease in the mean cohort height of suppressed trees for each unit increase in DMR (Figure 5).

Conversely, there was no evidence of an effect of DMR on the natural logarithm of cohort diameter of suppressed trees ($F_{1,25} = 0.419$, $p = 0.5235$) (Table 2, Figure 6). Although other models were equally preferred by BIC, no significant relationships were found within the set of candidate models (Table S12). However, the BIC preferred model (Table S13) showed there was strong evidence of an effect of DMR on the natural logarithm of cohort diameter of dominant/codominants after accounting for stand density ($F_{1,24} = 8.563$, $p = 0.0074$) (Table 2). There was an estimated 4.2% (95% CI: 0.3%, 8.0%) decrease in the median cohort diameter of dominant/codominants for each unit increase in DMR after accounting for stand density (Figure 6). The BIC indicated that the natural logarithm of cohort diameter of intermediate trees was best explained by stand density alone ($F_{1,25} = 16.212$, $p = 0.0005$) (Table 2, Figure 6), with no evidence of a significant effect of DMR (Table S14).

Discussion

Effects of Dwarf Mistletoe on Stand Structure and Cohort Distribution

Dwarf mistletoe is associated with reduced canopy volume, greater representation of the suppressed cohort, reduced representation and average diameter of the dominant/codominant cohort, and reduced average height of the suppressed cohort in lodgepole pine forests. This parasitic plant has a profound influence on the structure of lodgepole pine forests in central Oregon. *Arceuthobium americanum* is common throughout the area, so it is apparent that mountain pine beetle does not eradicate it by killing the host trees. Conversely, it appears that mountain pine beetle activity may exacerbate the stand-level infection by removing the largest trees in the stand and leaving the suppressed, dwarf mistletoe infected trees as described in previous research [10]. Based on our understanding of the epidemiology of *A. americanum* [30], we conclude that dwarf mistletoe then spreads onto the remaining uninfected trees and new regeneration in the stand with increased light availability and space in the mid and

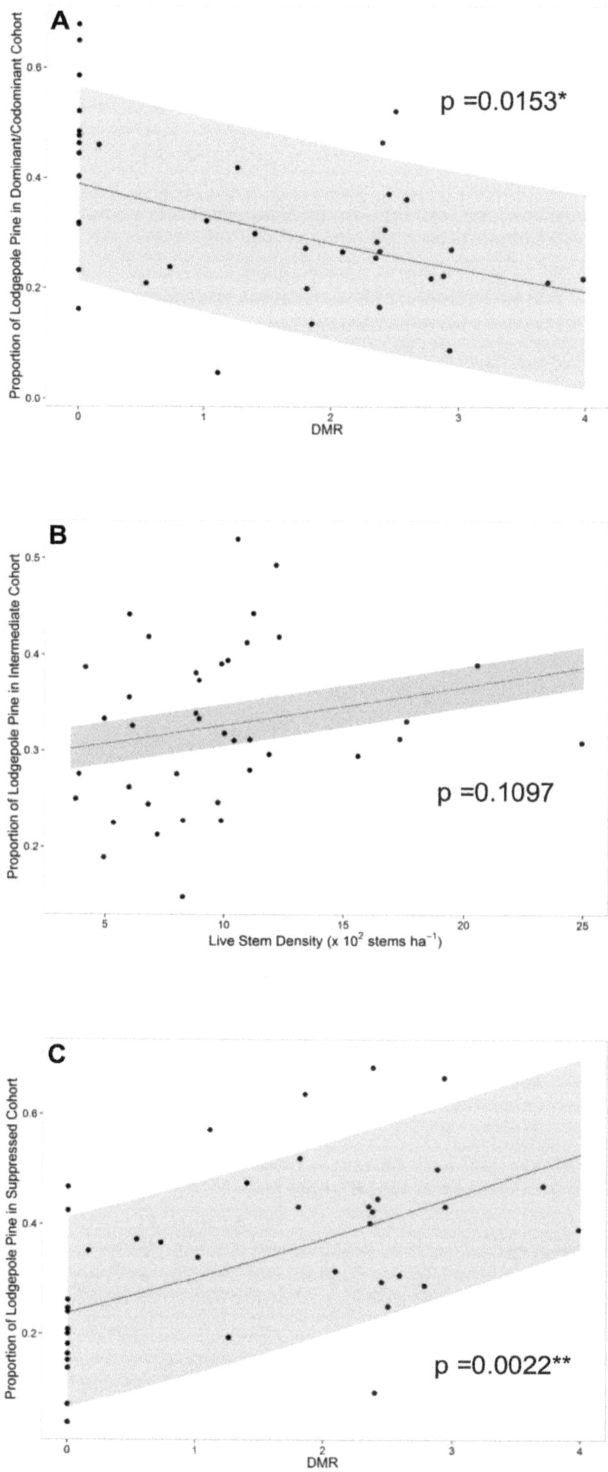

Figure 4. Generalized linear mixed models of proportion of lodgepole pine in cohorts on BIC preferred explanatory variables. Scatterplots of generalized linear mixed models of A) proportion of lodgepole pine in the dominant cohort on dwarf mistletoe rating (DMR), B) proportion of lodgepole pine in the intermediate cohort on stand density and C) proportion of lodgepole pine in the suppressed cohort on DMR with 95% confidence intervals.

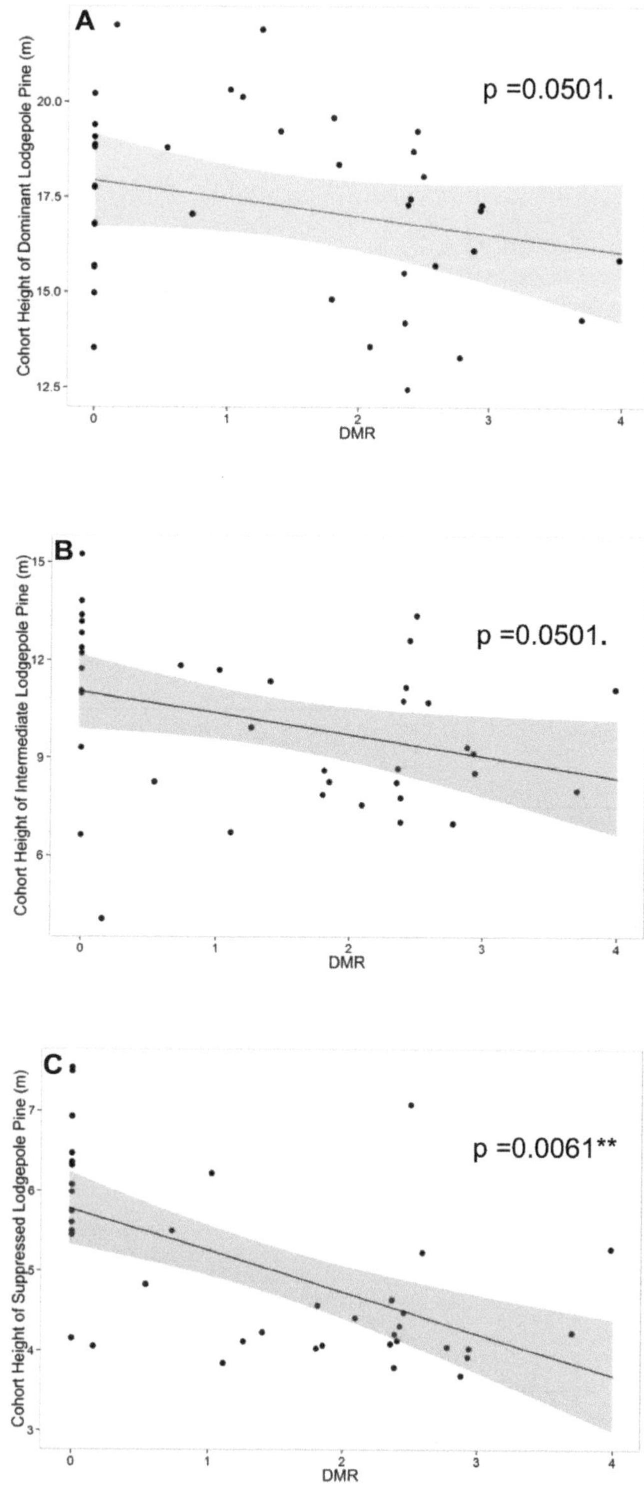

Figure 5. Linear mixed models of cohort height on dwarf mistletoe rating. Scatterplots of linear mixed models of A) cohort height of dominant lodgepole pine (m) with stand density fixed at its mean, B) cohort height of intermediate lodgepole pine (m), and C) cohort height of suppressed lodgepole pine (m) on dwarf mistletoe rating (DMR) with 95% confidence intervals.

lower canopy due to the removal of large trees [61], thereby reducing overall productivity and growth. Although this may be thought of as a negative influence dwarf mistletoe has other ecosystem influences in many forest types including lodgepole pine forests. Dwarf mistletoes provide both food (from the dwarf mistletoe plants themselves) and nesting resources (from the

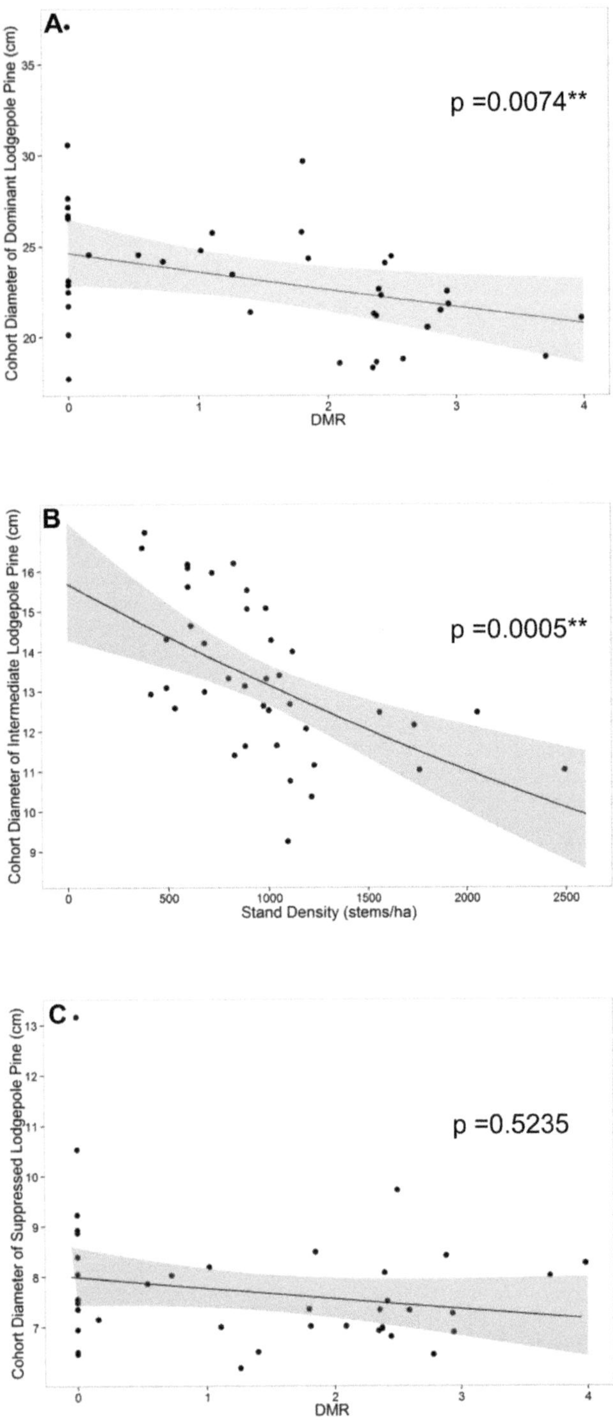

Figure 6. Linear mixed models of log cohort diameter on BIC preferred explanatory variables. Scatterplots of backtransformed linear mixed models of A) natural logarithm of cohort diameter of dominant lodgepole pine (cm) on dwarf mistletoe rating (DMR) with stand density fixed at its mean, B) natural logarithm of cohort diameter of intermediate lodgepole pine (cm) on stand density and C) natural logarithm of cohort diameter of suppressed lodgepole pine on DMR with 95% confidence intervals.

witches' brooms induced by many dwarf mistletoe species) for numerous bird, mammal, and insect species [30], [62]. The multiple effects of dwarf mistletoes on forest stands indicate that the overall effect of these organisms in a forest is complex and may be viewed as either negative or positive depending on management objectives [63].

The proportion of lodgepole pine in the dominant/codominant cohort significantly decreased as dwarf mistletoe severity increased in plots 21–28 years TSB. Conversely, the proportion of lodgepole pine in the suppressed cohort significantly increased with increased dwarf mistletoe severity. Both findings are consistent with previous work in dwarf mistletoe infected lodgepole pine [34], [36] but our findings are more broadly applicable due to our randomly selected sites and intensive sampling of stands in our study area. The shift in proportion of trees in each cohort indicates that high levels of dwarf mistletoe may keep residual trees in the suppressed cohort after a mountain pine beetle mortality event longer than in stands little or no dwarf mistletoe. However, this shift associated with increased DMR was not reflected in the proportion of trees in the intermediate cohort. This may be attributed to an equivalent proportional shift of both the dominant/codominant cohort and the intermediate cohort to the next smallest size class, reflecting zero net change of the proportion of the stand in the intermediate cohort.

The reduction in cohort height of suppressed trees and the reduced diameter of dominant/codominant trees is consistent with previous work at the individual tree-scale [34], [35], [36], but our findings suggest that this process is also happening at a stand-scale. Increased dwarf mistletoe infection in a stand is associated with increased representation by the suppressed cohort as well as decreased height within all cohorts and decreased diameter within the dominant/codominant cohort. This suggests that the presence of dwarf mistletoe has the ability to slow stand recovery after a mountain beetle epidemic as compared with that of uninfected stands. Although the presence of dwarf mistletoe likely alters the trajectory of stand development, published studies regarding post-mountain pine beetle epidemic stand structure and ecosystem recovery have not accounted for its effects [6], [11], [15]. Dwarf mistletoe is impacting overall stand development post-mountain pine beetle outbreak and its effects should be incorporated to accurately project recovery of stands experiencing mortality.

Total canopy volume indicates the relative amount of available space that is occupied by any given forest. Our data supports the contention that dwarf mistletoe slows the recolonization of available space in these low diversity lodgepole pine ecosystems in the decades following mountain pine beetle mortality. Previous research in central Oregon lodgepole pine showed that the volume of individual crowns were reduced in lodgepole pine trees with increased DMR [39], but sampling occurred in stands in which total canopy volume did not change with infection level [64]. Our results may differ due to our random sampling method and control of previous mountain pine beetle activity, which has not been accounted for in previous work.

Previous work on dwarf mistletoe in central Oregon lodgepole pine has found various results with respect to stand basal area and density, which were not influenced by DMR in this study. One study found no relationship between stand basal area and DMR [36], while another study found a significant reduction in stand basal area associated with increased dwarf mistletoe severity [33]. Both studies found that stand density increases with DMR. The inconsistency of results is likely a result of a disparity in mountain pine beetle legacy, as previous studies of dwarf mistletoe effects did not account for this factor. Previous research has shown that stand basal area and stand density change significantly over time after a mountain pine beetle mortality event [11], [12], [13], [15], which may confound the structural effects of other disturbances. The discrepancy in the understanding of the relationship between dwarf mistletoe and these stand metrics in lodgepole pine forests indicates that interpretation of these relationships must be made within the context of the time since the previous mountain pine beetle epidemic.

Potential Effects of Structural Heterogeneity Associated with Dwarf Mistletoe

Lodgepole pine dwarf mistletoe's influences on canopy volume, proportion of the stand in each cohort, and average diameter and height within cohorts show that its presence at various severities on the landscape introduces structural heterogeneity to lodgepole pine forests. Although dwarf mistletoe decreases vigor of individual trees [30], heterogeneity of stand structure leads to higher overall landscape resistance and resilience to various disturbances in many systems [11], [19], [27]. Diverse structure introduced by dwarf mistletoe may actually increase landscape resistance and resilience to disturbances, such as mountain pine beetle epidemics. To reach epidemic populations, mountain pine beetle needs densely stocked dominant and codominant lodgepole pine in which to lay their brood [9]. These conditions are more likely to be found in stands uninfected or lightly infected with dwarf mistletoe, given our results of increased proportion of the suppressed cohort, as well as decreased diameter of the dominant/codominant cohort in stands with severe dwarf mistletoe. Although this study did not directly address landscape-level processes, we hypothesize that heterogeneity of stand structure associated with dwarf mistletoe may influence the pattern and extent of mountain pine beetle mortality on the landscape. We hypothesize that dwarf mistletoe may create patches of mountain pine beetle habitat of varying suitability at a given time. The presence of dwarf mistletoe in lodgepole pine forests post-mountain pine beetle should be addressed at a landscape scape to better understand this relationship.

Furthermore, the relationship between decreased vigor at the individual tree scale associated with dwarf mistletoe and mountain pine beetle susceptibility is poorly understood. Several studies have suggested that during the early phase of an epidemic, trees which have low vigor may be preferred by mountain pine beetle due to lowered defense capabilities [65], [66]. Others have shown that more successful mountain pine beetle brood production occurs in trees with thick phloem [10], [67] which is related to strong tree vigor [68]. This has led to the hypothesis that dwarf mistletoe-infected trees are less susceptible to mountain pine beetle attack because they have thinner phloem than trees without dwarf mistletoe [69], [70], [71]. However, the evidence that dwarf mistletoe infection decreases phloem thickness in lodgepole pine is inconclusive. One study found that infected lodgepole pine have significantly lower phloem thickness than uninfected trees, and concluded that there was a negative relationship between dwarf mistletoe infection and mountain pine beetle brood production [10]. Conversely, another study found no relationship between dwarf mistletoe infection and phloem thickness in lodgepole pine [71]. It is possible that the presence of dwarf mistletoe has some influence on individual trees' ability to support a mountain pine beetle brood, thereby further intertwining the effects of each disturbance on lodgepole pine forest structure. However, further research is required to discern the nature of this relationship.

An additional factor which confounds the understanding of the ecology of dwarf mistletoe-infected lodgepole pine forests is the effect of mountain pine beetle epidemics on overall stand-level dwarf mistletoe rating. This interaction is not well understood. A study of post-mountain pine beetle dwarf mistletoe severity in British Columbia found a higher level of dwarf mistletoe in unattacked trees than in recently attacked trees, suggesting that mountain pine beetle may prefer uninfected trees to infected trees [72]. They hypothesize that dwarf mistletoe will intensify in stands post-mountain pine beetle epidemic, rather than decrease. We do not have data regarding dwarf mistletoe in these stands pre-mountain pine beetle epidemic, so it is unclear whether or not the epidemic intensified or reduced dwarf mistletoe infection in these stands. The frequency with which we observed dwarf mistletoe in our study area supports the hypothesis that mountain pine beetle epidemics do not remove dwarf mistletoe from attacked stands. Given our understanding of mountain pine beetle behavior, we hypothesize that an epidemic leads to an increase in the stand level DMR due to the removal of the largest and most vigorous trees, while leaving the dwarf mistletoe infected and suppressed trees. However, further research regarding host selection by mountain pine beetle and survival of attacked hosts is necessary to fully understand this relationship in lodgepole pine.

Conclusions

The mountain pine beetle influences many stand attributes, so it is important to interpret structural effects within the context of the previous mountain pine beetle epidemic. Conversely, when attempting to understand the influence of mountain pine beetle on stand structure and ecosystem processes in lodgepole pine, it is imperative to incorporate dwarf mistletoe effects. Structural differences associated with dwarf mistletoe create heterogeneous structure in this forest type and may have a significant influence on the productivity, resistance, and resilience of these stands in both positive and negative ways. Research at multiple spatial and temporal scales should be conducted to understand the complexity of the disturbance ecology of lodgepole pine forests. Furthermore, dwarf mistletoe may be similarly influencing ecosystem structure and function of other forest types and investigations of disturbance ecology should include dwarf mistletoes where they occur.

Supporting Information

Table S1 Plot locations.

Table S2 Cohort characteristics of 39 lodgepole pine plots within 13 stands in the Deschutes National Forest, Oregon.

Table S3 BIC table for the natural logarithm of stand density model.

Table S4 BIC table for stand basal area model.

Table S5 BIC table for the natural logarithm of canopy volume model.

Table S6 BIC table for the proportion of lodgepole pine in the dominant/codominant cohort model.

Table S7 BIC table for the proportion of lodgepole pine in the suppressed cohort model.

Table S8 BIC table for the proportion of lodgepole pine in the intermediate cohort model.

Table S9 BIC table for the cohort height of intermediates model.

Table S10 BIC table for the cohort height of suppressed model.

Table S11 BIC table for the cohort height of dominant/ codominants model.

Table S12 BIC table for the natural logarithm of cohort diameter of suppressed model.

Table S13 BIC table for the natural logarithm of cohort diameter of dominant/codominants model.

Table S14 BIC table for the natural logarithm of cohort diameter of intermediates model.

Acknowledgments

We would like to thank Stephanie Pickert for assistance with field work; Ariel Muldoon for valuable analytical insight and statistical advice; Andris Eglitis for his central Oregon mountain pine beetle expertise and feedback on this manuscript; Jay Pscheidt and two anonymous reviewers for providing helpful comments on earlier versions of this manuscript. We would also like to thank Richard Strachan for supporting and encouraging graduate research in forestry.

Author Contributions

Conceived and designed the experiments: MCA DCS TJW. Performed the experiments: MCA DCS MEQB. Analyzed the data: MCA. Contributed reagents/materials/analysis tools: MCA DCS TJW. Wrote the paper: MCA DCS TJW MEQB.

References

1. Lotan JE, Critchfield WB (1990) *Pinus contorta* (Dougl. ex. Loud.) lodgepole pine. In: Burns RM, Honkala BH, technical coordinators. Silvics of North America. Volume 1. Conifers. USDA Agricultural Handbook 654. Washington, DC, USA. 302–315.
2. Raffa KF, Aukema BH, Bentz BJ, Carroll AL, Hicke JA, et al. (2008) Cross-scale drivers of natural disturbances prone to anthropogenic amplification: the dynamics of bark beetle eruptions. Bioscience 58(6): 501–517.
3. Bright BC, Hicke JA, Hudak AT (2012) Landscape-scale analysis of aboveground tree carbon stocks affected by mountain pine beetles in Idaho. Environ Res Lett 7(4): 6.
4. Mikkelson KM, Bearup LA, Maxwell RM, Stednick JD, McCray JE, et al. (2013) Bark beetle infestation impacts on nutrient cycling, water quality and interdependent hydrological effects. Biogeochemistry 115: 1–21.
5. Hicke JA, Johnson MC, Hayes JL, Preisler HK (2012) Effects of bark beetle-caused tree mortality on wildfire. For Ecol Manage 271: 81–90.
6. Kayes LJ, Tinker DB (2012) Forest structure and regeneration following a mountain pine beetle epidemic in southeastern Wyoming. For Ecol Manage 263: 57–66.
7. Mitchell RG, Preisler HK (1998) Fall rate of lodgepole pine killed by the mountain pine beetle in central Oregon. West J Appl For 13(1): 23–26.
8. Simard M, Romme WH, Griffin JM, Turner MG (2011) Do mountain pine beetle outbreaks change the probability of active crown fire in lodgepole pine forests? Ecol Monogr 81(1): 3–24.
9. Amman GD (1977) The role of the mountain pine beetle in lodgepole pine ecosystems: impact on succession. In: Mattson WJ, editor. The role of arthropods in forest ecosystems: Springer-Verlag. 3–8.
10. Roe AL, Amman GD (1970) The mountain pine beetle in lodgepole pine forests. USDA Forest Service Research Paper INT-71. Rocky Mountain Research Station, Ogden, Utah, USA.
11. Diskin M, Rocca ME, Nelson KN, Aoki CF, Romme WH (2011) Forest developmental trajectories in mountain pine beetle disturbed forests of Rocky Mountain National Park, Colorado. Can J For Res 41(4): 782–792.
12. Klutsch JG, Negrón JF, Costello SL, Rhoades CC, West DR, et al. (2009) Stand characteristics and downed woody debris accumulations associated with a mountain pine beetle (*Dendroctonus ponderosae* Hopkins) outbreak in Colorado. For Ecol Manage 258(5): 641–649.
13. Pelz KA, Smith FW (2012) Thirty year change in lodgepole and lodgepole/ mixed conifer forest structure following 1980s mountain pine beetle outbreak in western Colorado, USA. For Ecol Manage 280: 93–102.
14. Sibold JS, Veblen TT, Chipko K, Lawson L, Mathis E, et al. (2007) Influences of secondary disturbances on lodgepole pine stand development in Rocky Mountain National Park. Ecol Appl 17(6): 1638–1655.
15. Collins BJ, Rhoades CC, Hubbard RM, Battaglia MA (2011) Tree regeneration and future stand development after bark beetle infestation and harvesting in Colorado lodgepole pine stands. For Ecol Manage 261(11): 2168–2175.
16. Varhola A, Coops NC, Bater CW, Teti P, Boon S, et al. (2010) The influence of ground- and lidar-derived forest structure metrics on snow accumulation and ablation in disturbed forests. Can J For Res 40(4): 812–821.
17. Koch AJ, Drever MC, Martin K (2011) The efficacy of common species as indicators: avian responses to disturbance in British Columbia, Canada. Biodivers Conserv 20(14): 3555–3575.
18. Mainwaring DB, Maguire DA (2004) The effect of local stand structure on growth and growth efficiency in heterogeneous stands of ponderosa pine and lodgepole pine in central Oregon. Can J For Res 34(11): 2217–2229.
19. Astrup RK, Coates D, Hall E (2008) Recruitment limitation in forests: lessons from an unprecedented mountain pine beetle epidemic. For Ecol Manage 256(10): 1743–1750.
20. Klutsch JG, Battaglia MA, West DR, Costello SL, Negrón JF (2011) Evaluating potential fire behavior in lodgepole pin-dominated forests after a mountain pine beetle epidemic in north-central Colorado. West J Appl For 26(3): 101–109.
21. Geiszler DR, Gara RI, Driver CH, Gallucci VF, Martin RE (1980) Fire, fungi, and beetle influences on a lodgepole pine ecosystem of south-central Oregon. Oecologia 46(2): 239–243.
22. Smith GD, Carroll AL, Lindgren BS (2011) Facilitation in bark beetles: endemic mountain pine beetle gets a helping hand. Agric For Entomol 13(1): 37–43.
23. Paine RT, Tegner MJ, Johnson EA (1998) Compounded perturbations yield ecological surprises. Ecosystems 1(6): 535–545.
24. D'Amato AW, Fraver S, Palik BJ, Bradford JB, Patty L (2011) Singular and interactive effects of blowdown, salvage logging, and wildfire in sub-boreal pine systems. For Ecol Manage 262(11): 2070–2078.
25. Buma B, Wessman CA (2012) Differential species responses to compounded perturbations and implications for landscape heterogeneity and resilience. For Ecol Manage 266: 25–33.
26. Seidl R, Fernandes PM, Fonseca TF, Gillet F, Jönsson AM, et al. (2011) Modelling natural disturbances in forest ecosystems: a review. Ecol Model 222: 903–924.
27. Turner MG (2010) Disturbance and landscape dynamics in a changing world. Ecology 91(10): 2833–2849.
28. Hull RJ, Leonard OA (1964a) Physiological aspects of parasitism in mistletoes (*Arceuthobium* and *Phoradendron*). I. The carbohydrate nutrition of mistletoe. Plant Physiol 39(6): 996–1007.
29. Hull RJ, Leonard OA (1964b) Physiological aspects of parasitism in mistletoes (*Arceuthobium* and *Phoradendron*). II. The photosynthetic capacity of mistletoe. Plant Physiol 39(6): 1008–1017.
30. Hawksworth FG, Wiens D (1996) Dwarf mistletoes: biology, pathology, and systematics. USDA Forest Service Agricultural Handbook 709, Fort Collins, Colorado, USA.
31. Johnson DW, Hawksworth FG, Drummond DB (1981) Yield loss of lodgepole pine stands to dwarf mistletoe in Colorado and Wyoming national forests. Plant Dis 65: 437–438.
32. Mathiasen RL, Nickrent DL, Shaw DC, Watson DM (2008) Mistletoes: pathology, systematics, ecology, and management. Plant Dis 92(7): 988–1006.
33. Godfree RC, Tinnin RO, Forbes RB (2002a) The effects of dwarf mistletoe, witches' brooms, stand structure, and site characteristics on the crown architecture of lodgepole pine in Oregon. Can J For Res 32(8): 1360–1371.
34. Hawksworth FG, Hinds TE (1964) Effects of dwarfmistletoe on immature lodgepole pine stands in Colorado. J Forest 62: 27–32.
35. Baranyay JA, Safranyik L (1970) Effect of dwarf mistletoe on growth and mortality of lodgepole pine stands in Alberta. Canadian Forestry Service, Department of Fisheries and Forestry Publication 1285. Ottawa, Ontario, Canada. 19 p.
36. Wanner JL, Tinnin RO (1989) Some effects of infection by *Arceuthobium americanum* on the population dynamics of *Pinus contorta* in Oregon. Can J For Res 19: 736–742.
37. Weir JR (1916) Mistletoe injury to conifers in the Northwest. U.S. Department of Agriculture Bulletin 306. Washington, DC., USA.
38. Wanner JL (1986) Effects of infection by dwarf mistletoe (*Arceuthobium americanum*) on the population dynamics of lodgepole pine (*Pinus contora*). Ph.D. dissertation, Portland State University, Portland, OR.
39. Godfree RC, Tinnin RO, Forbes RB (2002b) Relationships between *Arceuthobium americanum* and the structure of *Pinus contorta* var. *murrayana* stands in central Oregon. Plant Ecol 165(1): 69–84.
40. Klutsch JG, Beam RD, Jacobi WR, Negrón JF (2014) Bark beetles and dwarf mistletoe interact to alter downed woody material, canopy structure, and stand characteristics in northern Colorado ponderosa pine. For Ecol Manage 315(1): 63–71.

41. Simpson M (2007) Forested plant associations of the Oregon East Cascades. USDA Forest Service Technical Paper R6-NR-ECOL-TP-03-2007.
42. Simard M, Powell EN, Raffa KF, Turner MG (2012) What explains landscape patterns of tree mortality caused by bark beetle outbreaks in Greater Yellowstone? Global Ecol Biogeogr 21: 556–567.
43. Hansen EM (2014) Forest development and carbon dynamics after mountain pine beetle outbreaks. Forest Sci 60(3): 476–488.
44. Axelson JN, Alfaro RI, Hawkes BC (2009) Influence of fire and mountain pine beetle on the dynamics of lodgepole pine stands in British Columbia, Canada. For Ecol Manage 257(9): 1874–1882.
45. Axelson JN, Alfaro RI, Hawkes BC (2010) Changes in stand structure in uneven-aged lodgepole pine stands impacted by mountain pine beetle epidemics and fires in central British Columbia. Forest Chron 86(1): 87–99.
46. Volland LA (1988) Plant associations of the central Oregon pumice zone. USDA Forest Service Technical Paper R6-ECOL-104-1985.
47. Franklin JF, Dyrness CT (1973) Natural vegetation of Oregon and Washington. USDA Forest Service General Technical Report PNW-GTR-008. Pacific Northwest Research Station, Portland, Oregon, USA. 427 p.
48. Western Regional Climate Center [WRCC] (2013) Monthly precipitation, minimum and maximum temperatures for Wickiup Dam climate station. Available: http://www.wrcc.dri.edu/. Accessed 2013 Jan 6.
49. USDA Forest Service [USFS] (2012) Forest insect and disease aerial survey data, Pacific Northwest region. Available: http://www.fs.usda.gov/detail/r6/forest-grasslandhealth/insects-diseases/. Accessed 2012 Feb 21.
50. Stevens DL, Olsen AR (2004) Spatially balanced sampling of natural resources. J Am Stat Assoc 99(465): 262–278.
51. ESRI (2007) ArcGIS 9.3. Environmental Systems Research Institute, Redlands, California, USA.
52. Van Pelt R, Nadkarni NM (2004) Development of canopy structure in *Pseudotsuga menziesii* forests in the southern Washington Cascades. Forest Sci 50(3): 326–341.
53. Hawksworth FG (1977) The 6-class dwarf mistletoe rating system. USDA Forest Service General Technical Report RM-48, Rocky Mountain Research Station, Fort Collins, Colorado, USA.
54. Shaw DC, Freeman EA, Mathiasen RL (2000) Evaluating the accuracy of ground-based hemlock dwarf mistletoe rating: a case study using the Wind River Canopy Crane. West J Appl For 15(1): 8–14.
55. Coder KD (2000) Crown shape factors & volumes. Tree biomechanics series. Extension publication FOR00-32. University of Georgia, Warnell School of Forest Resources, Athens, Georgia, USA.
56. Burnham KP, Anderson DR (2002) Model selection and multimodel inference: a practical information-theoretic approach. Second edition. New York City: Springer.
57. Ramsey FL, Schafer DW (2013) The Statistical Sleuth: A course in methods of data analysis. Third edition. Boston: Brooks/Cole.
58. Browne WJ, Subramanian SV, Jones K, Goldstein H (2005) Variance partitioning in multilevel logistic models that exhibit overdispersion. J Roy Stat Soc A Sta 168(3): 599–613.
59. Nakagawa S, Schielzeth H (2013) A general and simple method for obtaining R^2 from generalized linear mixed-effects models. Methods in Ecol Evol 4(2): 133–142.
60. R Development Core Team (2009) R: A Language and Environment for Statistical Computing. Version 2.12.0. R Foundation for Statistical Computing, Vienna, Austria. Available: http://www.R-project.org.
61. Parmeter JR (1978) Forest stand dynamics and ecological factors in relation to dwarf mistletoe spread, impact, and control. In: Scharpf RF, Parmeter JR editors. Proceedings of the symposium on dwarf mistletoe control through forest management. 11–13 April 1978, Berkeley, California, USA. 16–30.
62. Shaw DC, Watson DM, RL Mathiasen (2004) Comparison of dwarf mistletoes (*Arceuthobium* spp., Viscaceae) in the western United States with mistletoes (*Amyema* spp., Loranthaceae) in Australia–ecological analogs and reciprocal models for ecosystem management. Aust J Bot 52(4): 481–498.
63. Winder RS, Shamoun SF (2006) Forest pathogens: friend or foe to biodiversity? Can J Plant Pathol 18: S221–S227.
64. Godfree RC, Tinnin RO, Forbes RB (2003) Relationships between dwarf mistletoe and the canopy structure of an old-growth lodgepole pine forest in central Oregon. Can J For Res 33: 997–1009.
65. Kulakowski D, Jarvis D (2013) Low-severity fires increase susceptibility of lodgepole pine to mountain pine beetle outbreaks in Colorado. For Ecol Manage 289: 544–550.
66. Amman GD (1972) Mountain pine beetle brood production in relation to thickness of lodgepole pine phloem. J Econ Entomol 65(1): 138–140.
67. Cole DM (1973) Estimation of phloem thickness in lodgepole pine. USDA Forest Service Research Paper INT-148, Ogden, Utah, USA. 10 p.
68. Amman GD (1978) Biology, ecology, and causes of outbreaks of the mountain pine beetle in lodgepole pine forests. In: Kibbee DL, Berryman AA, Amman GD, Stark RW, editors. Theory and practice of mountain pine beetle management in lodgepole pine forests: proceedings of the symposium. 25–27 April 1978, Washington State University, Pullman, WA, USA. 39–53.
69. McGregor MD (1978) Management of mountain pine beetle in lodgepole pine stands in the Rocky Mountain area. In: Kibbee DL, Berryman AA, Amman GD, Stark RW, editors. Theory and practice of mountain pine beetle management in lodgepole pine forests: proceedings of the symposium. 25–27 April 1978, Washington State University, Pullman, WA, USA. 129–139.
70. Stevens RE, Hawksworth FG (1984) Insect-dwarf mistletoe associations: an update. In: Hawksworth FG, Scharpf RF, technical coordinators. Biology of dwarf mistletoes: proceedings of the symposium. 8 August 1984, Colorado State University, Fort Collins, Colorado, USA. 94–101.
71. Hawksworth FG, Lister CK, Cahill DB (1983) Phloem thickness in lodgepole pine: its relationship to dwarf mistletoe and mountain pine beetle (Coleoptera: Scolytidae). Environ Entomol 12: 1447–1448.
72. Shore TL, Andrews RJ, van Sickle GA (1982) Survey for dwarf mistletoe infection in lodgepole pine regeneration under mountain pine beetle-attacked stands in the Chilcotin. Special Report. Environment Canada, Canadian Forestry Service, Victoria, British Columbia, Canada.

18

Tree Spatial Structure, Host Composition and Resource Availability Influence Mirid Density or Black Pod Prevalence in Cacao Agroforests in Cameroon

Cynthia Gidoin[1,2]*, Régis Babin[2,3], Leïla Bagny Beilhe[2,4], Christian Cilas[2], Gerben Martijn ten Hoopen[2,4], Marie Ange Ngo Bieng[5]

1 Supagro, UMR System, Montpellier, France, **2** CIRAD, UPR Bioagresseurs, Montpellier, France, **3** icipe, Plant Health Division, Coffee Pest Project, Nairobi, Kenya, **4** IRAD, Yaoundé, Cameroon, **5** CIRAD, UMR System, Montpellier, France

Abstract

Combining crop plants with other plant species in agro-ecosystems is one way to enhance ecological pest and disease regulation mechanisms. Resource availability and microclimatic variation mechanisms affect processes related to pest and pathogen life cycles. These mechanisms are supported both by empirical research and by epidemiological models, yet their relative importance in a real complex agro-ecosystem is still not known. Our aim was thus to assess the independent effects and the relative importance of different variables related to resource availability and microclimatic variation that explain pest and disease occurrence at the plot scale in real complex agro-ecosystems. The study was conducted in cacao (*Theobroma cacao*) agroforests in Cameroon, where cocoa production is mainly impacted by the mirid bug, *Sahlbergella singularis*, and black pod disease, caused by *Phytophthora megakarya*. Vegetation composition and spatial structure, resource availability and pest and disease occurrence were characterized in 20 real agroforest plots. Hierarchical partitioning was used to identify the causal variables that explain mirid density and black pod prevalence. The results of this study show that cacao agroforests can be differentiated on the basis of vegetation composition and spatial structure. This original approach revealed that mirid density decreased when a minimum number of randomly distributed forest trees were present compared with the aggregated distribution of forest trees, or when forest tree density was low. Moreover, a decrease in mirid density was also related to decreased availability of sensitive tissue, independently of the effect of forest tree structure. Contrary to expectations, black pod prevalence decreased with increasing cacao tree abundance. By revealing the effects of vegetation composition and spatial structure on mirids and black pod, this study opens new perspectives for the joint agro-ecological management of cacao pests and diseases at the plot scale, through the optimization of the spatial structure and composition of the vegetation.

Editor: Morag McDonald, Bangor University, United Kingdom

Funding: The authors thank the West African Scientific Partnership Platform on diversity of production systems and agroecological management of pests and diseases of the French Agricultural Research Centre for International Development (CIRAD) and the research project "Search for trade-offs between production and other ecosystem services provided by tropical agroforestry systems" (SAFSE) for funding. The funders had no role in study design, data collection and analysis, decision to publish, or preparation of the manuscript.

Competing Interests: The authors have declared that no competing interests exist.

* Email: gidoin@supagro.inra.fr

Introduction

Maintaining production and improving ecosystem services of agro-ecosystems while simultaneously reducing dependence on external inputs such as pesticides and fertilizers is a major challenge in agriculture today [1]. To achieve this goal, two concepts based on the "intensification in the use of the natural functionalities that ecosystems offer" are ecological intensification and agro-ecology [2]. A central idea behind these concepts is that associated biodiversity could provide a large range of ecosystem services that may have the same effect as chemical inputs, including pesticides, for pest regulation [3]. With regard to pest and disease regulation services, combining crop plants with other plant species can impact pest and disease occurrence through divers and often complex mechanisms [4]–[5]. However, the regulation of pests and diseases depends much more on specific characteristics of plant associations than on species richness *per se*, yet these characteristics are largely unknown.

Variability in resource availability is one of the main hypotheses proposed to explain the impact of plant diversity on pest and disease occurrence in agro-ecosystems [5]–[6]. Two mechanisms are involved in this hypothesis: (i) the "resource dilution" mechanism, which consists of a reduction in pest and disease occurrence through a decrease in host abundance [7] and (ii), the "alternate resource introduction" mechanism, which consists of an increase in pest and disease occurrence through the introduction of alternative hosts [8]. In general, "resource dilution" holds true for specialist pests and diseases that have only one or a few host plants [9], while the "alternate resource" mechanism generally holds true for more generalist pests and diseases.

Another mechanism involved in pest and disease regulation in multispecies agro-ecosystems is linked to variations in microclimatic conditions [6]. For example, shade trees associated with a crop reduce light availability and wind speed, buffer temperature, and increase relative humidity [10]. These microclimatic variations, which can be grouped under the term shading [11], are determined by several factors, including the spatial structure of shade trees. Tree spatial structure can be described by the characteristics of its vertical and horizontal structure including the number of strata, the density of individual trees in a stratum, and the horizontal distribution of shade trees [4]. These characteristics determine the mean and variance of sunlight transmitted to the understory and can directly affect processes related to the life cycles of pests and pathogens of the crop [6]–[8]–[12].

Under natural conditions, the microclimatic and resource alteration mechanisms are interlinked and interact. For example, microclimatic variations affect the vegetative growth of host plants and hence the availability of sensitive tissue resources for pests and diseases [13]–[14]. However, microclimatic and resource alteration mechanisms are rarely studied together and, as a result, their independent effects and relative importance in pest and disease occurrence at the plot scale are still not known. Although these mechanisms have been confirmed by both empirical research on plant pests and diseases [15] and by epidemiological models [16], the present study is one of the first to try to assess the relative importance of these different mechanisms in a real complex agro-ecosystem. Furthermore, depending on their biology, pests and diseases are differently impacted by each of these mechanisms [17]. In a previous work in cacao agroforest plots in Costa Rica, we studied the independent and joint effects of variables related to the two mechanisms on the impact of one cacao fungal disease (frosty pod rot caused by *Moniliophthora roreri*) [18]. The results of this study revealed the significant influence of the spatial structure of shade trees in the regulation of cacao disease.

In this paper, we focus on cacao (*Theobroma cacao* L.) agroforests in Cameroon. An agroforest is defined as an agro-ecosystem that is structurally close to a natural forest ecosystem with diverse vegetation composition and complex spatial structure [19]. Cameroon is currently the fifth cocoa producer of the world [20], but cocoa production is seriously affected by the mirid bug *Sahlbergella singularis* Hagl. [21] and black pod disease, caused by *Phytophthora megakarya* Brasier & Griffin [22]. Cocoa mirids are not specific to the crop and can be found on several plant species belonging to the Malvaceae family (e.g. *Ceiba pentandra*, *Cola acuminata*, *Cola lateritia* and *Sterculia rhinopetala*) [23], some of which are often associated with cacao trees in agroforests [21]. Black pod disease (BP) caused by *Phytophthora megakarya* can be considered as a specialized pathogen of cacao trees since it does not damage other plant species present in cacao based agroforests and has not been isolated from diseased plant material outside cacao based agroforests [24]–[25]. *Phytophthora megakarya* zoospores are dispersed by rain splash between 5–6 meters [26] and may spread further when free flowing water is available. Adult mirids are able to fly 1–2 km on average [27] but the larvae cannot fly.

The balance of shading effects on pest and disease regulation can be complex [4]–[8], as is the case in the regulation of mirids and black pod disease. Firstly, heavy shade reduces mirid density, probably by limiting cacao vegetative growth, i.e. the availability of young shoots (also named flushes), which are a primary food resource for mirids [21]. At the same time, the introduction of shade trees belonging to the Malvaceae family, known to include alternative host plants for mirids, could reduce the positive impact of shading on mirid density [28]. In the case of black pod, shading is considered to provide microclimatic conditions favorable to the spread and development of the disease [29]. However, the introduction of shade trees may result in resource dilution through a decrease in cacao tree abundance and hence in the availability of potentially sensitive tissue (mainly cacao fruits, also known as cacao pods), which could then reduce the negative impact of shading on the regulation of black pod. Therefore, assessing the respective effects of host composition, the availability of sensitive tissue, and shade tree spatial structure is particularly important as it could improve the joint integrated management of mirids and black pod disease in cacao agroforests.

The question we addressed in this study was "Which part of the variation in mirid density and black pod prevalence among agroforests is explained by host composition, the spatial structure of the vegetation, and resource availability?" To answer this question, mirid density, black pod prevalence and key variables related to spatial structure and composition of the vegetation and the amount of available sensitive tissue were characterized in 20 cacao agroforest plots in the region Centre in Cameroon. We examined the independent contributions of key variables explaining mirid density and disease prevalence using a hierarchical partitioning protocol. We then discuss the implications of our results for the joint agro-ecological management of mirids and black pod disease in cacao agroforests through the optimization of the spatial structure and composition of the vegetation.

Materials and Methods

Study site and sampling plots

The Lékié department in Cameroon was chosen because it represents an intermediate situation on the north–south gradient characterizing natural conditions in the cocoa growing region of central Cameroon [30]. In this department, cacao agroforests are located in a mosaic of cacao based agroforestry plantations owned by different farmers and of forest zones with substantial human activity. Individual farmers generally own one or two cacao plantations with an average size of 2.69 ha (SD = 0.09, n = 687) [30]. Twenty 2,500 m^2 sampling units (50×50 m, see Figure 1) were selected in 20 cacao agroforest plantations belonging to different farmers', based on the criterion that the unit contained more than 10 shade trees associated with the cacao trees. Each sampling unit (hereafter referred to as 'plot') was representative of the cacao agroforest plantation as a whole. The 20 plots were located in a total area of 10×10 km spread over three localities: Nkolobang (9 plots), Zima (7 plots) and Mbakomo (4 plots). The rainfall pattern is bimodal with between 1,500 and 1,600 mm of rainfall per year [30]. Data were collected between September 2011 and November 2012.

Characterization of vegetation structure

In September 2011, vegetation composition and spatial structure were characterized in each plot. The x,y cartesian coordinates of each plant (2 m tall or over) were recorded in each plot using a theodolite (Leica Builder 409, Leica Geosystems, Heerbrugg, Switzerland). Plants were identified to either species or family level, and in a few cases, were not identified. The trees associated with the cacao trees, whether identified or not, were classified in two categories: forest trees or fruit trees. This classification is based on the management of trees in agroforests and is that used in the literature, it states that the fruit trees are mainly introduced by farmers whereas the forest trees are mainly left standing in agroforests when the cacao plantation is established [19]–[31].

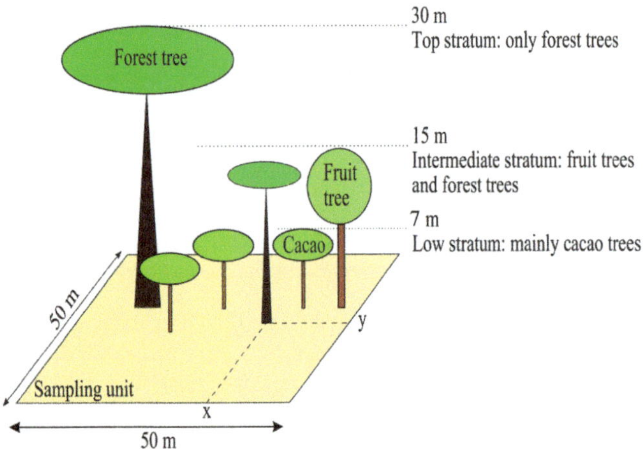

Figure 1. Schematic representation of the vertical structure of cacao agroforest in the Lékié department in Cameroon.

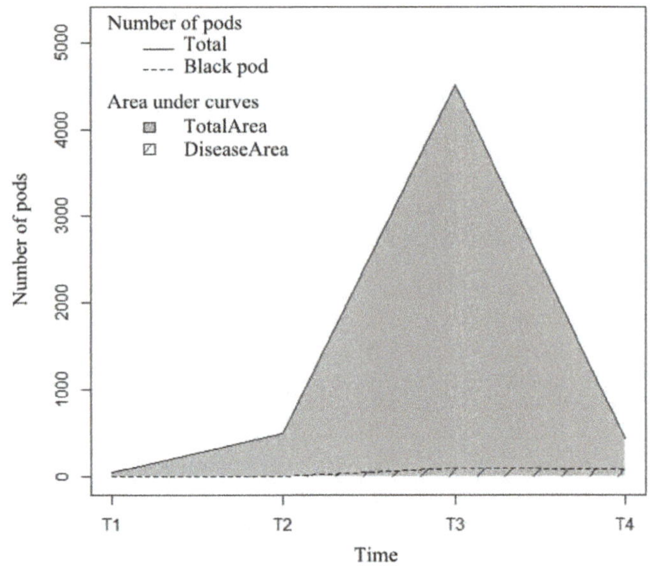

Figure 3. Number of total and damaged pods on the 80 cacao trees sampled in one plot. Number of pods from T1 (May 2012) to T4 (November 2012) on the 80 cacao trees sampled in the plot studied and mapped in Figure 2.

Pest and disease resources were described using variables related to host composition. The cacao tree was the main host for mirids and black pod in the agroforests we studied. For that reason, the relative abundance of cacao trees (*Abca*, i.e. their abundance with respect to all the vegetation mapped in the plot) was assumed to be a valid proxy to study the resource dilution mechanism. For

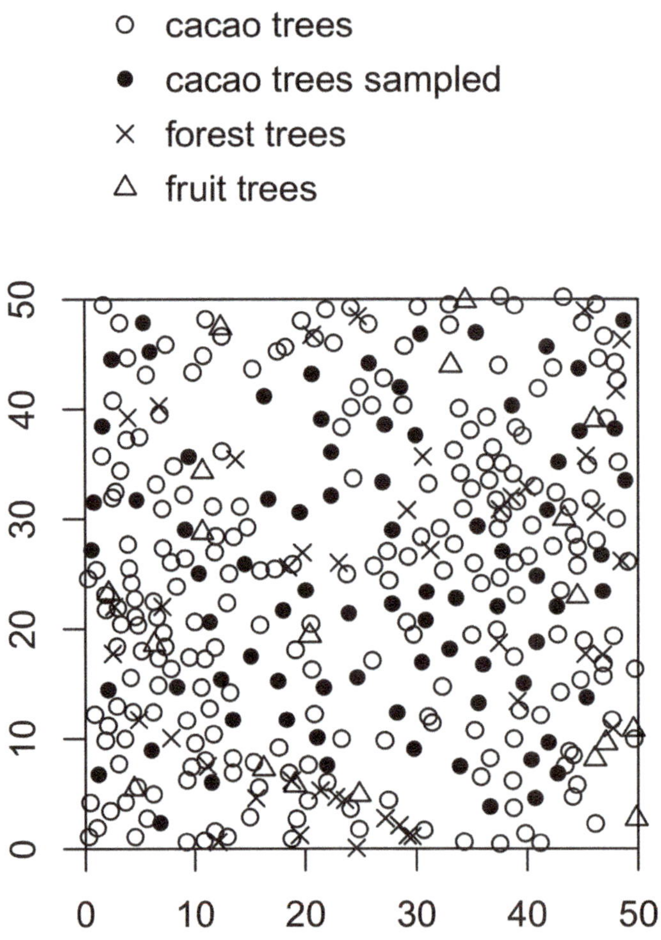

Figure 2. A sample map of a cacao agroforest plot studied.

Table 1. List of variables used to describe the 20 cacao agroforests.

Categories	Variables	Code	Unit or modalities	Min	Max	Mean	Transf.
Mirid pest	Mirid density 2011/12	*Dmir*	number of ind./ha	0	842	117	log
Black pod disease	Black pod prevalence	*BPP*	%	0	5	1	sqrt
Host composition "resource Hypothesis"	Cacao abundance	*Abca*	%	74	94	84	-
	Alternative host	*Alter*	Presence				
			Absence				
Sensitive tissue "resource Hypothesis"	Number of pods 2011/12	*Prod1*	number of pod/ha	78	22178	8497	sqrt
	Number of pods 2012	*Prod2*	number of pod x day	80.10^4	31.10^5	21.10^5	
	Flush presence	*Flush*	-	0.2	0.5	0.3	sqrt
Spatial structure "microclimate Hypothesis"	Total plant density	*Dtot*	number of trees/ha	580	1660	1129	-
	Density of associated shade trees	*Dsha*	number of shade trees/ha	40	168	97	log
	% intermediate trees	*%Inter*	%	18	92	51	sqrt
	Spatial structure of forest trees	*HSFo*	Low density				
			Aggregated				
			Random				
	Spatial structure of fruit trees	*HSFu*	Low density				
			Random				
			Regular				

Variables for density and number of pods are presented at the hectare scale but are used at the plot scale (1/4 ha) in statistical analyses.

Table 2. Information on plots concerned by the variable forest tree horizontal structure (*HSFo*).

HSFo modalities	Low density	Aggregated	Random
Number of plots	5	7	8
Density of forest trees	<10 ind./plot	>10 ind./plot	>10 ind./plot
L(r) curves *outside confidence envelope	No calculation of *L(r)* due to low forest tree densities	*L(r)*>0	*L(r)*=0
Log(*Dmir*) ANOVA: F = 3.9* Tukey HSD[1]	3.5 a	2.7 ab	2.1 b

[1]means with different letters are significantly different, P<0.05.

mirids, alternative hosts to cacao trees were present in the agroforests concerned [23] so the presence/absence of alternative mirid hosts (*Alter*) was also taken into account. Concerning black pod, in our study, the cacao tree was considered to be the only host.

Microclimatic conditions were described using variables related to spatial structure of the vegetation. The vertical and horizontal structure of the cacao trees and associated shade trees (forest trees and fruit trees) affect mean and variance of the light transmitted to the pest and disease present in the understory. The vertical spatial structure was characterized using four strata: the top stratum (at a distance of approximately from 15 to 30 m from the ground); the intermediate stratum (approximately 7 to 15 m from the ground), the low stratum with vegetation in the same stratum as cacao trees (approximately 3 to 7 m from the ground) and the lowest stratum with vegetation lower than cacao trees (approximately 2 to 3 m from the ground) (Figure 1). This classification is justified by field observations to distinguish the amount of shade received by the cacao trees (the top and intermediate canopy strata provided shade for the cacao trees) and the self-shading produced by cacao trees and the vegetation in the same strata. Each plant was recorded as belonging to one of these strata. The total cover (*Dtot*) produced by the cacao trees (self-shading) and associated shade trees was estimated by the total tree density from the cacao to the top canopy strata (the lowest stratum was not included in this calculation). The shade tree cover (*Dsha*) was estimated by the sum of tree densities in the top and intermediate canopy levels. The proportion of shade tree cover provided by the intermediate stratum was estimated by the proportion of shade trees in the intermediate stratum compared with the total number of shade trees (%*Inter*).

The horizontal spatial structure, which is rarely taken into account in agro-ecological studies, was characterized using the *L*(r) function (Besag in Ripley [32]). Although this function is often used in forest research [14] it has rarely been used for agro-ecosystems. The *L*(r) function is based on the calculation of the

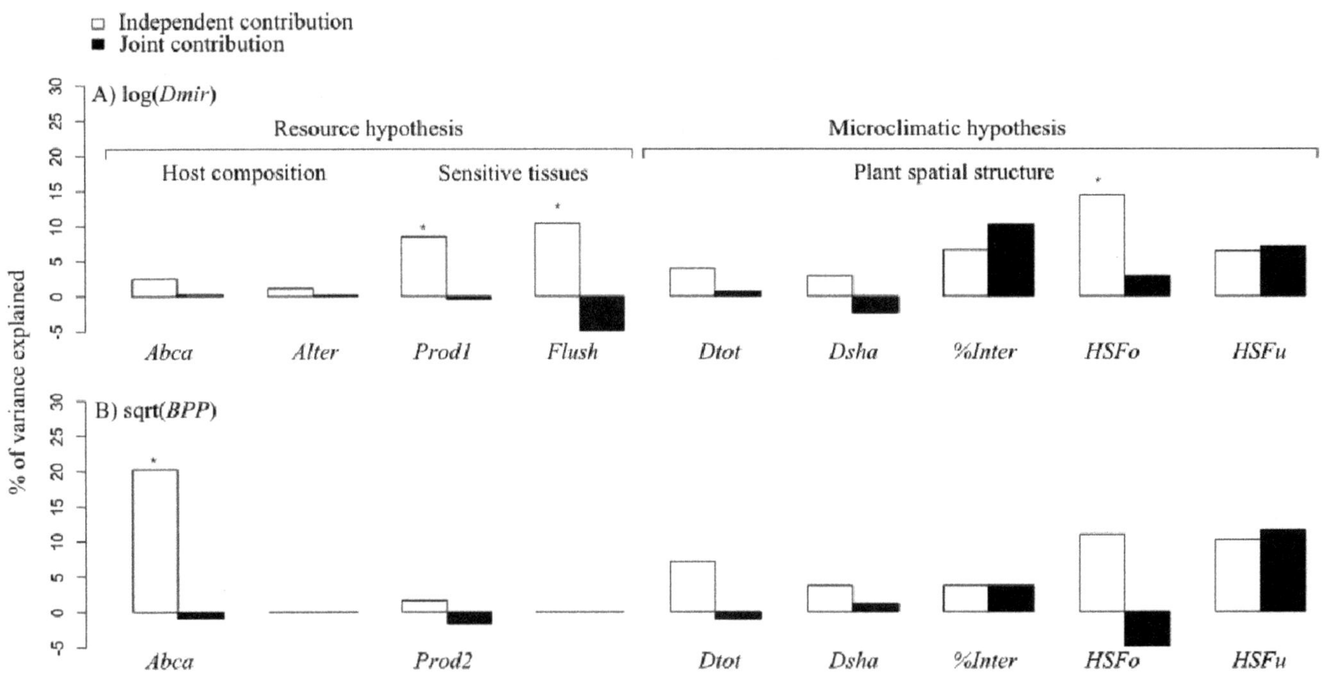

Figure 4. Independent and joint contribution (% of variance explained) of explanatory variables on mirid density and black pod prevalence. Results of the hierarchical partitioning analyses of A) mirid density (*Dmir*) and B) black pod prevalence (*BPP*). Significant independent contributions of explanatory variables are indicated by *(Z-score value>1.65, determined by randomization tests with 1,000 iterations). See Table 1 for the definition of the variables.

Table 3. Table of correlations.

Categories	Codes	Modalities	*Dmir*	*BPP*
Host composition	*Abca*		−0.02	−0.36
	Alter		0.00	-
		Presence	0.04	-
		Absence	−0.05	-
Sensitive tissues	*Flush*		0.25	-
	Prod1		0.34	-
	Prod2		-	−0.01
Spatial structure	*Dtot*		0.39	0.39
	Dsha		0.10	−0.02
	%Inter		0.33	0.20
	HSFo		0.13	0.21
		Low density	0.64	0.35
		Aggregated	0.03	−0.27
		Random	−0.43	0.02
	HSFu		0.11	0.17
		Low density	−0.44	−0.24
		Random	0.02	0.04
		Regular	0.48	0.24

Pearson coefficients and R^2 ANOVA values respectively, between continuous or categorical explanatory variables and residuals of models for mirid density (*Dmir*) or black pod prevalence (*BPP*) after controlling for the effects of the other hypotheses. The mean of residual values are indicated for modalities of categorical variables. See Table 1 for code and definition of variables.

expected number of neighbors within a distance below or equal to r of any point of the point pattern, i.e. in our study, of tree Cartesian positions in a plot. The L(r) function makes it possible to distinguish three types of tree distribution patterns: regular (L(r)$<$ 0), random (L(r)$\approx$0) or aggregated (L(r)$>$0). The statistical significance of the regular or aggregated pattern compared to a random pattern was determined by comparing the observed $L(r)$ value to a 95% confidence envelope based on simulated L(r) values of a complete spatial randomness (CSR) of tree horizontal structure, simulated by 1,000 Monte Carlo simulations. We applied the spatial structure analysis separately to forest trees and fruit trees. Indeed, the $L(r)$ function is applied to a set of individuals subjected to the same process of spatial distribution [33]. We assumed that this was the case for the two types of shade trees in the stands: forest trees and fruit trees. To mitigate edge effects (taken into account in our analyses by the isotropic edge correction as proposed in Ripley [32]), the outcome of the L(r) function should be interpreted for r values less than half or a quarter of the width of the plot [33]. Thus, L(r) was calculated for r values less than 12 m, i.e. less than 1/4 of 50 m. For statistical reasons, the L(r) function cannot be used on patterns at low stand densities [34]. Therefore, the L(r) function was calculated only in plots with at least 10 forest trees or 10 fruit trees per plot. The 12 values of L(r) calculated for forest trees (r from 1 m to 12 m) and the 12 values of L(r) calculated for fruit trees were used in two separate Principal Component Analyses (PCA). With respect with the Kaiser rule, we retained the PCA axes that explained at least 80% of the inertia [35]. Subsequently, the observation coordinates on these axes were used in two different hierarchical cluster analyses based on the Euclidean distance and Ward's criterion [36]. This resulted in clusters of plots with a similar horizontal spatial structure (regular, random, or aggregated) for forest trees (*HSFo*) and for fruit trees (*HSFu*). Plots not included in the typologies due to low densities of forest or fruit trees were attributed to a modality denoted "Low density" in the *HSFo* and *HSFu* categorical variables.

Pest and disease assessment

In each plot, 80 adult cacao trees were selected for observations of the presence of sensitive tissues and for pest and disease monitoring. The trees were selected so as to be homogeneously distributed in the plot (Figure 2).

Plots were not treated with insecticide by farmers for the two years of the study. Mirid density was evaluated in September 2011 and August 2012, i.e. during the usual period of mirid outbreaks in the region [21]. For sampling, the 80 cacao trees were sprayed with an insecticide composed of 20 g/L of lambda cyhalothrin and 20 g/L of imidaclopride using a motorized mist blower (Stihl 420, Stihl AG, Waiblingen, Germany) equipped with a formulation pump and a 0.8 restrictor, and fitted with the baffle plate provided by Stihl adjusted to 25 mL/ha. Treatments were carried out at 6:00 AM, when mirids are not yet very active to limit the risk of winged adults escaping. Seven hours after the treatment, dead insects were collected on 4×4 m plastic sheets positioned under the cacao trees before spraying. The collected insects were preserved in glass hemolysis tubes containing 70% alcohol. The French Agricultural Research Centre for International Development (CIRAD) and the Agricultural research institute for the development (IRAD) have granted permission. In the laboratory, the individual mirids belonging to *Sahlbergella singularis* collected under the 80 selected cacao trees in each plot were counted. For each year, mirid density per plot (*Dmir*) was estimated from the mean number of mirids per tree and the density of cacao trees per plot.

Data on disease damage were collected at four dates: T1: May 2012, T2: July–August 2012, T3: October 2012, T4: November

2012. At each date, healthy and diseased pods (infected by black pod) were counted on the 80 selected cacao trees. Only pods at least 10 cm in length were considered. For each plot, the mean of the total number of pods (healthy and diseased pods) and the mean of the number of diseased pods per tree were calculated for the 80 selected cacao trees and multiplied by the number of cacao trees present in the plot. Based on these values, for each plot, we plotted the curves of the total number of pods from T1 to T4 and the number of diseased pods over the same period (Figure 3). Next, black pod prevalence (*BPP*) was calculated using the formula: $BPP = \frac{DiseaseArea \times 100}{TotalArea}$, where *TotalArea* is the area under the total number of pods curve and *DiseaseArea* is the area under the number of diseased pods curve [18]. Fungicide treatments by farmers were homogeneous between plots with a recommended rate of one 50 g bag of Ridomil Gold plus 66 WP (Active ingredients: metalaxyl-M 6% and copper(1)oxide 60%) applied with a side lever knapsack sprayer containing 15 L water, four times between T1 to T4. Recorded differences between farmers' practices did not have a significant effect on black pod prevalence (data not shown) and were therefore not taken into account in our analysis.

Assessment of available sensitive tissue

We also describe pest and disease resources using variables that account for the amount of sensitive tissue available. Since black pod only attacks cacao tree pods and mirids preferentially attack the young shoots and pods, the attractiveness of a plot varied as a function of the amount of these sensitive tissues it contained. The amount of tissues sensitive to mirids was evaluated by the proportion of cacao trees with young shoots (called flushes) and the number of pods. For each plot, the flush presence variable (*Flush*) was the mean proportion of the 80 selected cacao trees with flushes during the study period, i.e. from T0 = September 2011 to T4 = November 2012. The number of pods available to mirids was estimated as the total number of pods, healthy and diseased, at the plot scale at T0 and T3 (the two periods when mirid outbreaks occur). The variable for pod production was denoted *Prod1* and is expressed as the number of pods.

The amount of tissues sensitive to *P. megakarya* was evaluated by the number of pods. For each plot, the number of pods was assessed using the value of the area under the curve of the total number of pods from T1 to T4 (Figure 3). The number of pods available for black pod was denoted *Prod2* and is expressed as fruit x day.

In all, 9 explanatory variables related to the host composition (*Abca*, *Alter*), shade tree spatial structure (*Dtot*, *Dsha*, *%Inter*, *HSFo*, *HSFu*), sensitive tissue amount (*Flush*, *Prod1* or *Prod2*) were calculated to explain mirid density (*Dmir*) or black pod prevalence (*BPP*) at the plot scale (see Table 1 for the list of variables).

Statistical models

Statistical models were developed to explain Ys variables, i.e. mirid density and black pod prevalence, by explanatory variables Xs for host composition, sensitive tissue availability and tree spatial structure. The approach we used comprised three main steps:

(1) Ys and Xs variables were log or square-root transformed to respect the normal distribution assumption (Table 1). Variation in mirid density between 2011 and 2012 was compared by ANOVA. No significant effect was found ($F = 2.10$, $P = 0.16$). Consequently, the data for 2011 and 2012 were pooled for subsequent analysis.

(2) Correlations between the Xs explanatory variables were tested using the Spearman rank correlation test for correlations between continuous variables and the Kruskal-Wallis test with the Dunn correction for pairwise comparisons for correlations between continuous and categorical variables. Some explanatory variables were not independent, with Spearman's correlation coefficient values ranging from −0.54 to 0.50. Therefore, a hierarchical partitioning protocol [37] was used to examine the independent effect of each explanatory variable on mirid and black pod variables. Two hierarchical partitioning protocols were used: (i) one to explain mirid density (*Dmir*) with the variables *Abca*, *Alter*, *Prod1*, *Flush*, *Dtot*, *Dsha*, *%Inter*, *HSFo* and *HSFu* and (ii) one to explain black pod prevalence (*BPP*) with the variables *Abca*, *Prod2*, *Dtot*, *Dsha*, *%Inter*, *HSFo* and *HSFu*. Hierarchical partitioning is a method that considers all possible models in a multiple regression setting to identify the most likely causal factors (explanatory approach). The independent effect (denoted I_i) of an explanatory variable X_i on Y is then obtained by averaging all the increases in fit generated by including the variable X_i in all the models in which X_i appears. R-squared, which expresses the Y variance explained by a model, was used as goodness-of-fit measure. The joint effect J_i of the X_i variable (effect caused jointly with the other Xs variables) is obtained by subtracting I_i from R_i, with R_i the R-squared of the model explaining Y with only X_i. For each X_i variable, the independent and joint contributions are expressed as a percentage of the total explained variance R_i, with $R_i = I_i + J_i$. The significance of independent effects is evaluated by Mac Nally's Z-scores using a randomization test with 1,000 iterations [38]. There is no measurement of the significance of joint effects. However, it is important to not only look at the independent "explanatory" power (I) of variables but also the ratio between I and J. Because, even if X_i has an apparently high independent effect on Y (great I_i), if X_i appeared to have complex inter-relationships with other explanatory variables (great J_i), then manipulating X_i alone may not have the desired outcomes [37].

(3) Hierarchical partitioning does not provide information on the direction of the relationship (positive or negative) between the Y variable and each explanatory variable. Therefore, to test the effect of microclimatic hypothesis variables on pest density or disease prevalence after controlling for the effect of resource hypothesis variables: (i) we built all the possible linear models that explain Y with the combination of variables of the resource hypothesis, i.e. the variables describing host composition and the sensitive tissue availability, (ii) the best model, with the smallest AIC (Akaike Information Criterion [39]) was chosen, (iii) we used the residuals of this model and we tested their relationship (positive or negative) with each variable of the microclimatic hypothesis, i.e. *Dtot*, *Dsha*, *%Inter*, *HSFo* and *HSFu*. To analyze the relationship, a Pearson correlation test (for continuous explanatory variables) or an ANOVA (for categorical explanatory variables) was performed. The same procedure was applied to study the effect of resource hypothesis variables after controlling for the effect of the microclimatic hypothesis [40]–[41].

Statistical analyses were performed with the packages "stats" for statistical hypothesis testing (e.g. ANOVA, Spearman rank correlation test, Kruskal-Wallis test), "ads" for the point pattern analyses (*L(r)* function), "FactoMineR" for the Principal Component Analyses and "hier.part" for the Hierarchical Partitioning protocol of the R software 2.15.0 [42]–[43].

Results

Vegetation composition and spatial structure

The relative abundance of cacao tree (*Abca*) ranged from 74% to 94% with a mean of 84% with respect to all plants 2 m-tall or over. In 11 plots, at least one of the alternative mirid hosts (*Alter*), i.e. *Ceiba pentandra*, *Cola acuminata*, *Cola lateritia* or *Sterculia rhinopetala*, was present (Table 1).

The total density of plants 2 m-tall or over (*Dtot*) ranged from 580 to 1,660 with a mean of 1,129 individuals per hectare (Table 1). Shade tree density (*Dsha*), i.e. only associated trees that were taller than the cacao trees, ranged from 40 to 168 with a mean of 97 shade trees per hectare (Table 1). The percentage of shade trees belonging to the intermediate stratum (*%Inter*) ranged from 18 to 92 with a mean of 51%. Concerning the horizontal structure of the forest trees (*HSFo*), five plots had a forest tree density <10 individuals per plot (40 trees per hectare, Table 2) and were classified as "low density". In the 15 other plots, forest tree density ranged from 48 to 156 trees per hectare. Seven of these plots displayed a significant aggregated structure of forest trees, with positive values of the $L(r)$ function > the 95% confidence envelope for at least one r value (Table 2). In the eight other plots, forest trees displayed a structure that did not significantly differ from a random distribution. Concerning horizontal structure of the fruit trees (*HSFu*), eight plots had a fruit tree density of less than 10 individuals per plot (40 trees per hectare) and were classified as "low density". In the 12 other plots, fruit tree density ranged from 40 to 104 trees per hectare. Three plots had significant negative values of the $L(r)$ function, and a trend towards a regular distribution was observed in four plots. In the five remaining plots, fruit trees displayed a structure that did not significantly differ from a random distribution.

Factors affecting mirid density

Mirid density ranged from 0 to 842 *Sahlbergella singularis* individuals per hectare with a mean of 117 mirids per hectare (Table 1). The variables sensitive tissue availability and shade tree spatial structure partially explained mirid density at the plot scale (Figure 4A).

The presence of flushes (*Flush*) and the number of pods (*Prod1*) had significant independent effects on mirid density (*Dmir*). They explained respectively 10.4% and 8.5% of mirid density variance independently of the other variables (Figure 4A). Flush presence and the number of pods were significantly positively correlated with mirid density (Table 3), after controlling for the effects of variables for tree spatial structure.

The horizontal structure of forest trees (*HSFo*) had a significant independent effect on mirid density (*Dmir*) and explained 14.5% of mirid density variance independently of the other variables (Figure 4A). Mean mirid density at the plot scale decreased when *HSFo* went from "Low density" to "Aggregated", and then from "Aggregated" to "Random" structure, after controlling for the effects of host composition and the availability of sensitive tissue (Table 2 and Table 3).

None of the host composition variables had a significant independent effect on mirid density (Figure 4A).

The joint effects of *Flush, Prod1* and *HSFo* with the other explanatory variables on mirid density were lower than their independent effects (Figure 4A).

Factors affecting black pod prevalence

Black pod prevalence (*BPP*) ranged from 0 to 5% infected pods of the total number of pods in 2012 at the plot scale with a mean of 1% (Table 1).

Among the host composition variables, cacao tree abundance (*Abca*) had a significant independent effect on black pod prevalence and explained 20.3% of the variance independently of the other variables (Figure 4B). Cacao tree abundance was negatively correlated with black pod prevalence (Table 3), after controlling for the effects of variables for tree spatial structure. The joint effect of *Abca* and the other explanatory variables on black pod prevalence was lower than its independent effect (Figure 4).

Neither sensitive tissue availability, i.e. the number of pods, nor the variables related to tree spatial structure had a significant independent effect on black pod prevalence (Figure 4B).

Discussion

The aim of this study was to assess the independent effects and the relative importance of different variables related to resource availability and microclimatic variation that explain pest and disease occurrence in real complex agro-ecosystems, at the plot scale.

Our study showed that some characteristics of vegetation structure related to the different study hypotheses influence pest and disease occurrence in the farmers' agroforests studied here.

Concerning mirids

Our hypothesis was that tree spatial structure influences mirid density through microclimatic variations. We assumed that the low density of shade trees or the microclimatic heterogeneity generated by the aggregation of shade trees could increase mirid density at the plot scale due to the higher density of mirids on cacao trees in direct sunlight than on cacao trees in the shade [21]. We also expected that when there were relatively more shade trees in the top stratum than in the intermediate stratum, mirid density would be reduced due to the decrease in mean transmitted light [21–12]. Concerning plant spatial structure variables, our results showed that only the variable horizontal structure of forest trees impacted mirid density independently of other variables. Mirid density decreased when a minimum number of randomly distributed forest trees were present compared with an aggregated distribution or when the density of forest trees was low. Babin et al. [21] found that mirid pockets, which are areas highly infested by mirids, were generally located in the sunniest zones of plots. In cacao agroforests, large forest trees, unlike fruit trees, tend to homogenize the distribution of the light resource and consequently limit the development of mirid pockets [21]. Our results confirm that the presence of forest shade trees is correlated with a decrease in mirid density. Several architectural characteristics distinguish forest trees from fruit trees. The most obvious difference is their height as only forest trees reached the top canopy level. However, this difference is probably not the only one involved, since otherwise the variable vertical structure (*%Inter*) would have been significant rather than the variable forest tree horizontal structure (*HSFo*). Thus, the difference in architecture of forest trees which generally have a larger and more porous canopy than fruit trees (a trend revealed by our data, data not shown) could thus also explain why forest trees are most likely to provide more uniform shading. Moreover, the reduction in mirid density is not only observed when forest trees are left standing in cacao agroforests, but this reduction appears to be more effective when forest trees are randomly distributed rather than aggregated. The interaction between the spatial organization of individuals and the distribution of the light resource has only rarely been studied in complex plant ecosystems. Using simulations, Martens et al. [12] showed that the variance in understory light is dependent on the spatial patterns of trees, with an increase in the variance of understory light when the horizontal structure of canopy trees progresses from random to aggregate. Thus, forest tree aggregation is likely to introduce heterogeneity in light distribution, with an alternation of shaded (under aggregates) and sunny (outside aggregates) environments, with sunny environments favoring the development of mirid pockets. Regular patterns of forest trees were not observed at our study site whereas they were observed for fruit trees. This is probably due to differences in the management of forest trees, which are left over from the previous forest ecosystem, compared with fruit trees, which are mainly planted by farmers [19]–[44]. In a previous study, we showed that compared with other horizontal structures or low density, regular forest tree structure reduced frosty pod rot (caused by *Moniliophthora roreri*) prevalence on cacao pods in an agroforest in the Talamanca region in Costa Rica [18]. The results we obtained concerning mirid density and the above result concerning frosty pod rot prevalence underline the potential of promoting forest tree horizontal structure for the agricultural management of cacao pests and diseases. It is interesting to note that in our study, the classical shade tree density or total tree density variables had no significant effect on pest density, whereas the variable tree horizontal structure, which is rarely taken into account, did. Our results thus open new perspectives for agro-ecological research on the regulation of pest and disease by shading through the tree spatial structure.

We also assumed that host composition and sensitive tissue availability influence mirid density through variations in resources. Agroforests are often described as complex agro-ecosystem with significant planned and spontaneous plant diversity [4]–[19]. We studied real agroforests and one of our observations was that cacao agroforests in the Lékié department varied little with respect to the abundance of cacao trees: cacao tree relative abundance ranged from 74% to 94%. This range may not be large enough to empirically test the resource dilution mechanism with the host abundance as proxy to describe the resource amount. Plus, the amount of available sensitive tissue is rarely taken into account in studies on the dilution mechanism. However, our study showed that the amount of sensitive tissue is probably a more valid proxy for describing the resources available for pests and diseases than host composition. Indeed, our results showed that, unlike cacao tree abundance, pod and flush production both had significant independent effects on mirid density. This effect is in agreement with the dilution mechanism since a decrease in the resource (here the amount of sensitive tissue) leads to a decrease in the pest density. In a previous study, we showed that the amount of sensitive tissue was more important than host composition in explaining variations in cacao tree disease [18]. Therefore, besides host abundance, it is important to include the availability of sensitive tissues to identify the effect of variations in resource availability on the pest and disease in real multispecies agro-ecosystems. This variable is particularly important since it makes it possible to determine whether microclimatic variation has a direct or indirect effect on the pest and disease, through variations in plant growth.

A major aim of this study was also to disentangle the effects of the mechanisms involved in the "microclimatic" and the "resource" hypotheses on mirid density in cacao agroforests. Our results showed that the variables forest tree spatial structure and the amount of sensitive tissue both had independent effects on mirid density, thus indicating that hypotheses concerning both the "microclimatic" and "resource" mechanisms may be involved in mirid infestation. Crucial life history traits of cocoa mirids could enlighten us: they are photophobic insects [23], yet at the plot level, they are often located in the sunniest environments [21]. To explain this contradiction, Babin et al. [21] suggested that higher resource availability related to flush intensity, which is greater in cacao trees exposed to sunlight, may attract and allow the development of larger mirid populations. From this assumption, we could conclude that, for cocoa mirids, "microclimatic" mechanisms may act indirectly through "resource" mechanisms. But our results also showed that the effects of the structure of shade trees and the amount of available resources were independent. Moreover, there was no significant difference in the proportion of cacao trees with flushes and in the number of pods between plots with different forest tree horizontal structure (unpublished data). Our study thus did not reveal a relationship between "resource" and "microclimatic" mechanisms. Consequently it is difficult to validate or invalidate the hypothesis that mirids are present in the sunniest areas because of the larger amount of sensitive tissues. A further study that considers the different variables at the cacao tree scale, (and not only at a plot scale as was the case in our study), would help answer this question. However, in terms of tradeoff between pest regulation and pod production services, our results are particularly interesting because we showed that mirid density could be reduced by optimizing the design of tree associations without affecting average pod production at the plot scale.

Concerning black pod

The low percentages of black pod prevalence (*BPP*) in our study plots, which did not reach more than 5% of the production, partly due to homogenized treatment with fungicides, preclude any definitive conclusions. Indeed, without treatment, yield losses as high as 50% are commonly observed and when conditions are favorable, yield losses can reach 80% [45]–[46]. We also presume that the year our observations were conducted was a year with relatively low losses and also that the four point observations of prevalence in the year may also underestimate real losses accumulated over entire year. Nevertheless, despite the fungicide treatment in the plots and the low prevalence rates observed, the host composition effect is probably a particularly strong effect. For that reason, we discuss this effect below, as we believe that it may be even more pronounced in cacao plots without fungicide treatments.

One of our hypotheses was that host composition and sensitive tissue availability influences black pod prevalence through resource variation. In our study plots, black pod prevalence was only explained by cacao tree abundance, independently of other variables, and black pod prevalence increased with a decrease in cacao tree abundance. Based on the "resource dilution" mechanism the opposite result would be expected, in other words a decrease in host tree abundance should have led to a decrease in disease prevalence [47]. Other hypotheses could be proposed to explain this result. Firstly, the relation that exists between the age of cacao agroforests and vegetation structure linked to the farmer's choice of management: On the one hand, old cacao trees that die are not systematically replaced with other cacao trees, or if they are replaced, the fact the cacao tree seedlings were small meant they were not taken into account in our mapping (which only included plants over 2 m in height). Moreover, cacao tree seedlings are sometimes planted together with Musaceae to provide shade during seedling growth [48] and also to provide an income for farmers from the sale of Musaceae fruits. This type of management can lead to a decrease in cacao tree density but also in cacao relative abundance linked to the age of the plantation. On the other hand, an increase in the age of the cacao plot may also lead to an increase in the amount of primary inoculum of *P. megakarya* in the plot, which has accumulated over the years. The negative relationship between cacao tree abundance and black pod prevalence could thus be partly explained by these two effects of the age of the plantation. Secondly, we could also expect that the increase in cacao tree abundance, which was negatively correlated with plant species richness in our plots (unpublished data), would lead to a decrease in fauna associated with non-cacao trees, like ants, at the plot scale. For example, ants of the genus *Crematogaster* are very common in cacao agroforests in our study region, and move in large numbers on cacao trees, where they tend honeydew hemiptera and prey on various arthropods [49]. These ants nest in dead wood or in hollow branches and trunks, so that they benefit from the presence of big forest trees [50]. These ants, like other insects, are assumed to play a determining role in the propagation of black pod by transporting soil particles that contain *Phytophthora* spores [51]–[52]. The potential decrease in the population of ants nesting in big shade trees, due to the increase in the relative abundance of cacao trees, could thus explain our results.

Another of our hypotheses was that the spatial structure of shade trees influences black pod prevalence by causing microclimatic variations. Concerning black pod prevalence, we assumed that the low density of self shading (*Dtot*) and of shade trees (*Dsha*) could reduce black pod prevalence at the plot scale due to aeration of the plot and a concomitant decrease in relative humidity [22]. We also expected that a decrease in the percentage of shade trees in the intermediate stratum (rather than in the top stratum) would decrease black pod prevalence due to more even rain interception and better plot aeration in the cacao tree stratum [22]. However, our results showed that, in our study, none of the variables related to plant spatial structure had an independent effect on disease prevalence. Mfegue [53] found no relationship between shade classes and black pod incidence either. In a previous study in Costa Rican cacao agroforests, we showed that only a regular spatial structure of forest trees reduced frosty pod rot prevalence (caused by *Moniliophthora roreri*) on cacao pods in comparison with plots with a low density of forest trees [18]. Like *M. roreri*, *P. megakarya* is primarily a cacao pod pathogen and the germination of these two pathogens is favored by high humidity, and hence by the presence of shade trees [52]. However, they have two different main modes of dispersal, *M. roreri* by wind and *P. megakarya* by rain. Consequently, shade reduces the spore dispersal of *M. roreri* while its impact on the spore dispersal of *P. megakarya* is not yet known. To control the development of frosty pod rot, it is generally advised to provide moderate and uniform shade at the plot scale, while to control black pod development, it is generally advised to reduce shade. In the Costa Rican study, in the case of *M. roreri*, we showed that although lower disease prevalence was not automatically achieved by conserving forest trees, this reduction was effective only if forest trees were regularly distributed throughout the plot. In the plots in the study area in Cameroon, the lack of impact of the variable tree spatial structure on black pod prevalence is certainly related to the generally high density of shade trees, more dense than the shade tree pattern that could reduce the development of the disease (Table S1).

This study also aimed to disentangle the effects of the mechanisms involved in the "microclimatic" and the "resource" hypotheses on the prevalence of black pod disease in cacao agroforests. One possible explanation for the surprising negative relationship between cacao tree abundance and black pod prevalence could be an increase in the relative abundance of shade trees due to the decrease in the abundance of the cacao trees. This increase in the relative abundance of shade trees could create a shadier environment, which supposedly increases black pod development. However, this explanation can be excluded, since if this were the case, one of the tree spatial structure variables, such as shade tree density or total tree density, would be significant in the hierarchical partitioning analysis. Due to the very low joint effect of the variable cacao tree abundance in the hierarchical partitioning analysis, to find an explanation for this result, we need to look for a variable that was not taken into account in our analysis and is correlated with cacao tree abundance, as proposed in the previous paragraph with regard to the age of the plot and the effect of inoculum accumulation.

Further studies, including studies involving classical variables known to influence the disease (e.g. inoculum potential) and plant structure variables, are needed to elucidate the natural mechanisms involved in the regulation of black pod. However, our results already show that optimizing host composition could be an effective way to decrease black pod prevalence.

Joint agro-ecological management implications for mirids and black pod disease and future outlook

One of the recommendations for the control of mirids is to maintain a uniform shade level in cacao tree plantations, while one of the recommendations for the control of black pod is to avoid excessive shade [21]–[22]. "Thus, to be effective, shade management strategies have to find a balance between shade conditions unfavourable for both mirids and black pod [21]". The present study provides new information on how tree structure characteristics affect the development of the pest and the disease. This is important, since knowledge of such characteristics will help improve pest and disease management strategies. Forest trees are expected to be better shade trees to reduce the occurrence of both mirids and black pod [21]–[54]. In our study, we confirmed that forest trees are better than fruit trees because they appear to provide better shade conditions for mirid control. What is more, our results suggest that reduced pest density is not automatically achieved by preserving forest trees. Indeed, we observed that aggregated forest trees are less efficient than a more randomised tree pattern probably because the latter homogenizes the light under the canopy. Our results also showed that variation in the spatial structure of forest trees did not significantly affect black pod prevalence at the plot scale. Consequently, in our study area, it appears that an agro-ecological approach to mirid and black pod management could be the preservation of natural resources, in this case native forest trees, but under farmer management. However, the preservation of forest trees in cacao agroforests could also favor the presence of a certain fauna, for example ants, that may be implicated in the dissemination of black pod. Thus a more detailed understanding is needed of the causes of the relationship between black pod prevalence and cacao tree abundance, along with studies that involve classical variables known to influence the disease and plant structure variables, are important.

Based on previous observations, we might expect similar results with respect to the impact of the spatial structure of canopy trees on biodiversity in general [55] and on other pests and diseases in particular. Indeed, we also assume that the importance of the spatial structure of shade trees for mirid or frosty pod rot [18] regulation also applies in coffee plantations, where leaf rust caused by *Hemileia vastatrix* is less severe under moderate shade [8]. However, the horizontal spatial structure of shade trees is a structural characteristic that has not received adequate attention to date, and the description of tree patterns and their impact on ecosystem functioning, particularly on pest and disease regulation services, need to be studied in more detail.

Many studies on the relationships between plant diversity and pest or disease occurrence have focused on the resource dilution mechanism. However, the main resource for the pests and diseases is often also the main financial resource for farmers, as is the case of the cacao tree and its fruits here. This study, which takes into account microclimatic and resource mechanisms, showed that the effect of the spatial structure of forest trees is independent of the amount of sensitive tissue (leaves and pods) and therefore provides interesting information on the tradeoff between pest regulation and pod production services.

Acknowledgments

The authors thank the cacao farmers of Nkolobang, Zima and Mbakomo for welcoming us in their plots. We thank Nga Marcus and IRAD-Cameroon technicians Mbarga Joseph Bienvenu, Owona Benoit and Petchayo Tigang Sandrine for their help with data collection. We also thank Daphne Goodfellow for reviewing the English. Finally, we acknowledge the two reviewers for their detailed and useful comments which enabled us to significantly improve the quality of this manuscript.

Author Contributions

Conceived and designed the experiments: CG MANB RB LBB GMTH CC. Performed the experiments: CG MANB RB LBB GMTH. Analyzed the data: CG. Contributed reagents/materials/analysis tools: CG. Contributed to the writing of the manuscript: CG MANB RB LBB GMTH CC.

References

1. Tilman D, Cassman KG, Matson PA, Naylor R, Polasky S (2002) Agricultural sustainability and intensive production practices. Nature 418: 671–677.
2. Chevassus-au-Louis B, Griffon M (2008) La nouvelle modernité: une agriculture productive à haute valeur écologique. In: Déméter 2008: économie et stratégie agricole. Paris: Club Déméter. 7–48.
3. Sabatier R, Wiegand K, Meyer K (2013) Production and Robustness of a Cacao agroecosystem: Effects of Two Contrasting Types of Management Strategies. PLoS ONE 8(12): e80352.
4. Malézieux E, Crozat Y, Dupraz C, Laurans M, Makowski D, et al. (2009) Mixing plant species in cropping systems: concepts, tools and models. A review. Agron Sustain Dev 29: 43–62.
5. Ratnadass A, Fernandes P, Avelino J, Habib R (2012) Plant species diversity for sustainable management of crop pests and diseases in agroecosystems: a review. Agro Sustain Dev 32: 273–303.
6. Avelino J, ten Hoopen GM, DeClerck F (2011) Ecological mechanisms for pest and disease control in coffee and cacao agroecosystems of the Neotropics. In: Rapidel B, DeClerck F, Le Coq J-F, Beer J, editors. Ecosystem Services from Agriculture and Agroforestry Measurement and Payment. London: Earthscan. 91–117.
7. Burdon JJ (1987) Diseases and plant population biology. Cambridge: Cambridge University Press. 222 p.
8. Schroth G, Krauss U, Gasparotto L, Aguilar JAD, Vohland K (2000) Pests and diseases in agroforestry systems of the humid tropics. Agroforest Syst 50: 199–241.
9. Mitchell CE, Tilman D, Groth JV (2002) Effects of grassland plant species diversity, abundance, and composition on foliar fungal disease. Ecology 83: 1713–1726.
10. Beer J, Muschler R, Kass D, Somarriba E (1997) Shade management in coffee and cacao plantations. Agroforest Syst 38: 139–164.
11. Somarriba E, Domínguez L, Harvey C (2005) ¿ Cómo evaluar y mejorar el dosel de sombra en cacaotales ? Agroforestería en las Américas 41: 120–128.
12. Martens SN, Breshears DD, Meyer CW (2000) Spatial distributions of understory light along the grassland/forest continuum: effects of cover, height, and spatial pattern of tree canopies. Ecol Modell 126: 79–93.
13. Calonnec A, Burie JB, Langlais M, Guyader S, Saint-Jean S, et al. (2013) Impacts of plant growth and architecture on pathogen processes and their consequences for epidemic behaviour. Eur J Plant Pathol 135: 479–497.
14. Ngo Bieng MA, Perot T, Coligny F, Goreaud F (2013) Spatial pattern of trees influences species productivity in a mature oak–pine mixed forest. Eur J For Res 132: 841–850.
15. Burdon JJ, Chilvers GA (1982) Host density as a factor in plant-disease ecology. Annual Review of Phytopathology 20: 143–166.
16. Anderson RM, May RM (1981) The population dynamics of microparasites and their invertebrate hosts. Philosophical Transactions of the Royal Society of London. Series B, Biological Sciences. 451–524.
17. Garrett KA, Mundt CC (1999) Epidemiology in mixed host populations. Phytopathology 89: 984–990.
18. Gidoin C, Avelino J, Deheuvels O, Cilas C, Ngo Bieng MA (2014) Shade tree spatial structure and pod production explain frosty pod rot intensity in cacao agroforests, Costa Rica. Phytopathology 104: 275–281.
19. Schroth G, Harvey CA, Vincent G (2004) Complex agroforests: Their structure, diversity, and potential role in landscape conservation. In: Schroth G, Da Fonseca G, Harvey C, Gascon C, Vasconcelos H, Izac A-M editors. Agroforestry and Biodiversity Conservation in Tropical Landscapes. Washington: Island Press. 227–260.
20. FAO (2013) Production. Part of FAOSTAT – FAO Database for food and agriculture. Rome: Food and Agriculture Organization of the United Nations (FAO). Available: http://faostat3.fao.org/faostat-gateway/go/to/home/E. Accessed 2013 Sept 1.
21. Babin R, ten Hoopen GM, Cilas C, Enjalric F, Yede, et al. (2010) Impact of shade on the spatial distribution of *Sahlbergella singularis* in traditional cocoa agroforests. Agric For Entomol 12: 69–79.
22. Cilas C, Despréaux D (2004) Improvement of Cocoa Tree Resistance to *Phytophthora* Diseases. Montpellier: CIRAD. 177 p.
23. Entwistle PF (1972) Pests of cocoa. London: Longman Group Limited. 779 p.
24. Holmes KA, Evans HC, Wayne S, Smith J (2003) *Irvingia*, a forest host of the cocoa black-pod pathogen, *Phytophthora megakarya*, in Cameroon. Plant Pathol 52: 486–490.
25. Opoku IY, Akrofi AY, Appiah AA (2002) Shade trees are alternative hosts of the cocoa pathogen *Phytophthora megakarya*. Crop Prot 21: 629–634.
26. Ten Hoopen GM, Sounigo O, Babin R, Yede, Dikwe G, et al. (2010) Spatial and temporal analysis of a Phytophthora megakarya epidemic in a plantation in the centre region of Cameroon. Proceedings of the 16th International Cocoa Research Conference. 683–687.
27. Leston D (1973) The Flight Behaviour of cocoa-capsids (Hemiptera: *Miridae*). Entomologia Experimentalis et Applicata 16: 91–100.
28. Mpé JM, Vos J, Neuenschwander P (2001) Integrated management of cocoa mirids in Cameroon. In: Vos J, Neuenschwander P, editors. West Africa Regional Cocoa IPM Workshop, Cotonou, Benin, 13–15 November, 2001. Proceedings CPL Press. 39–43.
29. Evans HC (1998) Disease and sustainability in the cocoa agroecosystem. In: Proceedings of the First International Workshop on Sustainable Cocoa Growing. Panama City, 30 March to 2 April 1998. p 12.
30. Jagoret P, Michel-Dounias I, Malézieux E (2011) Long-term dynamics of cocoa agroforests: a case study in central Cameroon. Agroforestry systems 81: 267–278.
31. Sonwa DJ, Nkongmeneck BA, Weise S, Tchatat M, Adesina A, et al. (2007) Diversity of plants in cocoa agroforests in the humid forest zone of Southern Cameroon. Biodivers Conserv 16: 2385–2400.
32. Ripley BD (1977) Modeling Spatial Patterns. J R Stat Soc Series B Stat Methodol 39: 172–212.
33. Goreaud F (2000) Apports de l'analyse de la structure spatiale en forêt tempérée à l'étude et la modélisation des peuplements complexes. Thèse de doctorat en Sciences forestières. ENGREF. 362 p.
34. Condit R (2000) Spatial Patterns in the Distribution of Tropical Tree Species. Science 288: 1414–1418.
35. Kaiser HF (1991) Coefficient alpha for a principal component and the Kaiser-Guttman rule. Psychol Rep 68: 855–858.
36. Lebart L (1994) Complementary use of correspondence analysis and cluster analysis. In: Greenacre M, Blasius, editors. Correspondence analysis in the social sciences. Academic Press. 162–178.
37. Mac Nally R (2000) Regression and model-building in conservation biology, biogeography and ecology: The distinction between and reconciliation of « predictive » and « explanatory » models. Biodivers Conserv 9: 655–671.
38. Mac Nally R (2002) Multiple regression and inference in ecology and conservation biology: further comments on identifying important predictor variables. Biodivers Conserv 11: 1397–1401.
39. Akaike H (1974) A new look at the statistical model identification. IEEE Trans Automat Contr 19: 716–723.
40. Gaba S, Chauvel B, Dessaint F, Bretagnolle V, Petit S (2010) Weed species richness in winter wheat increases with landscape heterogeneity. Agric Ecosyst Environ 138: 318–323.
41. Leprieur F, Beauchard O, Blanchet S, Oberdorff T, Brosse S (2008) Fish invasions in the world's river systems: when natural processes are blurred by human activities. PLoS Biol 6: e28.
42. R Development Core Team (2013) R: A Language and Environment for Statistical Computing. Vienna, Austria: R Foundation for Statistical Computing.
43. Walsh C, Mac Nally R (2009) Package "hier.part". Documentation for R (R. project for statistical computing). Available: http://cran.r-project.org/web/packages/hier.part/hier.part.pdf/.

44. Ngo Bieng MA, Gidoin C, Avelino J, Cilas C, Deheuvels O, et al. (2013) Diversity and spatial clustering of shade trees affect cacao yield and pathogen pressure in Costa Rican agroforests. Basic Appl Ecol 14: 329–336.
45. Berry D, Cilas C (1994) Etude génétique de la réaction à la pourriture brune des cabosses chez des cacaoyers (*Theobroma cacao* L.) issus d'un plan de croisement diallèle. Agronomie 14: 599–609.
46. Despréaux D, Cambrony D, Clément D, Partiot M (1988) Etude de la pourriture brune des cabosses du cacaoyer au Cameroun: définition de nouvelles méthodes de lutte. In: Proc. 10th Int. Cocoa Res. Conf., Santo Domingo, Dominican Republic. Cocoa Producers' Alliance (COPAL), UK. 407–412.
47. Keesing F, Holt RD, Ostfeld RS (2006) Effects of species diversity on disease risk. Ecol Lett 9: 485–498.
48. Jagoret P (2011) Analyse et évaluation de systèmes agroforestiers complexes sur le long terme: Application aux systèmes de culture à base de cacaoyer au Centre Cameroun. Thèse de doctorat. Montpellier Supagro. 288 p.
49. Tadu Z, Djiéto-Lordon C, Yede, Messop Youbi E, Aléné CD, et al. (2014) Ant mosaics in cocoa agroforestry systems of Southern Cameroon: influence of shade on the occurrence and spatial distribution of dominant ants. Agroforest Syst, doi:10.1007/s10457-014-9676-7.
50. Richard FJ, Fabre A, Dejean A (2001) Predatory behavior in dominant arboreal ant species: the case of *Crematogaster* sp. (Hymenoptera: Formicidae). J Insect Behav 14: 271–282.
51. Konam JK, Guest DI (2004) Role of beetles (Coleoptera: Scolytidae and Nitidulidae) in the spread of Phytophthora palmivora pod rot of cocoa in Papua New Guinea. Australas Plant Pathol 33: 55–59.
52. Evans HC (1998) Disease and sustainability in the cocoa agroecosystem. In Proceedings of the First International Workshop on Sustainable Cocoa Growing, Panama City.
53. Mfegue CV (2012) Origine et mécanismes de dispersion des populations de *Phytophthora megakarya*, pathogène du cacaoyer au Cameroun. Thèse de doctorat Montpellier SupAgro. 186 p.
54. Akrofi AY, Appiah AA, Opoku IY (2003) Management of *Phytophthora* pod rot disease on cocoa farms in Ghana. Crop Prot 22: 469–477.
55. Bartels SF, Chen HYH (2010) Is understory plant species diversity driven by resource quantity or resource heterogeneity? Ecology 91: 1931–1938.

Above- and Below-Ground Carbon Stocks in an Indigenous Tree (*Mytilaria laosensis*) Plantation Chronosequence in Subtropical China

Angang Ming[1,2,3]⁹, Hongyan Jia[1,3]⁹, Jinlong Zhao[4], Yi Tao[1], Yuanfa Li[5*]⁹

1 Experimental Center of Tropical Forestry, Chinese Academy of Forestry, Pingxiang, Guangxi, China PR, 2 Institute of Forest Ecology, Environment and Protection, Chinese Academy of Forestry, Beijing, China PR, 3 Guangxi Youyiguan Forest Ecosystem Research Station, Pingxiang, Guangxi, China PR, 4 College of Forestry, Beijing Forestry University, Beijing, China PR, 5 College of Forestry, Guangxi University, Nanning, Guangxi, China PR

Abstract

More than 60% of the total area of tree plantations in China is in subtropical, and over 70% of subtropical plantations consist of pure stands of coniferous species. Because of the poor ecosystem services provided by pure coniferous plantations and the ecological instability of these stands, a movement is under way to promote indigenous broadleaf plantation cultivation as a promising alternative. However, little is known about the carbon (C) stocks in indigenous broadleaf plantations and their dependence on stand age. Thus, we studied above- and below-ground biomass and C stocks in a chronosequence of *Mytilaria laosensis* plantations in subtropical China; stands were 7, 10, 18, 23, 29 and 33 years old. Our assessments included tree, shrub, herb and litter layers. We used plot-level inventories and destructive tree sampling to determine vegetation C stocks. We also measured soil C stocks by analyses of soil profiles to 100 cm depth. C stocks in the tree layer dominated the above-ground ecosystem C pool across the chronosequence. C stocks increased with age from 7 to 29 years and plateaued thereafter due to a reduction in tree growth rates. Minor C stocks were found in the shrub and herb layers of all six plantations and their temporal fluctuations were relatively small. C stocks in the litter and soil layers increased with stand age. Total above-ground ecosystem C also increased with stand age. Most increases in C stocks in below-ground and total ecosystems were attributable to increases in soil C content and tree biomass. Therefore, considerations of C sequestration potential in indigenous broadleaf plantations must take stand age into account.

Editor: Dafeng Hui, Tennessee State University, United States of America

Funding: Our research was financially supported by the Fundamental Research Funds for the Central Non-profit Research Institution of CAF (No. CAFYBB2014QA033), Nature Science Foundation of Guangxi (No 2014jjBA30073), The Ministry of Science and Technology (2012BAD22B01), National Science Foundation of China (No 31400542), and the Director Foundation Project of the Experimental Centre of Tropical Forestry, Chinese Academy of Forestry (No. RL2011-02). The funders had no role in study design, data collection and analysis, decision to publish, or preparation of the manuscript.

Competing Interests: The authors have declared that no competing interests exist.

* Email: 43294659@qq.com

⁹ These authors contributed equally to this work.

Introduction

Biomass and carbon (C) stocks in forest ecosystems play important roles in the global C cycle [1,2,3,4]. Trees and soils are components of forest ecosystems that provide the largest potential for C storage [3,5,6,7,8,9]. Increasing global C sequestration through enlargement of the proportion of forested land on the planet has been suggested as an effective measure for mitigating elevated concentrations of atmospheric carbon dioxide [10,11,12]. As the area of natural stands has decreased in recent decades, tree plantations have become increasingly important components of the planet's forest resources. Commercial plantations are now a central issue in sustainable forest management across the globe. Well-designed, multi-purpose plantations can reduce pressure on natural forests, restore some ecological services provided by natural forests and mitigate climate change through direct C sequestration [13].

China's large plantation programme is assuming an increasingly significant role in C sequestration from the atmosphere. The total land area under tree plantations has reached 6.2×10^7 ha and now accounts for 31.8% of the total forested landscape in the country [14]. The largest proportion (63%) of the total plantation area in China is located in subtropical regions, which provide hot and humid conditions appropriate for tree growth [15]. Most of these subtropical plantations consist of stands containing either a single coniferous species or an exotic tree (e.g. *Pinus massoniana*, *Cunninghamia lanceolata*, *Eucalyptus*) [15]. The creation of monospecific stands of trees that are not native to a landscape carries a high risk of consequential ecological damage, such as decrease in ecosystem stability and outbreak of diseases and insect pests [16,17,18,19,20]. As a result, alternative plantations of indigenous broadleaf species are spreading in this region of China and in neighbouring countries [21,22,23,24, 25,26,27].

Mytilaria laosensis, an indigenous broadleaf tree species, has potential in the afforestation of subtropical China. It grows rapidly and is strongly adaptable; the trunk is straight and the wood has desirable properties for the economic production of high-value timber. The species occurs naturally in western Guangdong, south-western Guangxi and south-eastern Yunnan. It is also indigenous to Vietnam and Laos. *M. laosensis* is expected to become a major afforestation species in subtropical China and beyond [28,29,30]. Its growth patterns, biomass production and wood properties, and the physical and chemical properties of the soils in which it grows have been reported in earlier literature [28,31,32,33]. However, information on biomass and C stocks in *M. laosensis* stands is still lacking. According to previous investigations, C stock size in plantations (especially biomass C) is related not only to tree species, site conditions and soil properties [34,35,36], but also to stand age. According to the previous studies, the C stock of *Castanopsis hystrix* plantations and *Erythrophleum fordii* plantation in subtropical China were increased with the increase in stand age [38,39]. And in the other region, stand age can also remarkably affect the C stocks of plantations' ecosystem [10,37]. However, Effects of diverse factors, including stand age, on C sequestration by *M. laosensis* plantation are poorly documented [33].

Here, we provide first measurements of the C stock across an age sequence of six *M. laosensis* plantation stands. Specifically, the objectives of this study were (1) to document the changes of the sizes and proportional contributions of plantation C pools as stands aged in the early decades following plantation establishment, and (2) to provide baseline information for forest biomass and C estimations focussing on indigenous broadleaf plantations in subtropical regions.

Materials and Methods

Study site and plot establishment

Ethics statement. This research was conducted in Experimental Center of Tropical Forestry, Chinese Academy of Forestry (ECTF for short), This study was also supported by this center. We confirmed that the location is not privately owned and the sampling of soils and plants was approved by ECTF. We also confirmed that the field studies did not involve endangered or protected species.

Study site. The study site is located in the Experimental Centre of Tropical Forestry at the Chinese Academy of Forestry location in Pingxiang City, Guangxi Zhuang Autonomous Region, China (22°02′–22°19′N, 106°43′–106°52′E). The region is within a semi-humid southern subtropical monsoon climate zone that has defined dry and wet seasons. The dry season extends from October to the following March, and the wet season from April to September. The annual mean precipitation at the site is 1200–1500 mm, annual average evaporation is 1261–1388 mm and the relative humidity is 80–84%. The annual mean temperature is 21°C, with a mean monthly minimum of 12.1°C and a mean monthly maximum of 26.3°C. The landscape is largely comprised of low mountains and hills at elevations of 350–650 m. The soils at the study site are categorised as red soils by the Chinese soil classification procedure; this category is equivalent to Oxisol in the USDA Soil Taxonomy. Most developed from granite and had a sandy texture [40,41,42].

Plot establishment. In 2013,six adjacent plantations with different stand ages were selected based on similarities in topography, soil texture, management methodology, environmental conditions and previous vegetation composition (dominated by *C. lanceolata*). We identified a chronosequence of *M. laosensis* stands that were 7, 10, 18, 23, 29 and 33 years of age. All six stands were located within 15 km of one another. They were established in 2006, 2003, 1995, 1990, 1984 and 1980, after the clear-cutting of previous *C. lanceolata* vegetation. Stand characteristics are summarised in Table 1.

During the summer of 2013, four sampling plots (each 30×20 m) were established at random locations in each of the six stands. In each of these plots, we measured the diameter at breast height (DBH, diameter at breast height) of individual trees using a diameter tape, and measured heights using a Hag-löf-VERTEX IV clinometer. Five subplots containing shrubs (each 2×2 m) and five subplots of herbaceous vegetation and litter (each 1×1 m) were established at random locations within each sampling plot (20 subplots in total in the stands of the same age). Plant species, numbers, heights and coverages were recorded; litter was collected from litter subplots (each 1×1 m). Environmental factors including altitude, slope, aspect and slope position were also recorded.

Measurements

Tree biomass. On the basis of the DBH and height measurements in the sampling plots, we selected and harvested six sample trees from different diameter classes in each of the six stands for biomass measurements (36 trees in total). The above-ground portions of the trees were divided into 2-cm sections for measurement. We measured the fresh weights of stems, bark,

Table 1. Site properties and vegetation characteristics of the six plantation stands studied (values are means ±SE; n = 4).

	7 yr old	10 yr old	18 yr old	23 yr old	29 yr old	33 yr old
Altitude (m)	360–520	450–530	450–540	550–650	450–550	350–500
Slope aspect	North-western	Northern	North-western	Northern	Northern	North-western
Slope gradient (°)	36.1±2.4	37.7±3.1	34.2±1.8	32.2±2.2	32.8±2.6	28.9±1.7
DBH (cm)	13.6±0.7	16.2±1.2	20.2±1.2	23.4±1.7	26.0±2.3	27.3±2.7
Tree height (m)	13.6±1.4	14.9±1.1	16.1±1.7	17.8±2.7	21.9±3.1	20.5±1.9
Stem density (trees ha^{-1})	1500±21	1224±18	911±12	721±11	675±8	671±10
Main understorey species	*M. laosensis, Cunninghamia lanceolata*	*M. laosensis, Thysanolaena maxima*	*M. laosensi, T. maxima*	*M. laosensis, C. lanceolata*	*M. laosensis, C. lanceolata*	*M. laosensis, T. maxima*

Table 2. Carbon contents in the vegetation components and litter layers of six differently aged plantation stands (values are means ±SE; n = 4 g kg^{-1}).

Layer	Components	7 yr old	10 yr old	18 yr old	23 yr old	29 yr old	33 yr old	Mean
Tree layer	Stem	527.8±14.1^{Eb}	536.9±12.7^{Cb}	545.6±16.4^{Bab}	520.7±29.3^{Cb}	556.7±15.2^{Ba}	569.1±34.7^{ABa}	542.8±20.4^{Cb}
	Bark	524.4±11.6^{Fa}	534.8±31.2^{Ca}	554.1 ±24.8^{ABa}	506.8±13.7^{Da}	543.2±19.4^{Ca}	517.9±18.8^{Da}	530.2±22.3^{Da}
	Branches	534.1±27.6^{Da}	512.5±21.8^{Ea}	532.2±31.5^{Ca}	498.8±33.2^{Eb}	516.9±19.7^{Ea}	542.3 ±41.2^{Ba}	522.8±28.6^{Ea}
	Leaves	573.6±44.7^{Aa}	525.9 ±31.2^{Ca}	547.1±28.8^{Ba}	560.2±42.9^{Aa}	568.3±44.4^{Aa}	586.1±36.1^{Aa}	560.2±38.4^{Aa}
	Fruit	512.4±17.2^{Ha}	498.8 ±21.8^{Fa}	505.2±15.4^{Ea}	489.7±17.7^{Fa}	504.3±21.9^{Fa}	488.4±20.4^{Ea}	499.8±18.8^{Ha}
	Stump roots	514.3±22.2^{Ga}	533.7±27.4^{Ca}	498.2±23.7^{Fb}	509.6±14.9^{Dab}	532.1±20.8^{Da}	508.1±24.6^{DEa}	516.0±23.8^{Fa}
	Coarse roots	543.4±31.8^{Ca}	554.1±37.1^{Aa}	527.6±30.4^{Ca}	538.7±28.6^{Bab}	520.8±27.3^{Eb}	558.4±40.2^{Ba}	540.5±33.6^{Cab}
	Medium roots	517.9±30.2^{Fa}	541.7±37.7^{Ba}	532.8±24.9^{Ca}	557.4±40.2^{Aa}	531.0±23.4^{Da}	520.2±28.7^{Ca}	533.5±32.6^{Da}
	Small roots	544.1±23.8^{Ca}	537.0 ±31.9^{BCa}	515.6±21.3^{Da}	520.4±19.4^{Ca}	536.3±33.1^{CDa}	525.4±29.2^{Ca}	529.8±25.2^{Da}
	Average of roots	529.9±26.8^{Ea}	541.6±34.2^{Ba}	518.6±21.7^{Da}	531.5±23.7^{BCa}	530.1±27.4^{Da}	533.7±28.9^{Ca}	530.9±28.3^{Da}
	Average of trees	538.0±25.5^{Da}	530.3±30.4^{Ca}	539.5±27.1^{BCa}	523.6±34.2^{Ca}	543.0±27.3^{Ca}	549.9 ±26.5^{Ba}	537.4±29.7^{CDa}
Shrub layer	Above ground	544.2±32.2^{Ca}	516.7±21.7^{Da}	545.2±31.8^{Ba}	534.8±25.5^{Ba}	541.7±27.4^{Ca}	522.2±18.9^{Ca}	534.1±25.8^{Da}
	Below ground	522.3±17.3^{Fa}	522.4 ±19.1^{CDa}	497.5±20.4^{Fa}	508.4±24.3^{Da}	516.8±19.4^{Ea}	519.2±9.7^{CDa}	514.4±18.3^{FGa}
	Average	533.3±28.3^{Da}	519.6±19.6^{Da}	521.4±25.3^{Ca}	521.6±14.8^{Ca}	529.3±25.4^{Da}	520.7±13.7^{Ca}	524.3±22.9^{DEa}
Herb layer	Above ground	513.3±31.7^{Ha}	532.8±28.8^{Ca}	527.9±35.4^{Ca}	508.4±31.0^{Da}	529.1±19.8^{Da}	511.7±27.6^{Da}	520.5±30.7^{Ea}
	Below ground	518.4±27.2^{Fa}	509.3±19.7^{Ea}	497.6±16.9^{Fa}	509.8±21.3^{Da}	524.2±24.7^{DEa}	510.5±18.8^{Da}	511.6±20.6^{Ga}
	Average	515.9±29.7^{Ga}	521.05 ±25.4^{Da}	512.8±27.3^{Ea}	509.1±26.8^{Da}	526.7±22.3^{Da}	511.1±25.2^{Da}	516.1±26.4^{Fa}
Litter layer	Undecomposed	557.8±32.2^{Ba}	578.4±40.4^{Aa}	562.1±34.2^{Aa}	550.9±27.9^{ABa}	564.2±40.1^{Aa}	578.5±31.2^{Aa}	565.3±35.7^{A}
	Semi-decomposed	554.2±29.9^{Ba}	543.8±31.0^{Ba}	542.2±27.6^{Ba}	537.1±30.4^{Ba}	529.7±27.7^{Da}	547.0±29.4^{Ba}	542.3±28.3^{C}
	Average	556.0±31.3^{Ba}	561.1±36.4^{Aa}	552.2±30.9^{Ba}	544.0±28.7^{Ba}	547.0±35.1^{Ca}	562.8±30.1^{Ba}	553.8±31.4^{B}

Note: Different Capital letters in the same list indicate significant pairwise differences within stand ages between components, and different lowercases indicate significant pairwise differences within components between stand ages (multiple comparisons test; $P<0.05$).

branches and leaves. The below-ground portions of the sample trees were dug out and examined using the open cut method. We measured the fresh weights of the stump roots, thick roots (diameter >2.0 cm), medium-thick roots (diameter 0.5–2.0 cm) and small roots (diameter <0.5 cm). Organ samples were collected (200 g of each organ) and oven-dried at a temperature of 65°C to constant weight to calculate the moisture contents and dry weights. We built regression models for the different organs to estimate tree biomass (using data from the 18 sample trees).

Understorey vegetation and litter biomass. We used a destructive harvesting method to measure the biomass in the above-ground and below-ground portions of the shrub and herbaceous layers [43]. The fresh masses of these two portions were obtained directly by weighing. After oven-drying to constant weight at 65°C, we weighed subsamples and calculated the respective dry weights from determinations of moisture contents. The components of understorey vegetation were then separated and measured. We weighed litter material that had not decomposed or was semi-decomposed at the same time. The litter samples were oven dried at 65°C and weighed.

C content. Samples of the above-ground and below-ground components of the sample trees (*M. laosensis*), shrubs, herbaceous plants and litter were dried, ground and sieved in the laboratory. These samples were then bottled for later chemical analysis. In total, 288 soil-sampling points were selected within the 24 sampling plots in each of the six stands. Soil pits were dug to a depth of 100 cm and samples were collected randomly from four depth horizons: 0–10 cm, 10–30 cm, 30–50 cm and 50–100 cm. Soil samples from the same depth horizon in the same stand were mixed in equal proportions and the mixtures were air-dried at room temperature (25°C). The samples were then ground and passed through a 2-mm-mesh sieve to remove coarse living roots and gravel; they were then ground in a mill in preparation for sieving through a 0.25-mm mesh before chemical analysis. A soil-sample cutting ring (100 cm^3) was used to collect samples of undisturbed soil from different horizons. These samples were taken to the laboratory for measurements of soil bulk density using the cutting ring method. We measured the C content of the tree component samples, understorey vegetation, litter by vario Macro Elemental Analyzer (Elementar Analyasensysteme GmbH, Germany), but the soil organic carbon was established by the oil-bath $K_2Cr_2O_7$ titration method.

C storage. The C stocks (C in biomass per unit area of land surface) in the vegetation and litter biomass were determined by multiplying C content by biomass (dry mass per unit area of land surface). The C stocks per unit area of land in each of the soil horizons were calculated by multiplying soil bulk density at a chosen soil depth by the C content at that depth. Total soil C stocks were computed by summing the stocks in each soil horizon.

Statistical analysis

We used one-way ANOVA to test for differences in the C content and C stock among plantations of different ages. The dependent parameters were normally distributed and homoscedastic. All analyses were performed using the following software: Microsoft Excel 2007 and SPSS (ver.13.0; SPSS, Chicago, IL) for Windows. Statistical significance was detected at $P<0.05$.

Results

C content in plantation stands

C content in the vegetation and litter layer. The C contents of the component organs differed significantly among the

Table 3. Carbon contents by soil depth in six differently aged *Mytilaria laosensis* plantation stands (values are means ±SE, n = 12; g kg^{-1}).

Soil depth (cm)	7 yr old	10 yr old	18 yr old	23 yr old	29 yr old	33 yr old	Mean
0–10	28.1 ± 4.2^{A}	27.4 ± 2.8^{A}	29.6 ± 2.9^{A}	31.7 ± 2.4^{AB}	39.7 ± 4.3^{C}	47.1 ± 5.7^{D}	33.9±3.4
10–30	23.2 ± 2.7^{A}	24.7 ± 2.1^{A}	26.3 ± 1.7^{B}	26.6 ± 2.3^{B}	27.7 ± 1.8^{B}	31.0 ± 2.5^{C}	26.6±2.4
30–50	18.9 ± 2.4^{A}	19.4 ± 2.7^{A}	18.6 ± 1.9^{A}	19.0 ± 2.6^{A}	19.4 ± 3.1^{A}	20.5 ± 3.4^{A}	19.3±3.0
50–100	10.9 ± 1.5^{A}	11.3 ± 1.7^{A}	13.2 ± 2.3^{B}	12.8 ± 2.0^{B}	14.4 ± 2.4^{C}	15.4 ± 2.7^{C}	13.0±2.3
Mean	20.3±2.5	20.7±2.3	21.9±2.7	22.5±2.4	25.3±3.2	28.5±3.7	23.2±2.7

Note: Different capital letters indicate significant pairwise differences within soil depths between stand ages (multiple comparisons test; $P<0.05$).

Table 4. Individual biomass regressions models for *Mytilaria laosensis* trees ($n = 18$ for all models); *W*, biomass; *D*, DBH; *H*, height.

Organ	Allometric equation	R^2	*F*-value	*P*
Stem	$W_s = 0.1740(D^2H)^{0.7661}$	0.9196	104.3812	<0.0001
Bark	$W_{ba} = 0.0220(D^2H)^{0.7081}$	0.7191	58.6375	<0.0001
Branches	$W_{br} = 0.0002(D^2H)^{1.2696}$	0.6291	11.2124	0.0065
Leaves	$W_l = 0.00003(D)^{1.2634}$	0.8091	41.0306	0.0001
Roots	$W_r = 0.0094(D^2H)^{0.9538}$	0.7247	14.7872	0.0027
Total tree	$W_t = 0.1536(D^2H)^{0.8268}$	0.9049	64.8073	<0.0001

six stands ($P<0.05$) and fell into the following rank order: leaf > stem > coarse root > medial root > bark > small root > branch > stump root > fruit. C content was not significantly different between medium roots and bark ($P<0.05$). The C contents of above-ground components of the shrub and herb layers in all six stands were higher than those of their below-ground components. In the litter layer, the C content of the undecomposed portion was higher than that of the semi-decomposed portion (Table 2).

The average C content did not differ significantly within plantation components among stands of different ages ($P>0.05$). No obvious pattern relationships were detected between C contents and increasing stand age.

C content in the soil layer. The C content changed markedly with increasing soil depth in all six stands ($P<0.01$; Table 3). The value in the topsoil (0–10 cm depth) was 46.1% higher than the average at 100 cm depth.

The C contents of the two shallowest soil layers (0–10 cm and 10–30 cm), the horizon at 50–100 cm and values obtained by summing across all soil layers increased significantly with increasing stand age ($P<0.05$). The C content in the 3050 cm soil layer, however, did not differ significantly among different ages ($P>0.05$).

C stocks in plantation stands

Biomass and C stocks in tree layers. The allometric relationship between the biomass of the tree organs (*W*) and DBH (*D*) andheight (*H*) were best-fitted with equations in the form $W = a(D^2H)^b$. The *F*-tests showed that all regressions were highly significant ($P<0.01$). Biomass calculations based on these allometric equations were used in the estimations of C stocks detailed in Table 4.

Fig. 1a depicts C stocks in stands of different ages and their allocation among component organs. The total C stocks in the trees were 77.3, 91.3, 104.1, 114.6, 153.0 and 156.2 t ha^{-1} in 7-, 10-, 18-, 23-, 29- and 33-year-old stands, respectively. Thus, stocks rapidly increased with age from 7 to 29 years, but plateaued thereafter. C stocks in stems made up 74.0, 72.1, 69.7, 67.7, 65.3 and 65.2% of the total tree content in 7-, 10-, 18-, 23-, 29- and 33-year-old stands, respectively. Furthermore, trends in the C stocks of stems, roots, bark, branches and leaves tracked those of total tree C stocks across all stand ages.

The allocation of C stocks to stems and bark decreased with stand age from 7 to 29 years; stem and bark allocations were similar in 29- and 33-year-old stands. In contrast, C allocations to branches, leaves and roots increased with stand age from 7 to 29 years, but were similar in 29- and 33-year-old stands.

C stocks in the shrub, herb and litter layers. The C stock levels of each layer of the six plantation stands are shown in Fig. 2. Small proportions of biomass and C were measured in the shrub, herb and litter layers. The summed C contents of the shrub, herb and litter layers made up 2.9%, 3.0%, 3.6%, 4.0%, 3.2% and 2.6% of the total C in 7-, 10-, 18-, 23-, 29- and 33-year-old stands, respectively. The above-ground C stocks in the shrub layers of the

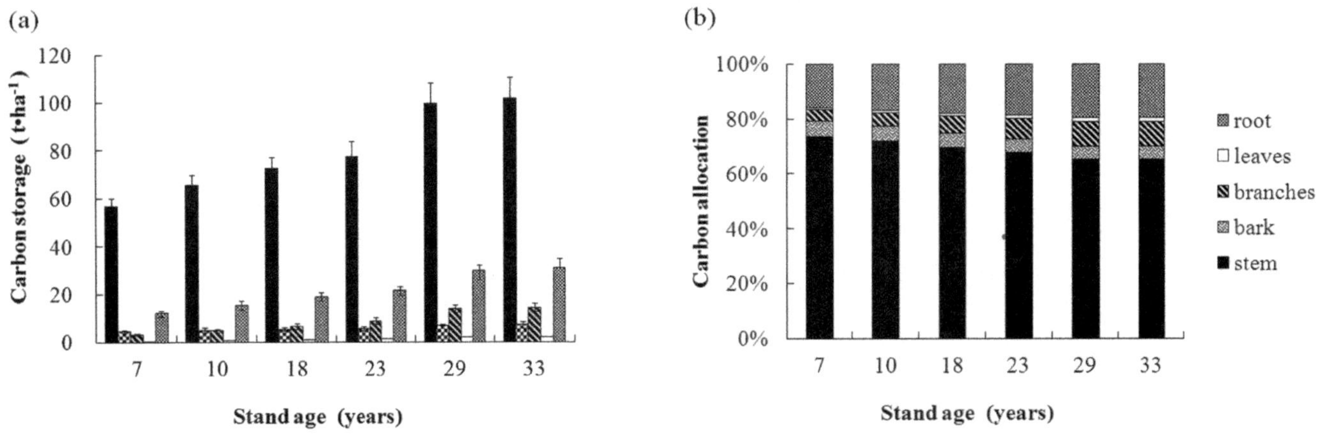

Figure 1. Carbon stocks and their allocation to tree components in six differently aged *Mytilaria laosensis* stands. Values in (a) are means ±SE, n = 4. Note: Different lowercase letters above the bars indicate significant differences between means (multiple comparisons test; $P<0.05$).

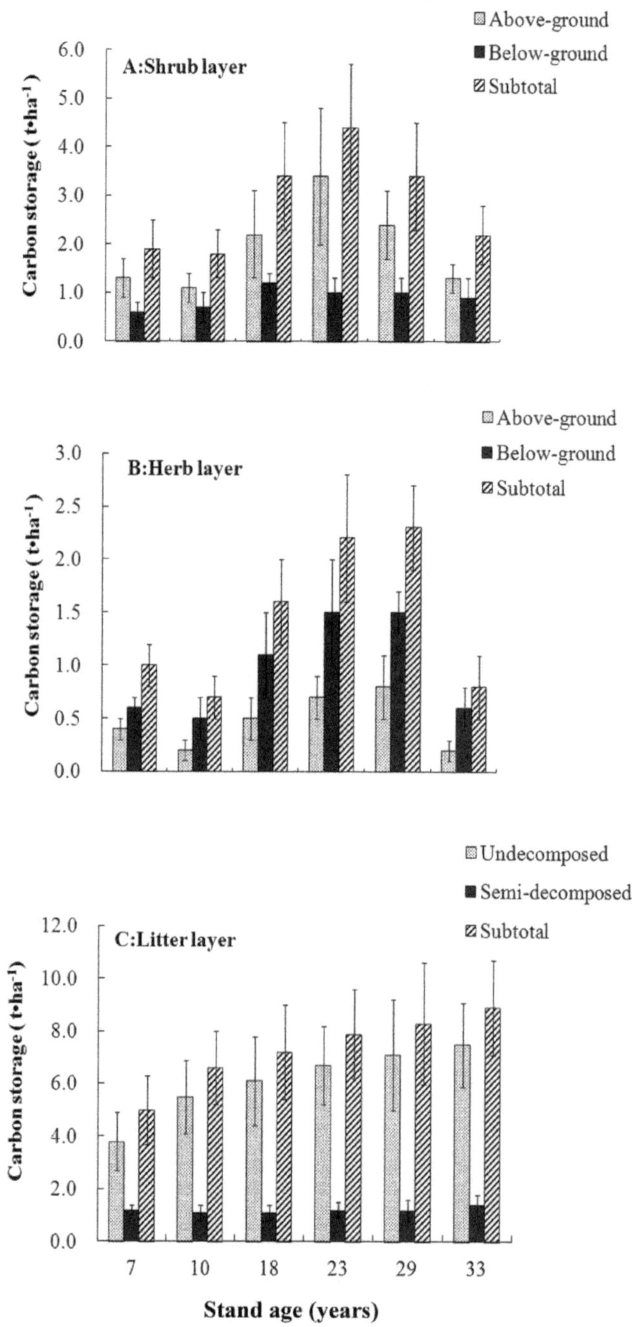

Figure 2. Carbon stocks in the shrub, herb and litter layers in six differently aged *Mytilaria laosensis* stands. Values are means ±SE, n = 4.

six stands were higher than below-ground shrub stocks. However, above-ground biomass and C stocks in the herb layer were lower than those below ground. The undecomposed biomass and C stocks in the litter layer were higher than those in the semi-decomposed portion; the highest average C stock in the undecomposed litter was 5.1-fold higher than that in the semi-decomposed portion.

Forest ground vegetation C stocks were correlated with stand age across the entire plantation chronosequence. The shrub C stocks increased with stand age from 7 to 23 years ($P<0.05$), but decreased with stand age from 23 to 33 years ($P<0.05$). Herb C stocks increased with stand age from 7 to 29 years, but decreased thereafter. C stocks in semi-decomposed portions of the litter layer were not related to stand age, but those in undecomposed portions and in the combined litter layer increased remarkably with increasing stand age ($P<0.05$). We therefore predict that the C stock in the litter layer will increase continually as the stands become older.

C stock in the soil layer. C stocks decreased with increasing soil depth even though the soil bulk density increased with depth. Fig. 3 depicts trends in the soil layer C stock across the *M. laosensis* stand age sequence. The top soil (0–10 cm) and deeper soil (50–100 cm) stocks followed an increasing trend with stand age ($P<0.05$), especially in the older stands (the difference between 29- and 33-year-old stands was especially significant; $P<0.01$). Although soil C stocks in the 10–30-cm and 30–50-cm horizons increased with age, the relationship was not significant (Fig. 3a).

Summed C stocks from 0 to 100 cm soil depth were 198.2, 199.6, 222.8, 227.9, 278.7 and 292.0 t ha^{-1} in the 7-, 10-, 18-, 23-, 29- and 33-year-old stands, respectively. This linear trend was significant (Fig. 3b).

C stock in the plantation ecosystem. Table 5 summarises individual ecosystem C stocks measured within each of the six stands. The rank order of C stock proportions across the six stands was as follows: soil layer (62.6–70.0%) > tree layer (27.3–33.9%) > litter layer (1.8–2.2%) > shrub layer (0.5–1.3%) > herb layer (0.2–0.6). Averaging across stands, 65.3% of the total C was in the soil and 31.5% in the trees (Fig. 4b).

Fig. 4a shows changes in C stocks in above-ground, below-ground and total ecosystem components with increasing stand age. Above-ground ecosystem C stocks increased as stand age increased from 7 to 29 years; thereafter, the C content changed little. Below-ground and total ecosystem C stocks increased across the entire age range.

The above-ground to below-ground ecosystem C stock ratios were 0.338, 0.388, 0.391, 0.417, 0.431 and 0.419 in the 7-, 10-, 18-, 23-, 29- and 33-year-old stands, respectively; the ratios increased gradually with age due to the accumulation of above-ground C in tree biomass.

Discussion

C content

C content in forest ecosystems varies by forest type. Tree species and site conditions are related to the C content [35,43]. Across six *M. laosensis* stands with different ages, we detected a significant difference in mean C contents among different tree organs. No significant effect of stand age was detected within individual tree organs in the *M. laosensis* plantations, which corroborates the findings of studies on other species [38,45].

The C contents of the various vegetation layers in the same stand fell into the following rank order: trees > shrubs > herbaceous plants (Table 2). Trees are probably high ranking because they can synthesise and accumulate more organic matter than can other types of vegetation [46].

The C content of litter varies with many factors, such as tree species, litter productivity, decomposition rate and microenvironment [35,44]. Litter reportedly decomposes at a considerably faster rate in broadleaf forests than in coniferous stands; thus, the standing stock of C is lower in broadleaf forests [35]. We measured a mean litter C content of 553.8 g kg^{-1} in *M. laosensis* stands. This value is considerably higher than that in *P. massoniana* stands (505.9 g kg^{-1}) in subtropical China [43], probably because

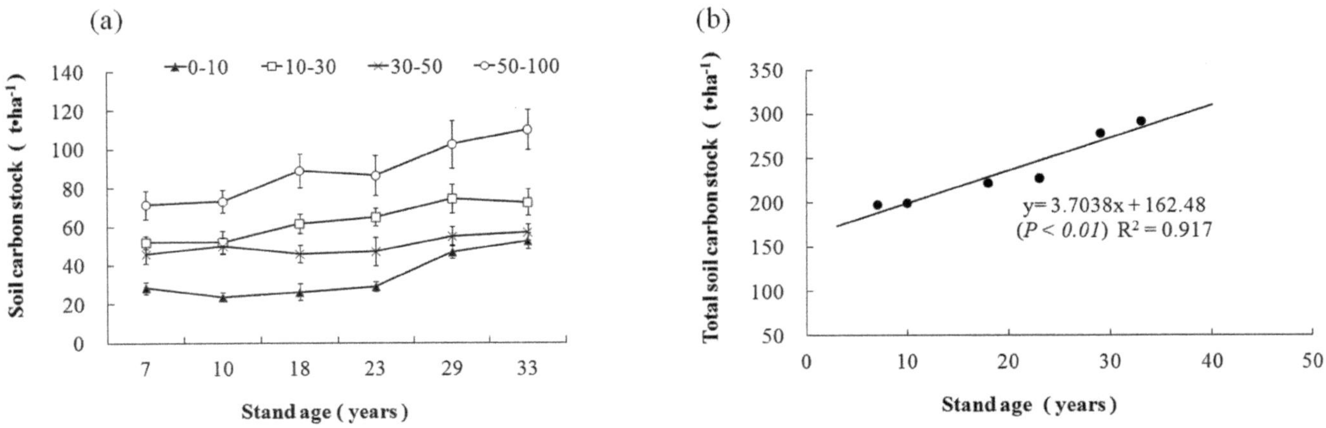

Figure 3. Soil layer carbon stocks in six differently aged *Mytilaria laosensis* stands (a) and the linear relationship between soil carbon stocks and stand age (b). Values are means ±SE n = 12. Soil depth ranges in the key to (a) are in cm units.

the leaves of *M. laosensis*, which were the main component of the litter layer, are very leathery and refractory [47].

The mean C content of the soil layers in each stand decreased as the depth increased. The C content in the topsoil was higher than that in the deeper soil because the organic C produced from the decomposition of litter and root systems near the ground surface had entered the topsoil first, as demonstrated in other studies [10,48,49]. The C contents of the top two soil layers (0–10 cm and 10–30 cm), the deepest layer (50–100 cm) and across all soil layers significantly increased with increasing stand age, probably due to the increasing litter productivity in older stands. This finding helps to explain the increase in below-ground C stocks as stands age.

C stocks in plantation stands

Estimating the C stock pools stored in age-sequenced plantations may contribute to forest management for C sequestration. C stocks in trees depend on stand density, biomass and relative C contents in the tissues. We found that C contents were positively correlated with C stocks (Table 5). The tree C stocks increased rapidly with age from 7 to 29 years, but plateaued thereafter, perhaps due to declining tree growth rates. Most previous studies have reported increases throughout the growth phases of trees [10,38]. The laws of growth may vary among tree species, For example, *Euclyptus urophylla ×E.grandis* grew faster in its early years, but turned slowly later, On the contrary, Castanopsis hystrix grew faster in the later stage than that in the early years [35,38,39,45], that is to say, different tree species show different growth charactersAs to *Mytilaria laosensis*, itneeds 25–30 years to reach the largest yield stage [28]. This variation may account for the difference between *Mytilaria laosensis* plantations and most other stands.

Only small portions of biomass and C were sequestered in the shrub, herb and litter layers, which accounted for only 2.6–4.0% of the total. We detected a correlation between forest ground vegetation C stock and stand age across the entire chronosequence. C stocks in ground vegetation increased in the early stages of tree stand development, but decreased in older stands in concert with changes in tree canopy cover and stand density. Nevertheless, no obvious common patterns have been detected in studies on ground vegetation C storage. In the litter layer, the undecomposed and the total litter C stock decreased markedly with increasing stand age, which probably accounts for increases in soil C content with increasing stand age.

Soil was the largest C pool in the six stands that we studied. Soil C stock depends on stand age, the physical and chemical properties of the soil, forest type, litter productivity and litter decomposition rate [6,35,43,50]. We found a significant linear relationship between total soil C stocks across the 0–100-cm-depth range and stand age. Hence, the proportion of C stock below ground will probably increase over protracted periods in the future life of the plantation ecosystems we studied.

Proportions of plantation ecosystem C stocks in the six stands fell into the following rank order: soil layer > tree layer > litter layer > shrub layer > herb layer. Soil and tree biomass harboured the largest C pools in the ecosystems, accounting for 96.8% of the total. These findings are congruent with previous studies [51,52,53,54].

Above-ground ecosystem C stocks increased during the early stages of stand development and plateaued after 29 years due to a deceleration in tree growth. However, below-ground and total ecosystem C stocks increased with the stand age during the whole chronosequence we studied.

Conclusions

Stand age is a major determinant of C stocks in plantations. Both the C stocks and their distributions among plantation ecosystem components were affected by stand age. We found no significant differences in C contents in above-ground components among stand ages, but the soil C content increased with increasing stand age. Tree C was the largest above-ground ecosystem fraction, which contributed 25.7% to the total ecosystem C stocks in all six stands. The soil fraction was the largest C pool across plantations. C stocks across the 0–100 cm soil depth range increased across the entire chronosequence. They were significantly linearly related to tree growth. The increase in above-ground tree biomass with increasing stand age significantly affected the above-ground ecosystem C stock size. The increases in below-ground ecosystem C stocks through a 7-year to 33-year stand chronosequence were mainly attributable to increases in soil organic C. Thus, one must take into account the successional development in forest ecosystem C pools when estimating C sink potentials over the complete life cycle of plantation stands.

Table 5. Carbon stocks and their allocation in six differently aged *Mytilaria laosensis* stands (values are means ±SE, n = 4; t ha^{-1}).

Layers	Components	7 yr old	10 yr old	18 yr old	23 yr old	29 yr old	33 yr old
tree layer	stem	56.8 ± 3.4^{F}	65.5 ± 4.2^{E}	72.3 ± 4.7^{D}	77.4 ± 6.3^{C}	99.8 ± 8.5^{B}	101.6 ± 9.1^{A}
	bark	4.5 ± 0.5^{D}	5.0 ± 1.0^{C}	5.4 ± 0.6^{B}	5.6 ± 0.8^{B}	7.1 ± 0.4^{A}	7.2 ± 1.3^{A}
	branches	3.2 ± 0.4^{E}	4.7 ± 0.7^{D}	6.7 ± 1.3^{C}	8.7 ± 1.4^{B}	13.9 ± 1.7^{AB}	14.4 ± 1.7^{A}
	leaves	0.5 ± 0.2^{E}	0.7 ± 0.1^{D}	1.0 ± 0.1^{C}	1.3 ± 0.1^{B}	2.1 ± 0.2^{AB}	2.2 ± 0.3^{A}
	total above-ground tree	65.0 ± 2.9^{E}	75.9 ± 4.5^{D}	85.4 ± 4.9^{C}	93.0 ± 5.8^{B}	122.9 ± 9.2^{AB}	$125.\ 4 \pm 10.4^{A}$
	root	12.3 ± 0.9^{E}	15.4 ± 1.7^{D}	18.7 ± 2.1^{C}	21.6 ± 1.4^{B}	30.1 ± 2.3^{AB}	30.8 ± 4.2^{A}
	subtotal	77.3 ± 4.4^{E}	91.3 ± 4.7^{D}	104.1 ± 5.1^{C}	114.6 ± 6.2^{B}	153.0 ± 8.4^{AB}	156.2 ± 10.3^{A}
shrub layer	above-ground	1.3 ± 0.4^{C}	1.1 ± 0.3^{D}	2.2 ± 0.9^{B}	3.4 ± 1.4^{A}	$2.4 \pm 0.7\ ^{B}$	1.3 ± 0.3^{C}
	below-ground	0.6 ± 0.2^{D}	0.7 ± 0.3^{C}	1.2 ± 0.2^{A}	1.0 ± 0.3^{AB}	$1.0 \pm 0.3\ ^{AB}$	0.9 ± 0.4^{B}
	subtotal	1.9 ± 0.6^{D}	1.8 ± 0.5^{D}	3.4 ± 1.1^{B}	4.4 ± 1.3^{A}	3.4 ± 1.1^{B}	2.2 ± 0.6^{C}
herb layer	above-ground	0.4 ± 0.1^{B}	0.2 ± 0.1^{C}	0.5 ± 0.4^{B}	0.7 ± 0.2^{A}	0.8 ± 0.3^{A}	0.2 ± 0.1^{C}
	below-ground	0.6 ± 0.1^{C}	0.5 ± 0.2^{C}	1.1 ± 0.4^{B}	1.5 ± 0.5^{A}	1.5 ± 0.2^{A}	0.6 ± 0.2^{C}
	subtotal	1.0 ± 0.2^{C}	0.7 ± 0.2^{D}	1.6 ± 0.4^{B}	2.2 ± 0.6^{A}	2.3 ± 0.4^{A}	0.8 ± 0.3^{D}
litter layer	undecomposed	$3.8 \pm 0.1.1^{D}$	5.5 ± 1.4^{C}	6.1 ± 1.7^{B}	6.7 ± 1.5^{B}	7.1 ± 2.1^{A}	7.5 ± 1.6^{A}
	semi-decomposed	1.2 ± 0.2^{B}	1.1 ± 0.3^{B}	1.1 ± 0.3^{B}	1.2 ± 0.3^{B}	1.2 ± 0.4^{B}	1.4 ± 0.4^{A}
	subtotal	5.0 ± 1.3^{E}	6.6 ± 1.4^{D}	7.2 ± 1.8^{C}	7.9 ± 1.7^{B}	8.3 ± 2.3^{AB}	8.9 ± 1.8^{A}
soil layer	0–10	28.5 ± 3.2^{E}	23.9 ± 2.4^{F}	26.3 ± 4.2^{D}	29.1 ± 2.5^{C}	46.9 ± 3.4^{B}	52.6 ± 4.1^{A}
	10–30	52.2 ± 3.4^{D}	52.1 ± 5.8^{D}	61.7 ± 4.8^{C}	65.1 ± 4.6^{B}	74.3 ± 7.4^{A}	72.5 ± 6.7^{A}
	30–50	46.1 ± 5.1^{C}	50.3 ± 4.4^{B}	45.9 ± 4.7^{C}	47.4 ± 7.4^{BC}	55.2 ± 4.6^{A}	57.0 ± 4.1^{A}
	50–100	71.4 ± 7.4^{E}	73.3 ± 5.7^{E}	88.8 ± 8.6^{D}	86.3 ± 10.4^{C}	102.3 ± 12.3^{B}	109.9 ± 10.3^{A}
	subtotal	198.2 ± 12.4^{E}	199.6 ± 10.3^{E}	222.7 ± 11.4^{D}	227.9 ± 13.2^{C}	278.7 ± 17.3^{B}	292.0 ± 16.4^{A}
	above-ground ecosystem	71.7 ± 14.2^{E}	83.8 ± 12.4^{D}	95.3 ± 14.2^{C}	105.0 ± 17.6^{B}	134.4 ± 18.8^{A}	135.8 ± 20.4^{A}
	below-ground ecosystem	211.7 ± 20.4^{F}	216.2 ± 17.8^{E}	243.7 ± 19.4^{D}	252.0 ± 21.3^{C}	311.3 ± 25.6^{B}	324.3 ± 27.3^{A}
	total ecosystem	283.4 ± 30.4^{F}	300.0 ± 34.1^{E}	339.0 ± 33.7^{D}	357.0 ± 29.8^{C}	445.7 ± 37.1^{B}	460.1 ± 36.4^{A}

Note: Different Capital letters in the same row indicate significant pairwise differences within components between stand ages; (multiple comparisons test; $P<0.05$).

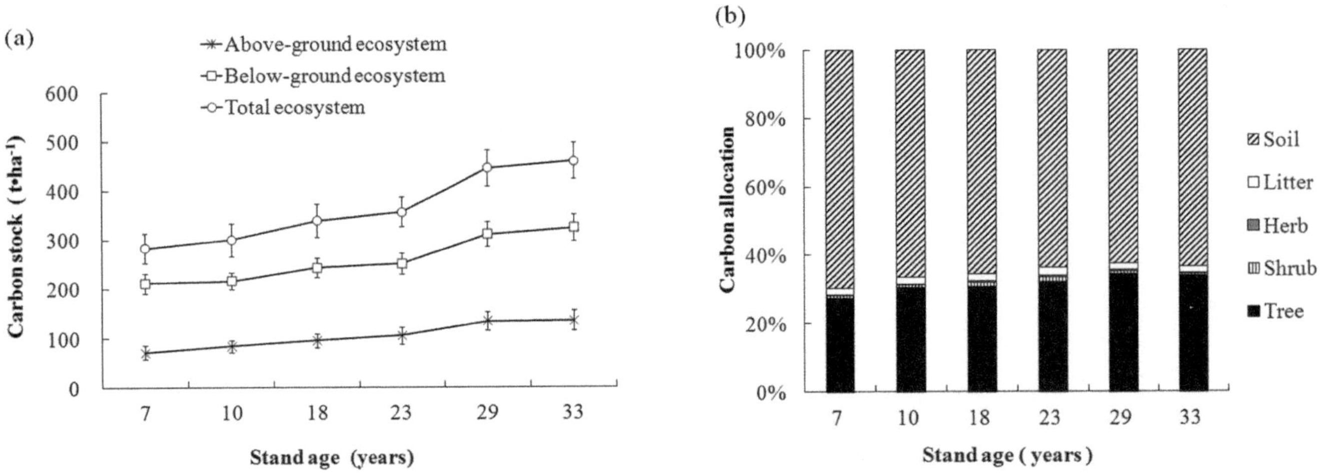

Figure 4. Carbon stocks in ecosystem components (a) and their proportional allocation (b) in six differently aged *Mytilaria laosensis* stands. Values in (a) are means ±SE, n=4.

Acknowledgments

The authors would like to thank Hui Wang, Dongjing Sun, Ji Zeng, You Nong, Haolong Yu, Jixin Tang, Henghui Wen, Yi Tao, Riming He, Hai Chen and Dewei Huang for their assistance in field sampling and data collection. We also thank Zhaoying Li, Lili Li and Bin He for their help in laboratory chemical analyses. We also acknowledge the helpful comments and suggestions of manuscript reviewers.

Author Contributions

Conceived and designed the experiments: AM YL. Performed the experiments: HJ. Analyzed the data: YT. Contributed reagents/materials/analysis tools: JZ. Wrote the paper: AM YL.

References

1. Choi SD, Lee K, Chang YS (2002) Large rate of uptake of atmospheric carbon dioxide by planted forest biomass in Korea. Global Biogeochemistry Cycle 16: 1089.
2. Goodale CL, Apps MJ, Birdsey RA, Field CB, HeathLS, et al. (2002) Forest carbon sinks in the Northern Hemisphere. Ecological Applications 12: 891–899.
3. Houghton RA (2005). Aboveground forest biomass and the global carbon balance. Global Change Biology 11: 945–958.
4. Litton CM, Raich JW, Ryan MG (2007). Carbon allocation in forest ecosystems. Global Change Biology 13: 2089–2109.
5. Brown S (2002) Measuring carbon in forests: current status and future challenges. Environmental Science and Pollution Research 116: 363–372.
6. Gower ST (2003) Patterns and mechanisms of the forest carbon cycle. Annual Review of Environment and Resource 28: 169–204.
7. Houghton RA (2007). Balancing the global carbon budget. Annual Review of Earth and Planetary Sciences 35: 313–347.
8. Kurz WA, Beukema SJ, Apps MJ (1996) Estimation of root biomass and dynamics for the carbon budget model of the Canadian forest sector. Canadian Journal of Forest Research 26: 1973–1979.
9. Vogt K (1991) Carbon budgets of temperate forest ecosystems. Tree Physiology 9: 69–86.
10. Peichl M, Arain MA (2006) Above- and belowground ecosystem biomass and carbon pools in an age-sequence of temperate pine plantation forests. Agricultural and Forest Meteorology 140: 51–63.
11. Peichl M, Arain MA (2007) Allometry and partitioning of above- and belowground tree biomass in an age-sequence of white pine forests. Forest Ecology and Management 253: 68–80.
12. Taylor AR, Wang JR, Chen HYH (2007) Carbon storage in a chronosequence of red spruce (*Picea rubens*) forests in central Nova Scotia. Canada. Canadian Journal of Forest Research 37: 2260–2269.
13. Paquette A, Messier C (2010) The role of plantations in managing the world's forests in the Anthropocene. Frontiers in the Ecology and Environment 8(1): 27–36.
14. Department of Forest Resources Management, SFA (2010). The 7th National forest inventory and status of forest resources. Forest Ecology and Management 1: 3–10.
15. SFA (State Forestry Administration) (2007) China's Forestry 1999–2005. China Forestry Publishing House, Beijing, China.
16. He Y, Li Z, Chen J, Liu Y, Liu D, et al. (2008) Sustainable management of planted forests in China: comprehensive evaluation, development recommendation and action framework. Chinese Forest Science and Technology 7: 1–15.
17. Jagger P (2003) The role of trees for sustainable management of less favored lands: the case of Eucalyptus in Ethiopia. Forest Policy and Economics 5: 83–95.
18. Peng SL, Wang DX, Zhao H, Yang T (2008) Discussion the status quality of plantation and near nature forestry management in China. Journal of Northwest Forestry University 23: 184–188.
19. Poore ME, Fries C (1985) The ecological effects of Eucalyptus. FAO Forestry p 59: 1–97, FAO, Rome, Italy.
20. Wang H, Liu SR, Mo JM, Wang JX, Makeschin F, et al. (2010a). Soil organic carbon stock and chemical composition in four plantations of indigenous tree species in subtropical China. Ecological Research 25: 1071–1079.
21. Borken W, Beese F (2005) Soil carbon dioxide efflux in pure and mixed stands of oak and beech following removal of organic horizons. Canadian Journal of Forest Research 35: 2756–2764.
22. Borken W, Beese F (2006) Methane and nitrous oxide fluxes of soils in pure and mixed stands of European beech and Norway spruce. European Journal of Soil Science 57: 617–625.
23. Kraenzel M, Castillo A, Moore T, Potvin C (2002) Carbon storage of harvest-age teak (Tectona grandis) plantations, Panama. Forest Ecology and Management 5863: 1–13.
24. Laclau P (2003) Biomass and carbon sequestration of ponderosa pine plantations and native cypress forests in northwest Patagonia. Forest Ecology and Management 180: 317–333.
25. Vesterdal L, Schmidt IK, Callesen I, Nilsson LO, Gundersen P (2008) Carbon and nitrogen in forest floor and mineral soil under six common European tree species. Forest Ecology and Management 255: 35–48.
26. Wang H, Liu SR, Mo JM, Zhang T (2010b) Soil–atmosphere exchange of greenhouse gases in subtropical plantations of indigenous tree species. Plant and Soil 335: 213–227.
27. Yuan SF, Ren H, Liu N, Wang J, Guo QF (2013) Can thinning of overstorey trees and planting of native tree saplings increase the establishment of native trees in exotic Acacia plantations in South China? Journal of Tropical Forest Science 25: 79–95.
28. Guo WF, Cai DX, Jia HY, Li YX, Lu ZF (2006) Growth laws of Mytilaria laosensis plantation. Forest Research 19: 585–589.
29. Li YX, Tan TY, Huang JG, Feng YQ (1988) A preliminary report on the densities of Mytilaria laosensis plantation. Forest Research 1: 206–212.
30. Liang RL (2007) Current situation of Guangxi indigenous broadleaf species resource and their development counter-measures. Guangxi Forest Science 36: 5–9.
31. Lin DX, Han JF, Xiao ZQ, Hong CF (2000) Mytilaria laosensis improved on the soil physical & chemical properties. Journal of Fujian College of Forestry 20: 62–65.

32. Lin JG, Zhang XZ, Weng X (2004) Effects of site conditions on growth and wood quality of Mytilaria laosensis plantations. Journal of Plant Resources and Environment 13: 50–54.
33. Ming AG, Jia HY, Tao Y, Lu LH, Su JM, et al. (2012) Biomass and its allocation in 28-year-old Mytilaria laosensis plantation in southwest Guangxi. Chinese Journal of Ecology 31: 1050–1056.
34. Jonard M, Andre F, Jonard F, Mouton N, Proces P, et al. (2007) Soil carbon dioxide efflux in pure and mixed stands of oak and beech. Annual of Forest Science 64: 141–150.
35. Kang B, Liu S, Zhang G, Chang J, Wen Y, et al. (2006) Carbon accumulation and distribution in Pinus massoniana and Cunninghamia lanceolata mixed forest ecosystem in Daqingshan, Guangxi of China. Acta Ecological Sinica 26: 1321–1329.
36. Kasel S, Bennett TL (2007) Land-use history, forest conversion, and soil organic carbon in pine plantations and native forests of south eastern Australia. Geoderma 137: 401–413.
37. Li X, Yi MJ, Son Y, Park PS, Lee KH, et al. (2011) Biomass and Carbon Storage in an Age-Sequence of Korean Pine (Pinus koraiensis) Plantation Forests in Central Korea. Journal of Plant Biology, 54 (1): 33–42.
38. Liu E, Wang H, Liu SR (2012) Characteristics of carbon storage and sequestration in different age beech (Castanopsis hystrix) plantations in south subtropical area of China. Chinese Journal of Applied Ecology 23: 335–340.
39. Ming AG, Jia HY, Tian ZW, Tao Y, Lu LH, et al. (2014) Characteristics of carbon storage and its allocation in Erythrophleum fordii plantations with different ages. Chinese Journal of Applied Ecology, 25(4): 940–946.
40. Soil Survey Staff of Usda (2006) Keys to Soil Taxonomy. United States Department of Agriculture (USDA), Natural Resources Conservation Service, Washington, DC, USA.
41. State Soil Survey Service Of China (1998) China Soil. China Agricultural Press, Beijing, China.
42. Wang WX, Shi ZM, Luo D, Liu SR, Lu LH, et al. (2013) Carbon and nitrogen storage under different plantations in subtropical south China. Acta Ecological Sinica 33: 925–933.
43. He YJ, Qin L, Li ZY, Liang XY, Shao MX, et al. (2013) Carbon storage capacity of monoculture and mixed-species plantations in subtropical China. Forest Ecology and Management 295: 193–198.
44. Zhou Y, Yu Z, Zhao S (2000) Carbon storage and budget of major Chinese forest types. Acta Phytoecology. Sinicea 24, 518–522.
45. Liang HW, Wen YG, Wen LH, Yin QC, Huang XZ, et al. (2009) Effects of continuous cropping on the carbon storage of Eucalyptus urophylla ×E. grandis short rotations plantations. Acta Ecological Sinica 29: 4242–4250.
46. Cleveland C, Townsend A, Taylor P, Alvarez-Clare S, Bustamante M, et al. (2011) Relationships among net primary productivity, nutrients and climate in tropical rain forests: a pan-tropical analysis. Ecology Letters 14: 939–947.
47. Lu LH, Cai DX, Jia HY, He RM (2009) Annual variations of nutrient concentration of the foliage litters from seven stands in the southern subtropical area. Scientia Silvae Sinicae 45: 1–6.
48. Tian D, Yin G, Fang X, Yan W (2010) Carbon density, storage and spatial distribution under different 'Grain for Green' patterns in Huitong, Hunan province. Acta Ecological Sinica 30: 6297–6308.
49. Zhang H, Guan DS, Song MW (2012) Biomass and carbon storage of Eucalyptus and Acacia plantations in the Pearl River Delta, South China. Forest Ecology and Management 277: 90–97.
50. Jandl R, Lindner M, Vesterdal L, Bauwens B, Baritz R, et al. (2007) How strongly can forest management influence soil carbon sequestration? Geoderma 137: 253–268.
51. Grigal DF, Ohmann LF (1992) Carbon storage in upland forests of the Lake States. Soil Science Society of America Journal 56: 935–943.
52. Wang H, Liu SR, Wang JX, Shi ZM, Lu LH, et al. (2013) Effects of tree species mixture on soil organic carbon stocks and greenhouse gas fluxes in subtropical plantations in China. Forest Ecology and Management 300: 43–52.
53. Wang FM, Xu X, Zou B, Guo ZH, Li ZA, et al. (2013) Biomass Accumulation and Carbon Sequestration in Four Different Aged Casuarina equisetifolia Coastal Shelterbelt Plantations in South China. PLOS ONE 8(10): e77449.
54. Zhao JL, Kang FF, Wang LX, Yu XW, Zhao WHH, et al. (2014) Patterns of Biomass and Carbon Distribution across a Chronosequence of Chinese Pine (Pinus tabulaeformis) Forests. PLOS ONE 9(4): e94966.

20

Effects of Climate Variability and Accelerated Forest Thinning on Watershed-Scale Runoff in Southwestern USA Ponderosa Pine Forests

Marcos D. Robles[1]*, Robert M. Marshall[1], Frances O'Donnell[2], Edward B. Smith[3], Jeanmarie A. Haney[1], David F. Gori[4]

1 The Nature Conservancy Center for Science and Public Policy, Tucson, Arizona, United States of America, 2 Northern Arizona University, Flagstaff, Arizona, United States of America, 3 The Nature Conservancy in California, Sacramento, California, United States of America, 4 The Nature Conservancy in New Mexico, Santa Fe, New Mexico, United States of America

Abstract

The recent mortality of up to 20% of forests and woodlands in the southwestern United States, along with declining stream flows and projected future water shortages, heightens the need to understand how management practices can enhance forest resilience and functioning under unprecedented scales of drought and wildfire. To address this challenge, a combination of mechanical thinning and fire treatments are planned for 238,000 hectares (588,000 acres) of ponderosa pine (*Pinus ponderosa*) forests across central Arizona, USA. Mechanical thinning can increase runoff at fine scales, as well as reduce fire risk and tree water stress during drought, but the effects of this practice have not been studied at scales commensurate with recent forest disturbances or under a highly variable climate. Modifying a historical runoff model, we constructed scenarios to estimate increases in runoff from thinning ponderosa pine at the landscape and watershed scales based on driving variables: pace, extent and intensity of forest treatments and variability in winter precipitation. We found that runoff on thinned forests was about 20% greater than unthinned forests, regardless of whether treatments occurred in a drought or pluvial period. The magnitude of this increase is similar to observed declines in snowpack for the region, suggesting that accelerated thinning may lessen runoff losses due to warming effects. Gains in runoff were temporary (six years after treatment) and modest when compared to mean annual runoff from the study watersheds (0–3%). Nonetheless gains observed during drought periods could play a role in augmenting river flows on a seasonal basis, improving conditions for water-dependent natural resources, as well as benefit water supplies for downstream communities. Results of this study and others suggest that accelerated forest thinning at large scales could improve the water balance and resilience of forests and sustain the ecosystem services they provide.

Editor: Julia A. Jones, Oregon State University, United States of America

Funding: The authors have no support or funding to report.

Competing Interests: The authors have declared that no competing interests exist.

* Email: mrobles@tnc.org

Introduction

As we transition into a warmer century, managing forests in the southwestern United States for resilience is more urgent and necessary than ever. Forest conditions across the region have declined significantly. In ponderosa pine (*Pinus ponderosa*) forests of central Arizona, stand densities range from 2 to 44 times greater than during pre-settlement conditions and total basal areas range from 2 to 4 times greater [1–3]. Severe wildfires and drought have caused high tree mortality rates across 14–18% of forests and woodlands in Arizona and New Mexico [4] and these results do not include mortality from two high-severity wildfires in 2011, the Wallow and the Las Conchas wildfires, which burned 217,000 and 63,500 ha (535,000 and 157,000 acres) in Arizona and New Mexico, respectively. Persistent drought conditions have also adversely impacted water-dependent habitats and species within and downstream of forest uplands, as evidenced by heightened mortality of a riparian tree species in central Arizona [5] and declines in number of species and densities of native fish in the Gila River in southwestern New Mexico [6], [7]. Researchers suggest that warming is amplifying the size and severity of wildfire- and drought-mortality events and decreasing snow pack levels [8–11]. Suggested mechanisms involve changes in the water cycle: increased evapotranspiration losses, extended water-stress periods, earlier snowmelt, and lengthened fire seasons. These changes and trends suggest a widespread drying of forests and an increasing risk of uncharacteristic fire and competition-induced water stress and mortality.

At the same time, many communities in the western U.S. are facing critical water shortages and river flows in some basins have diminished or been lost altogether due to unsustainable water management practices [12–15]. Recent studies at the scale of both the Colorado River basin and the state of Arizona concluded that water supplies for many communities will be inadequate to meet future projected demands [16],[17]. Moreover, elevated temperatures are predicted to directly impact river flows. A recent study estimated that for every one-degree Celsius rise in temperature, Colorado river flows will decline by 3–10% [18].

In this study, we explored the ways in which a new era of forest management, including accelerating the pace and scale of forest thinning, could be used to address declining conditions in forests in central and northern Arizona. At about 6,000 hectares per year (15,000 acres/year), the current pace of mechanical thinning in these forests does not begin to match the magnitude of recent forest disturbances. However, a new congressionally funded program called the Four Forest Restoration Initiative (hereafter "4FRI") will accelerate the use of mechanical thinning and prescribed burns across four national forests, treating 238,000 ha (588,000 acres) in the first analysis area over the next 10 years. The objective of the 4FRI project is "...to re-establish forest structure, pattern, and composition, which would lead to increases in forest resiliency and function," including reductions in fire risk and improved watershed function [19]. Mechanical thinning alone or in combination with prescribed fires reduced fire risk, increased runoff, and improved tree water-stress at the plot scale [20–23]. The 4FRI project provided the opportunity to evaluate the effectiveness of forest thinning to improve forest resilience and ecosystem functioning at larger scales.

The objectives of this study were to: (1) investigate how variation in climate and the pace and extent of thinning of ponderosa pine forests affects runoff in the Salt and Verde watersheds of central Arizona; and (2) explore the management implications of using accelerated thinning to improve forest resilience in a time of declining forest and water conditions. We chose the Salt-Verde watersheds because forests from these watersheds provide water for 1.5 million people in Phoenix, which is the 6th largest city in the United States. We focused on ponderosa pine because this forest type produces 50% of the runoff in these watersheds even though it accounts for only 20% of the area [24]. We modified a forest runoff model developed from historical paired watershed experiments in the Beaver Creek sub-watershed within our study area [20] and ran the model in multiple scenarios to estimate additional annual runoff from mechanical thinning. The scenarios simulated forest treatment at two scales (a) landscape-scale thinning planned within the first analysis area of the 4FRI project and (b) watershed-scale thinning in forests across the Salt-Verde watersheds (Figure 1). Scenarios were designed to account for variability in winter precipitation, as well as the pace and extent of forest treatments. We estimated runoff in periods with below-average precipitation, herein referred to as droughts, and in periods with above-average precipitation, herein referred to as pluvials, using a model of 20th century precipitation in ponderosa pine forests in the region [25]. We believe this is the first attempt to estimate the influence of mechanical thinning on runoff over multi-year broad-scale restoration projects that accounts for the effects of climate variability.

Materials and Methods

The subsequent sections provide detailed information on the study area, the development of the modified runoff model, the method we used to simulate variability in winter precipitation, the 4FRI runoff scenarios, and the Salt-Verde runoff scenarios. All statistical analyses and graphics were completed using the statistical software SigmaPlot 11 (Systat software Inc, San Jose, CA), except for regression model fitting which was performed using Matlab 2013 (Mathworks, Natick, MA). Methods and results are reported in SI units, but for accessibility to forest and water managers, we reported English units in parentheses within the text and included all study figures and tables in English units as File S1. Area and water volume amounts were rounded off to three significant figures.

Study Area

Ponderosa pine forests occupy approximately 0.68 million ha (1.68 million acres) of the 3.46 million ha (8.56 million acres) within the Salt-Verde watersheds in central Arizona (Figure 1). A portion of these forests, all within the Verde watershed, will be mechanically thinned in the 4FRI project in the next 10 years. Within the Salt-Verde watersheds, ponderosa pine forests grow at mid-elevations, from 1,800 to 2,600 m (5,900–8,600 feet) [26], [27], on soils derived from basalt (59%) and sedimentary rock (41%) [28]. Mean annual precipitation for ponderosa forests in the southwestern United States ranges from 510 to 760 mm (20 to 30 inches) and is highly variable [29]. Sixty percent of annual precipitation falls primarily as snow in the winter and spring months from October to April, and the melting of the resulting snowpack accounts for 80 to 90% of annual stream flow [29].

Forest Treatment Runoff Model

For the purposes of this study, we were interested in estimating the additional runoff that becomes available due to forest thinning on an annual basis, not absolute values of watershed runoff. We define "additional runoff due to thinning" as the portion of precipitation that appears as surface water at the sub-watershed outlet and that is directly attributable to mechanical thinning treatments. This additional runoff can be considered as "in-place" or "in-situ" as we did not attempt to model runoff accumulation, routing, groundwater outflows or inflows, or channel losses in downstream watersheds.

We modified the Baker-Kovner logistical regression equation from the historical Beaver Creek sub-watershed experiments to calculate annual runoff in our study [20]. We selected this model because the Beaver Creek sub-watershed is actually within our study area (Figure 1), and because the model was based upon measured empirical data. These experiments and others like it demonstrated that reductions in forest basal area of 30–100% in ponderosa forests on basalt-derived soils increased runoff between 15–40% for up to six years after the initial treatment [29]. Runoff increases were negligible for basal area reductions below 30%, six years after treatments, and for years with winter precipitation below 230 mm (9 inches) [20], [29]. The original model predicted annual watershed runoff based on three independent variables: winter precipitation (Oct-Apr), forest basal area, and a solar insolation index, and explained a fair amount of variability in total watershed runoff [20].

To derive increases in runoff directly associated with thinning using the paired watershed method, researchers in the Beaver Creek experiments first established a relationship between stream flows in the control and treated watersheds before treatments [30]. This relationship was used to predict what would have been baseline flows in the treated watershed. The difference between measured flows in the treated watershed and this predicted value were attributed to thinning effects. We compared data calculated in this fashion in the Beaver Creek experiments [30], [31] and one additional nearby site, Castle Creek [32], to output generated by the Baker-Kovner model. The output from the model was calculated as the difference in post-treatment and pre-treatment watershed runoff using post- and pre-treatment basal areas respectively and holding all other variables constant. In this comparison, we found the fit of the model output to the portion of runoff due to thinning to be much poorer ($r^2 = 0.43$, Figure 2a)

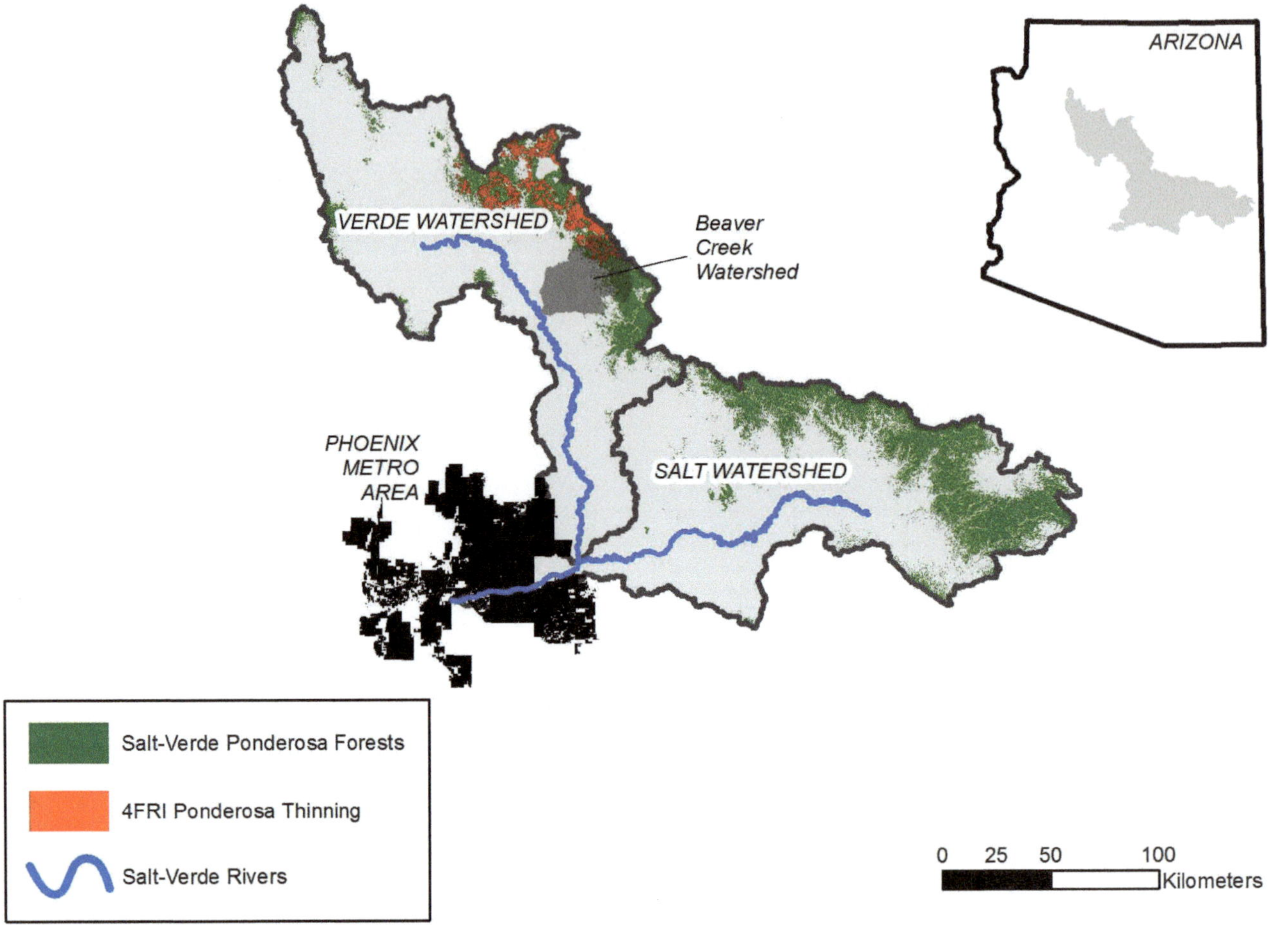

Figure 1. Map of Study Area. Map showing ponderosa pine forests in Salt-Verde watersheds in central Arizona, including those forests that are slated for mechanical thinning within the 4FRI project. Runoff from snowmelt in these forests is primary source of flow to Salt-Verde rivers which in turn are major sources of water for communities in the Phoenix Metro Area. Study used runoff model developed from experimental studies conducted in Beaver Creek watershed [20]. *Inset*: Location of study area in Arizona.

than model fit to total watershed runoff ($r^2 = 0.69$). We concluded that the original regression model was relatively insensitive to the direct effects of forest treatments, including time since treatment, on runoff.

We used a stepwise regression procedure to develop a revised model, beginning with an initial model that included the independent variables (winter precipitation and basal area) that were found to be significant in historical experiments [20]. Input data for all independent variables used to develop the model were derived from empirical data collected during the Beaver Creek and Castle Creek historical experiments and archived in databases or published reports [30–32]. We added independent variables and used an F-test to test the null hypothesis that the coefficient of each term was zero at the 95% significance level. To ensure that the model was not over-fitted, we repeated the stepwise regression procedure with 70% of the original data, selected randomly. We determined that the structure of the model did not change and provided a good fit to the remaining 30% of the data. We examined the residuals from the model and found no discernible patterns of dispersion or bias. Including terms for slope, percent of watershed area with a south-facing aspect, and mean, minimum, and maximum winter temperatures did not significantly improve the model.

The final model included three terms: winter precipitation, an interaction term between winter precipitation and years since treatment, and an interaction term between winter precipitation and basal area. It can be expressed as:

$$R_{treatment} = -28.464 + 0.148 * P - 0.015 * P * Y - 0.092 * P * [\exp(-BA_1/10.33) - \exp(-BA_2/10.33)]$$

$r^2 = 0.67$

where

$R_{treatment}$ = Increase in Annual Runoff attributed to Forest Thinning in mm

P = Total Winter Precipitation (Oct-Apr) in mm

Y = Years since Treatment

BA_1 = Basal Area before Treatment in m^2/ha

BA_2 = Basal Area after Treatment in m^2/ha

Figure 2b shows that the revised model explained more of the variability in predicting additional runoff associated with forest treatments ($r^2 = 0.67$) than did the original model. The model predicted runoff increases in a linear fashion with winter precipitation and basal area reductions, and decreases linearly

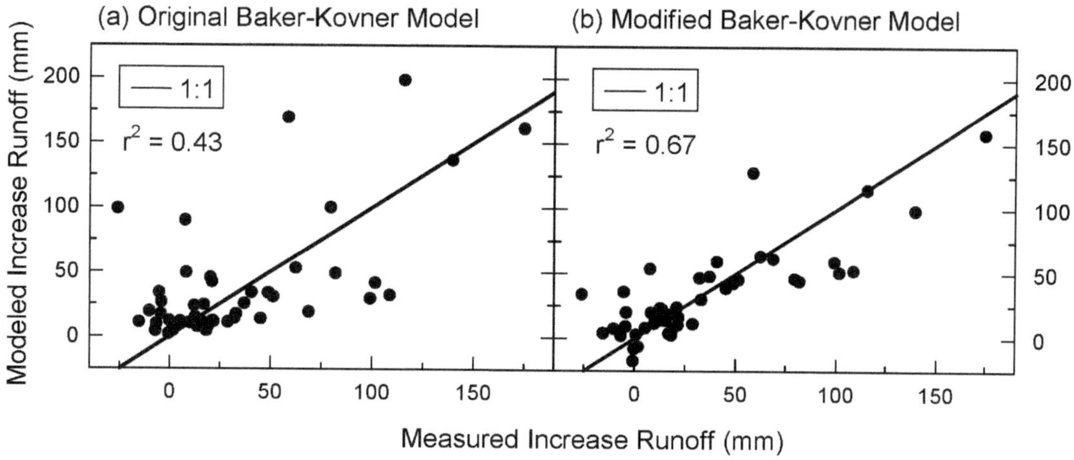

Figure 2. Comparison of models to observed runoff. Fit of (a) original [20] and (b) modified Baker-Kovner regression model output to increases in runoff associated with forests treatments in central Arizona, from Beaver Creek [30],[31] and Castle Creek [32] watersheds.

with years since treatment (Figure 3). To compare the original and modified Baker-Kovner models in English Units, see File S2.

We did use the original Baker-Kovner regression equation in this study to calculate total watershed runoff for forests in their current condition so that we could estimate the percentage increase in runoff associated with treatments. For these calculations, we used the average value for the solar insolation index, r = 0.7, reported from the Beaver Creek experiments [20]. In a sensitivity analysis, we found that calculating absolute runoff based on the low (r = 0.66) and high values (r = 0.74) of this index changed the estimated increase from treatments by only 1–2 percentage points.

Winter Precipitation

In order to account for inter-annual and decadal variability of winter precipitation, we used the PRISM model [25] to extract pluvials and droughts from the 20th century and inserted levels of winter precipitation from these periods into our runoff scenarios. We first extracted total winter precipitation, summed across October to April, for every year from 1900–2012 from the PRISM modeled dataset [25] across 2 scales: (a) ponderosa pine forests within the Verde watershed and (b) these forests within the Salt-Verde watersheds. We selected these geographies to be most representative of conditions for the 4FRI and Salt-Verde runoff scenarios, respectively. From 1900–2012, mean winter precipitation in Verde ponderosa pine forests was 394 mm (15.5 inches) with a range of 99–815 mm (3.9–32.1 inches) (Figure 4). For Salt-Verde forests, the mean was 368 mm (14.5 inches) and the range was 89–747 mm (3.5–29.4 inches) for this same period of time.

To test the accuracy of these modeled data, we compared values from the Verde ponderosa pine PRISM modeled data to measurements of total winter precipitation recorded during the original Beaver Creek experiments [31]. Given that we found the distribution of the data to be non-normal, we ran the non-parametric Spearman's rank order correlation. We found a strong association between modeled and observed data (Rho = 0.9577, $p<0.001$) although the model underestimated measured winter precipitation over the time period by approximately 25 mm (1 inch) and consistently under-estimated precipitation in wet winters (Figure 4 inset).

We calculated mean winter precipitation across 15-, 25-, and 35-year periods from this modeled data and used this information to select droughts and pluvials. For each scenario, we moved the start and end years that thinning was simulated to occur. For example, time periods for our 15-year scenarios included 1944–1958, 1945–1959, and 1946–1960 (examples illustrated as shaded areas in Figure 4). Using this method, we found two instances of pluvials – early 20th century and late 1970s to late-1990s – and two droughts – during the 1950s and the current drought.

Estimating runoff from 4FRI project

We used the modified regression model to calculate additional annual runoff from individual forests stands that will be thinned in the first analysis area of the 4FRI project, established an annual thinning schedule, and then summed increases in runoff from individual stands across treatment years. We ran a total of 26 scenarios where in each scenario we inserted a different 15-year winter precipitation sequence.

We obtained a Geographic Information System (GIS) dataset of alternative C in the draft environmental impact statement for the first analysis area of the 4FRI project [19]. For each stand to be mechanically thinned, this dataset contained estimates of current basal areas and desired post-treatment basal areas. Forest prescriptions for "group-selection" stands called for two post-treatment basal areas, one for "open" areas that would be evenly thinned and another for "group" areas where thinning would be minimal to enhance wildlife habitats by retaining greater tree densities. For these stands, we calculated a post-treatment basal area that was an average of these two basal areas, weighted by the proportional area of the open and group areas. We selected 4,064 ponderosa pine stands within the Verde watershed slated for mechanical thinning, where the prescription will reduce basal areas by at least 30%. They ranged in size from 0.4 to 217 ha (1 to 536 acres), for a total of 61,900 ha (153,000 acres). The prescriptions within these stands called for reductions of basal area between 30% and 70% with a mean reduction of 48% (Figure 5).

Based on consultation with US Forest Service staff, we found that it was reasonable to assume that an equal number of hectares will be mechanically thinned every year across the 10-year treatment period. So for the purposes of our 4FRI scenarios, we assigned each of these stands to one of ten cohorts until the area of each cohort equaled 1/10th of total area or approximately 6,190 ha (15,300 acres). Figure 6a illustrates the treatment schedule for

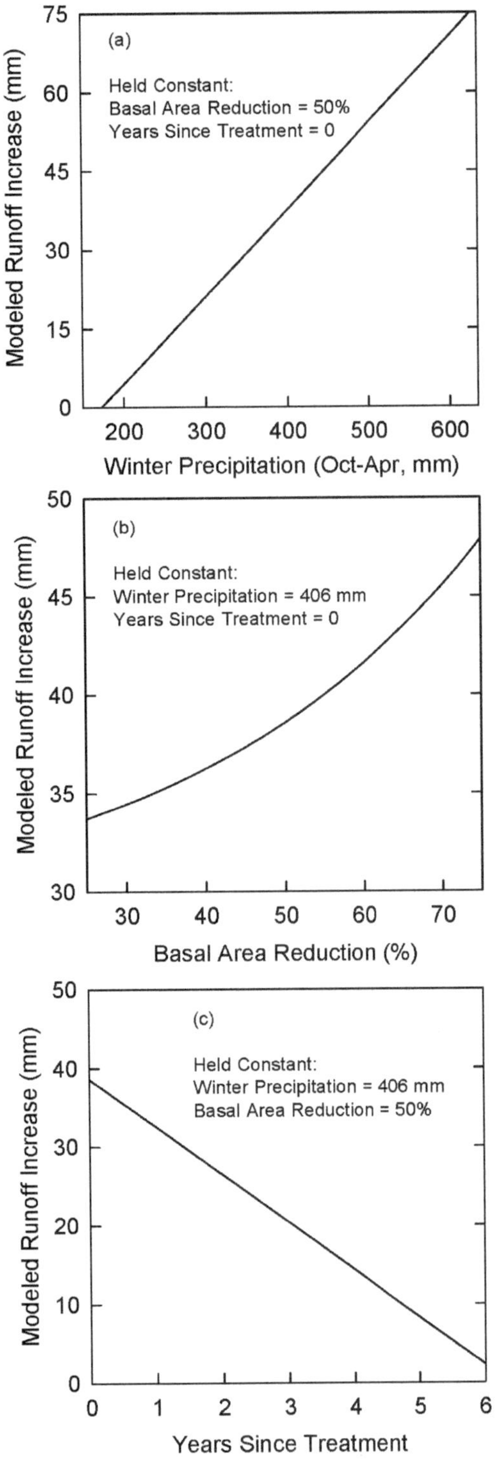

Figure 3. Influence of independent variables on runoff model. Relationships between model output to values for independent variables, including (a) winter precipitation, (b) percent basal area reduction, and (c) years since treatment. In all cases, other independent variables are held constant in order to view relationship of independent variable plotted on X-axis to model output.

the 15-year 4FRI scenarios reported in this study. Cohorts were treated consecutively in the first 10 years and it was assumed that cohorts influenced runoff for 6 years. So, for example, stands treated in cohort 1, contributed to additional runoff in scenario years 1-6 and stands treated in cohort 10 contributed from years 10–15.

We calculated additional runoff associated with thinning and total watershed runoff by inserting values of the independent variables into the revised and original Baker-Kovner regression models, respectively. Pre-treatment and post-treatment basal area values were derived from the 4FRI stand data. Years since treatment ranged from y = 0 to 5. As described previously, winter precipitation values were droughts and pluvials drawn from the PRISM model for ponderosa forests in the Verde watershed. We set runoff increase to zero in years when winter precipitation was less than 230 mm (9 inches). We transformed the unit of runoff from mm to volume (million m^3) using the area of each stand. We estimated annual increases in runoff at the landscape scale by summing stand level amounts for each scenario year, and calculated summary statistics (mean, median, max, cumulative) that allowed for a comparison of scenarios.

Estimating runoff from Salt-Verde watersheds

Unlike the first analysis area of the 4FRI project, there was not a planned forest restoration project across the larger geography of the Salt-Verde watersheds. We developed a range of estimates for the extent, pace, and intensity of forest thinning that could be conducted over this larger geography, grouped these estimates into runoff scenarios, and ran the scenarios using the revised and original regression models to estimate additional runoff from treatments and total watershed runoff. To estimate the potential areal extent of thinning, we subtracted from the total forested areas those land uses that are typically considered unsuitable for mechanical thinning. We bracketed this initial area estimate with lower values to account for the lack of comprehensive spatial information. Assuming that the pace of treatments at this scale could take longer, we constructed 15-, 25-, and 35-year runoff scenarios. Finally, we assumed that the intensity of forest thinning at this scale would be similar to the range of basal area reductions planned for the first analysis are of the 4FRI project. The remaining parts of this section describe these steps in greater detail.

Based on a methodology developed in a previous study that estimated wood supply on four National Forests in northern Arizona [33], we subtracted from the total area of ponderosa pine in the watersheds – 0.68 million ha (1.68 million acres) – those areas that are typically excluded from mechanical thinning projects due to steepness, restrictions in land management, previous treatments, or considerations of soil, habitat, or wildlife conditions (Table 1). We adopted 6 of the 7 "exclusion" criteria from that study and added one additional factor. We removed one criterion – Northern goshawk nest areas –because the mean basal area reduction for these areas in the 4FRI project was greater than 30% [19] suggesting that thinning in these areas could result in additional runoff based on the results of the Beaver Creek experiments [20]. We added one criterion – high severity burn patches – because a nearby study on the recovery of ponderosa pine forests after wildfire demonstrated that mature forests suitable for thinning would not likely develop in areas that had burned with high severity within the timeframe of this study [34]. While two of the three 4FRI prescriptions for Mexican Spotted Owl Protected Activity Centers (MSO PAC) would result in negligible changes in basal area, one would allow for basal area reductions greater than 30% [19]. However, we were not able to differentiate among these prescriptions from the data we obtained, so we opted for a conservative estimate for this factor by excluding MSO PAC areas from our analysis. Compiling the publicly available geospatial data

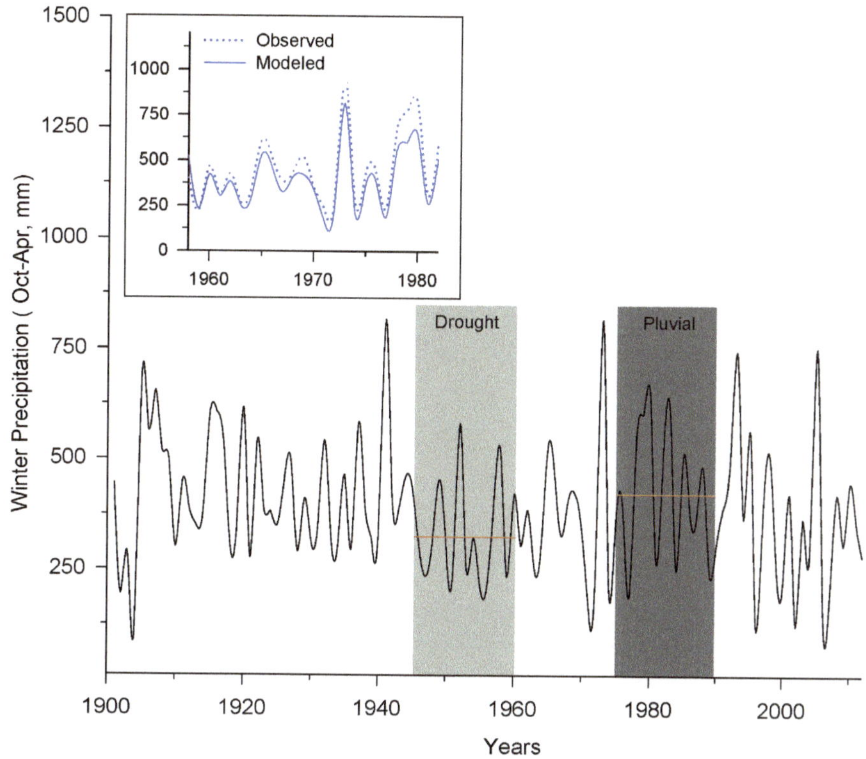

Figure 4. Variability in winter precipitation in ponderosa pine forests. Estimates of historical winter precipitation from 1900–2012 in ponderosa pine forests within Verde watershed from PRISM model [25]. Shaded areas are examples of 15-year droughts and pluvials that were used in study scenarios; horizontal red lines represent mean winter precipitation within these shaded areas. *Inset*: Comparison of measured winter precipitation observed during the historical Beaver Creek watershed experiments [20] from 1958–1982 versus modeled winter precipitation data shown in main figure.

for these factors [26], [27], [33], [35–38], we found that 401,000 ha (992,000 acres), or 59% of ponderosa pine forests, were available for mechanical thinning after accounting for these exclusion areas. The percentage of forests excluded in this estimate – 41% – compared favorably with the average of 37% of forests that were excluded in dozens of restoration projects evaluated in the northern Arizona [33]. We bracketed this high estimate with a moderate estimate – 45% or 301,000 ha (743,000 acres) – and a low estimate – 30% or 204,000 ha (505,000 acres) – because quantitative geo-spatial data on some exclusion areas was missing or incomplete. As illustrated graphically for the moderate estimate in Figure 6b-d, we divided these areas into 10-, 20-, and 30-year treatment cohorts assuming that equal areas would be thinned every year.

We scaled up the forest condition information contained in the 4FRI stand data using a proportion metric, assuming that these stands were representative of stand-level forest conditions across the Salt-Verde watersheds. For example, 4FRI stands that comprised 1,500-ha in a 15,000-ha 4FRI treatment cohort were estimated to be 3,000-ha in a 30,000-ha Salt-Verde cohort. If the thinning prescription for these stands called for a reduction in basal area from 33 to 17 m^2/ha, then the Salt-Verde scenarios calculated additional runoff from 3,000-ha that are thinned from 33 to 17 m^2/ha. After scaling up the area of each stand to the watershed scale, the procedure for estimating runoff in the Salt-Verde scenarios was the same as described for the 4FRI scenarios. We ran 147 scenarios in total, 67 fifteen-year, 52 twenty-five year, and 28 thirty-five year scenarios.

Results

Four Forest Restoration Initiative Runoff Scenarios

Mechanical thinning of ponderosa pine forests in the first analysis area of the 4FRI project – 6,190 ha/year for ten years totaling 61,900 ha (15,300 ac/year; 153,000 acres total) – increased mean annual runoff from 3.13 million m^3 (2,540 acre-feet) in a simulated drought to 7.27 million m^3 (5,890 acre-feet) in a pluvial. Differences in winter precipitation over the 15-year simulation periods significantly influenced runoff gains. A difference in mean winter precipitation of only 130 mm (5 inches), from 330 mm (13 inches) in drought scenario to 460 mm (18 inches) in a pluvial scenario, resulted in a doubling of the annual increase in runoff from treatments (Figure 7).

Inter-annual variability in winter precipitation was also important. Figure 8 illustrates year to year increases in runoff for those scenarios that resulted in the lowest and highest gains in runoff. Years with high winter precipitation played a disproportionate role in additional runoff in both droughts and pluvials. In the majority of scenarios ran (15 out of 26), winter precipitation from only 5 of the 15 years accounted for at least 75% of the increased runoff. Cumulative increases in runoff across 15-year periods resulted in a total increase from 54.3 to 111 million m^3 (44,000–89,800 acre-feet) (Figure 9). Runoff from thinned forests was approximately 20% greater than unthinned forests (as estimated using original Baker-Kovner regression model) in both droughts and pluvials (data not shown).

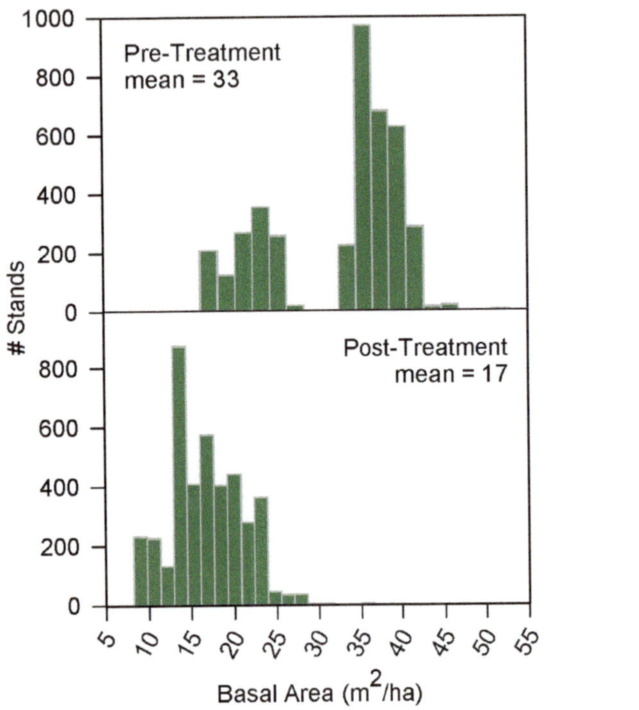

Figure 5. Ponderosa pine basal area reductions in 4FRI project. Histograms showing (top) pre-thinning and (bottom) desired post-thinning basal areas (in m^2/ha) of ponderosa pine stands in the first analysis area of the 4FRI project (excluding stands where basal area reduction <=30%).

Salt-Verde Runoff Scenarios

Depending on winter precipitation and the forest treatment schedule, mean annual increases in runoff from thinning of ponderosa forests across the Salt-Verde watersheds ranged from 4.76 to 15.0 million m^3 (3,860–12,200 acre-feet) over a 35-year treatment period, 6.18 to 23.4 million m^3 (5,010 to 19,000 acre-feet) over 25 years, and 9.23 to 42.8 million m^3 (7,480 to 34,700 acre-feet) over 15 years (Table 2). Similar to the 4FRI scenarios, additional runoff in the Salt-Verde watersheds was 1.6–2.3 times greater in pluvials than in droughts. Regardless of whether the scenarios occurred in a drought or pluvial, cumulative runoff gains in thinned forests were 20-26% greater than unthinned forests. Cumulative gains ranged from 167 to 525 million m^3 (135,000–426,000 acre-feet) in the 35-year scenarios, 154 to 585 million m^3 (125,000–474,000 acre-feet) in 25-year scenarios, and 138 to 643 million m^3 (112,000–521,000 acre-feet) in 15-year scenarios. In both droughts and pluvials, runoff increased in a positive linear fashion with increases in the pace and the extent of forest thinning (Figure 10). See File S1 to see study figures and tables in English units.

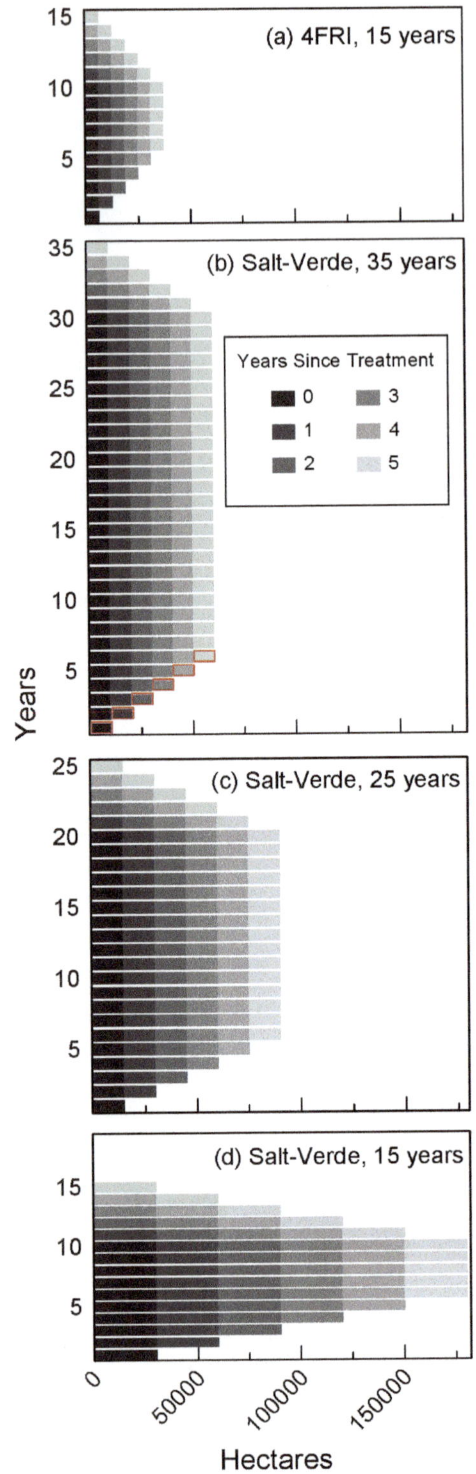

Figure 6. Forest treatment schedules for study scenarios. Graphical depiction of mechanical thinning treatment schedules for (a) 15-year 4FRI scenarios and (b-d) 35-year, 25-year, and 15-year Salt-Verde moderate thinning scenarios (total thinned area was 301,000 ha or 743,000 acres). Scenarios assumed consecutive treatments for 10-, 20-, and 30-year treatment periods shown as black bars in the bottom left portion of each of the figures. Bars outlined in red in (b) show the contribution of one cohort of stands through six years in the scenario.

Discussion

This study demonstrated that the pace, extent, and intensity of forest thinning that is planned under the first analysis area of 4FRI and at larger scales could measurably increase runoff in ponderosa pine forests in central Arizona. Modeled runoff from thinned forests was approximately 20% greater than runoff from unthinned forests which is within the range of 10–40% increases in runoff demonstrated in the Beaver Creek experiments [20], [29]. The Salt-Verde runoff scenarios showed that levels of

Table 1. Estimate of hectares of ponderosa pine forests available to mechanical thinning in Salt-Verde watersheds.

Category	Hectares	%	*Sources*
Salt-Verde Watersheds Ponderosa Pine Forests	*681,000*	-	[27]
Exclusions			
Steep Slopes greater than or equal to 40%	68,000	10%	[26]
Hi-Severity Burn Patches within Wallow and Rodeo-Chedeski Fires	43,700	6%	[35]
Specially Designated Areas including Wilderness Areas, Research Natural Areas, State Wildlife Management Areas[a]	21,800	3%	[36],[37]
Streamside Management Zones within 100 feet perennial reaches in national forests and 200 feet in tribal areas	8,050	1%	[38]
Mexican Spotted Owl Protected Activity Centers where owls have been found to be nesting[b]	57,500	8%	[33]
Erodible Soils where thinning is unsuitable due to erosion risk or rocky conditions[c]	93,100	14%	[33],[37]
Completed Treatments that have been thinned in last 10 years[d]	35,200	5%	[37]
Sub-Total Exclusions (accounting for overlap between layers)	*280,000*	*41%*	
Sub-Total Forest Available Mechanical Thinning	*401,000*	*59%*	

[a] Estimate derived from USDA Forest Service Southwestern Region Wilderness Status layer and Forest Level Special Interest Management Areas layers [37].
[b] Data available for Coconino, Apache-Sitgreaves, Kaibab, and Tonto National Forests only. Added qualitative estimate for this exclusion factor for tribal lands and Prescott National Forest assuming land excluded in these land units would be proportional to lands excluded for this factor on the four national forests above.
[c] List of excluded soils for Coconino, Apache-Sitgreaves, Kaibab, and Tonto National Forests based on Hampton et al. [33]. This list of excluded soils was augmented based on consultation with USDA Forest Service southwestern region soil scientists. Geospatial data for these soils derived from USDA Forest Service Forest Level Terrestrial Ecological Units layers [37]. Added qualitative estimate for this exclusion factor for tribal lands and Prescott National Forest where geospatial data was unavailable, assuming land excluded in these land units would be proportional to lands excluded for this factor on the four national forests above.
[d] Geospatial data for this factor derived from USDA Forest Service Forest Level Activities Layer joined to FACTs table for Coconino, Apache-Sitgreaves, Kaibab, Tonto, and Prescott National Forests [37]. Added qualitative estimate for this exclusion factor for tribal lands based on consultation with Bureau of Indian Affairs forestry personnel.
Summary of hectares and percentages of ponderosa pine forests within Salt-Verde watersheds in relation to potential for thinning treatments, including total hectares, those areas typically excluded from mechanical thinning, and the remaining hectares available for mechanical thinning. Estimates rounded off to three significant digits.

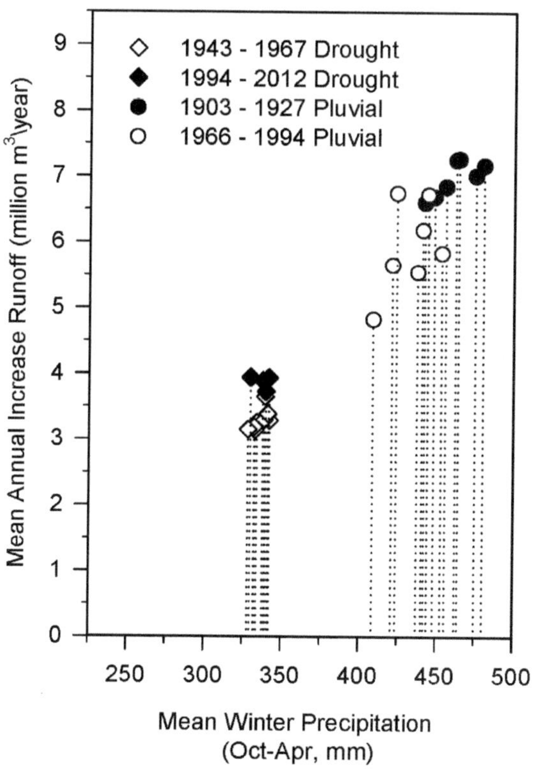

Figure 7. Increases in mean annual runoff from thinning in 4FRI project. Results from 26 scenarios with varying levels of winter precipitation showing increases in mean annual runoff associated with mechanical thinning of ponderosa pine forests in the first analysis area of the 4FRI project. In order to compare scenarios, only increases in *mean* annual runoff are shown. Annual variability in runoff for two of these scenarios is shown in Figure 8.

additional runoff increased proportionally with increases in the pace and extent of thinning. Runoff gains occurred in droughts and pluvials, a surprising outcome that underscored the importance of inter-annual and decadal variability in precipitation. These increases in runoff would likely improve conditions for water-dependent natural resources, such as cienegas, riparian areas, and aquatic habitats, which are vulnerable to low flows that are experienced seasonally, especially in summer months, and also during droughts [5–7]. Runoff gains also could provide incidental benefits to the water supply of downstream users, but the increases were more modest when compared to total runoff from the Salt-Verde watersheds. They would comprise a 0–3% increase from mean annual surface flows from the Salt-Verde rivers of 1.39 billion m^3 (1.13 million acre-feet) and approximately a 1-9% increase of Salt-Verde flows supplied to municipal users in the Phoenix Metro Area on an annual basis [39].

Consistent with the historical experiments, thinning effects on runoff were temporary in this study. Additional runoff would cease six years after the multi-year thinning schedules in our scenarios. Beaver Creek investigators speculated that thinning effects were short-lived because of subsequent regrowth of understory species [29]. We note that prescribed fires are planned as subsequent treatments for all stands that are initially thinned within the first analysis area of 4FRI [19]. These maintenance treatments should provide additional information as to how effective subsequent actions are in terms of removing understory vegetation and sustaining runoff gains.

Our study adds to previous research on the potential for increases in runoff associated with thinning of ponderosa pine forests in the Southwest. Using a simulation model, Brown and Fogel [40] found that thinning approximately 9,300 ha (23,000 acres) of ponderosa pine per year over a 10-year period produced increases in runoff that ranged from 0.28 to 11.8 million m^3/year (230–9,600 acre-feet/year). This range overlapped with but was

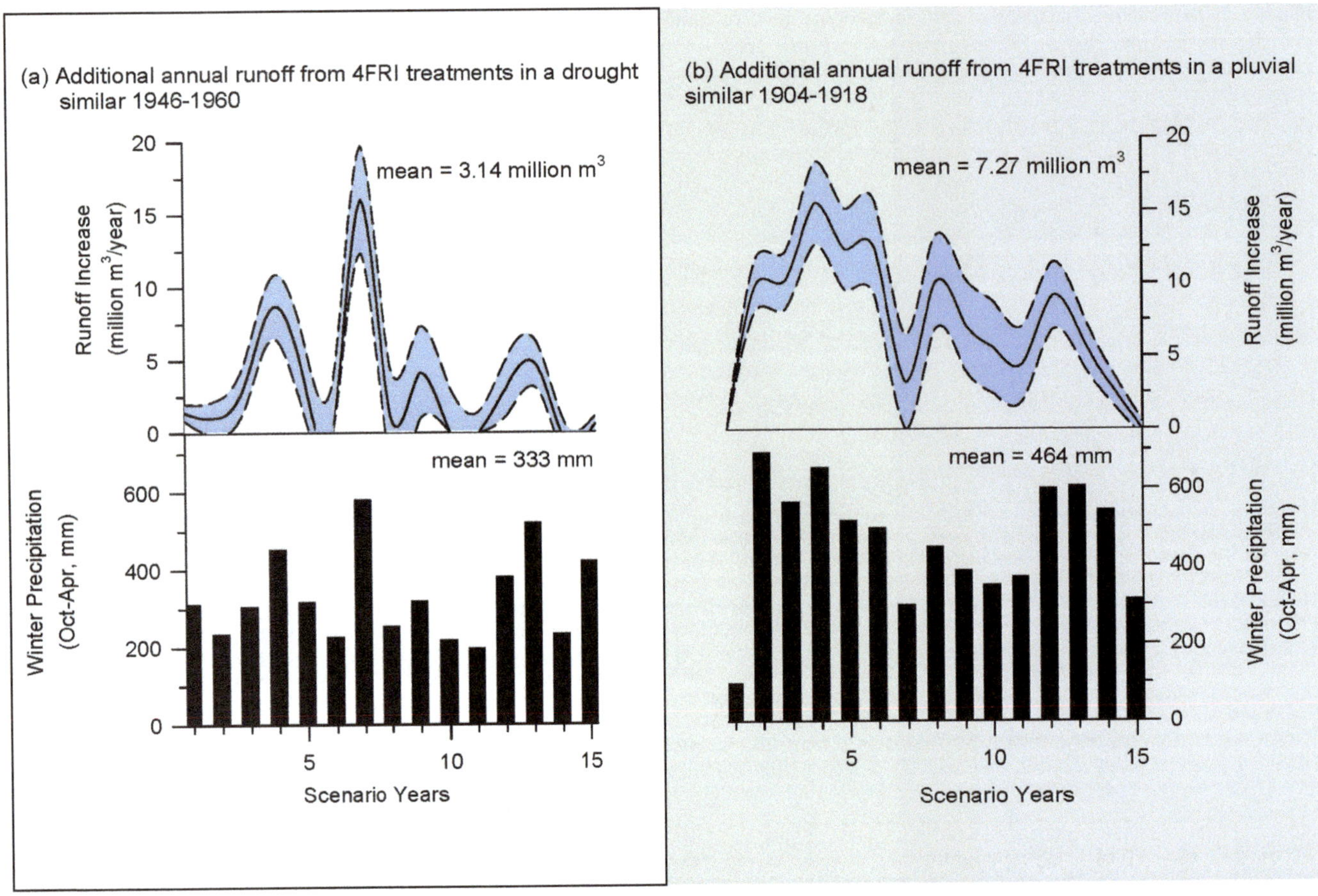

Figure 8. Year to year variability in runoff increases from thinning in 4FRI project. Modeled increases in annual runoff associated with mechanical thinning of ponderosa pine forests in the first analysis area of the 4FRI project during the (a) drought and (b) pluvial that produced lowest and highest gains in runoff respectively. Top panes show increases in annual runoff in million m^3/year. Solid black lines are output values from regression model; dotted lines and blue areas represent 90% confidence intervals. Bottom panes show corresponding winter precipitation values (Oct-Apr, mm) used as one of the independent variables to calculate runoff.

lower than the 6.18 to 11.9 million m^3/year (5,010–9,660 acre-feet/year) increases in runoff for the two Salt-Verde runoff scenarios with annual thinning of approximately 10,000 ha (25,000 acres) (Table 2). One potential reason for this difference was that Brown and Fogel [40] modeled the treatment-runoff relationship to be constant over time, 18 mm (0.71 inches), and did not evaluate variability in precipitation but instead assumed average winter precipitation of 396 mm (15.6 inches). Using similar values for winter precipitation and basal area reduction, our model simulated the runoff response to vary depending on years since treatment, from a high of 38 mm (1.5 inches) in the year immediately following treatments to a low approaching zero mm after six years (Figure 3c).

Study Limitations

In this study, we revised a statistical runoff model from the historical Beaver Creek watershed experiments to explore how variations in climate and forest thinning would affect runoff. To the extent possible, we limited the application of the runoff model to the range of climatic and environmental conditions under which the original data were collected and model developed. However, our modeling approach was constrained by several uncertainties that will require further investigation. The model was limited by the scope of forest practices conducted in the Beaver Creek watershed experiments [20], [30–32]. These experiments used thinning techniques such as strip-thinning and patch clearing that had a similar range of basal area reductions but a different spatial configuration from contemporary thinning prescriptions. They did not measure the effects of maintenance treatments on runoff. They measured "in-place" runoff at sub-watershed outlets and did not measure other aspects of the water budget, such as evapotranspiration, soil moisture, surface routing, and groundwater recharge. The revised model represented statistical relationships between annual runoff and winter precipitation, forest basal area reductions and time since treatment, but these relationships were not necessarily related to physical processes or functions that control water balance [20]. We summarized how important factors that are not directly modeled may increase or decrease the range or runoff due to thinning that is documented in this study in File S3.

Of all the factors where uncertainties in forest hydrology and runoff remain, perhaps the most significant were the effects of soil types, long-term climate variability, and climate change. In terms of soil types, the revised regression model in this study was derived from experimental thinning on basalt-derived soils that support 59% of ponderosa pine forests in the Salt-Verde watersheds. The runoff effects of thinning on the remaining 41% of forests on sedimentary-derived soils are unknown. Two studies demonstrated that runoff from untreated ponderosa forests on sedimentary-derived soils can be

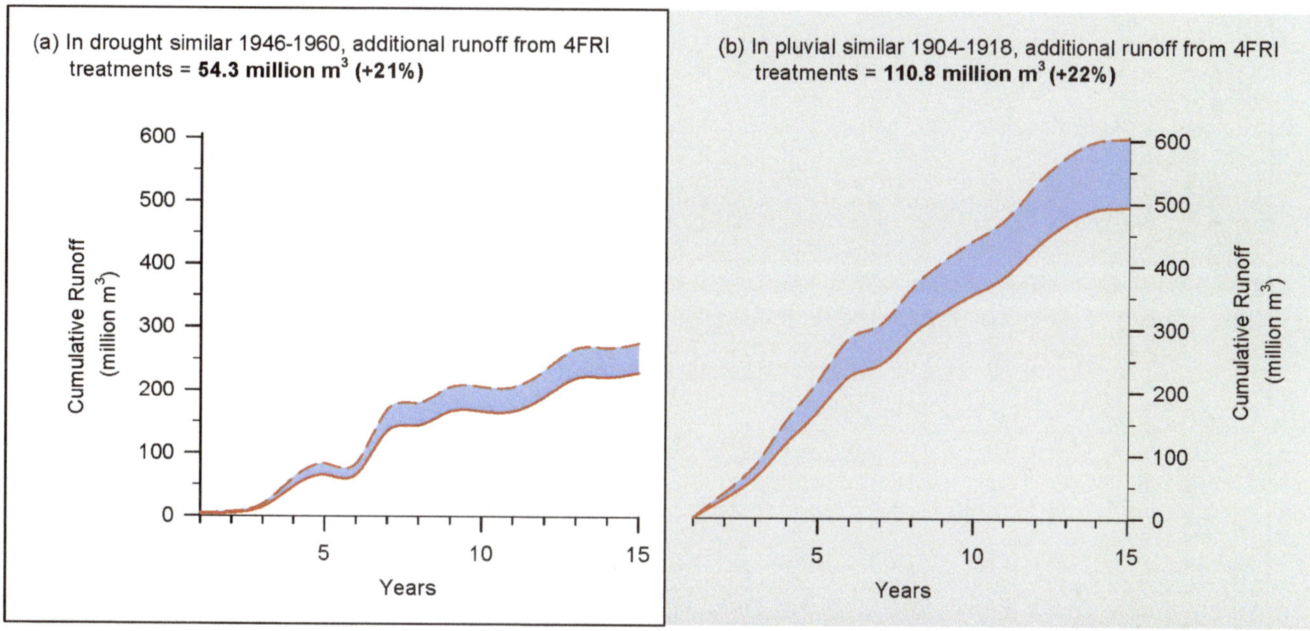

Figure 9. Cumulative runoff increases from thinning in 4FRI project. Estimates of cumulative increases in runoff (million m^3) from planned mechanical thinning of ponderosa pine forests in the first analysis area of the 4FRI project under (a) drought and (b) pluvial conditions. Solid red lines are estimates of cumulative runoff under current forest conditions using original Baker-Kovner regression model [20]. Dotted red lines represent increases in cumulative runoff associated with 4FRI treatments using modified Baker-Kovner regression model. Difference between these two values, shown with blue shading, is additional runoff from forest thinning treatments. Estimated increases in runoff ceased after 15 years.

substantially less than runoff from volcanic soils [41], [42]. Several properties of sedimentary-derived soils relative to basalt-derived soils may lead to greater water permeability and infiltration, including lower clay content, extensive fracturing, and deeper soil horizons [30], [43], [44]. Depending on a number of factors, increased infiltration may lead to increased recharge, which we would expect to result in increased discharge to surface water at some location downstream.

A nearby paired-watershed study in mixed conifer forests on sedimentary-derived soils found that overstory removal of 30-80% led to a slightly higher range of increased runoff, 30–110%, that persisted for more years than the Beaver Creek experiments [45],[46]. Direct comparisons with ponderosa pine forests are not possible because this study was conducted within a higher-elevation mixed-conifer forest that had higher initial basal areas and higher mean annual precipitation.

Relative to climate, this study's scenarios captured the high end of variability associated with pluvials but probably not the low end of variability associated with droughts. Specifically, the early 20th century pluvial may have been the wettest pluvial that has occurred in the last 12 centuries [47], whereas the two occurrences of 20th century drought were eclipsed in terms of duration and severity by several multi-decade "mega-droughts" [48–51]. Output from global circulation models indicates that climate variability will continue to be an important characteristic of the region in the future [52], but that climate change may increase the risk of extreme climatic events such as multi-decade droughts and extreme winter precipitation [53], [54]. Some global circulation models also project that mean winter precipitation in the Southwest will decline by up to 10% [52], but it may take many years to detect effects on stream flows because of precipitation variability [55]. The net effect of changes in precipitation on forest condition and hydrology still needs to be resolved and will require more investigation.

Unlike precipitation, the effects of anthropogenic warming on forest hydrology are clearer. Studies have found that warmer temperatures in recent decades help explain a downward trend in snowpack in the western United States, even after patterns of natural climate variability have been considered [9], [56]. In our revision of the historical runoff model, we attempted to add several parameters of temperature – seasonal means, maximums, and minimums – as explanatory variables to the revised regression model but none were significant. This suggests that the temporal and spatial range of winter temperatures that occurred during the years when the watershed experiments were conducted was not sufficient to add explanatory value to predicting runoff. In a separate trend analysis of winter temperatures (Oct-Apr), we found that temperatures in the months of March and April in the last 25 years, 1988–2012, were significantly warmer by 2 and 1 degrees Celsius (3.5 and 2 degrees Fahrenheit), respectively, than temperatures for the same months in the 25 years when the experiments were conducted, 1958–1982 (see File S4 for more detailed information). The magnitude of observed declines in snowpack in the Southwest, in the range of 20%, is similar to the increases in runoff associated with thinning from this study, suggesting that accelerated thinning may at least offset or ameliorate runoff losses due to climate change.

Management Implications: Accelerated Forest Thinning to Improve Resilience

On balance, we believe there is enough evidence for managers to explore the ways in which accelerated forest thinning can restore forest resilience and functioning to ameliorate the drying of ponderosa pine forests associated with warming and past management activities. Below we summarize the management implications of this and related studies.

Table 2. Annual runoff increases from thinning in Salt-Verde watersheds.

Forest Management: Thinning[a]				Winter Precipitation Regime			Runoff: Annual Increase			Runoff: Cumulative Increase	
Scenario Name	*Total Hectares*	*Hectares per Year*	*Effective Hectares per Year*	*Type*	*Similar to 20th Century Period*	*Mean (mm/yr)*	*Mean (million m^3)*	*Median (million m^3)*	*Maximum (million m^3)*	*Total (million m^3)*	*%*
35-low	204,000	6,800	35,000	Drought	1942–1976	340	4.76	2.99	17.5	167	23%
				Pluvial	1965–1999	396	7.65	5.85	25.7	268	26%
35-mid	301,000	10,000	51,400	Drought	1942–1976	340	7.01	4.39	25.8	245	23%
				Pluvial	1965–1999	396	11.3	8.61	37.7	393	26%
35-high	401,000	13,400	68,800	Drought	1942–1976	340	9.35	5.86	34.4	327	23%
				Pluvial	1965–1999	396	15.0	11.5	50.4	525	26%
25-low	204,000	10,200	49,000	Drought	1942–1966	338	6.18	4.06	26.3	154	20%
				Pluvial	1975–1999	411	11.9	9.65	33.6	299	22%
25-mid	301,000	15,100	72,000	Drought	1942–1966	338	9.09	5.97	38.7	227	20%
				Pluvial	1975–1999	411	17.5	14.2	49.3	438	22%
25-high	401,000	20,100	96,300	Drought	1942–1966	338	12.1	7.98	51.7	303	20%
				Pluvial	1975–1999	411	23.4	19.0	66.0	585	22%
15-low	204,000	20,400	81,700	Drought	1945–1959	315	9.23	5.95	52.3	138	20%
				Pluvial	1975–1989	422	21.8	8.31	60.1	327	22%
15-mid	301,000	30,100	120,000	Drought	1945–1959	315	13.6	8.75	77.0	204	20%
				Pluvial	1975–1989	422	32.1	12.2	88.4	481	22%
15-high	401,000	40,100	161,000	Drought	1945–1959	315	18.1	11.7	103	271	20%
				Pluvial	1975–1989	422	42.8	16.3	118	643	22%

[a] 'Scenario Name' indicates number of years simulated in each scenario (15, 25, 35 years) and extent of forest thinned (low = 30% of forest thinned; mid = 45%; high = 60%). 'Total Hectares' are total number of hectares thinned in each scenario. 'Hectares per Year' are number of hectares thinned every year for 10 consecutive years (15-year scenarios), 20 consecutive years (25-year scenarios), and 30 consecutive years (35-year scenarios). 'Effective Hectares per Year' are the average number of thinned hectares across all years of the scenario that contribute to additional runoff (see figure 6).

Estimates of additional runoff associated with thinning of ponderosa pine forests across Salt-Verde watersheds accounting for variation in treatment periods, total forest extent treated, and drought and pluvial periods of winter precipitation (Oct-Apr). Estimates rounded off to three significant digits.

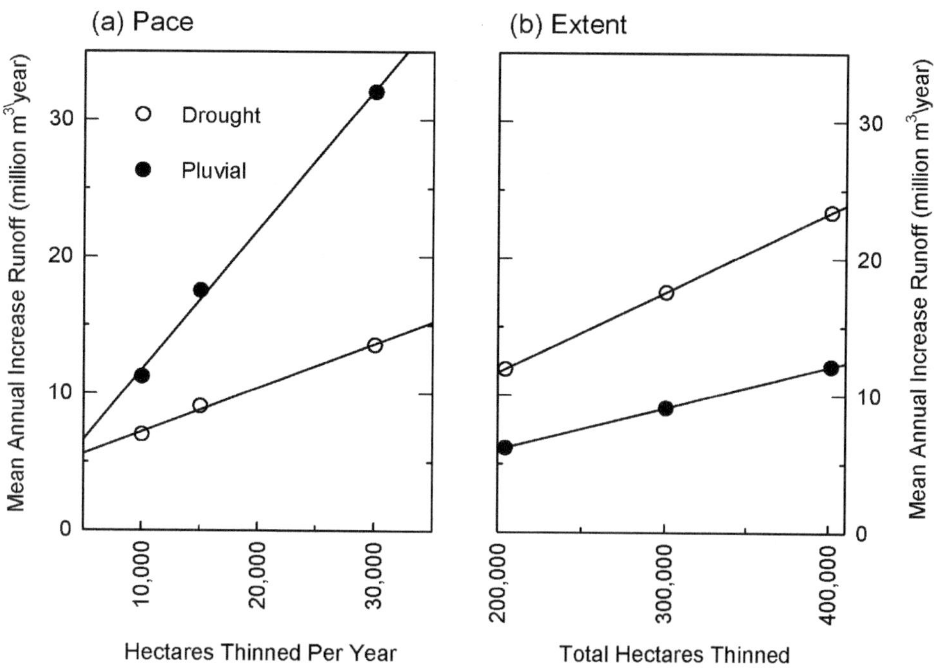

Figure 10. Scale effects of thinning on runoff in Salt-Verde watersheds. Effects of increasing (a) pace and (b) extent of thinning treatments in ponderosa pine forests in Salt-Verde watersheds on increases in mean annual runoff (million m^3/year). In (a) total area thinned is held constant at 301,000 ha (743,000 acres) (scenarios: 35mid, 25mid, 15mid) to show influence of increasing the area thinned per year. In (b) duration of thinning treatments is held constant at 25 years (scenarios: 25low, 25mid, 25high) to show influence of increasing the total area thinned across the scenario. In order to illustrate scale effects, only increases in *mean* annual runoff are shown. Statistics describing annual variability in runoff gains are shown in Table 2 and illustrated graphically for 4FRI scenario in Figure 8.

1. *Maintain & improve ecosystem function* – The current paradigm of forest management has moved away from a debate about managing forests to address one societal need, such as downstream water supply, to recognition that management strategies need to address multiple challenges [29], [57]. The 4FRI restoration project demonstrates this approach with planned benefits that include "...improved vegetation biodiversity, wildlife habitat, soil productivity, and watershed function" [19]. The impaired functioning and heightened vulnerability of southwestern forests to large-scale disturbances increases the urgency to improve ecosystem functioning, and to measure the effectiveness and trade-offs associated with these management activities using robust adaptive management and monitoring protocols. Such a focus may result in more options and flexibility in subsequent years, whereas failure to maintain functioning ecosystems may eliminate future options if soil and hydrologic processes are irreversibly altered.

Our study demonstrated that accelerated thinning can improve surface water runoff, a key ecosystem function, and provided indirect evidence of benefits for soil moisture and productivity. Another study found that average soil water content on low density ponderosa forests (250 trees/ha) was substantially higher than on high density forests (2,710 trees/ha), although these differences in forest density were not a result of mechanical thinning [58]. Researchers in the historical Beaver Creek experiments hypothesized that lower tree densities associated with thinning would reduce evapotranspiration losses, and thereby allow more water to be available for soil moisture, groundwater recharge, and surface water runoff [29]. Although we were unable to evaluate thinning effects on evapotranspiration, Dore et al. [59] found that light thinning (40% basal area reduction) in a ponderosa pine forest reduced stand-level evapotranspiration by 12% over four years, but the effect diminished over time and was not detectable in the fourth post-harvest year. This same study found that removal of all trees in a wildfire-burned site reduced annual evapotranspiration by 20%.

Other studies demonstrated that mechanical thinning can reduce the vulnerability of forests to uncharacteristic crown fires. Using a fire simulation model, Cochrane et al. [60] found that landscape-scale mechanical thinning that preceded actual wildfires could have reduced the average size of six wildfires in ponderosa pine and mixed conifer forests in California and the southwestern United States by an average of 18% (range 0.3 to 65%). A meta-analysis of 54 experimental studies showed that mechanical thinning significantly reduced the susceptibility of western ponderosa and Jeffrey pine forests to crown fires [23]. These studies describe the primary mechanism by which thinning reduces fire risk as a redistribution of fuels, thereby reducing fire spread rates and crowning behavior.

2. *Apply management tools at scale* – Managing forests at a scale that is comparable to recent forest disturbances will be another key factor for improving resilience. At 238,000 ha (588,000 acres) of ponderosa pine to be treated with mechanical thinning and prescribed fire, the 4FRI project is approaching the scale of these amplified disturbances. Our study demonstrated that potential increases in runoff associated with thinning were scale dependent: runoff increased in a linear fashion as the pace and extent of thinning increased in the Salt-Verde runoff scenarios (Figure 10). Recent efforts to reduce fire risk with landscape-scale treatments of ponderosa pine forests have been successful. For example, treatments completed before the largest wildfire in Arizona history, the Wallow Fire in 2011, effectively protected communities and towns by reducing fire severity before it reached key residential areas [61].

Further, accelerated thinning may open up opportunities to use other management tools at an accelerated pace. For example, wildland fire use has been an effective and inexpensive tool for maintaining forest resilience across large areas in remote forests and woodlands – the Gila Wilderness in New Mexico and Grand Canyon National Park in northern Arizona [62], [63] – but has been used less frequently in forests near population centers and towns. Completion of treatments in the first analysis area of the 4FRI project could have a multiplying effect on maintaining forest resilience if the thinning itself allows for increased use of wildland fire or prescribed fire over many more hectares.

3. *Leverage knowledge of climate variability* – The Beaver Creek watershed experiments and others like it demonstrated that winter precipitation was the best predictor of additional runoff from forest thinning [20], [30]. We add to this knowledge by examining how variability in winter precipitation could influence management effectiveness. Our scenarios demonstrated that gains in runoff from thinning could occur even under drought conditions. In simulated droughts, 1 out of 3 years had above-average precipitation, and these years accounted for 75% of the runoff gains in these scenarios. Even though these gains were substantially less than runoff gains during pluvials, additional runoff under drought conditions could help lessen the impact of predicted climate-driven losses in hydrologic connectivity important to native fish in the Salt and Verde watersheds [64].

This study also demonstrated that the same thinning schedule could result in a doubling in additional runoff if conducted in a pluvial typical of 20th century variability as compared to a drought. Scientists have developed decision support tools to help managers understand the likelihood that precipitation regimes will shift from wet to dry conditions or vice-versa [65], [66]. Knowledge of whether management activities are taking place in a pluvial or a drought could improve the ability of managers to predict the likelihood of success for various objectives. For example, thinning in a pluvial might meet substantive objectives to improve runoff, whereas thinning in a drought might be more significant for reducing risks of catastrophic crown fires or drought mortality.

With the notable exception of a mega-drought, the droughts and pluvials examined in this study were characteristic of climate variability in the region. An examination of hydro-climatic trends in the Salt-Verde watersheds revealed that temperatures are non-stationary, increasing significantly in recent decades, but the same is not true for winter precipitation or resulting stream flows [55]. Rather, winter precipitation has fluctuated between droughts and pluvials in these watersheds in the last century with no evidence of directional change, a pattern that has reoccurred in the southwestern United States for the last 1000 years [50]. These precipitation cycles have been linked to persistent anomalies in sea surface temperatures that vary across inter-annual and decadal time steps [67], [68]. If past patterns of precipitation variability remain stable in the near term, then it is probable that precipitation and flows in the Salt-Verde watersheds will shift into wetter conditions within the timeframe examined in this study [66].

4. *Building Knowledge to Reduce Uncertainties* – Landscape-scale restoration projects like 4FRI present the opportunity to learn about the influence of accelerated thinning on forest water budgets and resilience using modern forestry techniques and under a changing and variable climate. Stakeholders in the 4FRI project are developing a robust adaptive management and monitoring program so that progress towards objectives can be measured and timely adjustments to management made. Additionally, a paired-watershed experiment to evaluate the effects of forest treatments, including prescribed fire maintenance treatments, on watershed properties is planned within the first analysis area of 4FRI. It will monitor runoff, precipitation, evapotranspiration, snow water storage, soil moisture, groundwater recharge, and sediment yield, thereby reducing our knowledge gaps regarding the effects of forest management on the water cycle.

Conclusions

The widespread pattern of forest drying, evidenced by ecosystem level moisture deficits and larger and more severe wildfires, indicates we are entering a new era of forest management. The 4FRI project is an example of a restoration effort that is addressing this challenge in two ways: establishing objectives to improve resilience and increasing the scale at which management tools are applied. Our study demonstrated that this type of project can increase runoff by approximately 20% compared to unthinned forests, even under simulated drought conditions. If treatments at this scale are completed and repeated over the next several decades, increases in runoff could help offset the current and projected declines in snowpack and stream flow due to warming while improving the resilience of forest stands. As an incidental benefit in an era of dwindling water supplies and projected water shortages, forest thinning could play a role in augmenting river flows on a seasonal basis, improving conditions for water-dependent ecosystems, and benefitting the water supplies of downstream communities. Accelerated forest thinning to reduce water stress and wildfire risk is one of the only management options under our control, and it is probably the most critical to apply over the short term. Because the effects are temporary, it is important to frame accelerated thinning as a reset of ecosystem resilience, not a permanent cure. Rather, accelerated forest thinning creates opportunities to apply other management tools that restore forest resilience at broader scales, whereas management actions at the current pace will likely not be sufficient to address recent and projected changes in forest conditions.

Supporting Information

File S1 Methods and results figures and tables in English units.

File S2 Original and revised Baker-Kovner regression models in English units.

File S3 Additional factors that influence runoff.

File S4 Temperature trend analysis.

Acknowledgments

We thank Daniel Majka for his early contributions and input into this study. We thank Dan Huebner and Karen Koestner for providing data from the historical Beaver Creek experiments and Shaula Hedwall and Mark Nigrelli for providing data and expertise from the Four Forest Restoration Initiative and the Northern Arizona Wood Supply Study. We thank Kit MacDonald, Chris Nelson, Estella Smith, Rory Steinke, Joe Ringelberg, and Andrea Thode for providing data and expertise of forest thinning exclusions. We thank Julio Bentancourt, Michael Crimmins, and Gregg Garfin for input on climate variability and climate change, Charlie Ester for input on water management in the Salt-Verde River watersheds, and Robert Steidl for advice on statistical analyses. We thank Pat Graham, Molly Hunter, Dale Turner and Sue Sitko for critical review of this paper.

Author Contributions

Conceived and designed the experiments: MR RM FO JH. Performed the experiments: MR FO. Analyzed the data: MR FO. Contributed reagents/materials/analysis tools: MR FO JH. Wrote the paper: MR RM FO ES JH DG.

References

1. Covington WW, Moore MM (1994) Southwestern ponderosa pine forest structure: Changes since Euro-American settlement. Journal of Forestry 92(1): 39–47.
2. Fule PZ, Heinlein TA, Covington WW, Moore MM (2003) Assessing fire regimes on Grand Canyon landscapes with fire-scar and fire-record data. International Journal of Wildland Fire 12: 129–145.
3. Cocke AE, Fule PZ, Crouse JE (2005) Forest change on a steep mountain gradient after extended fire exclusion: San Francisco Peaks, Arizona, USA. Journal of Applied Ecology 42(5): 814–823.
4. Williams AP, Allen CD, Millar CI, Swetnam TW, Michaelsen J, et al. (2011) Forest responses to increasing aridity and warmth in the southwestern United States. Proceedings of the National Academy of Sciences 107(50): 21289–21294.
5. Gitlin AR, Sthultz CM, Bowker MA, Stumpf S, Paxton KL, et al. (2006) Mortality gradients within and among dominant plant populations as barometers of ecosystem change during extreme drought. Conservation Biology 20(5): 1477–1486.
6. Propst DL, Gido KB, Stefferud JA (2008) Natural flow regimes, nonnative fishes, and native fish persistence in arid-land river systems. Ecological Applications 18(5): 1236–1252.
7. Stefferud JA, Gido KB, Propst DL (2011) Spatially variable response of native fish assemblages to discharge, predators and habitat characteristics in an arid-land river. Freshwater Biology 56: 1403–1416.
8. Breshears DD, Cobb NS, Rich PM, Price KP, Allen CD, et al. (2005) Regional vegetation die-off in response to global-change-type drought. Proceedings of the National Academy of Sciences 102(42): 15144–15148.
9. Mote PW (2006) Climate-driven variability and trends in mountain snowpack in western North America. Journal of Climate 19: 6209–6220.
10. Westerling AL, Hidalgo HG, Cayan DR, Swetnam TW (2006) Warming and earlier spring increase Western US forest wildfire activity. Science 313: 940–943.
11. van Mantgem PJ, Stephenson NL, Byrne JC, Daniels LD, Franklin JF, et al. (2009) Widespread increase of tree mortality rates in the Western United States. Science 323: 521–524.
12. Thomas BE, Pool DR (2006) Trends in Streamflow of the San Pedro River, Southeastern Arizona, and Regional Trends in Precipitation and Streamflow in Southeastern Arizona and Southwestern New Mexico. U.S. Geological Survey Professional Paper 1712. Reston, VA: U.S. Geological Survey. 79 p.
13. Timmons D, Springer A (2006) Prescott AMA groundwater flow model update report. Available: http://www.azwater.gov/AzDWR/Hydrology/Modeling/documents/Prescott_Model_Update_Report.pdf. Accessed 6 May 2014.
14. Webb RH, Turner RM, Leake SA. (2007) The Ribbon of Green: Change in Riparian Condition in the Southwestern United States. Tucson, AZ: The University of Arizona Press. 480 p.
15. Marshall RM, Robles MD, Majka DR, Haney JA (2010) Sustainable Water Management in the Southwestern United States: Reality or Rhetoric? PLoS ONE 5(7): e11687. doi:10.1371/journal.pone.0011687.
16. Water Resources Development Commission (2011) Water Resources Development Commission Final Report: Volume II Committee Reports. Phoenix, AZ: Water Resources Development Commission. 685 p.
17. United States Department of Interior Bureau of Reclamation (2012) Colorado River Basin Water Supply and Demand Study: Study Report. Washington DC: U.S. Department of Interior, Bureau of Reclamation. 95p.
18. Vano JA, Udall B, Cayan DR, Overpeck JT, Brekke LD, et al. (2014) Understanding uncertainties in future Colorado River streamflow. Bulletin of the American Meteorological Society 95(1): 59–78.
19. United States Department of Agriculture Forest Service (2013) Draft environmental impact statement for the Four-Forest Restoration Initiative. Flagstaff, AZ: USDA Forest Service, Southwestern Region, Coconino and Kaibab National Forests. 744 p.
20. Brown HE, Baker MB, Rogers JJ, Clary WP, Kovner JL, et al. (1974) Opportunities for increasing water yields and other multiple uses on ponderosa pine forest lands. USDA Forest Service Research Paper RM-129. Fort Collins, CO: U.S. Department of Agriculture, Forest Service, Rocky Mountain Forest and Range Experiment Station. 36 p.
21. Zausen GL, Kolb TE, Bailey JD, Wagner MR (2005) Long-term impacts of stand management on ponderosa pine physiology and bark beetle abundance in northern Arizona: A replicated landscape study. Forest Ecology and Management 218: 291–305.
22. Stephens SL, Moghaddas JJ, Edminster C, Fiedler CE, Haase S, et al. (2009) Fire treatment effects on vegetation structure, fuels, and potential fire severity in western US forests. Ecological Applications 19(2): 305–320.
23. Fule PZ, Crouse JE, Roccaforte JP, Kalies EL (2012) Do thinning and/or burning treatments in western USA ponderosa pine or Jeffrey pine-dominated forests help restore natural fire behavior? Forest Ecology and Management 269: 68–81.
24. Barr GW (1956) Recovering rainfall. Tucson, AZ: Department of Agricultural Economics, University of Arizona. 33 p.
25. Daly C, Halbleib M, Smith JI, Gibson WP, Doggett MK, et al. (2008) Physiographically-sensitive mapping of climatological temperature and precipitation across the conterminous United States. International Journal of Climatology 28(15): 2031–2064.
26. Gesch D, Oimoen M, Greenlee S, Nelson C, Steuck M, et al. (2002) The National Elevation Dataset. Photogrammetric Engineering and Remote Sensing 68(1): 5–11.
27. US Geological Survey, Gap Analysis Program (2010) National Land Cover, Version 1. US Geological Survey. Available: http://gapanalysis.usgs.gov/gaplandcover/viewer/. Accessed 2012 Dec 23.
28. Arizona Geological Survey (2011) Geologic Map of Arizona. Available: http://www.azgs.az.gov/services_azgeomap.shtml. Accessed 2013 Apr 9.
29. Baker MB (2003) Hydrology. In: P . Friederici, editor. Ecological Restoration of Southwestern Ponderosa Pine Forests. Washington DC: Ecological Restoration Institute, Northern Arizona University, Island Press. pp. 161–174.
30. Baker MB (1986) Effects of ponderosa pine treatments on water yield in Arizona. Water Resources Research 22(1): 67–73.
31. Neary D (2011) Managing Arid and Semi-Arid Watersheds: Beaver Creek Watershed. U.S. Department of Agriculture, Forest Service, Rocky Mountain Research Station. Available: http://www.rmrs.nau.edu/watersheds/highlands/watersheds/beavercreek/. Accessed 2012 Oct 29.
32. Rich LR (1972) Managing a ponderosa pine forest to increase water yield. Water Resources Research 8(2): 422–428.
33. Hampton HM, Sesnie SE, Bailey JD, Snider GB (2011) Estimating regional wood supply based on stakeholder consensus for forest restoration in Northern Arizona. Journal of Forestry: 15–26.
34. Haire SL, McGarigal K (2010) Effects of landscape patterns of fire severity on regenerating ponderosa pine forests (*Pinus ponderosa*) in New Mexico and Arizona, USA. Landscape Ecology 25: 1055–1069.
35. Eidenshink J, Schwind B, Brewer K, Zhu Z, Quayle B, et al. (2007) A project for monitoring trends in burn severity. Journal of Association of Fire Ecology 3(1): 3–21.
36. Conservation Biology Institute (2010) Protected Areas Database of the United States, Version 2. Available: http://consbio.org/products/projects/pad-us-cbi-edition. Accessed 2013 Apr 9.
37. United States Department of Agriculture Forest Service (2013) GIS datasets for the Southwestern Region, United States Department of Agriculture, Forest Service. Available: http://www.fs.usda.gov/detail/r3/landmanagement/gis/?cid=stelprdb5202474. Accessed 2013 Apr 9.
38. Turner DS, List MD. 2007. Habitat mapping and conservation analysis to identify critical streams for Arizona's native fish. Aquatic Conservation: Marine and Freshwater Systems 17: 737–748.
39. Arizona Department of Water Resources (2010) Arizona Water Atlas Volume 8: Active Management Areas Water Atlas. Available: http://www.azwater.gov/azdwr/StatewidePlanning/WaterAtlas/ActiveManagementAreas/default.htm. Accessed 2014 May 5.
40. Brown TC, Fogel MM (1987) Use of streamflow increases from vegetation management in the Verde River Basin. Water Resources Bulletin 23(6): 1149–1160.
41. Campbell RE, Baker MB, Ffolliott PF, Larson FR, Avery CC (1977) Wildfire effects on a ponderosa pine ecosystem: An Arizona case study. USDA Forest Service Research Paper RM-191. Fort Collins, CO: U.S. Department of Agriculture, Forest Service, Rocky Mountain Forest and Range Experiment Station. 12 p.
42. Ffolliott PF, Baker MB (1977) Characteristics of Arizona ponderosa pine stands on sandstone soils. USDA Forest Service General Technical Report RM-44. Fort Collins, CO: U.S. Department of Agriculture, Forest Service, Rocky Mountain Forest and Range Experiment Station. 7 p.
43. Pearson GA (1931) Forest types in the Southwest as determined by climate and soil. Technical Bulletin No. 247. Washington DC: U.S. Department of Agriculture, Forest Service, Southwestern Forest and Range Experiment Station. 144 p.
44. Bosch JM, Hewlett JD (1982) A review of catchment experiments to determine the effect of vegetation changes on water yield and evapotranspiration. Journal of Hydrology 55: 3–23.
45. Rich LR, Gottfried GJ (1976) Water yields resulting from treatments on the Workman Creek experimental watersheds in Central Arizona. Water Resources Research 12(5): 1053–1060.
46. Hibbert AR, Gottfried GJ (1987) Streamflow responses to forest treatments on two Arizona mixed conifer watersheds. In: Troendle CA, Kaufmann MR, Hamre RH, Winokur RP editors. Management of subalpine forest: Building on 50 years of research. USDA Forest Service General Technical Report RM-149. Silver Creek, CO: U.S. Department of Agriculture, Forest Service, Rocky Mountain Forest and Range Experiment Station. pp. 189–194.

47. Woodhouse CA, Kunkel KE, Easterling DR, Cook ER (2005) The twentieth-century pluvial in the western United States. Geophysical Research Letters 32: L07701, doi:10.1029/2005GL022413.
48. Woodhouse CA, Overpeck JT (1998) 2000 years of drought variability in the central United States. Bulletin of the American Metereological Society 79(12): 2693–2714.
49. Stahle DW, Cook ER, Cleaveland MK, Therrell D, Meko DM, et al. (2000) Tree-ring data document 16th century megadrought over North America. Eos 81(12): 121–125.
50. Ni F, Cavazos T, Hughes MK, Comrie AC, Funkhouser G (2002) Cool-season precipitation in the southwestern USA since AD 1000: Comparison of linear and nonlinear techniques for reconstruction. International Journal of Climatology 22: 1645–1662.
51. Cook ER, Seager R, Cane MA, Stahle DW (2007) North American drought: Reconstructions, causes, and consequences. Earth-Science Reviews 81: 93–134.
52. Garfin G, Jardine A, Merideth R, Black M, LeRoy S, editors (2013) Assessment of Climate Change in the Southwest United States: A Report Prepared for the National Climate Assessment. A report by the Southwest Climate Alliance. Washington DC: Island Press. 531 p.
53. Ault TR, Cole JE, Overpeck JT, Pederson GT, Meko DM (2014) Assessing the risk of persistent drought using climate model simulations and paleoclimate data. Journal of Climate e-View doi:http://dx.doi.org/10.1175/JCLI-D-12-00282.1.
54. Dominguez F, Rivera E, Lettenmaier DP, Castro CL (2012). Changes in winter precipitation extremes for the western United States under a warmer climate as simulated by regional climate models. Geophysical Research Letters (39) L05803, doi:10.1029/2011GL050762.
55. Murphy KW, Ellis AW (2014) An assessment of the stationarity of climate and stream flow in watersheds of the Colorado River Basin. Journal of Hydrology 509: 454–473.
56. Pierce DW, Barnett TP, Hidalgo HG, Das T, Bonfils C, et al. (2008) Attribution of declining Western U.S. snowpack to human effects. Journal of Climate 21: 6425–6444.
57. Grant GE, Tague CL, Allen CD (2013) Watering the forest for the trees: An emerging priority for managing water in forest landscapes. Frontiers in Ecology and the Environment 11: 314–321.
58. Zou CB, Breshears DD, Newman BD, Wilcox BP, Gard MO, et al. (2008) Soil water dynamics under low-versus high-ponderosa pine tree density: Ecohydrological functioning and restoration implications. Ecohydrology 1: 309–315.
59. Dore S, Montes-Helu M, Hart SC, Hungate BA, Koch GW, et al. (2012) Recovery of ponderosa pine ecosystem carbon and water fluxes from thinning and stand-replacing fire. Global Change Biology 18: 3171–3185.
60. Cochrane MA, Moran CJ, Wimberly MC, Baer AD, Finney MA, et al. (2012) Estimation of wildfire size and risk changes due to fuel treatments. International Journal of Wildland Fire 21(4): 357–267.
61. Kennedy MC, Johnson MC (2014) Fuel treatment prescriptions alter spatial patterns of fire severity around the wildland-urban interface during the Wallow Fire, Arizona, USA. Forest Ecology and Management 318: 122–132.
62. Fule PZ, Laughlin DC (2007) Wildland fire effects on forest structure over an altitudinal gradient, Grand Canyon National Park, USA. Journal of Applied Ecology 44(1): 136–146.
63. Hunter ME, Iniguez JM, Lentile LB (2011) Short- and long-term effects on fuels, forest structure, and wildfire potential from prescribed fire and resource benefit fire in southwestern forests, USA. Fire Ecology 7(3): 108–121.
64. Jaeger KL, Olden JD, Pelland NA (2014) Climate change poised to threaten hydrologic connectivity and endemic fishes in dryland streams. Proceedings of the National Academy of Sciences. doi: 10.1073/pnas.1320890111.
65. Enfield DB, Cid-Serrano L (2006) Projecting the risk of future climate shifts. International Journal of Climatology 26: 885–895.
66. Gangopadhyay S, McCabe GJ (2010) Predicting regime shifts in flow of the Colorado River. Geophysical Research Letters 37: L20706, doi:10.1029/2010GL044513.
67. Sheppard PR, Comrie AC, Packin GD, Angersbach K, Hughes MK (2002) The climate of the Southwest. Climate Research 21: 219–238.
68. Mantua NJ, Hare SR (2002) The Pacific Decadal Oscillation. Journal of Oceanography 58: 35–44.

A High-Resolution Chronology of Rapid Forest Transitions following Polynesian Arrival in New Zealand

David B. McWethy[1]*, Janet M. Wilmshurst[2,4], Cathy Whitlock[1,3], Jamie R. Wood[2], Matt S. McGlone[2]

1 Department of Earth Sciences, Montana State University, Bozeman, Montana, United States of America, **2** Landcare Research, Lincoln, New Zealand, **3** Institute on Ecosystems, Montana State University, Bozeman, Montana, United States of America, **4** School of Environment, University of Auckland, Auckland, New Zealand

Abstract

Human-caused forest transitions are documented worldwide, especially during periods when land use by dense agriculturally-based populations intensified. However, the rate at which prehistoric human activities led to permanent deforestation is poorly resolved. In the South Island, New Zealand, the arrival of Polynesians c. 750 years ago resulted in dramatic forest loss and conversion of nearly half of native forests to open vegetation. This transformation, termed the Initial Burning Period, is documented in pollen and charcoal records, but its speed has been poorly constrained. High-resolution chronologies developed with a series of AMS radiocarbon dates from two lake sediment cores suggest the shift from forest to shrubland occurred within decades rather than centuries at drier sites. We examine two sites representing extreme examples of the magnitude of human impacts: a drier site that was inherently more vulnerable to human-set fires and a wetter, less burnable site. The astonishing rate of deforestation at the hands of small transient populations resulted from the intrinsic vulnerability of the native flora to fire and from positive feedbacks in post-fire vegetation recovery that increased landscape flammability. Spatially targeting burning in highly-flammable seral vegetation in forests rarely experiencing fire was sufficient to create an alternate fire-prone stable state. The New Zealand example illustrates how seemingly stable forest ecosystems can experience rapid and permanent conversions. Forest loss in New Zealand is among the fastest ecological transitions documented in the Holocene; yet equally rapid transitions can be expected in present-day regions wherever positive feedbacks support alternate fire-inhibiting, fire-prone stable states.

Editor: Christopher Carcaillet, Ecole Pratique des Hautes Etudes, France

Funding: This research was supported in part by National Science Foundation Grants OISE-0966472 (DBM, CW), and BCS-1024413 (DBM, CW), New Zealand Foundation for Research Science and Technology funding to JMW, JRW and MSM, and Royal Society of New Zealand Marsden funding to MSM, JMW, and CW. The funders had no role in study design, data collection and analysis, decision to publish, or preparation of the manuscript.

Competing Interests: The authors have declared that no competing interests exist.

* Email: dmcwethy@montana.edu

Introduction

The composition and structure of forests worldwide are being altered by intensification and expanding land-use, changing climatic conditions and increasing incidence of large fires [1,2]. Sudden transitions between forest, savanna and grassland biomes can be linked to feedbacks between land-use practices, climate change and disturbances such as fire [3]. Research suggests that the extent of forest cover reinforces conditions that maintain or inhibit fire (e.g., high forest cover often promotes greater fuel moisture), and consequently, the stability of forested, savanna or grassland vegetation [4]. Recent deforestation has been shown to initiate positive feedbacks replacing forested landscapes that rarely burned with fire-prone shrubland, savanna and grasslands [5]. Human-mediated forest transitions are not a new phenomenon, however. Humans have effected forest transitions for millennia through their use of fire, and widespread rapid transitions in the past are often linked to expanding human settlements and cultivation [6,7].

New Zealand provides a prime example of a relatively recent and major prehistoric forest loss. Here, the arrival of humans c. 750 years ago [8] instigated a period of deliberate burning, termed the Initial Burning Period (IBP), that led to the loss of nearly 50% of New Zealand's forests [9,10] and conversion of podocarp dominated forests to open shrubland vegetation. The conversion of forests in the South Island, presents a striking paradox, that extremely small populations (by any estimate – human mtDNA sequences suggest an initial founding population of between 50–100 females and a low population growth rate of c. 1% [11]) were able to instigate widespread and permanent forest transitions. Paleoenvironmental records from New Zealand suggest that ecosystem feedbacks during the initial burning period accelerated the rapid rate of forest demise and help explain this paradox [5,9]. In this paper, we examine the rate at which relatively stable forest biomes in New Zealand were converted to open shrublands in the 13th century, and how forest transitions following early human-set fires in the South Island of New Zealand compares with other regions in the world currently experiencing deforestation.

During the IBP, nearly all pollen and charcoal records derived from the drier parts of New Zealand show a distinct decline in pollen from native podocarp tree taxa (and to a lesser extent beech forest trees), an increase in pollen from seral taxa particularly bracken fern (*Pteridium esculentum*) and grasses (Poaceae), and an increase in charcoal (e.g., [9,12,13]. Previous results from radiocarbon dated high-resolution microscopic charcoal analyses from a network of sites across the central South Island showed a distinct increase in charcoal accumulation rates during the IBP within 200 years following Polynesian (Māori) settlement, although the increase is diachronous among sites [9]. At all but the most remote and wet sites, paleofire reconstructions suggest that repeated fires (2 or more) were responsible for the watershed conversion of native forest to a fern/scrub or grassland. [9,10,14]. However, the age-depth chronologies described in [9,14] were based on a small number of radiocarbon dates selected prior to and after the IBP, and consequently, the rate of forest loss remains poorly constrained.

To precisely determine the rate at which prehistoric forest transitions occurred following initial human arrival in the South Island, New Zealand, we developed high-resolution reconstructions of vegetation (pollen) and fire (macroscopic charcoal) from radiocarbon dated lake-sediment records from two small, closed-basin lakes, Lake Kirkpatrick and Dukes Tarn (Fig. 1). We specifically choose two sites that represent different vulnerabilities to human-set fires: one drier lowland site and a second, wetter high-elevation site. The chronology of these reconstructions and rates-of-change calculations are based on a stratigraphic series of Accelerator Mass Spectrometry (AMS) radiocarbon dates from the sediment cores (Table S1).

Results

Chronology model results were highly convergent, suggesting forest transitions occurred within decades of the first human-set fires (Figs. S1–S5). Significant (>15% decline in native taxa) forest loss occurred within 17 yrs (SD = 7) at Lake Kirkpatrick and 48 yrs (SD = 19) at Dukes Tarn (Fig. 2). At Lake Kirkpatrick, the IBP was punctuated by two fire episodes at c. AD 1367 and AD 1391 (Fig. 3, Bchron chronology). Following the initial increase in charcoal accumulation rates (CHAR; particles cm^{-2} yr^{-1}), fire episodes occurred periodically (every 50–100 yrs) until c. AD 1600 when fire activity decreased. A second increase in fire activity coincided with European arrival ca. AD 1800. Vegetation assemblages changed dramatically following the first fires associated with human arrival. Native trees (e.g., *Nothofagus menziesii*, Nothofagus spp., *Podocarpus* spp. and *Prumnopitys* spp.) declined from 99 to 47% of the total terrestrial pollen percentage from 1331 to 1391 yr AD, whereas disturbance related-taxa (e.g. Poaceae and *Pteridium*) increased from up to 25 and 27%, respectively, over the same period (see Fig. S6 for additional pollen information). Pollen of native trees increased slightly to 64% at c. 1642 yr AD at the expense of grass pollen. Tree pollen percentages then declined, recovered again at c. 1792 yr AD before declining to <30% in the late 20th century. Pollen of exotic taxa (Pinaceae, *Rumex* spp. and *Taraxacum* spp.) reached 2% by the mid 19th century and increased to >10% at c. 1947 yr AD.

At Dukes Tarn, charcoal data register initial fire events at c. AD 1337, and 1387, and the pollen data indicate that forests recovered within decades even though fire events occurred periodically (once every 50–100 yrs) from the IBP until the last century (Fig. 3). Pollen of some native trees (*Nothofagus fuscospora* type) collectively declined from 97 to 83% whereas disturbance-related taxa, notably Poaceae and *Pteridium*, increased from 1 to 9% and 0 to 7% respectively during the interval from c. AD 1355 to 1414 (see Fig. S7 and [9] for additional pollen information). In contrast to Lake Kirkpatrick, pollen of native trees increased to 94% by c. AD 1502 at the expense of Poaceae, then fluctuated between 85–94%, and accounted for 97% at c. 1694 yr AD (back to pre-deforestation levels). Tree percentages decreased to 83% in the last century. Non-native plant taxa (Pinaceae, *Rumex* spp. and *Taraxacum* spp.) account for >1% of the pollen at c. 1928 yr AD and 5% at c. 1980 yr AD before declining to <1% in recent decades. Fires that were recorded during the same time period at both sites (Fig. 4) likely represent a pattern of burning associated with humans moving through the landscape to enhance resource conditions and facilitate travel to nearby greenstone quarries [15] and could have been facilitated by extreme fire weather (e.g., hot, dry, windy conditions). Compared to Lake Kirkpatrick where forest transitions led to open vegetation that persists to the present, native vegetation at Duke's Tarn experienced relatively minor modification and showed partial recovery within several decades of the first IBP fires and full recovery 350 years after the IBP. Post IBP fires at Duke's Tarn seem to have had less of a lasting impact on the long-term structure and composition of vegetation.

Statistical analyses of rates of change reflect the rapid transition from native trees to grasses and bracken, evident in changes in pollen percentages (Fig. 3). Sørensen's index quantifying dissimilarity in pollen taxa prior to and during the IBP shows sharp increases in the dissimilarity in vegetation composition and dominance, with transitions in vegetation taxa peaking during the IBP at both Lake Kirkpatrick and Dukes Tarn. The Sørensen's index (dissimilarity between each sample and the previous sample) was greatest at c. 1363 yr AD (10^{-2}) at Lake Kirkpatrick and at c. 1375 AD (10^{-2}) at Dukes Tarn, coinciding with the first IBP fires at both sites. Dissimilarity between pre-human vegetation assemblages and post-IBP vegetation taxa increased from present to the pre-IBP period at both sites but was much more pronounced at Lake Kirkpatrick where dramatic vegetation transitions from forest to open shrublands persisted until present. Changes in vegetation assemblages at Dukes Tarn are characterized by an initial forest loss, a late recovery of native trees, followed by an increase in grasses and the presence of introduced taxa associated with European colonization.

Discussion

Human-caused forest transitions are documented worldwide, especially during periods when land use by dense agriculturally-based populations intensified [6,16,17]. However, the rate at which prehistoric human activities led to a persistent biome shift is commonly poorly resolved [18]. In the South Island of New Zealand, the arrival of Polynesians resulted in dramatic forest loss, with conversion of nearly half of native forests to open vegetation by a small number of sparsely distributed populations [10]. Paleoenvironmental records suggest that this conversion occurred within decades to centuries of human arrival but the speed at which site conversion took place once burning commenced has been unclear until now. Chronological models (Bchron, Oxcal and MCAgeDepth) that explicitly estimate joint uncertainty of all radiocarbon calibration dates suggest that the forest loss during IBP was rapid, occurring in 10–24 years at the drier site, Lake Kirkpatrick (median = 17) and 29–67 years at the wetter site, Dukes Tarn (median = 48). The speed at which vegetation conversion occurred is likely even more rapid than estimates based on age-depth chronologies alone because they incorporate joint uncertainty originating from both AMS radiocarbon dating and radiocarbon calibration distributions. Despite inherent limitations in calculating the rate of past vegetation change from lake-sediment records, forest loss in New Zealand was rapid by any measure. These rates of deforestation parallel those made for the eastern North Island where dry, lowland podocarp forests

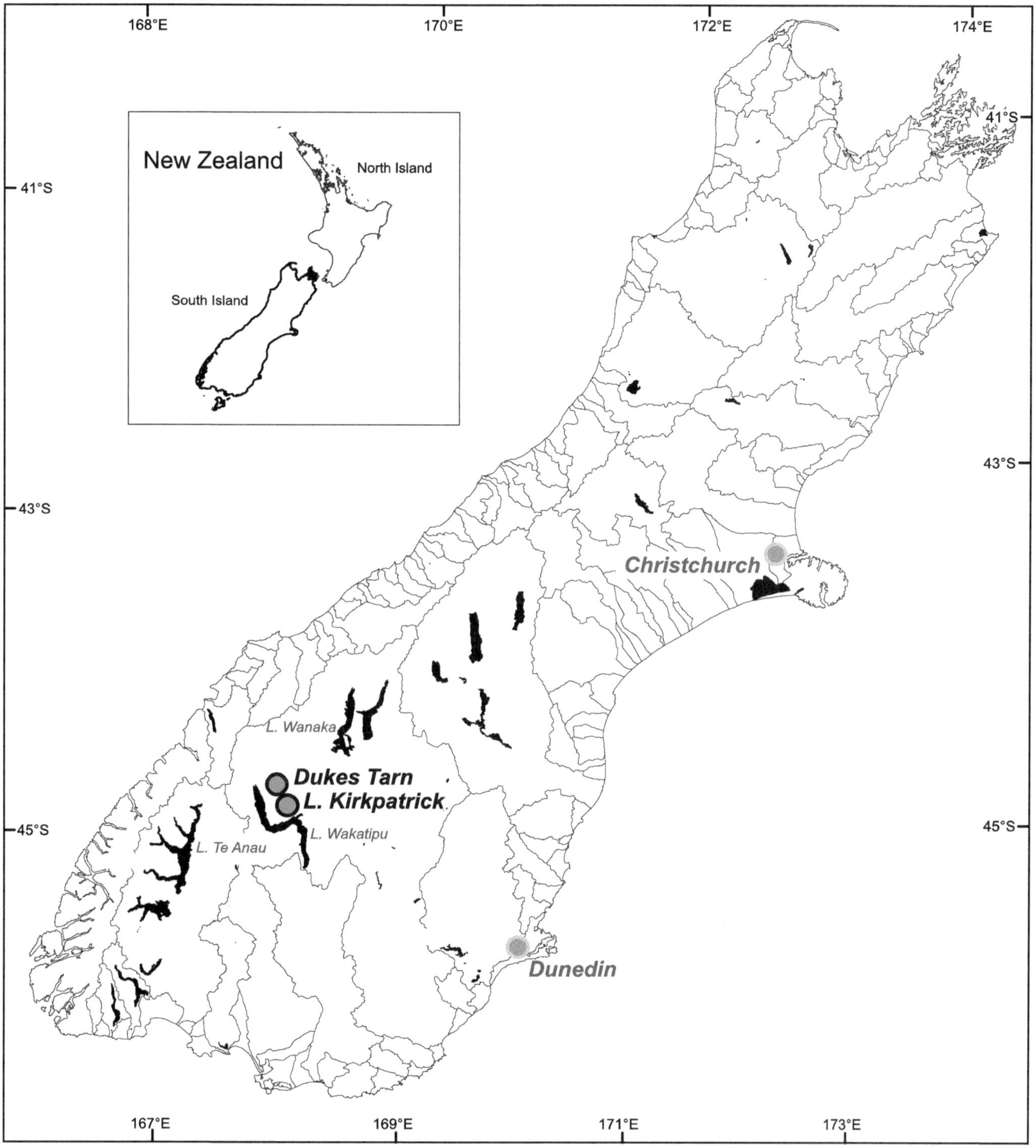

Figure 1. Location of study sites, South Island, New Zealand.

transitioned to bracken fern shrubland within decades of the first human-set fires whereas loss of more mesic, upland forests took a century or longer [19].

For most of the South Island, fire impacts on forested ecosystems were ecologically insignificant throughout the Holocene, in that changes in the structure and composition of vegetation were poorly related to the occurrence of fire [20]. The widespread and mostly one-way shift from forest to open shrubland that occurred with the arrival of Polynesians, and then following European arrival, was unprecedented in the Holocene, even after extreme disturbance events (e.g., large volcanic eruptions). This was the case for Lake Kirkpatrick and many other sites throughout similarly dry forests of eastern New Zealand including the eastern North Island [21]. The higher-elevation site,

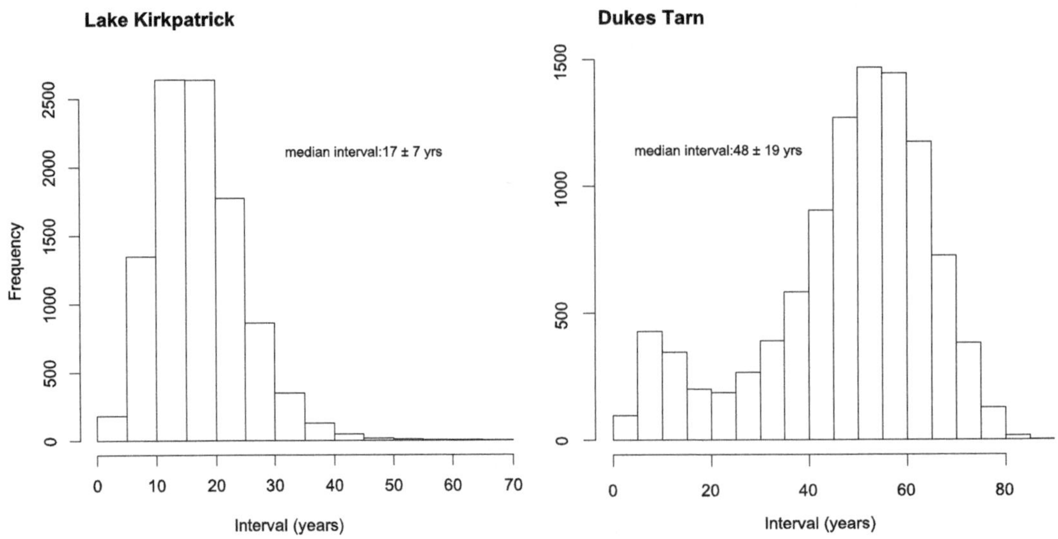

Figure 2. Bchron posterior frequency distribution of the interval (yrs) associated with forest transitions at Lake Kirkpatrick and Dukes Tarn, South Island, New Zealand. The interval associated with forest transitions was identified using two criteria: 1) visible change in core lithology, and 2) a rapid decline in pollen percentages of native forest taxa and a coincident increase in pollen percentages of open vegetation taxa.

Dukes Tarn, experienced equally rapid vegetation change, but declines in forest taxa at this site were less pronounced. This is most likely because this site is wetter and because the deeply-incised forested drainage basin south of the tarn has provided an ongoing refuge from fire, contributing significant wind-transported pollen to the tarn sediments. Differences in the extent to which fire activity and vegetation changed following human arrival at Lake Kirkpatrick and Dukes Tarn may be explained by variation in the intensity of human activity at these two sites; the most parsimonious explanation, however, centers on differences in rainfall and inferred fuel moisture at these two sites and is supported by previous results comparing a network of sites from the South Island ([9]). Paleofire records at even wetter sites (>1600 mm/yr) near or west of the Southern Alps of the South Island indicate only limited fire activity and minor associated vegetation change [9,22]. Similarly, along the western coast of the North Island moist conditions limited the extent to which Māori were able to burn forests [23]. Here, Māori gardens were established by burning small patches of forests but the scale of forest clearing was insignificant compared to deforestation that occurred in drier regions of the North and South Island sites [23]. What is surprising is the extent and rapid rate of forest clearance that occurred in the South Island where prehistoric human populations were at their lowest compared to the North Island, and cultivation was limited to a few coastal sites north of latitude 44° South.

Dynamic Global Vegetation Models that estimate rates of prehistoric deforestation at a regional to global scale suggest that the extent and pattern of preindustrial deforestation in most regions was strongly linked to population densities, inherent landscape productivity and lifestyles and technological innovation [16,17]. Estimates of population density and reconstructions of environmental change in these models are derived from archeological and paleoenvironmental records, and in most regions, rates of deforestation are poorly resolved. Our high-resolution chronology of environmental change from the South Island, New Zealand indicates rapid forest transitions occurred during the IBP during a time of exceedingly low population numbers and limited use of technological innovation (i.e., intensive cultivation). Why then were New Zealand's forests particularly vulnerable to fire? The lowland evergreen conifer/angiosperm forests were characterized as being both fire sensitive and ignition limited [20]. Low- to mid-elevation forests experiencing moderate to low levels of annual rainfall (300–1600 mm/yr) persisted for millennia simply because ignitions were rare and fuels were typically wet [20,24]. It was only when human-set fires expanded the proportion of flammable early-seral vegetation on the landscape that widespread forest transitions were possible [5]. Positive feedbacks between the fire sensitivity of the vegetation, introduction of human-set ignitions, and changes in landscape flammability with the spread of early-seral vegetation offer an explanation for these rapid transitions. Perry et al. [5,22] provide empirical support for these feedbacks, showing that dry, low- to mid-elevation evergreen conifer/angiosperm forests in the South Island New Zealand were vulnerable to fire, and that targeted burning created positive feedbacks where burning of forest resulted in more flammable seral vegetation which was more vulnerable to ignitions. The model estimated that conversion of the forest to shrubland could have occurred in 50–200 years if fires were targeted both spatially (in highly-flammable early-seral vegetation) and temporally (every few decades to maintain early-seral vegetation).

Forest transitions occurring at present in a number of regions around the world bring new attention to conditions and feedbacks that drive rapid transitions between seemingly stable states [25–28]. Analyzing global tree cover and fire activity data, Staver et al. [4] show that mesic landscapes (>2500 mm/y rainfall) with >60% tree cover are resistant to changes in fire activity and biome shifts, whereas areas with intermediate levels of rainfall (1000–2500 mm/y) and <50% tree cover are vulnerable to rapid transitions to savanna. Hirota et al. [29] suggest that landscapes with intermediate levels of tree cover are extremely rare because the interaction between climate, edaphic conditions and fire activity interact to attract vegetation to relatively stable states of higher (forest) and lower (savanna) levels of tree cover. A number of mechanisms create positive feedbacks that trigger and maintain these multiple stable states. Landscapes with high canopy cover of mature forests are resilient to increased fire activity because they maintain fuel moisture levels that suppress fire and often lack

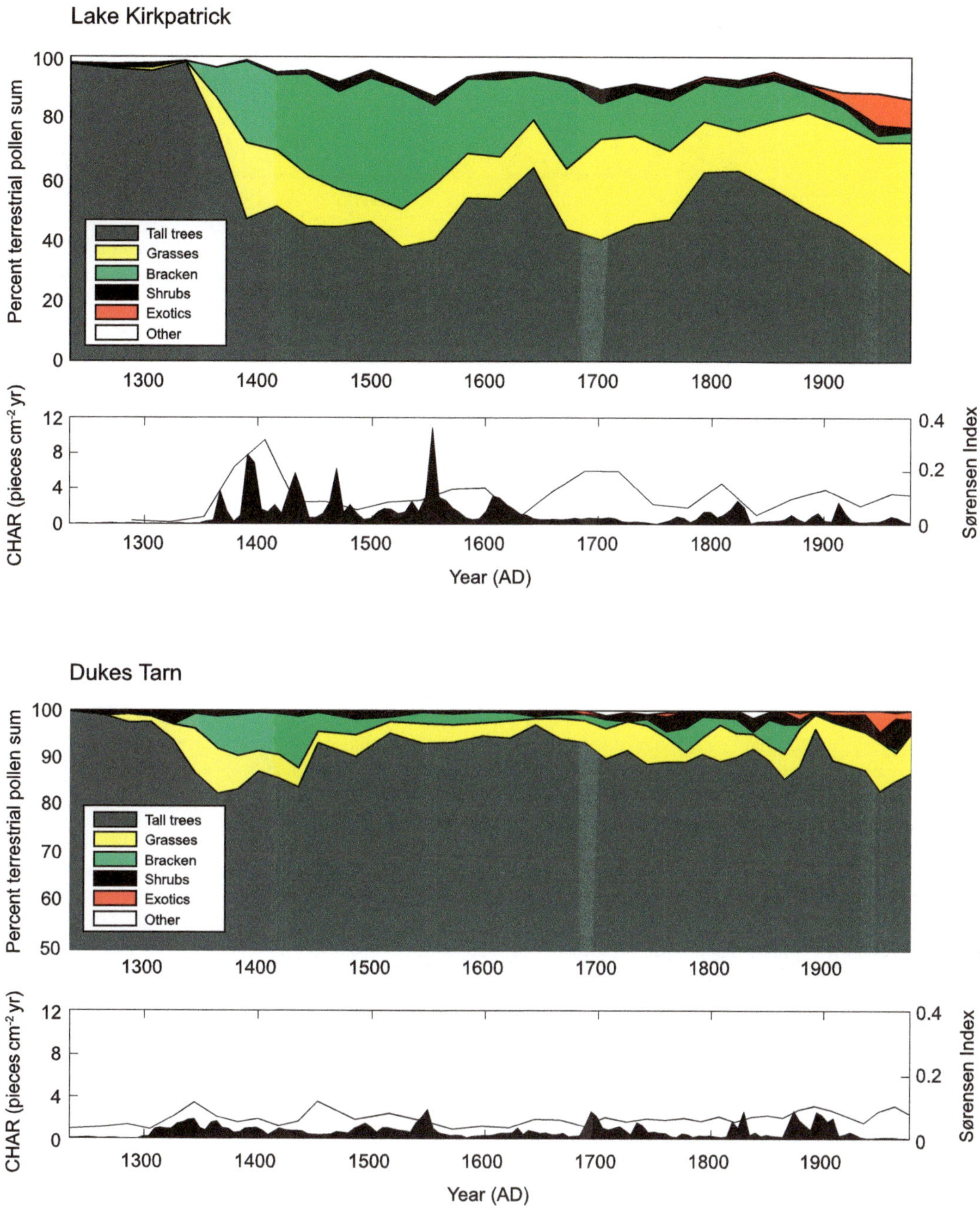

Figure 3. Statistical analysis of vegetation change and rate of vegetation change (based on Sørensen's similarity index) and changes in fire activity (based on CHAR, charcoal particles cm^{-2} yr^{-1}, and fire event determination), Lake Kirkpatrick and Dukes Tarn. Colored panels show change in percent of total terrestrial pollen percentages for native trees and disturbance-associated taxa (e.g., Poaceae, *Pteridium*) and non-native taxa (Pinaceae, *Rumex*, *Taraxacum*-type) introduced by Europeans for Lake Kirkpatrick (top) and Dukes Tarn (bottom). Black and white panels show charcoal accumulation rates and Sørensen's distance between each pollen sample and the next oldest pollen sample.

continuous understory fuels for carrying fire [3]. Dry landscapes reinforce open conditions by supporting low fuel moisture and species with plant functional traits that promote fire.

A consequence of these attracting states is the potential for human activities to reduce forest density and/or canopy cover past a tipping point where they become increasingly vulnerable to rapid transitions to more fire-prone states [30]. This has been shown to be particularly true for forests where flammability decreases with forest age. Kitzberger et al. [3] demonstrate that even small increases in the frequency of ignitions and/or climate variability in these forests can result in dramatic shifts in fire regimes at large scales. The New Zealand example suggests that targeted ignitions alone where sufficient to initiate feedbacks promoting high landscape flammability and increased fire activity.

Today, the vulnerability of temperate forests to rapid transitions is evident in a number of regions. For example, alterations to fuel and microclimate conditions following logging activities in evergreen forests of the northwestern US increase post-fire

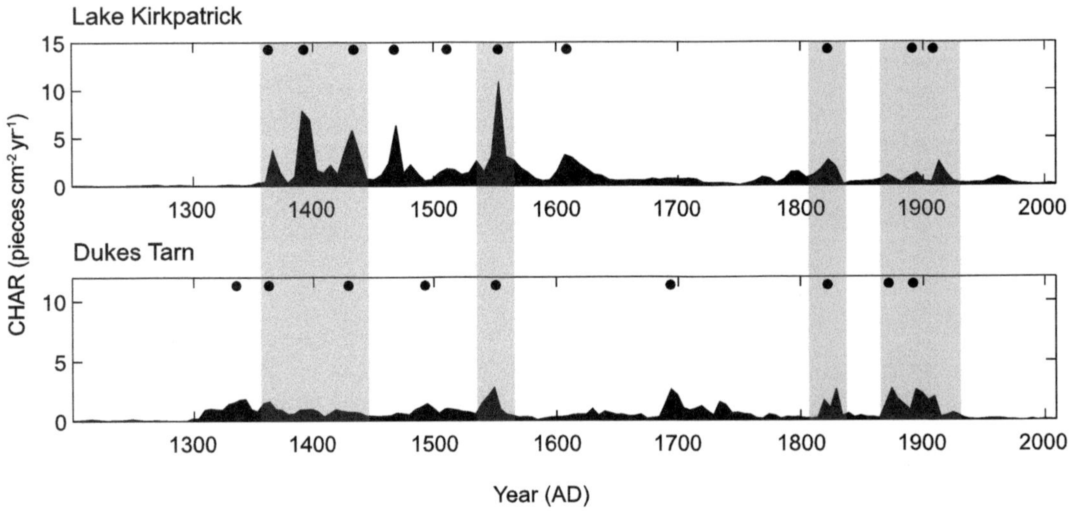

Figure 4. Variation in CHAR (charcoal particles cm^{-2} yr^{-1}), and peak identification for Lake Kirkpatrick and Dukes Tarn. Black circles identify peaks in CHAR that are statistically significant from background variation in CHAR and are thus likely to be local fires (within 1–3 kms of lake). Shaded areas highlight time periods when fire events were recorded at both lakes.

flammability, ultimately increasing fire severity such that forest recovery is delayed or non-existent [31,32]. Lindenmayer et al. [27] describe a similar feedback in mountain ash forests in Victoria, Australia where logging activity results in an abundance of fine fuels, slash and more flammable early-seral vegetation. While fire occurred in these forests in the past, it was typically infrequent, and eventually followed by conditions that promoted robust recruitment of mountain ash seedlings. In recent decades, the increase in fine fuels, slash and persistently drier microclimatic conditions following logging activity and subsequent increased fire activity interact to create these new fire-prone 'landscape traps' [27]. Likewise, Cochrane et al. [33] demonstrate that previously burned forests in Amazonia are highly vulnerable to subsequent fires that further promote more open, highly-flammable vegetation. In southern South America, the combination of increased anthropogenic fires and herbivory of native seedlings by introduced species is rapidly facilitating the replacement of mesic *Nothofagus* forests with fire-prone shrubs [34]. On Great Barrier Island North Island, New Zealand, Perry et al. [35] found that conversion of mesic broadleaf forests to shrublands initiated positive feedbacks in which an increase in fire activity not only reinforced the spread and persistence of highly flammable early-seral woody vegetation but also promoted flammable invasive species and degraded soil conditions, especially on drier aspects. Lacking fire, these forests would likely recover but Perry et al. [35] observe that, along with a continued loss of topsoil, seed trees and dispersal agents (e.g. avian frugivores) are becoming increasingly rare such that forest recovery becomes ever more improbable. These recent forest transitions demonstrate the strong influence of human-initiated positive feedbacks that increase landscape flammability and highlight the inherent vulnerability of seemingly stable temperate forests to rapid biome and fire regime shifts. Widespread forest transitions are predicted for the future in regions with similar age-flammability and low-ignition characteristics [36].

Conclusions

The timing of an abrupt change from forest to open shrubland following human-set fires at two sites in the South Island of New Zealand is well-constrained in sediment cores using high-resolution chronology models, suggesting widespread deforestation occurred within decades of the first anthropogenic fires in the most vulnerable drier regions. The fire history of New Zealand is an example of how a seemingly stable ecosystem can pass natural tipping points and experience persistent transformations. The rate at which anthropogenic forest loss occurred was rapid, taking place within decades following the initial arrival of a small founding population. The rapidity of the response was partly facilitated by positive feedbacks created by the introduction of a new and frequent ignition source by humans. In addition, the dominant forest taxa lacked natural adaptations to fire (e.g., serotiny, resprouting) which made them highly vulnerable to increased ignitions [24]. The newly created fire-prone landscapes persisted even when fire activity decreased after the IBP. These shrublands will likely be maintained for centuries if not millennia unless fire is removed from the system for long periods of time (>100 years). As with temperate forests experiencing similar transitions today, simply removing fire from New Zealand landscapes could eventually lead to forest recovery. However, recent examples suggest that a confluence of factors that degrade biophysical conditions, impede seedling regeneration, remove source populations and dispersal agents and introduce fire-adapted weedy species will continue to hinder native forest recovery in New Zealand and elsewhere. Even in the absence of fire, management of landscapes to promote open vegetation, agriculture, and exotic forestry plantations promises to reinforce conditions that prevent regeneration of native species. The rapid rate of forest transitions witnessed in New Zealand 750 years ago portends potentially rapid transitions in biomass-rich regions experiencing similar land-use-disturbance interactions and feedbacks and highlights challenges to native forest recovery.

Materials and Methods

Study Sites

The study regions lie in the southeastern hill country and southern alpine region of the South Island of New Zealand, an area that ranges between 250 and 1900 m elevation (Fig. 1). The climate is cool year-round and moderately wet with most of the precipitation occurring in winter. There is a strong steep east/west gradient both in precipitation and elevation. During the Holocene the vegetation of the region was dominated by closed-canopy broadleaf evergreen forests [20]. Natural fires occurred in areas of forest in the dry southeastern region of New Zealand but were characterized by long intervals between fires [37]. The original forest extent and composition are inferred from pollen records, isolated surviving forest stands and remnant wood [38,39]. Permission to conduct fieldwork at the two study sites was granted by the New Zealand Department of Conservation (DOC).

Lake Kirkpatrick (lat. 45.03°S, long. 168.57°E; 567 masl). The Lake Wakatipu catchment is a mid-elevation (570 masl) site receiving moderate levels of rainfall (1077 mm/yr) and was part of a trade route for highly prized greenstone that the indigenous population (Māori) used and traded [15]. Lake Kirkpatrick lies within the Lake Wakatipu catchment. The small lake (~3.0 ha) is surrounded by pastureland with browntop (*Agrostis capillaris*), Poa spp. and other introduced grasses, clover (*Trifolium* spp.), *Hypericum* spp.; pine plantations (*Pinus radiata*); and small areas of native silver beech forest (*Nothofagus menziesii*).

Dukes Tarn (lat. 44.96°S, long. 168.49°E; 825 masl). Dukes Tarn is approximately 9.5 km northwest of Lake Kirkpatrick at 830 masl and also located within the Lake Wakatipu catchment. Dukes Tarn (~1.5 ha) receives more annual rainfall (1340 mm/yr) than Lake Kirkpatrick and vegetation surrounding the lake consists of patches of black beech (*Nothofagus solandri*), small trees and shrubs (e.g., *Coprosma* spp. *Phyllocladus, Dracophyllum longifolium, Hebe* spp., *Gaultheria* spp.), native tussock (*Chionochloa* spp.) and introduced pasture grasses (*Festuca* spp.). Deeply-incised drainages supporting native forest remnants border Dukes Tarn to the south whereas open vegetation surrounds the tarn on rolling terrain to the north.

Lake Sediment Cores

Lake sediment cores were obtained from Lake Kirkpatrick and Dukes Tarn using a polycarbonate tube (Klien core). Cores were split at University of Minnesota's LacCore facility for charcoal and pollen analysis. A 195 cm sediment core was retrieved from Lake Kirkpatrick in 2009 and a 173 cm sediment core was retrieved from Dukes Tarn in 2008. Core lithology for Lake Kirkpatrick consisted of light green gyttja 0–27 cm (depth below surface), gray silty clay 27–29 cm, dark green gyttja 29–105 cm, light gray silty clay 105–115 cm, and light green gyttja 115–195 cm. Core lithology for Dukes Tarn consisted of light green gyttja 0–42 cm, silty clay 42–43 cm, light green gyttja 43–128 cm, dark silty clay with charcoal 128–136 cm, and light green gyttja 138–173 cm.

Reconstruction of Vegetation and Fire Histories

Pollen analysis followed the preparation methods of Moore et al. [40]. Pollen counts exceeded 250 terrestrial pollen grains and bracken (*Pteridium*) spores. Percentage calculations were based on this terrestrial sum (excluding other ground ferns and tree fern spores as they tend to be over-represented). We followed the protocol of McGlone and Wilmshurst [10] to detect initial human impact on the vegetation and grouped pollen taxa into tall forest taxa (predominantly *Fuscospora* spp. and podocarps); disturbance-related taxa including ferns (*Pteridium esculentum*), *Coriaria* spp. and grasses (Poaceae); and introduced taxa, including Pinaceae and *Rumex* spp. Remaining taxa, mostly herbaceous species, were labeled "other". These groupings were used to identify the transition from closed forest to open shrubland and grassland.

High-resolution charcoal analysis followed methods of Whitlock and Larsen [41]. Charcoal particles greater (>125 µm in diameter) were examined to reconstruct local fire events based on changes in charcoal accumulation rates (CHAR; particles cm^{-2} yr^{-1}). Decomposition of the charcoal time series employed a Gaussian mixture model to identify the mean and variance of the background CHAR (BCHAR) distribution [42]. The 99th percentile of this distribution is defined as the threshold value separating peak fire events from "noise". Significant charcoal peaks, determined by CHAR values greater than a locally-defined threshold value, identified specific fire events. Because each 1-cm vertical interval in the lakes encompasses 1–10 years at both sites, a charcoal peak is assumed to represent one or more fires occurring within the time span of a charcoal peak. The statistical significance of each peak was evaluated by comparing the original charcoal counts against the values in samples occurring 35 years before the peak. If the maximum count of a peak had a >5% chance of coming from the same Poisson-distributed population as the minimum charcoal count within the preceding 35 years, then a "peak" was not identified. Fire size and/or intensity were inferred from the magnitude of individual charcoal peaks (particles cm^{-2} yr^{-1}) [43].

Chronology Models

We obtained 22 radiocarbon dates at two sites to develop high-resolution chronologies (Table S1). Radiocarbon dates were obtained from macrofossils found during the interval of the first IBP fires where changes in charcoal accumulation rates and pollen percentages indicate dynamic changes in fire activity and vegetation. The chronology of these reconstructions and rates-of-change calculations are based on a stratigraphic series of AMS radiocarbon dates from the sediment cores. We incorporated this suite of radiocarbon dates into three readily accessible chronology models designed to estimate lake-sediment age-depth chronologies and chronological uncertainty: Bchron [44], Oxcal [45] and McAgeDepth [43]. Model outputs were highly convergent for all three models (Figs. S1–S5). We highlight results from Bchron because the chronology model has an environment event analysis function which explicitly estimates the duration of the interval at which significant environmental events occurred (Fig. 2).

Accelerator mass spectrometry (AMS) ^{14}C dates obtained on twig charcoal and terrestrial plant macrofossils, provided the chronology for precisely estimating rates of past environmental change. Thirteen radiocarbon dates were obtained from Lake Kirkpatrick and nine from Dukes Tarn. Eleven radiocarbon dates were obtained from across 20 centimeters of sediment where charcoal and pollen percentages identify the IBP at Lake Kirkpatrick and six radiocarbon dates were obtained across the comparable 10-centimeter interval at Dukes Tarn. We used this set of radiocarbon dates in a suite of age-depth chronology models to estimate the duration at which forest transitions occurred.

To derive age-depth chronologies we used three statistical age-depth chronology models: Bchron [44], Oxcal [45] and McAge-Depth [43] which all address two primary types of uncertainty: uncertainty in radiocarbon dates and their calibration, and variability in the sedimentation process itself. These models share a number of features that are desirable for developing a continuous age-depth chronology for sediment cores including, monotonicity (deeper sediments are likely older), variable sedimentation (rates are assumed to be variable), probabilistic interpolation based on uncertainty associated with calibrated probability distributions (posterior samples or chronologies are based on joint estimates of uncertainty of all dates), and increased uncertainty at depths far from radiocarbon-dated material [46]. Each model uses a Markov-Chain Monte-Carlo algorithm to estimate model parameters, which provides a distribution of the chronologies that are most likely given all radiocarbon-dated probability distributions. Most importantly, they explicitly account for and estimate joint uncertainty of all of the radiocarbon dates and consider the probability density functions of calibrated age ranges in chronology interpolation. All high-resolution chronology models achieved convergence and identified no outliers at Dukes Tarn, but two outliers at Lake Kirkpatrick (macrofossil dated material at depths 110 and 111 cm), which were subsequently left out of modeled chronologies. The radiocarbon material at these two depths were identified as having low mass during AMS analyses (KCCAMS lab), which likely increased their age uncertainty and inaccuracy.

Rate of vegetation change

We used the Sørensen's similarity statistic [47] to assess the rate of vegetation change between pollen samples. Sørensen's index, similar to the Bray-Curtis dissimilarity [48] is a similarity index that measures compositional similarity between two sites, or in this case, compositional similarity in pollen taxa at two depths where pollen percentages are measured. Sørensen's distance statistic is calculated between each pollen sample and the preceding pollen sample providing a statistical measure of vegetation change over time.

Supporting Information

Figure S1 Comparison of chronology model estimates (circles and bars represent median and 95% Confidence Intervals) for the interval associated with forest transitions that followed the first human-set fires, Lake Kirkpatrick and Dukes Tarn, South Island, New Zealand. Median (and 95% confidence estimate ranges) for the interval associated with forest transitions for Lake Kirkpatrick were 17 yrs (3–13, Bchron), 27 yrs (7–47, MCAgeDepth), and 31 yrs (16–46, Oxcal). Median and 95% confidence estimate ranges (in parentheses) for the interval associated with forest transitions for Dukes Tarn were 47 yrs (35–59, Bchron), 44 yrs (24–64, MCAgeDepth), and 25 yrs (14–36, Oxcal).

Figure S2 MCAgeDepth age depth model for Lake Kirkpatrick with sedimentation rates and sample resolution. Chronologies developed using a weighted cubic smoothing spline which takes into account the number and uncertainty of age estimates. Gray shading represent 95% confidence intervals reflect the combined uncertainty of all age estimates in a model and are derived from 1000 bootstrapped chronologies.

Figure S3 MCAgeDepth age depth model for Dukes Tarn with sedimentation rates and sample resolution. Chronologies developed using a weighted cubic smoothing spline which takes into account the number and uncertainty of age estimates. Gray shading represent 95% confidence intervals reflect the combined uncertainty of all age estimates in a model and are derived from 1000 bootstrapped chronologies.

Figure S4 Oxcal age-depth model for Lake Kirkpatrick using the P_Sequence deposition model which assumes deposition to be random (*k* parameter set at 1 – which assumes a postulated event spacing of approximately 1 cm). Light grey distributions represent likelihoods for single calibrated dates and in darker grey shading represents the marginal posterior distributions which take into account the depth model; the depth model curves show envelopes for the 95% and 68% highest probability density (HPD) ranges.

Figure S5 Oxcal age-depth model for Dukes Tarn using the P_Sequence deposition model which assumes deposition to be random (*k* parameter set at 1 – which assumes a postulated event spacing of approximately 1 cm). Light grey distributions represent likelihoods for single calibrated dates and in darker grey shading represents the marginal posterior distributions which take into account the depth model; the depth model curves show envelopes for the 95% and 68% highest probability density (HPD) ranges.

Figure S6 Pollen diagram for Lake Kirkpatrick, South Island, New Zealand.

Figure S7 Pollen diagram for Dukes Tarn, South Island, New Zealand.

Table S1 AMS radiocarbon dating determined age information for Lake Kirkpatrick and Dukes Tarn, New Zealand.

Acknowledgments

We thank Mairie Fromont for conducting pollen analyses for Dukes Tarn, John Southon (University of California-Irvine, Keck-Carbon Cycle AMS Facility) for radiocarbon dates and Philip Higuera for help with Figure 3 and the application of Sørensen's similarity index to pollen data.

Author Contributions

Conceived and designed the experiments: DBM CW JMW. Performed the experiments: DBM CW JMW JRW. Analyzed the data: DBM CW JMW JRW. Contributed reagents/materials/analysis tools: DBM CW JMW JRW. Wrote the paper: DBM CW JMW JRW MSM.

References

1. Nepstad DC, Stickler CM, Soares B, Merry F (2008) Interactions among Amazon land use, forests and climate: prospects for a near-term forest tipping point. Philosophical Transactions of the Royal Society B-Biological Sciences 363: 1737–1746.
2. Mayer AL, Khalyani AH (2011) Grass Trumps Trees with Fire. Science 334: 188–189.
3. Kitzberger T, Araoz E, Gowda J, Mermoz M, Morales J (2011) Decreases in fire spread probability with forest age promotes alternative community states, reduced resilience to climate variability and large fire regime shifts. Ecosystems: 1–16.
4. Staver AC, Archibald S, Levin SA (2011) The global extent and determinants of savanna and forest as alternative biome states. Science 334: 230–232.
5. Perry GLW, Wilmshurst JM, McGlone MS, McWethy DB, Whitlock C (2012) Explaining fire-driven landscape transformation during the Initial Burning Period of New Zealand's prehistory. Global Change Biology 18: 1609–1621.
6. Pongratz J, Reick C, Raddatz T, Claussen M (2008) A reconstruction of global agricultural areas and land cover for the last millennium. Global Biogeochem Cycles 22: GB3018.
7. Kaplan JO, Krumhardt KM, Ellis EC, Ruddiman WF, Lemmen C, et al. (2011) Holocene carbon emissions as a result of anthropogenic land cover change. The Holocene 21: 775–791.
8. Wilmshurst JM, Anderson AJ, Higham TFG, Worthy TH (2008) Dating the late prehistoric dispersal of Polynesians to New Zealand using the commensal Pacific rat. Proceedings of the National Academy of Sciences 105: 7676–7680.
9. McWethy DB, Whitlock C, Wilmshurst JM, McGlone MS, Fromont M, et al. (2010) Rapid landscape transformation in South Island, New Zealand, following initial Polynesian settlement. Proceedings of the National Academy of Sciences 107: 21343–21348.
10. McGlone MS, Wilmshurst JM (1999) Dating initial Maori environmental impact in New Zealand. Quaternary International 59: 5–16.

11. Murray-McIntosh RP, Scrimshaw BJ, Hatfield PJ, Penny D (1998) Testing migration patterns and estimating founding population size in Polynesia by using human mtDNA sequences. Proceedings of the National Academy of Sciences of the United States of America 95: 9047–9052.
12. Newnham R, Lowe DJ, McGlone MS, Wilmshurst JM, Higham TFG (1998) The Kaharoa Tephra as a critical datum for earliest human impact in northern New Zealand. Journal of Archaeological Science 25: 533–544.
13. Horrocks M, Deng Y, Ogden J, Alloway B, Nichol S, et al. (2001) High spatial resolution of pollen and charcoal in relation to the c. 600 yr BP Kaharoa Tephra: implications for Polynesian settlement of Great Barrier Island, northern New Zealand. Journal of Archaeological Science 28: 153–168.
14. McWethy DB, Whitlock C, Wilmshurst JM, McGlone MS, Li X (2009) Rapid deforestation of South Island, New Zealand by early Polynesian fires. The Holocene 19: 883–897.
15. CINZAS (2008) Central Index of New Zealand Archaeological Sites. nzarchaeologyorg.
16. Kaplan JO, Krumhardt KM, Zimmermann N (2009) The prehistoric and preindustrial deforestation of Europe. Quaternary Science Reviews 28: 3016–3034.
17. Pfeiffer M, Spessa A, Kaplan JO (2013) A model for global biomass burning in preindustrial time: LPJ-LMfire (v1.0). Geosci Model Dev\ 6\: 643\-685\.
18. Williams M (2000) Dark ages and dark areas: global deforestation in the deep past. Journal of Historical Geography 26: 28–46.
19. Wilmshurst JM, McGlone MS, Partridge TR (1997) A late Holocene history of natural disturbance in lowland podocarp/hardwood forest, Hawke's Bay, New Zealand. New Zealand Journal of Botany 35: 79–96.
20. Ogden J, Basher L, McGlone M (1998) Fire, forest regeneration and links with early human habitation: Evidence from New Zealand. Annals of Botany 81: 687–696.
21. Wilmshurst JM (1997) The impact of human settlement on vegetation and soil stability in Hawke's Bay, New Zealand. New Zealand Journal of Botany 35: 97–111.
22. Perry GLW, Wilmshurst JM, McGlone MS, Napier A (2012) Reconstructing spatial vulnerability to forest loss by fire in pre-historic New Zealand. Global Ecology and Biogeography: no-no.
23. Wilmshurst JM, Higham TFG, Allen H, Johns D, Phillips C (2004) Early Maori settlement impacts in northern coastal Taranaki, New Zealand. New Zealand Journal of Ecology 28: 167–179.
24. Perry G, Wilmshurst J, McGlone M (2014) The ecology and long-term history of fire in New Zealand's ecosystems. New Zealand Journal of Ecology 38.
25. Kitzberger T, Brown PM, Heyerdahl EK, Swetnam TW, Veblen TT (2007) Contingent Pacific-Atlantic ocean influence on multi-century wildfire synchrony over western North America. Proceedings of the National Academy of Sciences 104: 543–548.
26. Field RD, Van der Werf G, Shen SSP (2009) Human amplification of drought-induced biomass burning in Indonesia since 1960. Nature Geoscience 2: 185–188.
27. Lindenmayer DB, Hobbs RJ, Likens GE, Krebs CJ, Banks SC (2011) Newly discovered landscape traps produce regime shifts in wet forests. Proceedings of the National Academy of Sciences 108: 15887–15891.
28. Odion DC, Moritz MA, DellaSala DA (2010) Alternative community states maintained by fire in the Klamath Mountains, USA. Journal Of Ecology 98: 96–105.
29. Hirota M, Holmgren M, Van Nes EH, Scheffer M (2011) Global Resilience of Tropical Forest and Savanna to Critical Transitions. Science 334: 232–235.
30. McWethy DB, Higuera PE, Whitlock C, Veblen TT, Bowman DMJS, et al. (2013) A conceptual framework for predicting temperate ecosystem sensitivity to human impacts on fire regimes. Global Ecology and Biogeography 22: 900–912.
31. Thompson JR, Spies TA, Ganio LM (2007) Reburn severity in managed and unmanaged vegetation in a large wildfire. Proceedings of the National Academy of Sciences 104: 10743–10748.
32. Odion DC, Frost EJ, Strittholt JR, Jiang H, Dellasala DA, et al. (2004) Patterns of fire severity and forest conditions in the western Klamath Mountains, California. Conservation Biology 18: 927–936.
33. Cochrane MA, Alencar A, Schulze MD, Souza CM Jr., Nepstad DC, et al. (1999) Positive Feedbacks in the Fire Dynamic of Closed Canopy Tropical Forests. Science 284: 1832–1835.
34. Veblen TT, Holz A, Paritsis J, Raffaele E, Kitzberger T, et al. (2011) Adapting to global environmental change in Patagonia: What role for disturbance ecology? Austral Ecology: no-no.
35. Perry GLW, Ogden J, Enright NJ, Davy LV (2010) Vegetation patterns and trajectories in disturbed landscapes, Great Barrier Island, northern New Zealand. New Zealand Journal of Ecology: 311–323.
36. Moritz MA, Parisien M-A, Batllori E, Krawchuk MA, Van Dorn J, et al. (2012) Climate change and disruptions to global fire activity. Ecosphere 3: art49.
37. McGlone MS, Wilmshurst JM, Leach HM (2005) An ecological and historical review of bracken (Pteridium esculentum) in New Zealand, and its cultural significance. New Zealand Journal of Ecology 29: 165–184.
38. McGlone MS (1995) Lateglacial landscape and vegetation change and the Younger Dryas climatic oscillation in New Zealand. Quaternary Science Reviews 14: 867–881.
39. Wardle P (2001) Distribution of native forest in the upper Clutha district, Otago, New Zealand. New Zealand Journal of Botany 39: 435–446.
40. Moore PD, Webb JA, Collinson ME (1991) Pollen Analysis. Oxford, UK: Blackwell Scientific.
41. Whitlock C, Larsen C. (2001) Charcoal as a Fire Proxy. In: Smol JP, Birks, H.J.B, Last, W M., editor.Tracking Environmental Change Using Lake Sediments: Volume 3 Terrestrial, Algal, and Siliceous indicators.Dordrecht: Kluwer Academic Publishers. pp. 75–97.
42. Higuera PE, Peters ME, Brubaker LB, Gavin DG (2007) Understanding the origin and analysis of sediment-charcoal records with a simulation model. Quaternary Science Reviews 26: 1790–1809.
43. Higuera PE, Brubaker LB, Anderson PM, Hu FS, Brown TA (2009) Vegetation mediated the impacts of postglacial climate change on fire regimes in the south-central Brooks Range, Alaska. Ecological Monographs 79: 201–219.
44. Haslett J, Parnell A (2008) A simple monotone process with application to radiocarbon-dated depth chronologies. Journal of the Royal Statistical Society: Series C (Applied Statistics) 57: 399–418.
45. Bronk Ramsey C (2008) Deposition models for chronological records. Quaternary Science reviews 27: 42–60.
46. Parnell AC, Buck CE, Doan TK (2011) A review of statistical chronology models for high-resolution, proxy-based Holocene palaeoenvironmental reconstruction. Quaternary Science Reviews 30: 2948–2960.
47. Sørensen T (1948) A method of establishing groups of equal amplitude in plant sociology based on similarity of species and its application to analyses of the vegetation on Danish commons. Kongelige Danske Videnskabernes Selskab 5: 1–34.
48. Bray JR, Curtis JT (1957) An ordination of upland forest communities of southern Wisconsin. Ecological Monographs 27: 325–349.

22

Ecological Implications of a Flower Size/Number Trade-Off in Tropical Forest Trees

Chris J. Kettle[1,2]*, Colin R. Maycock[2], Jaboury Ghazoul[1], Pete M. Hollingsworth[3], Eyen Khoo[4], Rahayu Sukmaria Haji Sukri[2], David F. R. P. Burslem[2]

1 Ecosystem Management, Institute of Terrestrial Ecosystems, ETH Zurich, Zurich, Switzerland, **2** Institute of Biological and Environmental Sciences, University of Aberdeen, Aberdeen, United Kingdom, **3** Royal Botanic Gardens Edinburgh, Edinburgh, United Kingdom, **4** Forest Research Centre, Sabah Forest Department, Sabah, Malaysia

Abstract

Background: In angiosperms, flower size commonly scales negatively with number. The ecological consequences of this trade-off for tropical trees remain poorly resolved, despite their potential importance for tropical forest conservation. We investigated the flower size number trade-off and its implications for fecundity in a sample of tree species from the Dipterocarpaceae on Borneo.

Methodology/Principal Findings: We combined experimental exclusion of pollinators in 11 species, with direct and indirect estimates of contemporary pollen dispersal in two study species and published estimates of pollen dispersal in a further three species to explore the relationship between flower size, pollinator size and mean pollen dispersal distance. Maximum flower production was two orders of magnitude greater in small-flowered than large-flowered species of Dipterocarpaceae. In contrast, fruit production was unrelated to flower size and did not differ significantly among species. Small-flowered species had both smaller-sized pollinators and lower mean pollination success than large-flowered species. Average pollen dispersal distances were lower and frequency of mating between related individuals was higher in a smaller-flowered species than a larger-flowered confamilial. Our synthesis of pollen dispersal estimates across five species of dipterocarp suggests that pollen dispersal scales positively with flower size.

Conclusions and Their Significance: Trade-offs embedded in the relationship between flower size and pollination success contribute to a reduction in the variance of fecundity among species. It is therefore plausible that these processes could delay competitive exclusion and contribute to maintenance of species coexistence in this ecologically and economically important family of tropical trees. These results have practical implications for tree species conservation and restoration. Seed collection from small-flowered species may be especially vulnerable to cryptic genetic erosion. Our findings also highlight the potential for differential vulnerability of tropical tree species to the deleterious consequences of forest fragmentation.

Editor: Jerome Chave, Centre National de la Recherche Scientifique, France

Funding: Financial support was provided by the British Ecological Society (http://www.britishecologicalsociety.org/) "Overseas Research Programme" and the Natural Environment Research Council (http://www.nerc.ac.uk/) grant numbers NER/A/S/2001/00835 and NE/D003822/1. The funders had no role in study design, data collection and analysis, decision to publish, or preparation of the manuscript.

Competing Interests: The authors have declared that no competing interests exist.

* E-mail: chris.kettle@env.ethz.ch

Introduction

Angiosperms manifest a wide array of floral displays, from species that produce only a single large flower per plant, to those that produce hundreds of tiny flowers on many inflorescences per individual. The attractiveness of these different floral displays is closely coupled to mating system, the type and breadth of pollinators and individual fecundity [1,2]. Studies across plant species show that species tend to produce fewer flowers per individual as the size of the flower increases [3,4,5,6,7,8,9]. The causal agents of this trade-off have been investigated empirically in a number of plant species, indicating that hierarchical resource allocation [10,11], pollen discounting via geitonogamy in obligate out-crossing species [1,4,11] and pollen limitation [12] are all important drivers in this trade-off.

Few studies have, however, determined whether a flower size/number trade-off occurs across, as well as within, species (but see 7) and examined its consequences for plant community structure. The potential implications could include partitioning of pollinators among species as a function of flower size, differential pollen dispersal as a function of pollinator size and mobility relationships [13], and differences in fecundity among species. Surprisingly, interactions between plants and their pollinators have received limited attention as a driver of species diversity in tropical tree communities [14,15,16]. Partitioning of pollinator services could limit competitive exclusion, and if these interactions can be shown to contribute to an equalisation of fitness, contribute to species coexistence [17].

We combine experimental ecology, field observation and molecular ecological approaches to investigate the ecological implications of a flower size/number trade-off in a clade of tropical forest canopy and emergent trees. The processes that determine the differential fecundity of individuals are poorly understood, and therefore the importance to demography of pre-dispersal stages of plant reproduction may have been underestimated. Studies that link floral traits to fecundity are also likely to reveal important implications of flower size vs number trade-offs for species' differential responses to habitat degradation [12,18]. For example, increased isolation of individual trees is of concern for the management of tropical tree species that have been heavily fragmented in recent years [19].

We used a group of 12 coexisting dipterocarp species ranging in flower size (calyx diameter from 1.2 mm to 10.2 mm) to explore the function of flower size and number and pollination systems in fruit production. Specifically, we combine measures of flower size, flower number and pollination success (as inferred from pollen tube growth data) with measures of paternity and genetic relatedness within mapped populations of two species to test the following explicit hypotheses.

1. Across species, flower size scales negatively with mean flower production per individual, which would be indicative of a flower size/number trade-off.
2. Across species, mean pollination success (as defined by the proportion of flower styles containing pollen tubes) increases with flower size.
3. Species with small flowers have small-bodied pollinators.
4. Species with smaller bodied pollinators have more limited pollen dispersal distances.
5. The proportion of flowers that give rise to fruit scales positively with flower size.
6. Spatial aggregation and limited pollen dispersal increase inbreeding in a small flowered species.

These hypotheses encapsulate a set of trade-offs between flower size, pollinator body size and pollen dispersal that might have consequences for fruit production and fecundity, which are important components of plant fitness (see Figure 1).

Tree species in the Dipterocarpaceae dominate lowland rain forests of Southeast Asia, accounting for up to 80 percent of the density and basal area of canopy trees [20]. Dipterocarps display variable flower size and morphology and are pollinated by insects including thrips, beetles, moths and giant honeybees [21,22,23,24,25,26,27]. Reproductively mature trees of many dipterocarps exhibit irregular mass flowering [20]. Dipterocarps produce hermaphroditic flowers and are considered predominately out-crossing, which is maintained through pre- and post- zygotic incompatibility mechanisms [28]. Apomixis has been reported in some species [24,29], but we found no evidence of this in the progeny arrays of our study species. Fruit dispersal occurs by gravity and wind assisted gyration for the majority of fruit falling beneath the mother tree or a within 20–30 m [30,31]. Limited seed dispersal leads to the aggregation of related individuals, and significant fine-scale spatial genetic structure has been observed in the majority of dipterocarp species studied [32].

Results

Flower and fruit production as a function of flower size

Analyses of absolute (unscaled) log-transformed values of maximum potential flower production yielded a single most likely model (Table S1a Supplementary Information), in which flower production declined as a function of flower size for the comparison of 11 species ($F_{1,7} = 35.88$, $P = 0.0005$, Fig. 2a). In this model canopy species had lower values of unscaled maximum flower production than emergent species ($F_{1,7} = 57.05$, $P = 0.0001$), and the interaction between life form and flower size was also significant ($F_{1,7} = 21.81$, $P = 0.0023$). Analyses of maximum flower production scaled to tree size yielded three most likely models (Table S1a Supplementary Information). In all three cases scaled maximum flower production declined significantly with increasing flower size (e.g. $F_{1,7} = 70.39$, $P < 0.0001$ for the model with the lowest AIC value, Fig. 2b)), but there was no difference between canopy and emergent species ($F_{1,7} = 1.71$, $P = 0.23$) and no interaction of life form and flower size ($F_{1,7} = 2.13$, $P = 0.18$).

In contrast to flower production (where small flowered species produce more flowers than larger flowered species), there was not a significant effect of flower size in any of the three most likely models analysing absolute values of maximum fruit production (Fig. 2c) or in the single most likely model of maximum fruit production scaled to tree size (Fig. 2d). Emergent species had a marginally greater value of maximum absolute fruit production than canopy species (Fig. 2c), and this term was significant in one of the three most likely models for this variable, but there was no difference between canopy and emergent species in maximum fruit production scaled to tree size (Fig. 2d). The nearest neighbour distance term was non-significant in all analyses of fruit production among species. The relationships between flower size and fruiting variables and pollination success (median number of pollen tubes per pollinated flower) were all supported by phylogenetically independent contrasts. Of the 12 species used in this study, only eight species exist in well resolved phylogenies [33,34]. With the exception of *S. xanthophylla* all the *Shorea* species in our sample fall within the Red Meranti clade [34], with no apparent phylogenetic signal of calyx size.

Pollination success

Analyses of the proportion of flowers pollinated identified five most likely models (Table S1c, Supplementary Information), but in all cases a positive effect of flower size was the only significant term ($F_{1,7} = 55.97$, $P < 0.001$, Fig. 2e). Similarly, flower size was the only significant term in all the five most likely models analysing interspecific differences in the mean number of pollen tubes per pollinated flower ($F_{1,9} = 45.72$, $P < 0.0001$, Fig. 2f). Differences in life form and in median nearest neighbour distance were not significant in these models of pollination success among species.

Pollinator body size

The experimental exclusion of pollinators based on body size resulted in a significant decline in pollination success with increasing flower size ($t = -6.18$, $P < 0.0001$, Fig. 2). In small-flowered species (1.0–2.0 mm), such as *Shorea xanthophylla*, restricting access to pollinators had very little effect on the proportion of fruit produced relative to that expected among open pollinated flowers. In contrast, in the largest flowered species *Dipterocarpus grandiflorus* fruit set within treatments that limited pollinator access to small bodied insects was substantially reduced under all mesh sizes (Fig. 3 see Dg). The response to mesh bag treatments was significant ($t = 2.76$, $P = 0.011$), but the interaction of flower size and mesh bag treatment was not significant.

Pollen dispersal distance and genetic relatedness between assigned parents

Paternity analysis revealed that short-distance mating events were far more common among the smaller flowered species *S. xanthophylla* (47% of matings within 50 m and 71% within 100 m of the mother tree) than the larger flowered species *P. tomentella* (13% within 50 m and 32% within 100 m) ($\chi^2 = 17.96$, $P < 0.001$). The median (Wilcox rank sum $= 7325$, $P < 0.0001$) and mean ($t = -4.609$, $P < 0.0001$) pollen dispersal distances for all assigned

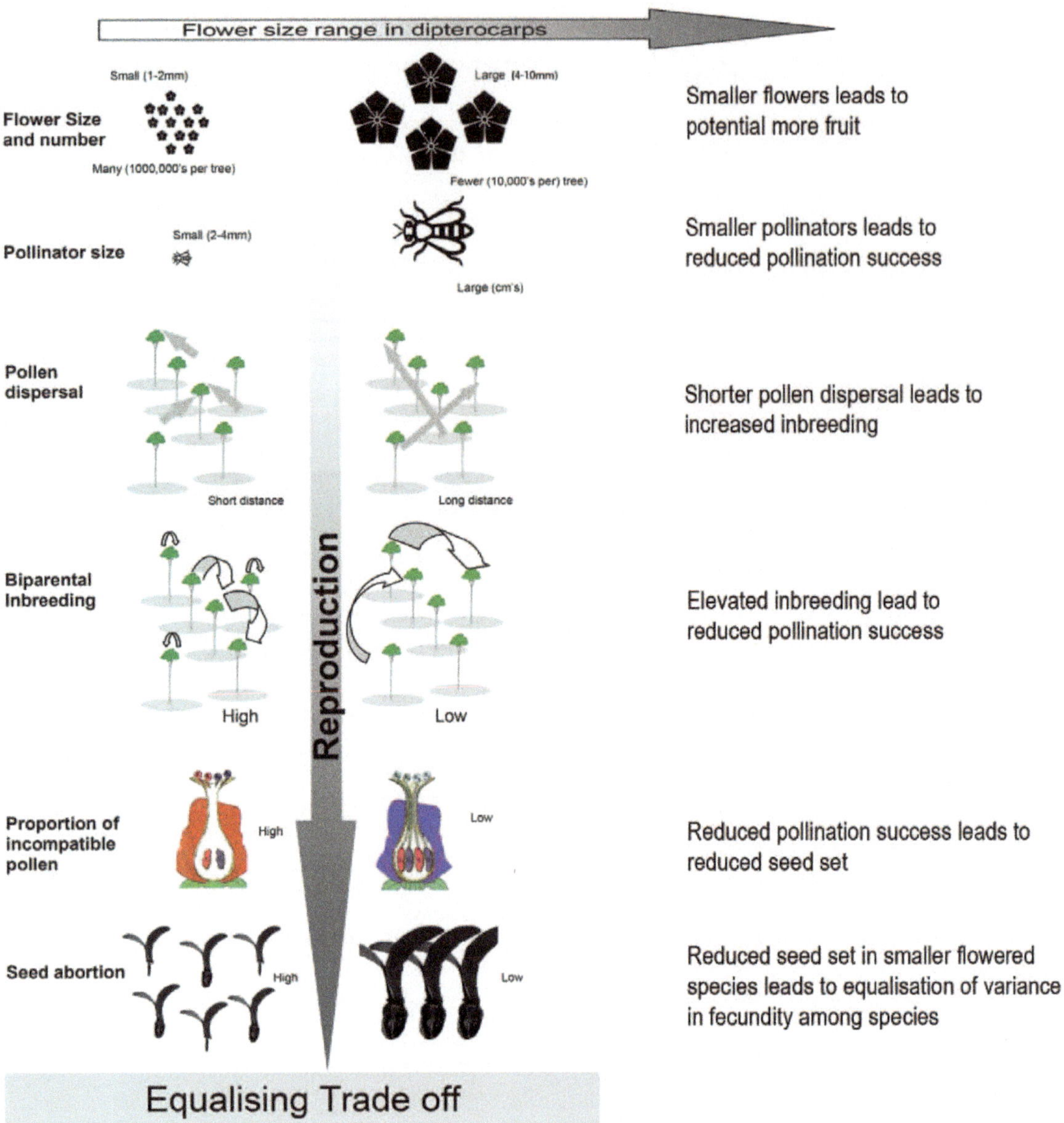

Figure 1. Schematic illustrating how interactions between flower size and number, pollination biology, pollen dispersal and fruit production might contribute to equalisation of fitness components among species of Dipterocarpaceae.

progeny showed a similar pattern: 63 m and 119 m in *S. xanthophylla* and 164 m and 177 m in *P. tomentella* respectively. Because the density of potential pollen donors within 100 m of mother trees was generally greater for *S. xanthophylla* than *P. tomentella*, we also tested for differences in pollen dispersal between the two species using a subset of mother trees that had equivalent distances to nearest flowering neighbours. In the subset of *S. xanthophylla* mothers, 32 percent of matings were assigned at less than 50 m and 34 percent at less than 100 m. The results of the TWOGENER analysis of pollen dispersal are summarised in Table 1. The differentiation in the pollen gene pool is significant in both species ($P<0.01$) and almost twice as great in *S. xanthophylla* than in *P. tomentella* (0.0624 and 0.0364 respectively). In this analysis the mean pollen dispersal distance for *S. xanthophylla* was estimated at 56 m (S.E. 2.88) and for *P. tomentella* 416 m (S.E. 29.93). The effective number of pollen donors per mother tree (N_{ep}) was 8.01 in *S. xanthophylla* and 13.73 in *P. tomentella* (Table 1). The effective pollen neighbourhood area (A_{ep}) ranged from 1.46 ha ($A_{ep}{}^{min}$) $-$14.62 ha ($A_{ep}{}^{max}$) in *S. xanthophylla* and from 8.43 ha ($A_{ep}{}^{min}$) $-$84.26 ha ($A_{ep}{}^{max}$) in *P. tomentella* (Table 1).

Genetic relatedness between mothers and assigned fathers was an order of magnitude higher in progeny of the small-flowered species *S. xanthophylla* than in the larger flowered species *P. tomentella* (mean pairwise kinship coefficient $=0.033$ (S.E. 0.008) and 0.002 (S.E. 0.01) respectively; $t=2.195$, df$=229$, $P=0.028$). The mating system parameters based upon the progeny arrays indicate that both species are extensively outcrossing ($t_m=0.996$ and $t_m=0.907$ in *S. xanthophylla* and *P. tomentella* respectively). However, the parental inbreeding coefficient and biparental inbreeding were both greater in *S. xanthophylla* (see supporting information Table S2).

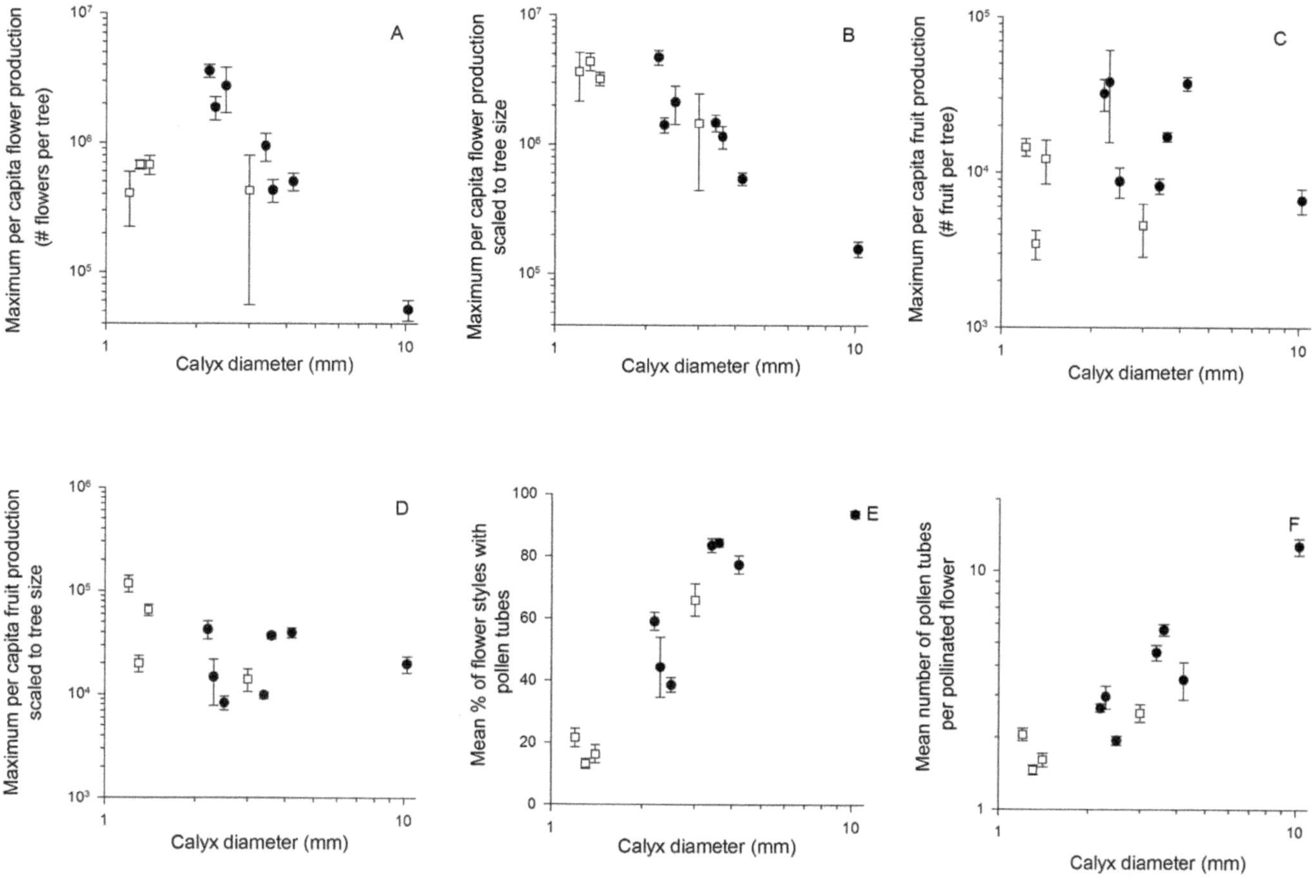

Figure 2. Effects of calyx tube diameter (flower size, mm) on estimated flower and fruit production in 11 dipterocarp tree species. Effects of calyx tube diameter on estimated (a) maximum per capita flower production, (b) maximum per capita flower production scaled to basal area and mean species tree height, (c) maximum per capita fruit production, (d) maximum per capita fruit production scaled to basal area and mean species tree height, (e) mean percentage of flower styles with pollen tubes and (f) mean number of pollen tubes per pollinated flower for four canopy (open symbols) and seven emergent (closed symbols) dipterocarp species growing in Sepilok Forest Reserve, Sabah. Maximum flower and fruit production were estimated from the mean of the upper quartile of values per species (n = 2 to 14 individuals per species).

Discussion

Flower size/number trade-off and fecundity

We confirmed that flower size scaled negatively with flower number across 11 dipterocarp species. Although canopy species produced fewer flowers for their flower size than predicted by the scaling relationship for emergent species, all species fit the same negative relationship when flower production was scaled to tree size. These relationships suggest that the trade-off between flower size and potential flower production was consistent across all species, but subject to allometric constraints. These observations are robust to phylogenetic comparison These results are consistent with other studies that suggest that a flower size vs number trade-off is common in angiosperms [7,35]. Maximum fruit production per tree, in contrast to flower production, did not scale negatively with flower size. This indicates that the potential fecundity offered by a large number of small flowers was offset by the reduced likelihood that each of these flowers would be pollinated and develop into a mature fruit. Our results suggest that processes operating between flower production and fruit dispersal contribute to equalising fecundity among dipterocarp species. Below we interpret our findings and discuss the range of alternative hypotheses that could lead to the observed patterns of fecundity.

Fruit production in larger fruited species could be resource limited, leading to larger fruits being produced in fewer numbers [11]. However, we exclude this hypothesis for two reasons. First, the scaling relationships were determined for trees representing the upper quartile for flower production, which are least likely to be resource-limited among the sample. Secondly, fruit size scales positively with flower size in dipterocarps (C.R. Maycock *et al.*, unpubl. data), therefore small-flowered species would have lower overall per capita fruit mass, even though fruit number remains the same, a scenario that is at odds with the expectations of resource limited fruit production.

Experimental exclusion of all potential pollinators except thrips reduced pollination success in large-flowered but not small-flowered species. This result suggests that the role of thrips in pollination is important for small-flowered species of dipterocarp such as *Shorea xanthophylla* but less so as flower size increases. Thrips are abundant in the flowers of all the dipterocarps examined in this study and are widespread in the flowers of many non-dipterocarps in the same forest (C.R. Maycock, *et al.* unpubl. data). Hence their abundance is unlikely to limit pollination success in the small-flowered species, and therefore we reject pollinator limitation as a determinant of the flower size-dependent reduction in pollination success.

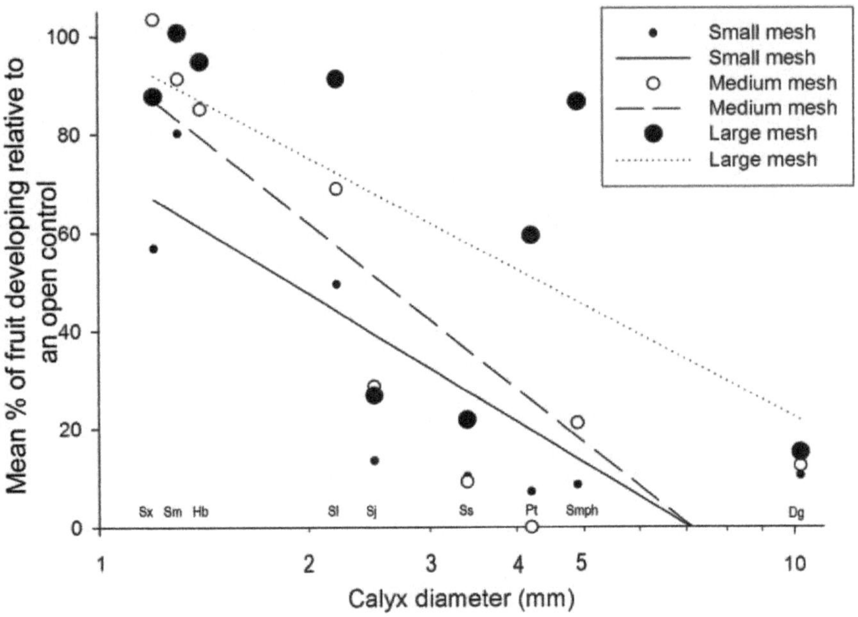

Figure 3. Effect of pollinator exclusion using bags with mesh apertures of 0.2 mm (solid line, small closed symbols), 2 mm (dashed line, open symbols) and 5 mm (dotted line, large closed symbols) on pollination success one month after anthesis. Pollination success is defined as the mean percentage of flowers that gave rise to immature fruit for 8 to 10 bags per treatment on 2 to 6 replicate trees per species, expressed as a percentage of the mean value for an equivalent sample of open-pollinated flowers on the same tree.

There are several concurring lines of evidence to support the hypothesis that not only do large-flowered species receive a higher proportion of pollen (as empirically demonstrated here for 11 species) but also that pollen received is of greater genetic quality (*i.e.*, more compatible). First, our examination of pollen tube growth among species showed that both the proportion of pollinated flowers and mean number of pollen tubes per pollinated flower was greater in larger-flowered species. This implies that either smaller-flowered species receive fewer per capita visits by pollinators than larger-flowered species, and/or that a higher proportion of pollination events result in the transfer of non-compatible self-pollen that is blocked at the stigmatic surface or at the early stages of pollen tube growth. Dipterocarp species are considered to be self-incompatible [28], and although there is evidence that selfing can lead to fruit production, these fruit are preferentially aborted at an early stage of maturation [36,37]. This

Table 1. Mean ± SEM calyx tube width (mm) for the 12 study species of Dipterocarpaceae sampled in Sepilok forest reserve (SFR).

					Number of sample trees (n)		
Species (code)	**Life form †**	**Mean tree height**	**Calyx tube width (mm)**	**Median NN distance (m)**	**Flower & Fruit production [a]**	**Pollination success [b]**	**Pollinator Exclusion [c]**
Shorea xanthophylla (Sx)	Canopy	40 m (25 m) - 35 m*	1.2±0.1	19.7	29	26	3
Shorea multiflora (Sm)	Canopy	40 m (No estimate) - 35 m*	1.3±0.2	37.9	35	31	6
Hopea beccariana (Hb)	Canopy	45 m (35 m*)	1.4±0.1	200	14	11	6
Shorea leprosula (Sl)	Emergent	60 m (60 m) - 65 m*	2.2±0.2	85.1	55	33	6
Shorea parvifolia (Sp)	Emergent	65 m* (60 m)	2.3±0.1	69.7	6	6	0
Shorea johorensis (Sj)	Emergent	50 m (65 m*)	2.5±0.1	66.7	32	29	6
Shorea macropteraa (Smac)	Canopy	40*- 50 m (60 m)	3.0±0.2	118.2	6	6	0
Shorea smithiana (Ss)	Emergent	60 m* (55 m)	3.4±0.2	91.8	22	10	3
Shorea beccariana (Sb)	Emergent	60 m (55 m*)	3.6±0.3	39.5	53	38	0
Parashorea tomentella (Pt)	Emergent	65 m (65 m*)	4.2±0.1	50.5	29	25	6
Shorea macrophylla (Smph)	Emergent	45 m (45 m)	4.9±0.2	N/A	N/A	N/A	2
Dipterocarpus grandiflorus (Dg)	Emergent	45 m * (45 m)	10.2±0.7	24.8	30	18	3

Mean tree heights are from Ashton (2004), except values in brackets which are from [91] and those marked *, which were measured directly at SFR or obtained from a published source equivalent to our own measure. Median two nearest neighbouring (NN) distance to flowering conspecific trees; number of sample trees (n) for a) quantification of flower and fruit production; b) pollination success, based on pollen tube analysis and c) trees for experimental pollinator exclusion.
† Following Ashton [89].

behaviour may have evolved as a predator avoidance strategy [38].

Second, the pollinator exclusion experiment also revealed that fruit set increased as access to larger pollinators was permitted by larger mesh sizes in larger flowered species. The exception to this pattern was the largest flowered species, *Dipterocarpus grandiflorus*, which had significantly reduced fruit set even with the largest mesh (5 mm). This outcome suggests that the importance of larger pollinators increases as flower size increases and, for the largest flowered species, *D. grandiflorus*, pollinators that exceed 5 mm in size, such as giant honey bees *Apis dorsata*, large scarabid beetles and Sphingidae and Noctuidae moths, play an important role. We conclude that the role of small-bodied insects in dipterocarp pollination, in particular thrips, declines with increasing flower size. The relevance of this finding in the context of putative trade-offs is that the larger bodied pollinators may transfer pollen over greater distances and may therefore be more effective agents for outcrossing [13,39].

Differential pollen dispersal. Our study provides evidence that larger bodied pollinators disperse pollen over greater distances than smaller bodied pollinators in dipterocarps. Our paternity analysis and indirect estimates of pollen dispersal demonstrated that short-distance pollen dispersal events were more frequent, and that the mean pollen dispersal distance were shorter, for the small-flowered species *Shorea xanthophylla* than for *Parashorea tomentella*, independent of density of flowering conspecifics. Progeny collected from a subset of *S. xanthophylla* mother trees with an equivalent density of flowering conspecifics to that of *P. tomentella* were assigned more frequently to nearer pollen donors (<100 m) than for *P. tomentella*. Other studies have estimated mean pollen dispersal distances in dipterocarp species to lie in the range 175–207 m [22,25,26]. For the five dipterocarp species where pollen dispersal distances have been quantified, representing flower sizes ranging over two orders of magnitude, mean pollen dispersal distance increases with flower size, with the lowest mean dispersal distances in *S. xanthophylla* and greatest dispersal distances in *Dipterocarpus tempehes* (Table 2). Such a pattern remains inconclusive, as it is based on only five species, and other factors may influence pollen-mediated gene flow, including differences in the density of flowering conspecifics. Nevertheless, the results are suggestive that increasing flower size corresponds to higher pollen flow distances, which we interpret as being mediated by the larger-bodied pollinators attracted by the larger flowers. Hence, all else being equal, these larger-flowered species experience higher outcrossing rates and are less susceptible to inbreeding. This corresponds to a higher quality of pollination service for large flowered species, which compensates for the relatively few flowers produced by these species.

Inbreeding and thus reduced fecundity through inbreeding depression are predicted to be greater in species where related individuals are highly aggregated. Seed dispersal limitation and habitat specialisation are common in dipterocarps [30,31], and both processes can lead to aggregation of related individuals [40,41,42]. This is measurable as fine-scale spatial genetic structure (SGS) within populations. A strong SGS signal would not be expected if gene flow by pollen or seed dispersal is extensive. Dipterocarp seed dispersal is known to occur over short distances [43,44,45,46,47], which suggests that long distance pollen flow is likely to be the primary agent undermining SGS. The evidence for fine-scale SGS observed in comparable studies of dipterocarp species [11,26,29,32,48,49] suggests that SGS is common in the Dipterocarpaceae, but weaker in larger flowered species. This interpretation is consistent with our pollinator exclusion experiments, and other studies [25,50,51] that identify the importance of large insects such as *Apis dorsata* and sphingid moths for pollination of large-flowered dipterocarps.

Significant fine-scale SGS combined with limited pollen dispersal may impose a fitness cost due to elevated inbreeding depression or incompatible matings in self-incompatible species (Fig. 1), and this is likely to be more substantial among smaller flowered species for the reasons given above. Such processes would lead to a genetic cost under limited pollen dispersal and is supported by evidence of a higher frequency of matings between more closely-related individuals in the smaller-flowered species *Shorea xanthophylla* than in *Parashorea tomentella*. High neighbourhood genetic relatedness has been shown to interact with density to increase the proportion of aborted fruit in the insect pollinated tropical tree *Jacaranda copaia* [41], which supports our hypothesis that genetic processes contribute to the trade-off between flower size and pollination success (Fig. 1).

Differential Fecundity. As a component of plant fitness, fecundity is represented in many models of population persistence and species coexistence [52,53,54,55]. However, empirical investigations of coexistence in tropical trees have focused more on plant responses to abiotic resource availability [56,57,58]. These studies do not support an unequivocal role for habitat partitioning in the maintenance of the diverse tropical tree communities observed in Borneo. While our study is limited to only a few species, it does highlight the potential importance of partitioning of the pollinator niche as an important biotic resource, and also provides the basis for postulating a series of trade-offs (depicted in Figure 1) linking fruit production to flower size, flower number and pollination systems within the Dipterocarpaceae. These mechanisms may well contribute to equalisation of fecundity among species, and warrant further investigation as a novel mechanism for promoting coexistence.

Conservation implications

Our findings have implications for the conservation of genetic and species level diversity. First, our results highlight the potential for differential vulnerability of species to inbreeding, genetic drift and pollen limitation as a result of habitat fragmentation [15,59,60]. Larger-flowered dipterocarps may be less vulnerable to fragmentation as larger-bodied pollinators have a greater

Table 2. Summary table of pollen dispersal from TWOGENER analysis in two species of dipterocarp with contrasting flower size.

Species	d/trees ha	Φft	δ (m)	N_{ep}	A_{ep}^{min} (ha)	A_{ep}^{max} (ha)
S. xanthophylla	5.48	0.0624**	56.27 (43.08–69.48)	8.01	1.46	14.62
P. tomentella	1.63	0.0364**	416.25 (292.84–539.67)	13.73	8.43	84.26

Density of adult trees (d); differentiation in pollen gene pool (Φ_{ft}); mean pollen dispersal distance(δ); effective number of pollen donors (N_{ep}); minimum effective pollen neighbourhood area (A_{ep}^{min}) and maximum effective pollen neighbourhood area (A_{ep}^{max}) following Smouse *et al* 2001.
** Significant at $P<0.01$, values in parentheses are 95% confidence intervals.

capacity to disperse between forest fragments [15,61,62] and thus maintain larger effective populations of forest trees. Second, the genetic processes invoked here also have clear implications for tropical forest restoration, and specifically the collection of seeds for generating planting stock. The genetic diversity of seed collections from small flowered species may be more vulnerable to cryptic processes, such as an elevated proportion of inbred progeny, which may compromise the long-term viability of forest restoration [63]. Hence we emphasise the importance of maintaining the integrity of pollinator communities and dispersal corridors between forest fragments for tropical forest conservation.

Conclusions

Our study suggests that a trade-off exists between flower size and number that, coupled with a positive scaling between flower size and pollination success, reduces the variance in per tree fruit production between species. We have shown that species with small flowers produce them in greater numbers, but these flowers are pollinated by small-bodied insects that are less likely to disperse pollen over long distances. This gives rise to limited gene flow by pollination and, coupled with adult populations with strong spatial genetic structuring, increases the likelihood of mating between related individuals in small-flowered species. As mean flower size increases across species, flower number per tree declines, but the likelihood of receiving pollen from unrelated (more distant) conspecifics increases because larger-bodied pollinators such as sphingid moths and *Apis dorsata* become increasingly important as pollinators. Thus a trade-off embedded in the relationship between flower size and the pollination ecology of co-occurring tropical trees contributes to reducing variance in per capita fecundity among species and life-forms. Although we cannot infer that fecundity relates directly to life-time mean fitness, seed production is an important component of plant reproductive success. Inbreeding has been shown to have direct implications on differential seed mass in dipterocarps [64] and germination rates and survival in other tropical tree species [65,66,67]. It is therefore plausible that these processes reduce fitness inequalities among species and thereby delay competitive exclusion. This perspective suggests that partitioning of pollinators may contribute to maintaining the diversity of species in this ecologically and economically important family of tropical trees, and that the pollination component of the regeneration niche may play a more substantial role in maintaining species diversity than has been recognised previously.

Materials and Methods

Study site and species

The study site was the Sepilok Forest Reserve (SFR: 5°47′–5°52′ N, 117°55′–118°03′ E), which is about 24 km west of the state capital of Sandakan on the east coast of Sabah, Malaysia (SI Fig. S1). SFR is a "Class VI virgin jungle reserve" gazetted in 1930, managed by the Forest Research Centre Sabah for forest protection and research. SFR supports lowland dipterocarp and heath forest (described by Fox [68]). Mean annual temperature falls in the range of 26.7–27.7°C and mean annual rainfall is 2929±134 mm (Malaysian Meteorological Service, unpubl. data).

Flower size, flower number and fruit production

We examined the trade-offs among flower size and total flower and fruit production across 12 dipterocarp species during two minor flowering events between July 2001 and November 2002 supplemented by information obtained during flowering events in 2006–2008 (Table 3). Mean flower size of these species, measured as calyx tube diameter on an average sample of 700 flowers per species from 6 to 55 individuals, varied in the range 1.2 mm–10.2 mm (Table 3). Flowering trees were located using binoculars along an extensive trail system, in an area of approximately 640 ha. A mean of 22 trees were sampled for each species (Table 3). Diameter at breast height (dbh) and crown width were measured for each tree, and their positions were recorded using a GPS (Magellen 315, Thales Navigation, Santa Clara, USA). Similarly, the nearest two flowering neighbours for each tree sampled were located and their positions recorded as above. Phylogenetically independent contrasts (PICs) were used to examine trade-offs between flower size and total flower production, pollination success, the mean number of pollen tubes per pollinated flower or maximum total fruit production to take account of phylogenetic inertia within closely related species [69]. We applied the relationships determined from a recent dipterocarp phylogeny using *PgiC* gene sequence [34] which includes seven of the 12 study species. The PIC analysis was conducted using the computer program Compare 4.6b on the transformed variables [70]. Following PIC analyses, the Y-variable contrasts were regressed on the X-variable contrasts as outlined in [71].

Four 1 m^2 mesh traps raised 1 m above the ground were installed under each tree to estimate flower production. The traps were placed approximately 5 m from the trunk, and at 90° intervals from a random azimuth. All traps were installed prior to the commencement of flower fall and their contents were collected

Table 3. Summary statistics of pollen dispersal from paternity analysis in dipterocarp species with a range of flower size, including the number of genotyped offspring (n); progeny type used in the paternity analysis: mature embryos directly collected from mother (ME); forest floor seedlings (FS), or Saplings (Sp); seedlings germinated from seeds collected below putative mothers (GS); proportion of total embryos assigned at 80% confidence level (% *asgn* 80%); mean pollen dispersal distance across all assigned embryos (*MPDd*), and sample area over which paternity analysis was conducted (PN area); N/A = not available.

Species	Calyx tube Width (mm)	n	Progeny type	% *asgn* 80%	*MPDd* (m)	S.E	PN area (ha)	source
S. xanthophylla	1.2	456	ME	68	119	8.4	100	This study
S. lumutensis	2.2	182	GS	63	175	N/A	8	[26]
P. tomentella	4.2	408	ME	26	176	14.67	100	This study
N. heimii	10.0	248	FS & Sp	37	191	N/A	42	[22]
D. tempehes	10.0	335	FS	88	207	N/A	70	[25]

approximately every three days until the end of fruit-fall. The samples were air dried and sorted into fractions representing buds, flowers and fruit, and the fractions were counted and weighed. Flower counts were based on calyxes, which we presume to be subjected to less lateral dispersal by wind than corollas. Total flower production was estimated by multiplying the total number of buds, flowers and fruit captured by the traps (expressed per m^2) by the projected area of the tree crown. The flowering events we sampled for this paper were not general flowering events sensu [72], and a high proportion of flowering individuals produced only a small number of flowers on relatively few branches. Such flowering events are thought to contribute very little to population-level fecundity [73]. Therefore we estimated mean maximum flower and fruit production from the upper quartile of values per species (n = 2–14 individuals per species). Because basal area alone is insufficient to account for differences in architecture and allometry among species and to account for effects of canopy exposure on flower production, we present values of maximum flower production scaled to tree size:

$$\textit{Scaled flower production} = \frac{\textit{Flower production}}{\textit{Basal area}} \times \frac{MHE}{MHS}$$

where MHE/MSE is a scaling factor for height calculated by dividing the maximum height of emergents at the study site (MHE: 65 m) by the maximum height of the study species (MHS). Basal area was determined by direct measurement for each tree sampled. Maximum height of the emergent trees in the SFR and maximum height of each species were estimated to the nearest 5 m using a measuring line dropped from the canopy. Total fruit production was estimated from the mean number of fruit captured per trap multiplied by the projected area of the tree crown.

Pollination success

We examined pollen tube growth as a measure of pollination success in 11 of the same 12 dipterocarp species using flower samples collected during 2001–2002 (*H. beccariana, S. beccariana, S. johorensis, S. leprosula, S. macroptera, S. multiflora, S. parvifolia, S. smithiana* and *S. xanthophylla,*) or 2006–2008 (*D. grandiflorus* and *P tomentella*) (Table 3). Flower samples were obtained using mixed rope-climbing techniques to gain access to the canopy. Inflorescences were collected directly from each tree at five randomly selected locations within the crown. Flowers that had recently shed their corollas were removed from the inflorescences. We collected a minimum of 500 flowers from each tree for the *Shorea* spp and *H. beccariana*, while the low flower densities of *D. grandiflorus* and *P. tomentella* limited the sample to approximately 50 and 100 flowers per tree, respectively. All flowers were stored in formalin-acetic-alcohol (FAA). Mean calyx diameters for each tree were determined from direct measurement on 25 randomly selected flowers. The styles of the flowers were stained with 1 percent aniline blue solution and the number of pollen tubes in each were determined using fluorescence microscopy as outlined in Ghazoul *et al.*, [21]. We examined pollen tubes within 25200 flowers sampled randomly from the flower collections taken from a total of 638 individuals of the 11 species.

Pollinator body size

Mesh bags were used to exclude visitors from inflorescences prior to anthesis on 2–6 trees from each of nine of the 12 study species that represented an order-of-magnitude range in flower size and flowered between May 2006 and May 2008 (Table 1). Ten replicate inflorescences per tree were used for each of the following treatments: (a) open pollination (no mesh); (b) exclusion of large pollinators (5 mm mesh); (c) exclusion of medium-sized and large pollinators (2 mm mesh); and (d) exclusion of all pollinators except thrips (0.2 mm mesh). Control large mesh bags containing fly trap paper confirmed that large mesh bags did not inhibit visitation by smaller pollinators. The bags were removed at the end of anthesis. Fruit initiation and development were monitored every five days throughout the first month, and then every 10 days until fruit maturity. The effect of mesh bag treatments is expressed as percent pollination success relative to the open pollination treatment one month after anthesis.

Pollen dispersal distances

Pollen dispersal distance was estimated using paternity analysis based on samples of mature fruit collected from known mother trees of *Shorea xanthophylla* (mean calyx diameter 1.2 mm) and *Parashorea tomentella* (mean calyx diameter 4.6 mm). These species were selected because this contrast in flower size was predicted to generate differences in pollen dispersal distance based on our sampling of their flower visitor communities (C.R. Maycock *et al.*, unpubl. data), and because their densities of flowering individuals maximised sampling effort for the available resources. We applied identical sampling strategies in both species, combining the genotypes of mother and progeny with genotypes from all potential pollen donors in mapped populations within 500 m of each mother trees. Because flowering tree density can have a high variance, both within and among species, our sampling design enables us to explicitly control for this.

DNA sampling, extraction and genotyping. All sampling was carried out between April 2006 and September 2008. 20–35 mature fruits were sampled from each of 20 and 17 mother trees of *Shorea xanthophylla* and *Parashorea tomentella* respectively. Fresh fruits were dissected and embryos removed and dried in Sigma™ silica gel prior to DNA extraction. In addition to sampling inner bark from each mother tree, inner bark was sampled from all flowering con-specifics within a 500 m radius of each mother and the location of all adult trees recorded with a GPS. For adult trees, fresh inner bark (cambium) was collected using a 2 cm diameter leather punch and a hammer following Colpaert *et al.* [74]. The fresh samples were placed in labelled tea bags and dried in Sigma™ silica gel within 12 hours of sampling, and stored under the same conditions prior to DNA extraction. Total genomic DNA was extracted from approx. 0.03 g of silica dried material (inner bark, embryo or leaf, depending on sample type) and ground to a fine powder using a Qiagen Mixer-mill™ with Retsch © stainless steel 'cone balls' and the Qiagen DNAeasy™ 96-well-plate plant extraction system. For the paternity analysis the genotype of each mother tree, candidate father and embryo was determined at nine polymorphic, genomic, nuclear microsatellite loci. Primers are summarised in Supplementary Information Table S3. The forward primer of each locus was 5- prime end labelled with one of four florescent-labelled dyes (FAM, NED, PET or VIC) to enable multiplexing of multiple loci during fragment analysis. All PCR products were quantified using an ABI 3730xl DNA Analyser (Applied Biosystems) and sample genotypes scored using Genemapper v.4.0™ software (Applied Biosystems).

To establish whether data from different microsatellite loci were independent, genotypic linkage disequilibrium was tested between all pairs of loci within each species adult sample using GENEPOP 4.0 [75] with the following settings; dememorization: 1000, batch: 100, iteration per batch: 5000. No evidence for significant linkage disequilibrium was detected.

Direct estimates of pollen dispersal. Mother trees were sampled based upon the abundance of fruits and accessibility. Using multilocus genotypes (9 loci, see Table S3 Supporting information) of progeny and known mother trees we applied a maximum likelihood exclusion analysis in CERVUS 3.0, to exclude candidate fathers [76,77]. Candidate fathers were defined as all flowering conspecifics within a 500 m radius of each mother tree. Pollen dispersal distance was deduced for each embryo based on the position of the mother tree and the assigned pollen parent (father) within the 500 m radius. Simulations of paternity were run using the allele frequencies of all adult reproductive trees and the following settings: 10000 cycles; minimum number of loci typed 5, mothers as known parents, test for selfing (*i.e.*, mothers also included as candidate fathers); the flowering conspecifics within 500 m of each mother were set as the candidate fathers for progeny from that mother only; 1 percent for proportion of loci mistyped, and 87 percent for proportion of loci genotyped. The proportion of candidate fathers sampled was set at 80 percent, which was considered a conservative estimate based upon the observed frequency of flowering conspecifics across the entire SFR. Assignment was based upon the 80% confidence level of the critical LOD score.

Based upon paternity assignment, 68 percent of progeny (310 of 456) from 20 mothers trees of *S. xanthophylla* and 26 percent of progeny (106 of 408) from 17 mother trees of *P. tomentella* could be assigned with an 80% confidence within 500 m of the mother tree. For all progeny that could not be assigned at this level of confidence the pollen donors were assumed to be beyond a 500 m radius of each mother tree. This assumption would be supported by evidence that the gene pool of the genotyped progeny differed significantly from that of the pollen donor pool within the 500 m radius. To test this we examined the frequency of rare alleles (abundance <0.01) in the embryo cohorts of each species. In *P. tomentella* 92 percent of the rare alleles (<0.01) are observed only in the embryos, not in the mothers or neighbouring adults, thus must have come from outside the 500 m radius we sampled. In contrast in *S. xanthophylla* only 49 percent of rare alleles were unique to the embryo cohort.

Indirect estimate of pollen dispersal. Pollen dispersal was also evaluated using the TWOGENER [78,79] approach with POLDISP [80]. TWOGENER estimates the heterogeneity of the pollen gene pool sampled within a set of maternal trees (Φ_{FT}) using AMOVA [81]. This has the added advantage that it provides an estimate of contemporary gene flow over the scale of the landscape which is less sensitive to flowering tree density and complementary to the more localised estimate of the paternity analysis. The mean effective pollen dispersal distance (δ) was calculated assuming the following dispersal models; normal, bivariate normal and anisotropic normal, following [79,82,83] in *S. xanthophylla* and *P. tomentella.* The density of adult trees was calculated using our detailed spatial distribution data of our focal species within SFR (Table 1). The effective number of pollen donors N_{ep} and effective pollination neighbourhood area ($\hat{A}_{ep}$) were calculated as follows: $N_{ep} = 1/2\Phi_{ft}$ and $\hat{A}_e = N_{ep}/d_e$ where d_e is the effective density of trees. We used a range of values of d_e including those derived from the TWOGENER analysis. Because the effective density of flowering adults may vary markedly from year to year we calculated N_{ep} assuming a range of densities based upon the census density of reproductive adults (d) and $d/2$, $d/4$ and $d/10$ following the approach of [84] after [79].

Inbreeding of progeny

The relatedness between mothers and putative fathers, derived from the paternity analysis, was estimated for each of the assigned progeny using SPAGEDI 1.2 g [85]. Here we calculated the individual and mean kinship coefficient F [86] for each parent pair assigned by CERVUS [76] within *S. xanthophylla* and *P. tomentella* (n = 229 and 106 respectively). Mean kinship coefficients F have been used to express the relatedness between mating individuals in ants [87] and are useful as they are directly comparable to measures of spatial genetic structure among species. We supplement the measures of relatedness between mating pairs with estimates of the number of pollen donors per seed crop (Nep) and multilocus outcrossing estimates using the program MLTR v 2.4 [88].

Statistical analysis

Linear models were used to investigate trade-offs among species' between log-transformed flower size and: (1) log-transformed maximum flower production; (2) arcsine-transformed percentage of flowers pollinated; (3) log-transformed mean number of pollen tubes per pollinated flower; and (4) log-transformed maximum fruit production. We included life-form (canopy or emergent following Ashton [89]), as an additional independent variable in all models. For analyses of pollination success and fruit production (2–4) we also included median values of the mean distance to the two nearest flowering trees (for pollination success) or fruiting trees (for fruit production) as an independent variable to account for possible effects of differences between species in the density of reproductive trees. We compared models containing all possible combinations of one or more explanatory variables and their interactions and used AIC values to select the most likely models. Models with AIC values that differed by less than two were judged equally valid [90] but F statistics and probability values reported in this paper are taken from the model with the lowest AIC value.

The effects of pollinator exclusion on pollination success were analysed using a generalised linear model with log-transformed calyx diameter and mesh bag treatment as fixed explanatory variables. The non-significant interaction term was removed from the model. Pollination success was expressed as the proportion of fruits within each treatment surviving to one month after the end of anthesis, relative to open pollination.

To test the null hypothesis of no difference in the frequency of short distance mating we conducted a chi-squared test on counts of pollen dispersal within distance classes <50 m, 50–99.9 m, 100–199.9 m and 200–500 m in *S. xanthophylla* and *P. tomentella.* Average pollen dispersal distances over all assigned embryos were compared between *S. xanthophylla* and *P. tomentella* using a Student *t*-test (arithmetic mean) and Wilcox test (median). To determine the significance of inbreeding of progeny as a function of flower size we used a Student *t*-test to test the null hypothesis of no difference in mean kinship coefficient. All statistics were conducted using R (R Development Core Team 2009).

Supporting Information

Figure S1 Location of Sepilok Forest Reserve, boundary demarked by orange line together with distribution of flowering trees of *Parashorea tomentella* and *Shorea xanthophylla* used for paternity analysis in. The two grey ellipses indicate the subset of mother trees from *S. xanthophylla* which had comparable local density of flowering conspecifics to *P. tomentella.*

Table S1 (**a**) AIC and Δ_{AIC} values from four candidate models of log-transformed absolute (unscaled) and scaled flower production for 11 dipterocarp species at Sepilok Forest Reserve, Sabah. The most likely models are shown in bold. (**b**) AIC and Δ_{AIC} values from 18 candidate models of arcsine square root – transformed proportion of flowers pollinated and log-transformed mean number of pollen tubes per pollinated flower for 11 dipterocarp species at Sepilok Forest Reserve, Sabah. The most likely models are shown in bold. logFS, log-transformed flower size; LF, life-form; NND, median values of the mean distance to the two nearest flowering trees. (**c**) AIC and Δ_{AIC} values from 18 candidate models of log-transformed absolute (unscaled) and scaled fruit production for 11 dipterocarp species at Sepilok Forest Reserve, Sabah. The most likely models are shown in bold. logFS, log-transformed flower size; LF, life-form; NND, median values of the mean distance to the two nearest flowering trees.

Table S2 Mating system statistics for progeny of *Shorea xanthophylla* and *Parashorea tomentella* based upon 9 microsatellites loci. Number of progeny genotypes (N); multilocus outcrossing rate (t_m); single locus outcrossing rate (t_s); biparental inbreeding as defined by the difference between multilocus and single locus outcrossing rates (t_m - t_s); Parental inbreeding coefficient. Values in parentheses are standard error (SE) based upon 100 bootstraps.

Table S3 Summary of the 11 microsatellite primers used for paternity analysis and quantification of relatedness between assigned parents in two dipterocarp species *Shorea xanthophylla* and *Parashorea tomentella*. Number of alleles (Na); observed heterozygosity (H_{obs}); expected heterozygosity (H_e); paternity non-exclusion probability at each locus (N-P_E) and total exclusion probability over all loci (P_E) given known mother. [a] Redesigned primers based on published primers. [b] Newly developed microsatellite primers. [c] Published primers.

Acknowledgments

We thank the governments of Malaysia and the state of Sabah for permission to conduct research; the Sabah Forest Department for logistic support; Jeisin bin Jumian and Mike Charkow for field assistance; Will Goodall-Copestake and Viola Huschauer for laboratory assistance and Soon Leong Lee for advice on primers. Fragment analysis was conducted using the sequencing facility of the Genetic Diversity Centre of ETH Zurich. The manuscript benefited from the comments of the academic editor Jerome Chave and anonymous reviewers.

Author Contributions

Conceived and designed the experiments: CK CM JG PH DB. Performed the experiments: CK CM EK RS. Analyzed the data: CK CM DB. Contributed reagents/materials/analysis tools: CK CM EK RS. Wrote the paper: CK CM JG PH DB.

References

1. Goodwillie C, Sargent RD, Eckert CG, Elle E, Geber MA, et al. (2010) Correlated evolution of mating system and floral display traits in flowering plants and its implications for the distribution of mating system variation. New Phytologist 185: 311–321.
2. Primack RB (1987) Relationships among flowers, fruits and seeds. Annual Review of Ecology and Systematics 18: 409–430.
3. Charlesworth D (1989) Evolution of low female fertility in plants: pollen limitation, resource allocation and genetic load. Trends in Ecology and Evolution 4: 289–292.
4. Harder LD, Barrett SCH (1995) Mating cost of large floral displays in hermaphrodite plants. Nature 373: 512–515.
5. Sakai S (2000) Size-number trade-off and allocation to flower production in animal-pollinated flowers. Plant Species Biology 15: 261–268.
6. Sakai S, Sakai A (1995) Flower size-dependent variation in seed size- theory and a test. American Naturalist 145: 918–934.
7. Sargent RD, Goodwillie C, Kalisz S, Rees RH (2007) Phylogenetic evidence for a flower size and number trade-off. American Journal of Botany 94: 2059–2062.
8. Worley AC, Barrett SCH (2000) Evolution of floral display in Eichhornia paniculata (Pontederiaceae): Direct and correlated responses to selection on flower size and number. Evolution 54: 1533–1545.
9. Worley AC, Barrett SCH (2001) Evolution of floral display in Eichhornia paniculata (Pontederiaceae): genetic correlations between flower size and number. Journal of Evolutionary Biology 14: 469–481.
10. Obeso JR (2002) The costs of reproduction in plants. New Phytologist 155: 321–348.
11. Richards S, Williams NM, Harder LD (2009) Variation in Pollination: Causes and Consequences for Plant Reproduction. The American Naturalist 174: 382–398.
12. Harder LD, Aizen MA (2010) Floral adaptation and diversification under pollen limitation. Philosophical Transactions of the Royal Society B-Biological Sciences 365: 529–543.
13. Greenleaf SS, Williams NM, Winfree R, Kremen C (2007) Bee foraging ranges and their relationship to body size. Oecologia 153: 589–596.
14. Stone GN, Willmer P, Rowe JA (1998) Partitioning of pollinators during flowering in an African Acacia community. Ecology 79: 2808–2827.
15. Ghazoul J (2005) Pollen and seed dispersal among dispersed plants. Biological Reviews 80: 413–443.
16. Sargent RD, Vamosi JC (2008) The influence of canopy position, pollinator syndrome, and region on evolutionary transitions in pollinator guild size. International Journal of Plant Sciences 169: 39–47.
17. Chesson P (2000) Mechanisms of maintenance of species diversity. Annual Review of Ecology and Systematics 31: 343.
18. Vamosi JC, Knight TM, Steets JA, Mazer SJ, Burd M, et al. (2006) Pollination decays in biodiversity hotspots. Proceedings of the National Academy of Sciences of the United States of America 103: 956–961.
19. Lowe AJ, Boshier D, Ward M, Bacles CFE, Navarro C (2005) Genetic resource impacts of habitat loss and degradation; reconciling empirical evidence and predicted theory for neotropical trees. Heredity 95: 255–273.
20. Ashton PS, Givnish TJ, Appanah S (1988) Staggered Flowering in the Dipterocarpaceae - New Insights into Floral Induction and the Evolution of Mast Fruiting in the Aseasonal Tropics. American Naturalist 132: 44–66.
21. Ghazoul J, Liston KA, Boyle TJB (1998) Disturbance-induced density-dependent seed set in *Shorea siamensis* (Dipterocarpaceae), a tropical forest tree. Journal of Ecology 86: 462–473.
22. Konuma A, Tsumura Y, Lee CT, Lee SL, Okuda T (2000) Estimation of gene flow in the tropical-rainforest tree Neobalanocarpus heimii (Dipterocarpaceae), inferred from paternity analysis. Molecular Ecology 9: 1843–1852.
23. Sakai S, Momose K, Yumoto T, Kato M, Inoue T (1999) Beetle pollination of Shorea parvifolia (section Mutica, Dipterocarpaceae) in a general flowering period in Sarawak, Malaysia. American Journal of Botany 86: 62–69.
24. Fukue Y, Kado T, Lee SL, Ng KKS, Muhammad N, et al. (2007) Effects of flowering tree density on the mating system and gene flow in Shorea leprosula (Dipterocarpaceae) in Peninsular Malaysia. Journal of Plant Research 120: 413–420.
25. Kenta T, Isagi Y, Nakagawa M, Yamashita M, Nakashizuka T (2004) Variation in pollen dispersal between years with different pollination conditions in a tropical emergent tree. Molecular Ecology 13: 3575–3584.
26. Lee SL, Ng KKS, Saw LG, Lee CT, Muhammad N, et al. (2006) Linking the gaps between conservation research and conservation management of rare dipterocarps: A case study of *Shorea lumutensis*. Biological Conservation 131: 72–92.
27. Naito Y, Konuma A, Iwata H, Suyama Y, Seiwa K, et al. (2005) Selfing and inbreeding depression in seeds and seedlings of Neobalanocarpus heimii (Dipterocarpaceae). Journal of Plant Research 118: 423–430.
28. Bawa KS (1998) Conservation of Genetic resources in the Dipterocarpaceae. In: Appanah S, Turnbull JM, eds. Review of the Dipterocarps: Taxonomy, Ecology and Silviculture. Bogor, Indonesia: CIFOR.
29. Ng KKS, Lee SL, Koh CL (2004) Spatial structure and genetic diversity of two tropical tree species with contrasting breeding systems and different ploidy levels. Molecular Ecology 13: 657–669.
30. Osada N, Takeda H, Furukawa A, Awang M (2001) Fruit dispersal of two dipterocarp species in a Malaysian rain forest. Journal of Tropical Ecology 17: 911–917.
31. Seidler TG, Plotkin JB (2006) Seed dispersal and spatial pattern in tropical trees. Plos Biology 4: 2132–2137.
32. Kettle CJ, Hollingsworth PM, Burslem DFRP, Maycock C, Khoo E, et al. Determinants of fine-scale spatial genetic structure in three co-occurring rain forest canopy trees in Borneo *Perspectives in Plant Ecology Evolution and Systematics*. (in press).
33. Gamage DT, de Silva MP, Inomata N, Yamazaki T, Szmidt AE (2006) Comprehensive molecular phylogeny of the sub-family Dipterocarpoideae (Dipterocarpaceae) based on chloroplast DNA sequences. Genes & Genetic Systems 81: 1–12.
34. Kamiya K, Harada K, Tachida H, Ashton PS (2005) Phylogeny of PgiC gene in Shorea and its closely related genera (Dipterocarpaceae), the dominant trees in Southeast Asian tropical rain forests. American Journal of Botany 92: 775–788.

35. Worley AC, Baker AM, Thompson JD, Barrett SCH (2000) Floral display in Narcissus: Variation in flower size and number at the species, population, and individual levels. International Journal of Plant Sciences 161: 69–79.
36. Dayanandan S, Attygolla DNC, Abeygunasekera AWWL, Gunatilleke IAUN, Gunatilleke CVS (1990) Phenology and floral morphology in relation to pollination of some Sri Lankan Dipterocarps. In: Bawa KS, Hadley M, eds. Reproductive Ecology of Tropical Forest Plants. Paris, France: UNESCO.
37. Momose K, Nagamitsu T, Inoue T (1996) The reproductive ecology of an emergent dipterocarp in a lowland rain forest in Sarawak. Plant Species Biology 11: 189–198.
38. Ghazoul J, Satake A (2009) Nonviable seed set enhances plant fitness: the sacrificial sibling hypothesis. Ecology 90: 369–377.
39. Haskell JP, Ritchie ME, Olff H (2002) Fractal geometry predicts varying body size scaling relationships for mammal and bird home ranges. Nature 418: 527–530.
40. Hardy OJ, Maggia L, Bandou E, Breyne P, Caron H, et al. (2006) Fine-scale genetic structure and gene dispersal inferences in 10 Neotropical tree species. Molecular Ecology 15: 559–571.
41. Jones FA, Comita LS (2008) Neighbourhood density and genetic relatedness interact to determine fruit set and abortion rates in a continuous tropical tree population. Proceedings of the Royal Society B-Biological Sciences 275: 2759–2767.
42. Loveless MD, Hamrick JL (1984) Ecological Determinants of Genetic-Structure in Plant-Populations. Annual Review of Ecology and Systematics 15: 65–95.
43. Bagchi R, Press MC, Scholes JD (2009) Evolutionary history and distance dependence control survival of dipterocarp seedlings. Ecology Letters 9999.
44. Blundell AG, Peart DR (1998) Distance-dependence in herbivory and foliar condition for juvenile *Shorea* trees in Bornean dipterocarp rain forest. Oecologia 117: 151–160.
45. Blundell AG, Peart DR (2004) Seedling recruitment failure following dipterocarp mast fruiting. Journal of Tropical Ecology 20: 229–231.
46. Curran LM, Webb CO (2000) Experimental tests of the spatiotemporal scale of seed predation in mast-fruiting Dipterocarpaceae. Ecological Monographs 70: 129–148.
47. Wells K, Bagchi R (2005) Eat in or take away: Seed predation and removal by rats (Muridae) during a fruiting event in a dipterocarp rainforest. Raffles Bulletin of Zoology 53: 281–286.
48. Ng KKS, Lee SL, Saw LG, Plotkin JB, Koh CL (2006) Spatial structure and genetic diversity of three tropical tree species with different habitat preferences within a natural forest. Tree Genetics & Genomes 2: 121–131.
49. Takeuchi Y, Ichikawa S, Konuma A, Tomaru N, Niiyama K, et al. (2004) Comparison of the fine-scale genetic structure of three dipterocarp species. Heredity 92: 323–328.
50. Harrison RD, Nagamitsu T, Momose K, Inoue T (2005) Flowering phenology and pollination of Dipterocarpus (Dipterocarpaceae) in Borneo. Malayan Nature Journal 57: 67–80.
51. Ghazoul J (1997) The pollination and breeding system of *Dipterocarpus obtusifolius* (Dipterocarpaceae) in dry deciduous forests of Thailand. Journal of Natural History 31: 901–916.
52. Chambers J, Aarssen LW (2009) Offspring for the next generation: most are produced by small plants within herbaceous populations. Evolutionary Ecology 23: 737–751.
53. Hubbell SP (2001) Monographs in Population Biology. The unified neutral theory of biodiversity and biogeography. Princeton, NJ: Princeton University Press. pp i-xiv, 1–375.
54. Kohyama T, Takada T (2009) The stratification theory for plant coexistence promoted by one-sided competition. Journal of Ecology 97: 463–471.
55. Verhulst J, Montana C, Mandujano MC, Franco M (2008) Demographic mechanisms in the coexistence of two closely related perennials in a fluctuating environment. Oecologia 156: 95–105.
56. Fine PVA, Miller ZJ, Mesones I, Irazuzta S, Appel HM, et al. (2006) The growth-defense trade-off and habitat specialization by plants in Amazonian forests. Ecology 87: S150–S162.
57. Kraft NJB, Valencia R, Ackerly DD (2008) Functional traits and niche-based tree community assembly in an amazonian forest. Science 322: 580–582.
58. Russo SE, Brown P, Davies SJ (2008) Interspecific demographic trade-offs and soil-related habitat associations of tree species along resource gradients (vol 96, pg 192, 2008). Journal of Ecology 96: 1330–1330.
59. Aguilar R, Quesada M, Ashworth L, Herrerias-Diego Y, Lobo J (2008) Genetic consequences of habitat fragmentation in plant populations: susceptible signals in plant traits and methodological approaches. Molecular Ecology 17: 5177–5188.
60. Honnay O, Jacquemyn H (2007) Susceptibility of common and rare plant species to the genetic consequences of habitat fragmentation. Conservation Biology 21: 823–831.
61. Dick CW, Etchelecu G, Austerlitz F (2003) Pollen dispersal of tropical trees (*Dinizia excelsa*: Fabaceae) by native insects and African honeybees in pristine and fragmented Amazonian rainforest. Molecular Ecology 12: 753–764.
62. Steffan-Dewenter I, Tscharntke T (1999) Effects of habitat isolation on pollinator communities and seed set. Oecologia 121: 432–440.
63. Kettle CJ, Ennos RA, Jaffre T, Gardner M, Hollingsworth PM (2008) Cryptic genetic bottlenecks during restoration of an endangered tropical conifer. Biological Conservation 141: 1953–1961.
64. Naito Y, Kanzaki M, Iwata H, Obayashi K, Lee SL, et al. (2008) Density-dependent selfing and its effects on seed performance in a tropical canopy tree species, Shorea acuminata (Dipterocarpaceae). Forest Ecology and Management 256: 375–383.
65. Aldrich PR, Hamrick JL, Chavarriaga P, Kochert G (1998) Microsatellite analysis of demographic genetic structure in fragmented populations of the tropical tree *Symphonia globulifera*. Molecular Ecology 7: 933–944.
66. Hufford KM, Hamrick JL (2003) Viability selection at three early life stages of the tropical tree, Platypodium elegans (Fabaceae, Papilionoideae). Evolution 57: 518–526.
67. Hufford KM, Hamrick JL, Rathbun SL (2009) Male reporductive success at three early life stages in the tropical tree Platypodium elegans. International Journal of Plant Sciences 170: 724–734.
68. Fox JED (1973) A handbook to Kabili-Sepilok Forest Reserve. Kuching, Sarawak, Malaysia.
69. Harvey PH, Pagel MD (1991) The Comparative Method in Evolutionary Biology. Oxford: Oxford University Press. Oxford: Oxford University Press.
70. Martin EP (2004) Compare, version 4.6b. Computer programs for the statistical analayis of comparative data.: Distributed by the author.
71. Purvis A, Rambaut A (1995) Comparative analysis by independent contrasts (CAIC): an Apple Macintosh application for analysing comparative data. Computer applications in the biosciences: CABIOS 11: 247–251.
72. Sakai S (2002) General flowering in lowland mixed dipterocarp forests of South-east Asia. Biological Journal of the Linnean Society 75: 233–247.
73. Maycock CR, Thewlis RN, Ghazoul J, Nilus R, Burslem D (2005) Reproduction of dipterocarps during low intensity masting events in a Bornean rain forest. Journal of Vegetation Science 16: 635–646.
74. Colpaert N, Cavers S, Bandou E, Caron H, Gheysen G, et al. (2005) Sampling tissue for DNA analysis of trees: Trunk cambium as an alternative to canopy leaves. Silvae Genetica 54: 265–269.
75. Rousset F (2008) GENEPOP ' 007: a complete re-implementation of the GENEPOP software for Windows and Linux. Molecular Ecology Resources 8: 103–106.
76. Kalinowski ST, Taper ML, Marshall TC (2007) Revising how the computer program CERVUS accommodates genotyping error increases success in paternity assignment. Molecular Ecology 16: 1099–1106.
77. Marshall TC, Slate J, Kruuk LEB, Pemberton JM (1998) Statistical confidence for likelihood-based paternity inference in natural populations. Molecular Ecology 7: 639–655.
78. Austerlitz F, Dutech C, Smouse PE, Davis F, Sork VL (2007) Estimating anisotropic pollen dispersal: a case study in Quercus lobata. Heredity 99: 193–204.
79. Smouse PE, Dyer RJ, Westfall RD, Sork VL (2001) Two-generation analysis of pollen flow across a landscape. I. Male gamete heterogeneity among females. Evolution 55: 260–271.
80. Robledo-Arnuncio JJ, Austerlitz F, Smouse PE (2007) POLDISP: a software package for indirect estimation of contemporary pollen dispersal. Molecular Ecology Notes 7: 763–766.
81. Excoffier L, Smouse PE, Quattro JM (1992) Analysis of Molecular Variance Inferred from Metric Distances among DNA Haplotypes - Application to Human Mitochondrial-DNA Restriction Data. Genetics 131: 479–491.
82. Cloutier D, Kanashiro M, Ciampi AY, Schoen DJ (2007) Impact of selective logging on inbreeding and gene dispersal in an Amazonian tree population of Carapa guianensis Aubl. Molecular Ecology 16: 797–809.
83. de Lacerda AEB, Kanashiro M, Sebbenn AM (2008) Long-pollen movement and deviation of random mating in a low-density continuous population of a tropical tree Hymenaea courbaril in the Brazilian Amazon. Biotropica 40: 462–470.
84. Born C, Hardy OJ, Chevallier MH, Ossari S, Atteke C, et al. (2008) Small-scale spatial genetic structure in the Central African rainforest tree species *Aucoumea klaineana*: a stepwise approach to infer the impact of limited gene dispersal, population history and habitat fragmentation. Molecular Ecology 17: 2041–2050.
85. Hardy OJ, Vekemans X (2002) SPAGEDi: a versatile computer program to analyse spatial genetic structure at the individual or population levels. Molecular Ecology Notes 2: 618–620.
86. Loiselle BA, Sork VL, Nason J, Graham C (1995) Spatial Genetic-Structure of a Tropical Understory Shrub, *Psychotria Officinalis* (Rubiaceae). American Journal of Botany 82: 1420–1425.
87. Hardy OJ, Pearcy M, Aron S (2008) Small-scale spatial genetic structure in an ant species with sex-biased dispersal. Biological Journal of the Linnean Society 93: 465–473.
88. Ritland K (2002) Extensions of models for the estimation of mating systems using n independent loci. Heredity 88: 221–228.
89. Ashton PS (2004) Dipterocarpaceae. In: Soepadmo E, Saw LG, Chung RCK, eds. Tree Flora of Sabah and Sarawak: Government of Malaysia. pp 63–388.
90. Burnham KP, Anderson DR (2002) Model selection and multimodel inference: a practical information-theoretic approach. New York, NY: Springer-Verlag.
91. Wood GHS, Meijer W (1964) Dipterocarps of Sabah. Sandakan: Forest Department.

PERMISSIONS

All chapters in this book were first published in PLOS ONE, by The Public Library of Science; hereby published with permission under the Creative Commons Attribution License or equivalent. Every chapter published in this book has been scrutinized by our experts. Their significance has been extensively debated. The topics covered herein carry significant findings which will fuel the growth of the discipline. They may even be implemented as practical applications or may be referred to as a beginning point for another development.

The contributors of this book come from diverse backgrounds, making this book a truly international effort. This book will bring forth new frontiers with its revolutionizing research information and detailed analysis of the nascent developments around the world.

We would like to thank all the contributing authors for lending their expertise to make the book truly unique. They have played a crucial role in the development of this book. Without their invaluable contributions this book wouldn't have been possible. They have made vital efforts to compile up to date information on the varied aspects of this subject to make this book a valuable addition to the collection of many professionals and students.

This book was conceptualized with the vision of imparting up-to-date information and advanced data in this field. To ensure the same, a matchless editorial board was set up. Every individual on the board went through rigorous rounds of assessment to prove their worth. After which they invested a large part of their time researching and compiling the most relevant data for our readers.

The editorial board has been involved in producing this book since its inception. They have spent rigorous hours researching and exploring the diverse topics which have resulted in the successful publishing of this book. They have passed on their knowledge of decades through this book. To expedite this challenging task, the publisher supported the team at every step. A small team of assistant editors was also appointed to further simplify the editing procedure and attain best results for the readers.

Apart from the editorial board, the designing team has also invested a significant amount of their time in understanding the subject and creating the most relevant covers. They scrutinized every image to scout for the most suitable representation of the subject and create an appropriate cover for the book.

The publishing team has been an ardent support to the editorial, designing and production team. Their endless efforts to recruit the best for this project, has resulted in the accomplishment of this book. They are a veteran in the field of academics and their pool of knowledge is as vast as their experience in printing. Their expertise and guidance has proved useful at every step. Their uncompromising quality standards have made this book an exceptional effort. Their encouragement from time to time has been an inspiration for everyone.

The publisher and the editorial board hope that this book will prove to be a valuable piece of knowledge for researchers, students, practitioners and scholars across the globe.

LIST OF CONTRIBUTORS

Nicola K. Abram
Durrell Institute for Conservation and Ecology, School of Anthropology and Conservation, University of Kent, Canterbury, Kent, United Kingdom
HUTAN/Kinabatangan Orang-utan Conservation Programme, Kota Kinabalu, Sabah, Malaysia
Living Landscape Alliance, Reading, Berkshire, United Kingdom
Borneo Futures initiative, People and Nature Consulting International, Jakarta, Indonesia

Panteleimon Xofis
Technological Education Institute of Kavala, Department of Forestry and Management of Natural Environment, Drama, Greece

Joseph Tzanopoulos and Douglas C. MacMillan
Durrell Institute for Conservation and Ecology, School of Anthropology and Conservation, University of Kent, Canterbury, Kent, United Kingdom

Marc Ancrenaz
HUTAN/Kinabatangan Orang-utan Conservation Programme, Kota Kinabalu, Sabah, Malaysia
Borneo Futures initiative, People and Nature Consulting International, Jakarta, Indonesia
North England Zoological Society, Chester, Cheshire, United Kingdom

Robin Chung
C H Williams Talhar and Wong (Sabah) Sdn Bhd, Sandakan, Sabah, Malaysia

Lucy Peter
Danau Girang Field Centre, c/o Sabah Wildlife Department, Kota Kinabalu, Sabah, Malaysia

Robert Ong
Forest Research Centre, Sabah Forestry Department, Sandakan, Sabah, Malaysia

Isabelle Lackman
HUTAN/Kinabatangan Orang-utan Conservation Programme, Kota Kinabalu, Sabah, Malaysia

Benoit Goossens
Danau Girang Field Centre, c/o Sabah Wildlife Department, Kota Kinabalu, Sabah, Malaysia
Organisms and Environment Division, School of Biosciences, Cardiff University, Cardiff, Wales, United Kingdom
Sabah Wildlife Department, Kota Kinabalu, Sabah, Malaysia

Laurentius Ambu
Sabah Wildlife Department, Kota Kinabalu, Sabah, Malaysia

Andrew T. Knight
Department of Life Sciences, Imperial College London, Ascot, Berkshire, United Kingdom
Department of Botany, Nelson Mandela Metrropolitan University, Port Elizabeth, Eastern Cape, South Africa

Daniel Magnabosco Marra
Spezielle Botanik und Funktionelle Biodiversität, Universität Leipzig, Leipzig, Deutschland
BGC Prozesse, Max-Planck-Institut für Biogeochemie, Jena, Deutschland
Laboratório de Manejo Florestal, Instituto Nacional de Pesquisas da Amazônia, Manaus, Brazil

Jeffrey Q. Chambers
Laboratório de Manejo Florestal, Instituto Nacional de Pesquisas da Amazônia, Manaus, Brazil
Geography Department, University of California, Berkeley, California, United States of America
Climate Sciences Department, Lawrence Berkeley National Laboratory, Berkeley, California, United States of America
Ecology and Evolutionary Biology Department, Tulane University, New Orleans, Louisiana, United States of America

Niro Higuchi, Gabriel H. P. M. Ribeiro and Joaquim dos Santos
Laboratório de Manejo Florestal, Instituto Nacional de Pesquisas da Amazônia, Manaus, Brazil

Susan E. Trumbore
BGC Prozesse, Max-Planck-Institut für Biogeochemie, Jena, Deutschland
Laboratório de Manejo Florestal, Instituto Nacional de Pesquisas da Amazônia, Manaus, Brazil

Robinson I. Negrón-Juárez
Climate Sciences Department, Lawrence Berkeley National Laboratory, Berkeley, California, United States of America
Ecology and Evolutionary Biology Department, Tulane University, New Orleans, Louisiana, United States of America

Björn Reu
Spezielle Botanik und Funktionelle Biodiversität, Universität Leipzig, Leipzig, Deutschland

Christian Wirth
Spezielle Botanik und Funktionelle Biodiversität, Universität Leipzig, Leipzig, Deutschland
BGC Prozesse, Max-Planck-Institut für Biogeochemie, Jena, Deutschland
German Centre for Integrative Biodiversity Research (iDiv) Halle-Jena-Leipzig, Leipzig, Deutschland

Marko Kovač
Department of Forest and Landscape Planning and Monitoring, Slovenian Forestry Institute, Večna pot 2, Ljubljana, Slovenia

Arthur Bauer
Department of Food, Agriculture and Forestry Cham - Division of Forestry, Bavarian State Ministry of Food, Agriculture and Forestry, Ölbergstrabe 3, Waldmünchen, Germany

Göran Ståhl
Department of Forest Resource Management, Swedish University of Agricultural Sciences (SLU), Skogsmarksgränd, Umeå, Sweden

Dewei Yang, Tao Luo, Tao Lin, Quanyi Qiu and Yunjian Luo
Key Lab of Urban Environment and Health, Institute of Urban Environment, Chinese Academy of Sciences, Xiamen, China
Xiamen Key Lab of Urban Metabolism, Xiamen, China

Alvaro G. Gutiérrez
Department of Ecological Modeling, Helmholtz Centre for Environmental Research (UFZ), Leipzig, Germany
Forest Ecology Group, Institute of Terrestrial Ecosystems, Department of Environmental Sciences, Swiss Federal Institute of Technology (ETH Zürich), Zürich, Switzerland

Juan J. Armesto
Instituto de Ecología y Biodiversidad (IEB), Santiago, Chile
Departamento de Ecología, Facultad de Ciencias Biológicas, Universidad Catolica de Chile, Santiago, Chile

M. Francisca Díaz
Departamento de Ciencias Biológicas, Facultad de Ciencias Biológicas, Universidad Andrés Bello, Santiago, Chile

Andreas Huth
Department of Ecological Modeling, Helmholtz Centre for Environmental Research (UFZ), Leipzig, Germany

Hao Chen
Key Laboratory of Vegetation Restoration and Management of Degraded Ecosystems, South China Botanical Garden, Chinese Academy of Sciences, Guangzhou, China
University of Chinese Academy of Sciences, Beijing, China

Geshere A. Gurmesa
Department of Geosciences and Natural Resource Management, University of Copenhagen, Copenhagen, Denmark
University of Chinese Academy of Sciences, Beijing, China
Sino-Danish Center for Education and Research, Aarhus, Denmark

Lei Liu
Key Laboratory of Vegetation Restoration and Management of Degraded Ecosystems, South China Botanical Garden, Chinese Academy of Sciences, Guangzhou, China
State Key Laboratory of Urban and Regional Ecology, Research Center for Eco-Environmental Sciences, Chinese Academy of Sciences, Beijing, China

Tao Zhang
Key Laboratory of Vegetation Restoration and Management of Degraded Ecosystems, South China Botanical Garden, Chinese Academy of Sciences, Guangzhou, China
Institute of Tropical Pratacultural Science, Zhanjiang Normal University, Zhanjiang, China

Shenglei Fu, Zhanfeng Liu, Shaofeng Dong, Chuan Ma and Jiangming Mo
Key Laboratory of Vegetation Restoration and Management of Degraded Ecosystems, South China Botanical Garden, Chinese Academy of Sciences, Guangzhou, China

John T. Longino
Department of Biology, The University of Utah, Salt Lake City, Utah, United States of America,

Michael G. Branstetter
Department of Entomology, National Museum of Natural History, Smithsonian Institution, Washington, D.C., United States of America,

Robert K. Colwell
Department of Ecology and Evolutionary Biology, University of Connecticut, Storrs, Connecticut, United States of America, and University of Colorado Museum of Natural History, Boulder, Colorado, United States of America

Edith Bai, Weiwei Dai, Ping Jiang and Shijie Han
State Key Laboratory of Forest and Soil Ecology, Institute of Applied Ecology, Chinese Academy of Sciences, Shenyang, China

Wei Li, Shanlong Li, Jianfei Sun and Bo Peng,
State Key Laboratory of Forest and Soil Ecology, Institute of Applied Ecology, Chinese Academy of Sciences, Shenyang, China
College of Resources and Environment,University of Chinese Academy of Sciences, Beijing, China

Martin M. Gossner and Wolfgang W. Weisser
Friedrich-Schiller-University of Jena, Institute of Ecology, Jena, Germany
Technische Universität München, Terrestrial Ecology Research Group, Department of Ecology and Ecosystem Management, Center for Food and Life Sciences Weihenstephan, Freising-Weihenstephan, Germany

Esther Pašalić and Patricia Lange
Friedrich-Schiller-University of Jena, Institute of Ecology, Jena, Germany

Markus Lange
Friedrich-Schiller-University of Jena, Institute of Ecology, Jena, Germany
Max Planck Institute for Biogeochemistry, Jena, Germany

Dominik Hessenmöller and Ernst-Detlef Schulze
Max Planck Institute for Biogeochemistry, Jena, Germany

Jörg Müller
University of Potsdam, Institute of Biochemistry and Biology, Potsdam, Germany
Stephanie A. Socher, Markus Fischer and Steffen Boch
University of Bern, Institute of Plant Sciences, Bern, Switzerland

Long Yang
Centre of Resource and Environment, Guangzhou Institute of Geography, Guangzhou, China
Key Laboratory of Vegetation Restoration and Management of Degraded Ecosystems, South China Botanical Garden, Chinese Academy of Sciences, Guangzhou, China

Jun Wang
Key Laboratory of Vegetation Restoration and Management of Degraded Ecosystems, South China Botanical Garden, Chinese Academy of Sciences, Guangzhou, China

Yuhui Huang
Guangdong Academy of Forestry, Guangzhou, China

Dafeng Hui
Department of Biological Sciences, Tennessee State University, Nashville, Tennessee, United States of America

Meili Wen
Centre of Resource and Environment, Guangzhou Institute of Geography, Guangzhou, China

Manuel A. Hernández-Ruedas, Víctor Arroyo-Rodríguez, Guillermo Ibarra-Manríquez and Miguel Martínez-Ramos
Centro de Investigaciones en Ecosistemas, Universidad Nacional Autónoma de México, Morelia, Michoacán, Mexico

Jorge A. Meave
Departamento de Ecología y Recursos Naturales, Facultad de Ciencias, Universidad Nacional Autónoma de México, Mexico City, Mexico

Esteban Martínez
Departamento de Botánica, Instituto de Biología, Universidad Nacional Autónomade México, Mexico City, Mexico

Gilberto Jamangapé
Ejido Loma Bonita, Ocosingo, Chiapas, Mexico

Felipe P. L. Melo
Departamento de Botánica, Universidade Federal de Pernambuco, Recife, Pernambuco,Brazil

Bráulio A. Santos
Departamento de Sistemática e Ecologia, Universidade Federal da Paraíba, João Pessoa, Paraíba, Brazil

You-xin Shen
Key Laboratory of Tropical Forest Ecology, Xishuangbanna Tropical Botanical Garden, Chinese Academy of Sciences, Menglun, Yunnan, China
Restoration Ecology Group, Xishuangbanna Tropical Botanical Garden, Chinese Academy of Sciences, Kunming, China

Wei-li Liu
Kunming Metallurgical Research Institute, Kunming, China

Yu-hui Li and Hui-lin Guan
Yunnan Normal University, Kunming, China

Wenfa Xiao, Zhilin Huang, Lixiong Zeng
State Forestry Administration Key Laboratory of Forest Ecology and Environment, Research Institute of Forest Ecology, Environment and Protection, Chinese Academy of Forestry, Beijing, PR China

Xiaogai Ge
State Forestry Administration Key Laboratory of Forest Ecology and Environment, Research Institute of Forest Ecology, Environment and Protection, Chinese Academy of Forestry, Beijing, PR China
Research Institute of Subtropical Forestry, Chinese Academy of Forestry, Fuyang, Zhejiang, PR China

Jingpin Lei
State Forestry Administration Key Laboratory of Forest Silviculture, Research Institute of Forestry, Chinese Academy of Forestry, Beijing, PR China

Benzhi Zhou
Research Institute of Subtropical Forestry, Chinese Academy of Forestry, Fuyang, Zhejiang, PR China

Maihe Li
Swiss Federal Research Institute WSL, Birmensdorf, Switzerland
State Key Laboratory of Forest and Soil Ecology, Institute of Applied Ecology, Chinese Academy of Sciences, Shenyang, PR China

Ole Petter Laksforsmo Vindstad, Rolf Anker Ims and Lauri Kapari
Department of Arctic and Marine Biology, University of Tromsø, Tromsø, Norway

Sabrina Schultze
Department of Ecology, Leuphana University Lüneburg, Lüneburg, Germany

Jane Uhd Jepsen and Martin Biuw
Norwegian Institute for Nature Research, Fram Centre, Tromsø, Norway

Anne Sverdrup-Thygeson
Norwegian University of Life Sciences, Department of Ecology and Natural Resource Management, Aas, Norway

Francesco Rovero
Tropical Biodiversity Section, MUSE - Museo delle Scienze, Trento, Italy
Udzungwa Ecological Monitoring Centre, Udzungwa Mountains National Park, Mangúla, Tanzania

Emanuel Martin
Sokoine University of Agriculture, Morogoro, Tanzania

Melissa Rosa and Jorge A. Ahumada
Tropical Ecology Assessment & Monitoring (TEAM) Network, Betty & Gordon Moore Center for Science and Oceans, Conservation International, Washington, D.C., United States of America

Daniel Spitale
Tropical Biodiversity Section, MUSE - Museo delle Scienze, Trento, Italy

Rodrigo S. Rios and Cristian Salgado-Luarte
Departamento de Biología, Universidad de La Serena, La Serena, Chile

Ernesto Gianoli
Departamento de Biología, Universidad de La Serena, La Serena, Chile
Departamento de Botánica, Universidad de Concepción, Concepción, Chile

Michelle C. Agne, David C. Shaw and Travis J. Woolley
Department of Forest Engineering, Resources, and Management, Oregon State University, Corvallis, Oregon, United States of America

Mónica E. Queijeiro-Bolaños
Departamento de Ecología y Recursos Naturales, Facultad de Ciencias, Universidad Nacional Autónoma de México, Mexico City, DF, Mexico

Cynthia Gidoin
Supagro, UMR System, Montpellier, France
CIRAD, UPR Bioagresseurs, Montpellier, France

Régis Babin
CIRAD, UPR Bioagresseurs, Montpellier, France
icipe, Plant Health Division, Coffee Pest Project, Nairobi, Kenya

Leïla Bagny Beilhe and Gerben Martijn ten Hoopen
CIRAD, UPR Bioagresseurs, Montpellier, France
IRAD, Yaoundé, Cameroon

Christian Cilas
CIRAD, UPR Bioagresseurs, Montpellier, France

Marie Ange Ngo Bieng
CIRAD, UMR System, Montpellier, France

Angang Ming
Experimental Center of Tropical Forestry, Chinese Academy of Forestry, Pingxiang, Guangxi, China PR
Institute of Forest Ecology, Environment and Protection, Chinese Academy of Forestry, Beijing, China
Guangxi Youyiguan Forest Ecosystem Research Station, Pingxiang, Guangxi, China PR, PR

Hongyan Jia
Experimental Center of Tropical Forestry, Chinese Academy of Forestry, Pingxiang, Guangxi, China PR
Guangxi Youyiguan Forest Ecosystem Research Station, Pingxiang, Guangxi, China PR, PR

Jinlong Zhao
College of Forestry, Beijing Forestry University, Beijing, China PR

Yi Tao and Yuanfa Li
College of Forestry, Guangxi University, Nanning, Guangxi, China PR

Marcos D. Robles, Robert M. Marshall and Jeanmarie A. Haney
The Nature Conservancy Center for Science and Public Policy, Tucson, Arizona, United States of America

Frances O'Donnell
Northern Arizona University, Flagstaff, Arizona, United States of America

Edward B. Smith
The Nature Conservancy in California, Sacramento, California, United States of America

David F. Gori
The Nature Conservancy in New Mexico, Santa Fe, New Mexico, United States of America

David B. McWethy
Department of Earth Sciences, Montana State University, Bozeman, Montana, United States of America

Janet M. Wilmshurst
Landcare Research, Lincoln, New Zealand
School of Environment, University of Auckland, Auckland, New Zealand

Cathy Whitlock
Department of Earth Sciences, Montana State University, Bozeman, Montana, United States of America
Institute on Ecosystems, Montana State University, Bozeman, Montana, United States of America

Jamie R. Wood and Matt S. McGlone
Landcare Research, Lincoln, New Zealand

Chris J. Kettle
Ecosystem Management, Institute of Terrestrial Ecosystems, ETH Zurich, Zurich, Switzerland
Institute of Biological and Environmental Sciences, University of Aberdeen, Aberdeen, United Kingdom

Colin R. Maycock, Rahayu Sukmaria Haji Sukri and David F. R. P. Burslem
Institute of Biological and Environmental Sciences, University of Aberdeen, Aberdeen, United Kingdom

Jaboury Ghazoul
Ecosystem Management, Institute of Terrestrial Ecosystems, ETH Zurich, Zurich, Switzerland

Pete M. Hollingsworth
Royal Botanic Gardens Edinburgh, Edinburgh, United Kingdom

Eyen Khoo
Forest Research Centre, Sabah Forest Department, Sabah, Malaysia Pharmaceuticals

Index